Environmental Science

Series editors: R. Allan · U. Förstner · W. Salomons

Springer

Berlin
Heidelberg
New York
Barcelona
Hong Kong
London
Milan
Paris
Singapore
Tokyo

Antonio Gianguzza · Ezio Pelizzetti
Silvio Sammartano (Eds.)

Chemical Processes in Marine Environments

With 194 Figures and 56 Tables

Springer

Editors

Prof. Antonio Gianguzza

Dipartimento di Chimica Inorganica, Sezione di Chimica Ambientale
Università di Palermo
Viale delle Scienze, Parco d'Orleans II
I-90128 Palermo, Italy

Prof. Ezio Pelizetti

Dipartimento di Chimica Analitica
Università di Torino
Via P. Giuria 5
I-10125 Torino, Italy

Prof. Silvio Sammartano

Dipartimento di Chimica Inorganica, Chimica Analitica e Chimica Fisika
Università di Messina
Salita Sperone 31
I-98166 Messina, Italy

ISSN 1431-6250
ISBN 978-3-642-08589-5

Die Deutsche Bibliothek – CIP-Einheitsaufnahme

Chemical processes in marine environments : with 56 tables / Antonio Gianguzza ... (ed.). - Berlin ; Heidelberg ; New York ; Barcelona ; Hong Kong ; London ; Milan ; Paris ; Singapore ; Tokyo : Springer, 2000
(Environmental science)

Printed in Germany

Cover Design: Struve & Partner, Heidelberg
Dataconversion: Büro Stasch · Uwe Zimmermann, Bayreuth

Preface

This book collects the scientific contributions from the lecturers at the 2nd edition of the "International School on Marine Chemistry" held in Ustica (Palermo, Italy) from 5 to 12 September 1998. The School was planned with the aim of giving an overview about the chemical processes occurring in the marine environment and the more recent analytical methodologies for their study.

The School was organised under the auspices of the Italian Chemical Society and with the financial support of the Marine Reserve of Ustica Island, the Committee for Environment of Italian CNR, the University of Palermo, the Provincia Regionale of Palermo and the Shimadzu-Italia Corporation. The book has been printed with the financial support of the "Assessorato Ambiente" of the Provincia Regionale of Palermo.

All the participants, about a hundred including the lecturers and the Organising Committee, are grateful for the generous support of the agencies. A particular acknowledgement from the Editors is to all the lecturers for their availability and courtesy and for the high quality of their scientific contributions.

Prof. Antonio Gianguzza
Prof. Ezio Pelizzetti
Prof. Silvio Sammartano

Contents

Contributors

Stefania Angelino
(PhD in Chemistry)
Dip. di Chimica Analitica, Università di Torino
Via P. Giuria 5, I-10125 Torino, Italy
Phone: +39 (0)11 6707626, Fax: +39 (0)11 6707615
E-mail: gennaro@al.unipmn.it

Renato Barbieri
(Professor of Inorganic Chemistry)
Dip. di Chimica Inorganica, Università di Palermo
Viale delle Scienze, Parco d'Orleans II
I-90128 Palermo, Italy
Phone: +39 (0)91 590578, Fax +39 (0)91 427584
E-mail: barbieri@mbox.unipa.it

Adriana Barbieri-Paulsen
(Ph.D. in Chemistry)
Institut für Physik, Medizinische Universität Lübeck
Ratzeburger Allee 160
D-23538 Lübeck, Germany
Phone: +49 (0)451 5004208
E-mail: barbieri@physik.mu-luebeck.de

Renato Barbieri
(Professor of Inorganic Chemistry)
Dip. di Chimica Inorganica, Università di Palermo
Viale delle Scienze, Parco d'Orleans II
I-90128 Palermo, Italy
Phone: +39 (0)91 590578, Fax +39 (0)91 427584
E-mail: barbieri@mbox.unipa.it

Giampaolo Barone
(Ph.D. in Chemistry)
Dip. di Chimica Inorganica, Università di Palermo
Viale delle Scienze, Parco d'Orleans II
I-90128 Palermo, Italy
Phone: +39 (0)91 590578, Fax +39 (0)91 427584
E-mail: gruisi@mbox.unipa.it

Rodolfo Cecchi
(Professor of Physical Oceanography)
Dip. di Scienze dell'Ingegneria, Università di Modena
Via Campi 213, I-41100 Modena, Italy
Phone: +39 (0)59 370703, Fax: +39 (0)59 373180
E-mail: cecchi@rainbow.unimo.it

Peter J. Craig
(Professor. BSc PhD Chem FRSC)
Department of Chemistry, De Monfort University
PO Box 143, Leicester LE8 OHL, UK
Phone: +44 (0)116 2577102, Fax: +44 (0)116 2577287
E-mail: pjcraig@dmu.ac.uk

Concetta de Stefano
(Professor of Analytical Chemistry)
Dip. di Chimica Inorganica, Chimica Analitica e Chimica Fisica, Università di Messina
Salita Sperone 31, I-98166 Messina, Italy
Phone: +39 (0)90 391354, Fax: +39 (0)90 392827
E-mail: destefano@chem.unime.it

Margaret Farago
(Professor of Environmental Chemistry)
Environmental Geochemistry Research Group
The T.H. Huxley School of Environment, Earth Science and Engineering Royal School of Mines
London SW7 2BP, UK
Phone: +44 (0)171 5946397, Fax: +44 (0)171 5946408
E-mail: m.farago@ic.ac.uk

Michele Forina
(Professor of Analytical Chemistry)
Dip. di Chimica e Tecnologie Farmaceutiche ed Alimentari, Università di Genova
Via Brigata Salerno s/n, I-16147 Genova, Italy
Phone +39 (0)10 3532630, Fax: +39 (0)10 3532684
E-mail: forina@dictfa.unige.it

Sarah N. Forster
(Dr.)
Deptartment of Chemistry, De Monfort University
PO Box 143, Leicester LE8 OHL, UK
Phone: +44 (0)116 2577102, Fax: +44 (0)116 2577287
E-mail: pjcraig@dmu.ac.uk

Claudia Foti
(Ph.D., Researcher of Analytical Chemistry)
Dip. di Chimica Inorganica, Chimica Analitica e Chimica Fisica, Università di Messina
Salita Sperone 31, I-98166 Messina, Italy
Phone: +39 (0)90 391354, Fax: +39 (0)90 392827
E-mail: foti@chem.unime.it

Roberto Frache
(Professor of Analytical Chemistry)
Dipartimento di Chimica e Chimica Industriale
Università degli Studi di Genova
Via Dodecaneso 31, I-16146 Genova, Italy
Phone: +39 (0)10 3536186, Fax: +39 (0)10 3625051
E-mail: frache@chimica.unige.it

Maria C. Gennaro
(Professor of Analytical Chemistry)
Dipartimento di Scienze e Tecnologie Avanzate
Corso Borsalino 54, I-15100 Alessandria, Italy
Phone: +39 (0)131 283806, Fax: +39 (0)131 283800
E-mail: gennaro@al.unipmn.it

Grazia Ghermandi
(Professor of Physics)
Dipartimento di Scienze dell'Ingegneria
Università di Modena
Via Campi 213, I-41100 Modena, Italy
Phone: +39 (0)59 370703, Fax: +39 (0)59 373180
E-mail: ghermand@rainbow.unimo.it

Antonio Gianguzza
(Prof. of Analytical and Environmental Chemistry)
Dip. di Chimica Inorganica, Sezione di Chimica Ambientale, Università di Palermo
Viale delle Scienze, Parco d'Orleans II
I-90128 Palermo, Italy
Phone: +39 (0)91 489409, Fax: +39 (0)91 427584
E-mail: giang@mbox.unipa.it

Monica Gulmini
(PhD, Researcher)
Dip. di Chimica Analitica, Università di Torino
Via P. Giuria 5, I-10125 Torino, Italy
Phone: +39 (0)11 6707614, Fax: +34 (0)11 6707615
E-mail: gulmini@ch.unito.it

Alessandra Irico
(PhD, Student)
Dip. di Chimica Analitica, Università di Torino
Via P. Giuria 5, I-10125 Torino, Italy
Phone: +39 (0)11 6707636, Fax: +34 (0)11 6707615

Richard O. Jenkins
(Dr.)
Dep. of Biological Sciences, De Monfort University
The Gateway, Leicester LE1 9BH, UK
Phone: +44 (0)116 2576306, Fax: +44 (0)116 257 7287
E-mail: roj@dmu.ac.uk

Silvia Lanteri
(Professor)
Dip. di Chimica e Tecnologie Farmaceutiche ed Alimentari, Università di Genova
Via Brigata Salerno s/n, I-16147 Genova, Italy
Phone: +39 (0)10 3532634, Fax: +39 (0)10 3532684
E-mail: silvia@dictfa.unige.it

David P. Miller
(Dr.)
Department of Chemistry, De Monfort University
PO Box 143, Leicester LE8 OHL, UK
Phone: +44 (0)116-257 7102, Fax: +44 (0)116 2577287
E-mail: pjcraig@dmu.ac

Frank J. Millero
(Professor of Marine and Physical Chemistry)
Rosenstiel School of Marine and Atmospheric Sciences, Dep. of Marine Chemistry
University of Miami
4600 Rickenbacker Causeway
Miami, Florida 33149, USA
Phone: +1 (0)305 3614707: Fax: +1 (0)305 3614144
E-mail: fmillero@rsmas.miami.edu

Claudio Minero
(Professor of Environmental Chemistry)
Dip. di Chimica Analitica, Università di Torina
Via P. Giuria 5, I-10125 Torino, Italy
Phone: +39 (0)11 6707632, Fax: +39 (0)11 6707615
E-mail: minero@ch.unito.it

Tracy-Ann Morris
(Dr.)
Department of Chemistry, De Monfort University
PO Box 143, Leicester LE8 OHL, UK
Phone: +44 (0)116 2577102, Fax: +44 (0)116 2577287
E-mail: pjcraig@dmu.ac.uk

Naaman Ostah
(Dr.)
Department of Chemistry, De Monfort University
PO Box 143, Leicester LE8 OHL, UK
Phone: +44 (0)116 2577102, Fax: +44 (0)116 2577287
E-mail: pjcraig@dmu.ac.uk

*Silvio Pantoja**
(Assistant Professor)
Dep. of Marine Chemistry and Geochemistry
Whods Hole Oceanographic Institution
Whods Hole, MA 02543, USA
Phone: +1 (0)508 2892740, Fax: +1 (0)508 4572164
E-mail: spantoja@whoi.edu
*current adress:
Dep. of Oceanography, University of Concepción
Casilla 160-C, Concepción, Chile
Phone: +56 (0)41 203499, Fax: +56(0)41 225400
E-mail: spantoja@udec.cl

Ezio Pelizzetti
(Professor of Analytical Chemistry)
Dip. di Chimica Analitica, Università di Torino
Via P. Giuria 5, I-10125 Torino, Italy
Phone: +39 (0)11 6707630, Fax: +39 (0)11 6707615
E-mail: pelizzett@ch.unito.it

Lorenzo Pellerito
(Professor of Inorganic Chemistry)
Dip. di Chimica Inorganica, Università di Palermo
Viale delle Scienze, Parco d'Orleans II
I-90128 Palermo, Italy
Phone: +39 (0)91 590367, Fax: +34 (0)91 427584
E-mail: bioinorg@mbox.unipa.it

Maurizio Pettine
(Dr.)
Istituto di Ricerca sulle Acque, Consiglio Nazionale delle Ricerche (CNR)
Via Reno 1, I-00198 Roma, Italy
Phone: +39 06 8841451, Fax: +39 06 8417861
E-mail: pettine@irsa1.irsa.rm.cnr.it

Salvatore Posante
(Ph.D in Chemistry)
Dip. di Chimica Inorganica, Università di Palermo
Viale delle Scienze, Parco d'Orleans II
I-90128 Palermo, Italy
Phone: +39 (0)91 489369, Fax: +39 (0)91 427584
E-mail: gruisi@mbox.unipa.it

Martin R. Preston
(Dr., BSc, PhD, MRSC, Chem)
Oceanography Laboratories
University of Liverpool
Liverpool L69 3BX, UK
Phone: +44 (0)151 7944093, Fax: +44 (0)151 7944099
E-mail: preston@liverpool.ac.uk

Paola Rivaro
(Ph.D. in Analytical Chemistry)
Dipartimento di Chimica e Chimica Industriale
Università degli Studi di Genova
Via Dodecaneso 31, I-16146 Genova, Italy
Phone: +39 (0)10 3536178, Fax: +39 (0)10 3626190
E-mail: rivaro@chimica.unige.it

Mario Rossi
(Dr.)
Dip. di Chimica Inorganica, Università di Palermo
Viale delle Scienze, Parco d'Orleans II
I-90128 Palermo, Italy
Phone: +39 (0)91 489369, Fax: +39 (0)91 427584
E-mail: gruisi@unipa.it

Giuseppe Ruisi
(Professor of Analytical Chemistry)
Dip. di Chimica Inorganica, Università di Palermo
Viale delle Scienze, Parco d'Orleans II
I-90128 Palermo, Italy
Phone: +39 (0)91 489369, Fax: +39 (0)91 427584
E-mail: gruisi@mbox.unipa.it

Silvio Sammartano
(Professor of Analytical Chemistry)
Dip. di Chimica Inorganica, Chimica Analitica e Chimica Fisica, Università di Messina
Salita Sperone 31, I-98166 Messina, Italy
Phone: +39 (0)90 393659, Fax: +39 (0)90-392827
E-mail: sammartano@chem.unime.it

Corrado Sarzanini
(Professor of Analytical Chemistry)
Dip. di Chimica Analitica, Università di Torino
Via P. Giuria 5, I-10125 Torino, Italy
Phone: +39 (0)11 6707628, Fax: +39 (0)11 6707615
E-mail: sarzanini@silver.ch.unito.it

Arturo Silvestri
(Professor of Inorganic Chemistry)
Dip. di Chimica Inorganica, Università di Palermo
Viale delle Scienze, Parco d'Orleans II
I-90128 Palermo, Italy
Phone: +39 (0)91 489369, Fax: +39 (0)91 427584,
E-mail: asilves@mbox.unipa.it

Louise M. Smith
(Dr.)
Dep. of Biological Sciences, De Monfort University
PO Box 143, Leicester LE8 OHL, UK
Phone: +44 (0)116 2576306, Fax: +44 (0)116 2577287
E-mail: roj@dmu.ac

Barbara Sulzberger
(Ph.D., Professor of Environmental Chemistry)
Swiss Federal Institute for Environmental Science and Technology
Überlandstrasse 133
CH-8600 Duebendorf, Switzerland
Phone: +41 (0)1 8235511, Fax: +41 (0)1 8235028
E-mail: sulzberger@eawag.ch

Roberto Todeschini
(Professor)
Dipartimento di Scienze dell'Ambiente e del Territorio, Università di Milano
Via L. Emanueli 15, I-20126 Milano, Italy
Phone: +39 (0)2 64474307, Fax: -39 (0)2 64474300
E-mail: roberto.todeschini@unimi.it

Constant van den Berg
(Ph.D.)
Oceanography Laboratories
University of Liverpool
Liverpool L69 3BX, UK
Phone: +44 (0)151 7944096, Fax: +44 (0)151 7944099
E-mail: c.m.g.van-den-berg@liverpool.ac.uk

Marco Vincenti
(Professor of Analytical Chemistry)
Dip. di Chimica Analitica, Università di Torino
Via P. Giuria 5, I-10125 Torino, Italy
Phone: +39 (0)11 6707636, Fax: +39 (0)11 6707615
E-mail: vincenti@ch.unito.it

Stuart Wakeham
(Ph.D., Professor)
Skidaway Institute of Oceanography
10 Ocean Science Circle, Savannah, GA 31411, USA
Phone: +01 (0)912 5982347, Fax: +01 (0)912598 2310
E-mail: stuart@skio.peachnet.edu

Vincenzo Zelano
(Professor of Analytical Chemistry)
Dip. di Chimica Analitica, Università di Torino
Via P. Giuria 5, I-10125 Torino, Italy
Phone: +39 (0)11 6707619, Fax: +39 (0)11 6707615
E-mail: zelano@ch.unito.it

Introduction – Environmental Analytical Chemistry as a Tool for Studying Chemical Processes in Marine Environments

A. Gianguzza · E. Pelizzetti · S. Sammartano

Introduction

The marine analytical chemist must be aware that the constituents – dissolved and suspended – of sea water have a three-dimensional distribution pattern. In contrast to distribution patterns in a mineral resource or in a batch of ore, distribution patterns in the marine environment are influenced by chemical, biological and physical processes, resulting in pronounced variability of individual patterns. Very often, only the general features of such a distribution pattern are known before seawater analysis is carried out. The optimal approach, therefore, presupposes an understanding of the marine environment and of the wide variety of chemical processes influencing the distribution of the compounds under examination. This means the marine analytical chemist is forced to have a reasonable knowledge of chemical oceanography.

A correct understanding of processes in the marine environment can only be achieved by unravelling the extreme physical and chemical complexity of the different compartments involved – atmosphere, water and sediments – and their interconnections. It is recognised that the determination of global parameters, such as total concentrations or average constants, is only a first step in the correct and rigorous representation of environmental properties. Environmental processes are affected by many factors that are still poorly evaluated or not evaluated at all. New approaches and methodologies are needed, such as the combination of analytical techniques (hyphenated techniques) or the use of mathematical procedures to extract detailed information and to build up useful predictive models, and finally a critical body of compilation data. By following the suggestions of the IUPAC Commission on Environmental Analytical Chemistry (1989), concerning the "Principal Activities" in which the evaluation of concepts and methodologies are of primary importance, the second meeting of the "International School on Marine Chemistry" (Ustica, Palermo 7–14 September 1998), was planned with the aim of presenting the most recent developments in the chemistry of the sea, with particular emphasis on chemical processes and analytical methodologies. Some aspects of the main topics presented during the School are briefly outlined, below.

Biogeochemical Cycles

The Carbon Cycle and Organic Matter. The biogeochemical processes occurring at the air-ocean interface are closely linked to the marine productivity and to earth's climate. Among these, the most important for the life on the earth is by far the assimilatory reduction of carbon and nitrogen by photosynthetic organisms. Therefore, the carbon and the nitrogen cycles cannot be considered as independent between them, but, on the contrary, as interconnected aspects of the whole living cycle. Despite their importance for marine life and the great interest by scientist, there are large gaps in their understanding. The greatest difficulty in studying the different aspects of biogeochemical cycles mainly is in the variability of rates and stoichiometry of the chemical reactions involved ("open systems"), which are dependent on the environmental conditions, such as light, temperature, oxidation state, etc., of the marine ecosystem. Therefore, the most convenient way of determining the overall amounts and fluxes of organic compounds in natural cycles is to use radioactive carbon tracers. For example, measurements of the variation of the $^{13}C/^{12}C$ ratio vs. time demonstrate that the ocean is receiving fossil fuel carbon (Quay et al. 1992; Hedges 1992). The measurements of *DOC* in sea water are matters of great controversy. The most used method, based on high-temperature combustion to CO_2, is not yet completely pointed out, as demonstrated by noticeable differences in the results of recent inter-laboratory comparisons (Sugimura and Suzuki 1988; Martin and Fitzwater 1992). The qualitative and the quantitative analysis of individual organic compounds present in sea water at nanomolar concentrations represents a challenge for the analytical chemist, and a further obstacle in the study of natural cycles (Baker and Hites 1999; Tittlemeier et al. 1999; Agrell et al. 1999; Harner et al. 1999). Current methodologies require large samples, time-consuming chromatographic separation, preconcentration (Yang et al. 1993) and other sample handling which are not always easy to carry out on ship cruise. Advances in mass spectrometry and in hyphenated HPLC-MS techniques that provide detection limits in the picomolar range (Vincenti 1997) allow an expansion in the number of detectable compounds.

The first and most important step of living cycles is by far the uptake of atmospheric CO_2 by the oceans. Whereas biologists have focused their attention on the role of carbon dioxide in photosynthetic processes (*Biological Pump*), chemists and chemical oceanographers were interested in the participation of carbon dioxide in the major geochemical cycles and also in the processes that led to the distribution of CO_2 and its related chemical forms (HCO_3^- and CO_3^{2-}) in the oceans (*Solubility Pump*), with the aim of building up chemical models to define the alkalinity and the buffer capacity of sea water.

During the last decades, interest in the carbon dioxide system comes also from its importance as a major greenhouse gas and its effect on the climate's global change.

Despite the fundamental importance of this chemical system, the CO_2 exchange rate between atmosphere and surface sea water is not yet well known, owing to the difficulty in measuring the net increase of inorganic carbon in the oceans and in distinguishing the fraction of inorganic carbon involved in living processes. Many efforts are in progress by scientists to develop procedures to detect CO_2 partial pressure by means optical sensors (Millero 1999) and, moreover, to realize routine measurements, to be carried out aboard ship, of alkalinity and pH of sea water within errors $<0.1\%$.

Metals and Organometallic Compounds

Analytical Methodologies. Natural concentrations of "heavy metals" are extremely low in the hydrosphere (Salomons and Forstner 1984; Bruland 1983). Their detection implies that the following problems have to be solved by an analytical chemist:

a. sampling, avoiding contamination
b. separation by matrix in order to remove interference
c. cleanup
d. selective preconcentration
e. accuracy and precision of the instrumental detection method

The development of these procedures and the use of chromatographic techniques (particularly HPLC) in the preconcentration and separation steps, together with the availability of very sensitive instrumental techniques, such as Atomic Absorption Spectroscopy with graphite furnace (AAS-ETA) or Inductively-Coupled Plasma Mass Spectrometry (ICP-MS, Alves et al. 1993), means that it is now possible to determine trace metal concentrations in ng l^{-1} rather than in µg l^{-1}, i.e. a significant achievement. However, efforts have still to be made to obtain better reproducibility of results and to set up faster analytical methods. As regards sensitivity, low accuracy risk and low cost requirements, the Cathodic and Anodic Stripping Voltammetry, (CSV and ASV, respectively) are the most versatile analytical methods for direct determination of the total concentration of some metals (Van Den Berg 1989). To carry out simultaneous detection of many trace elements, the PIXE (Photo Inductive X-ray Emission) analysis can be also used.

In spite of their low concentration, metals are generally extremely reactive and generally widely dispersed in nature in association with or as part of a variety of solid phases. The natural pathways of heavy metals are largely controlled by two major processes: incorporation to or into various solid phases and chemical complexation in solution. Different factors regulate these processes in the marine environment: pH, pE, the oxidation state of the metal and its acidity hardness. These parameters act to respectively restrict and promote the solubilization of the metal, but, owing to pH and pE values and to remarkable concentrations of complexing organic matter in sediments, immobilisation to the various solid phases generally predominates in the marine environment, so that sediments behave as a "reservoir sink" for metals in the environment (Bloom et al. 1999). Kinetics studies can give useful information about the absorption-release mechanism and contribute to defining the equilibrium distribution of metals between different phases (Millero 1996).

Chemical Speciation. Bare-metal ions cannot exist in water. Their basic unit is the hydrated aquo-ion which behaves like Lewis acid and, by a series of proton transfer (hydrolysis) reactions, a sequence of hydroxo- and possibly oxo-complexes may be produced (Stumm and Morgan 1996; Morel 1983). In complexation studies of metal ions in aqueous solutions, hydrolytic species cannot be neglected (Baes and Mesmer 1976). Moreover, when dealing with a multicomponent solution, such as sea water, other possible interactions of metals with inorganic (chloride, sulfate, carbonate, phosphate) and organic low and high molecular weight ligands must be taken into account (Stumm and Brauner 1975). Then, a network of interactions is established and metal species are present in different chemical forms (*Chemical Speciation*), each of them often showing different trends in bioavailability, accumulation, toxicity, absorption and/or release, etc. (Sadiq 1992; Tessier and Turner 1995).

Some metals, such as lead, tin, mercury, arsenic and antimony, are often present in sea waters as organic derivatives deriving both from anthropic inputs and from the enzymatic bioalkylation processes of inorganic metal ions (Craig and Miller 1997; Farago 1997). Particular attention must be paid to the study of organomercury and organotin compounds; the latter are used in antifouling paints for ships and are very widespread in sea water (Craig 1986; Champ and Seligman 1996). Since different R alkyl groups in the R_3SnX class determine different toxicity levels towards living organisms, it becomes relevant to evaluate the distribution of mono-, di- and trialkyltin derivatives, both in water and in sediments. For this purpose different extraction and detection methodologies using hyphenated techniques (HPLC-AAS, GLC-AAS) have been set up. Moreover, the interactions of soluble organometallic cations with sea water components have to be considered in the general picture of chemical speciation of trace elements, first of all by taking into account their hydrolysis processes. In this case, speciation analysis can be successfully performed by potentiometry. Mössbauer spectroscopy can be used to determine the structure of some complexes of organotin moieties with some important biological molecules (such as DNA and its derivatives), contributing to an understanding of the biochemical interactions of organometallic compounds with living organisms.

Chemometric Analysis. All the information obtained from the analysis of the parameters of the systems under study can be exploited with a better rationality by Chemometrics, with the aim of finding a correct correlation between them. Chemometric analysis plays a very important role in the studies of natural systems, such as marine waters, where many and variables parameters have to be controlled contemporary. In these cases, Chemometrics allows the definition of the better and more representative sampling methodology and the prediction of some general trends of chemical processes in the natural systems.

The School and this book, its outcome, do not presume to give a whole picture of the complex subject on the chemical processes in the marine environment. The aim was to give the analytical chemist a sense of the enormous challenge facing the chemical oceanographer, showing the role of the analytical chemistry in the knowledge of chemical aspects of the global biogeochemical cycle. Thus, it will be possible to avoid monitoring becoming a goal in itself and to set the conceptual basis on which analytical protocols and oceanographic cruises are programed. Then, hopefully, the power of environmental analytical chemistry to make a contribution to the understanding, preservation and correct management of the ocean and the earth as a whole should emerge.

References

Agrell C, Okla L, Larsson P, Blacke C, Wania F (1999) Evidence of latitudinal fractionation of polychlorinated byphenil congeners along the Baltic Sea region. Environ Sci Technol 33:1149–1156
Alves LC, Allen LA, Houk RS (1993) Measurements of vanadium, nickel and arsenic in seawater and urine reference materials by inductively-coupled plasma mass spectrometry with cryogenic desolvation. Anal Chem 65:2468
Baes CF, Mesmer RE (1976) The hydrolysis of cations. Interscientia, New York
Baker JI, Hites RA (1999) Polychlorinated dibenzo-p-dioxins and dibenzofurans in the remote North Atlantic marine atmosphere. Environ Sci Technol 33:14–20
Bloom NS, Gill GA, Cappellino S, Dobbs C, Mcshea L, Driscoll C, Mason R, Rudd J (1999) Speciation and cycling of mercury in Lavaca Bay, Texas, sediments. Environ Sci Technol 33:7–13

Bruland KW (1983) Trace elements in sea water. In: Riley JP, Chester R (eds) Chemical oceanography, 2nd ed, vol VIII. Academic Press, London, pp 157–220
Champ MA, Seligman, PF (eds.) (1996). Organotin. Environmental fate and effects. Chapman & Hall, London
Craig PJ (ed) (1986) Organometallic compounds in the environment. Principles and reactions. Longman, Harlow Essex
Craig PJ, Miller D (1997) Metal ions and organometallic compounds in sea water and in sediments: Biogeochemical cycles. In: Gianguzza A, Pelizzetti E, Sammartano S (eds) Marine chemistry – an environmental analytical chemistry approach. Kluwer Academic, Dordrecht (Water Science and Technology Library, vol XXV, pp 85–98)
Farago ME (1997) Arsenic in the marine environment. In: Gianguzza A, Pelizzetti E, Sammartano S (eds) Marine chemistry – an environmental analytical chemistry approach. Kluwer Academic, Dordrecht (Water Science and Technology Library, vol XXV, pp 275–292)
Harner T, Kylin H, Bidleman TF, Strachan WMJ (1999) Removal of alpha- and gamma-hexachlorocycloexane and enantiomers of alpha-hexachlorocycloexane in the Eastern Artic Ocean. Environ Sci Technol 33:1157–1164
Hedges J (1992) Global biogeochemical cycles: progress and problems. Mar Chem 39:67–93
Martin JH, Fitzwater SE (1992) Dissolved organic carbon in the Atlantic, Southern and Pacific Oceans. Nature 356:699–700
Morel FMM (1983) Principles of aquatic chemistry. John Wiley &Sons, New York
Millero FJ (1996) Chemical oceanography, 2nd edn. CRC Press, Boca Raton
Millero FJ (1999) Physical chemistry of natural waters. CRC Press, Boca Raton
Quay PD,Tilbrook B, Wong CS (1992) Oceanic uptake of fossil fuel CO_2: carbon-13 evidence. Science 256:74
Sadiq M (1992) Toxic metal chemistry in marine environments. Marcel Dekker Inc., New York
Solomons W, Forstner U (1984) Metals in the hydrocycle. Springer Verlag, München
Stumm W, Brauner PA (1975) Chemical speciation. In: Riley JP, Skirrow G (eds) Chemical oceanography, 2nd edn, vol I. Academic Press, London, pp 173–240
Stumm W, Morgan JJ (1996) Aquatic chemistry, 3rd edn. John Wiley & Sons, New York
Sugimura Y, Suzuki Y (1988) High temperature catalytic oxidation method for the determination of non volatile dissolved organic carbon in seawater by direct injection of a liquid sample. Mar Chem 24:105–112
Tessier A, Turner DR (1995) Metal speciation and bioavailability in aquatic systems. John Wiley & Sons, Chichester (IUPAC Series on Analytical and Physical Chemistry of Environmental Systems, vol III)
Tittlemeier SA, Simon M, Jarman WM, Elliott JE, Norstrom RJ (1999) Identification of a novel $C_{10}H_6N_2Br_4C_{l2}$ heterocyclic compound in seabird eggs. A bioaccumulation marine natural products? Environ Sci Technol 33:26–33
Van Den Berg CMG (1989) Electroanalytical chemistry of sea water. In: Riley JP (ed) Chemical oceanography, vol IX. Academic Press, London, pp 197–245
Vincenti M (1997) Application of mass spectrometric techniques to the detection of natural and anthropogenic substances in the sea. In: Gianguzza A, Pelizzetti E, Sammartano S (eds) Marine chemistry – an environmental analytical chemistry approach. Kluwer Academic, Dordrecht (Water Science and Technology Library, vol XXV, pp 189–210)
Yang XH, Lee C, Seranton MI (1993) Determination of nanomolar concentrations of individual dissolved low molecular weight amines and organic acids in seawater. Anal Chem65:572

Part I

Biogeochemical Cycles

Chapter 1

The Carbonate System in Marine Environments

F.J. Millero

1.1 Introduction

The major portion of carbon in the oceans occurs in the carbonate system. This system involves the following equilibria

$$CO_{2(g)} \rightleftharpoons CO_{2(aq)} \tag{1.1}$$

$$CO_{2(aq)} + H_2O \rightleftharpoons H^+ + HCO_3^- \tag{1.2}$$

$$HCO_3^- \rightleftharpoons H^+ + CO_3^{2-} \tag{1.3}$$

$$Ca^{2+} + CO_3^{2-} \rightleftharpoons CaCO_{3(s)} \tag{1.4}$$

The carbonate system is very important since it regulates the pH of sea water, and controls the circulation of CO_2 between the biosphere, the lithosphere, the atmosphere and the oceans. Recent interest in the carbonate system in the oceans has resulted from the "greenhouse effect" of CO_2. The concentration of CO_2 in the atmosphere has increased in the twentieth century (Fig. 1.1) (Keeling and Whorf 1994; Neftel et al. 1994). Since CO_2 can absorb infrared (IR) energy, this increase may cause the temperature of the earth to increase and could eventually melt the polar ice caps. The increase in CO_2 is related to the burning of fossil fuels (coal, petroleum and natural gas) and the production of cement. The atmospheric CO_2 enters the oceans across the air-sea interface and participates in the equilibrium processes outlined by Eq. 1.1 to 1.4. It also can be used by ocean plants in primary productivity

$$CO_2 + H_2O \longrightarrow CH_2O + O_2 \tag{1.5}$$

The adsorption of CO_2 by the oceans is quite complicated, since the rates of movement of the gas across the interface and from surface to deep waters varies with latitude, time, and season. Diurnal and seasonal variations in the carbonate system are caused by the removal of CO_2 by photosynthesis and solar heating. If the oceans were well-mixed and in equilibrium with the atmosphere, most of the increased CO_2 would be absorbed. This, however, is not the case, and the ocean response to increases in CO_2 is slow due to physical and chemical factors. The exchange involves the hydration of CO_2, which is a slow process relative to ionization. Approximate time scales for the mixing process can be determined using radioactive tracers to gain some idea of the mixing times. To use these estimates it is necessary to have some idea of the total car-

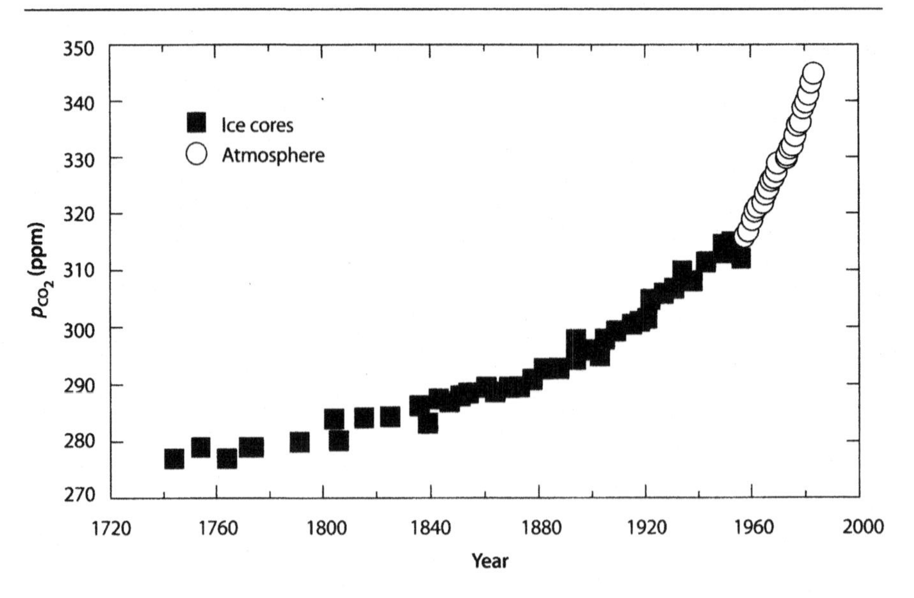

Fig. 1.1. The increase of CO_2 in the atmosphere with time (Keeling and Whorf 1994; Neftel et al. 1994)

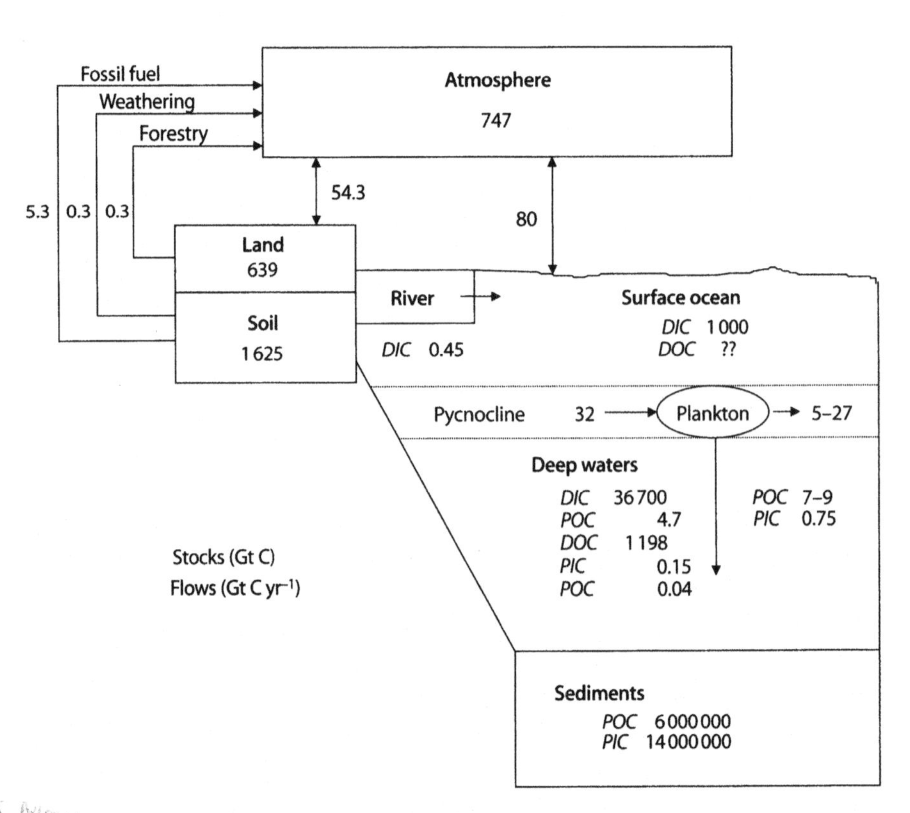

Fig. 1.2. The global carbon cycle (Millero 1996)

bon in various reservoirs and the global carbon cycle. The most recent estimates are shown in Fig. 1.2. The inorganic carbon estimates are reasonably accurate, but estimates for the carbon in the marine biosphere and humus are less precisely known. Most of the carbon in the oceans resides below the thermocline. The amount of carbon stored in carbonate rocks and sediments is a lot larger than the cycled CO_2, but is not important on short time scales (years).

The recent interest in the distribution of CO_2 in the oceans is related to the need to understand how the increase of CO_2 in the atmosphere and the expected increase in temperature will affect the climate. The classical measurements of p_{CO_2} in the atmosphere were made by Keeling and Whorf (1994) at the Mauna Loa Observatory in Hawaii, started in 1958. More recent measurements have been made throughout the world and on the air trapped in ice cores. These measurements clearly demonstrate that the CO_2 in the atmosphere is increasing due to the burning of fossil fuels. Although the rates of increase are the same as the increase in the use of fossil fuels, the amounts in the atmosphere are only 40% of the expected values (Table 1.1). This is thought to be due to the uptake by the oceans. The estimates of the sources and sinks of CO_2 in the atmosphere are given in Table 1.1. The difference between the sources (7.0 Gt yr^{-1}) and sinks (5.4 Gt yr^{-1}) of 1.6 Gt yr^{-1} is close to the overall uncertainty (1.4 Gt yr^{-1}). The estimated ocean sink of 2 Gt yr^{-1} has largely been determined using ocean models. Little direct proof is available to support this estimate. Atmospheric modelling (Tans et al. 1990) gives a lower value (1.0 Gt yr^{-1}). The value estimated from the penetration of ^{13}C into the oceans by Quay et al. (1992) supports the ocean modelling value. The fossil fuel carbon has a higher amount of ^{13}C than the atmospheric CO_2. Over time this $^{13}CO_2$ penetrates into the deep ocean. The modelling of the changes in the $^{13}CO_2$ has been used to determine a penetration rate of 25 to 35 m yr^{-1} and an oceanic uptake of sink of 2.1 Gt yr^{-1}.

The difference between the CO_2 produced and the amount put into the atmosphere has changed with time, which means that the natural sources and sinks have also changed. The global CO_2 measurements made recently will improve our estimate of the ocean sink and provide data that can be used to test the relibility of global CO_2 models used to predict the effect that future increases have on the climate. As pointed out recently by Sarmiento (1993), the CO_2 added to the atmosphere will eventually come

Table 1.1. Budget for the global CO_2 system (1980–1989)

	Average perturbation (10^{15} g C yr^{-1})
Sources	
Fossil fuel combustion	5.4 ±0.5
Deforestation	1.6 ±1.0
Total	7.0 ±1.2
Sinks	
Atmosphere	3.4 ±0.2
Oceans (models)	2.0 ±0.8
Total	5.4 ±0.8
Unaccounted for sinks	1.6 ±1.4

to equilibrium with seawater. Although this process is slow, his calculations indicate that the addition of 1 095 molecules of CO_2 to the atmosphere, will decrease to 15 molecules after 1 000 years. Most of the added CO_2 (985 molecules) will end up as part of the inorganic pool as bicarbonate and carbonate ions.

The CO_2 produced from the burning of fossil fuel can be taken up by the oceans by two methods:

1. the Solubility Pump and
2. the Biology Pump.

A sketch of the solubility pump is shown in Fig. 1.3. The driving force for the flux of CO_2 across the air-sea interface is the difference between the concentrations in the atmosphere and oceans given by

$$\text{Flux} = k(p_{CO_2(sw)} - p_{CO_2(atm)}) = k\Delta p_{CO_2} \quad (1.6)$$

where the value k is called the transfer velocity. The transfer velocity increases with increasing wind speed. The values of k (Wanninkhof 1992), determined in wind tunnel measurements (3–4 mol m^{-2} yr^{-1} μatm^{-1}), are much smaller than the value estimated from ^{14}C measurements, making it difficult to use Eq. 1.6 to calculate global fluxes of CO_2. When Δp_{CO_2} is positive, the oceans are a source of CO_2, and when it is negative the oceans are a sink for CO_2. To take up the missing CO_2, the value of Δp_{CO_2} world wide would have to be about 8 ppm. If rapid exchange takes place, one would expect the p_{CO_2} in the atmosphere to be equal to the values in the surface waters. If the exchange is sluggish, the p_{CO_2} in surface waters will be higher in upwelling areas and lower in colder waters than the values in the atmosphere.

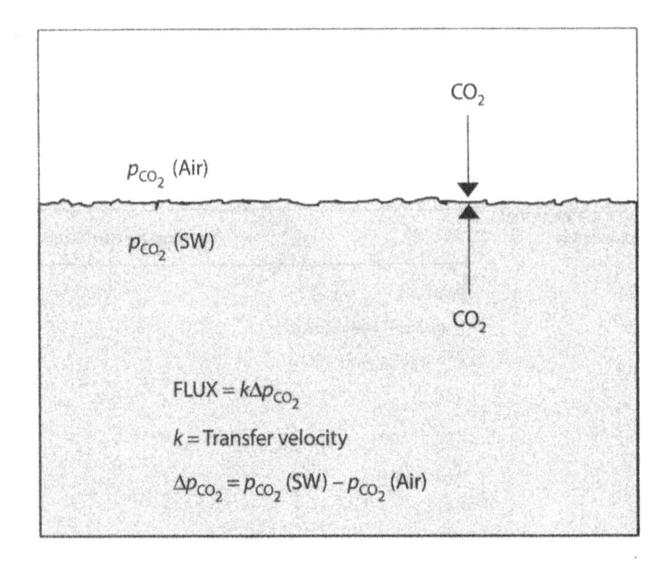

Fig. 1.3. The solubility pump

Fig. 1.4. The biological pump

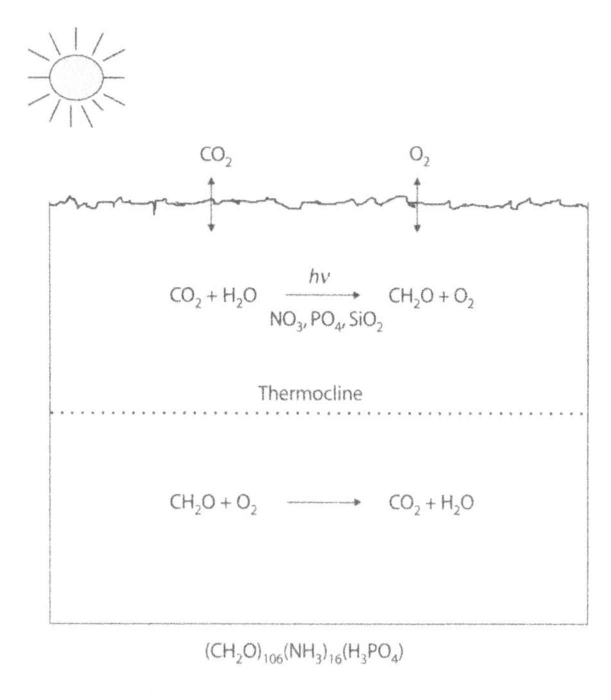

A sketch of the biological pump is shown in Fig. 1.4. The biological pump is primed in surface waters by the formation of plant material. This can be represented by the reaction (Redfield et al. 1963)

$$106\ CO_2 + 122\ H_2O + 16\ HNO_3 + H_3PO_4 \xrightarrow{hv} (CH_2O)_{106}(NH_3)_{16}H_3PO_4 + 138\ O_2 \quad (1.7)$$

The plants need nitrate, phosphate and silicate to grow, as well as some metals such as Fe and Mn (Millero 1997). When the plants die or are eaten by zooplankton, they can be removed from the surface and transported to deep waters below the thermocline. Bacteria can oxidize this material back to the starting material

$$(CH_2O)_{106}(NH_3)_{16}H_3PO_4 + 138\ O_2 \longrightarrow 106\ CO_2 + 122\ H_2O + 16\ HNO_3 + H_3PO_4 \quad (1.8)$$

This returns the CO_2 back to the waters that will not return to the surface for 500 to 1 000 years. Over short periods of time (100 years) the biological pump is thought to be in a steady state. Over geological periods, it is thought to be the cause of the changes in CO_2 that have occurred in the atmosphere and thus changed the temperature of the earth. It should be pointed out that the composition of phytoplankton given above (Redfield et al. 1963) has been shown by a number of workers to vary somewhat on location and species (Takahashi et al. 1985; Anderson and Sarmiento 1994; Steinberg et al. 1998).

Before we examine the distribution of CO_2 in the oceans, we will discuss the chemical concepts which were needed to understand the carbonate system in the oceans.

1.2 Equilibria of Carbonate Species

When carbon dioxide is in contact with water, equilibria, as defined by Eq. 1.1 to 1.4, will be established. Kinetics can affect the features of these reactions. Reaction 1.2 is first order with respect to CO_2, and has a first-order rate constant $k_1 = 0.03\ s^{-1}$ or a half time $t_{1/2} = \ln 2 / k_1 = 23$ s. The dehydration reaction, $H_2CO_3 \longrightarrow CO_2 + H_2O$, is first order with respect to $[H_2CO_3]$ with rate constant $k_{-1} = 20\ s^{-1}$ and $t_{1/2} = 0.03$ s. The values for the forward and backward reactions

$$CO_2 + H_2O \underset{k_{-1}}{\overset{k_1}{\rightleftharpoons}} H_2CO_3 \tag{1.9}$$

can be used to determine the equilibrium ratio of $K = k_1 / k_{-1} = 0.03 / 20 = 1 / 670$. This indicates that at equilibrium, the concentration of CO_2 is 670 times higher than H_2CO_3. This has led workers to use the so-called hydration convention to define the first ionization of carbonic acid.

The thermodynamics of the carbonic acid system have recently been reviewed (Millero 1995). The dissociation constants have been determined in natural waters using four pH scales (Dickson 1984; Millero 1986):

1. The activity scale $pH_{NBS} = -\log a_H$
2. The free (F) proton scale $pH_F = -\log [H^+]_F$
3. The total (T) proton scale $pH_T = -\log [H^+]_T$
4. The seawater proton scale $pH_{SWS} = -\log [H^+]_{SWS}$

where a_H is the activity of the proton and $[H^+]_F$ is the concentration of the free proton. The total seawater concentrations are related to the free concentrations by

$$[H^+]_T = [H^+]_F + [HSO_4^-] \tag{1.10}$$

$$[H^+]_{SWS} = [H^+]_F + [HSO_4^-] + [HF] \tag{1.11}$$

where $[HSO_4^-]$ and $[HF]$ are the concentration of the proton complexing with SO_4^{2-} and F^- in the solution. The relationships between the various scales are given in the Appendix along with equations for the dissociation constants of HSO_4^- and HF in seawater.

Since variations in the liquid junction potentials of various reference electrodes are different, it is better to use pH_{SWS}, pH_T or pH_F scales (Millero 1986). Seawater buffers are available that can be used to calibrate electrodes on these scales at a given temperature and salinity (Dickson 1993; Millero et al. 1993a). Although emf measurements are normally used to measure pH, it is also possible to use indicators that absorb light to measure pH. Clayton and Byrne (1993) have developed an indicator that can be used to measure the pH of seawater solution to a precision of 0.0004 and an accuracy of 0.003.

The stoichiometric dissociation constant for the first ionization is defined by

$$K_1^* = [H^+]_T [HCO_3^-]_T / [CO_2^*] \tag{1.12}$$

where $[H^+]_T = [H^+]_F + [HSO_4^-] + [HF]$, $[CO_2^*] = [CO_2] + [H_2CO_3]$ and the subscripts F and T are used to denote free and total concentrations. The concentration of dissolved CO_2 is related to the pressure by

$$[CO_2^*] = p_{CO_2} K_0 \tag{1.13}$$

where K_0 is the Henry's law constant similar to the values described for other gases (Weiss 1974). The solubility of CO_2 decreases with increasing temperature and salinity as for other gases. The solubility of CO_2 is greater than that of O_2 or N_2. The air ratios are $N_2 / O_2 / CO_2 = 630 / 240 / 1$, while the solution ratios are 28 / 19 / 1. Henry's law is not obeyed at high pH due to the formation of HCO_3^- and CO_3^{2-}. The stoichiometric value of K_1^* is related to the thermodynamic value by

$$K_1 = a_H a_{HCO_3} / a_{CO_2} a_{H_2O} = K_1^* \gamma_H \gamma_{HCO_3} / (\gamma_{CO_2} a_{H_2O}) \tag{1.14}$$

with $\gamma_{CO_2} = [CO_2]^0 / [CO_2]$ where the superscript zero denotes the solubility in pure water. The activity coefficients given above are the total values and include the effects due to the formation of ion pairs.

The stoichiometric constant for the second ionization of carbonic acid is defined by

$$K_2^* = [H^+]_T [CO_3^{2-}]_T / [HCO_3^-]_T \tag{1.15}$$

which is related to the thermodynamic value, K_2, by

$$K_2 = a_H a_{CO_3} / a_{HCO_2} = K_2^* \gamma_H \gamma_{CO_3} / \gamma_{HCO_3} \tag{1.16}$$

where the activity coefficients are total values which include the effect due to ionic interactions. The values of K_2^* in sea water can be determined from equations given in the appendix. The effect of pH on the various forms of carbonate in sea water are shown in Fig. 1.5.

The solubility product of $CaCO_3$ in its two major forms, calcite and aragonite, is also needed when studying the carbonate system. The stoichiometric solubility product is given by

$$K_{SP}^* = [Ca^{2+}]_T [CO_3^{2-}]_T \tag{1.17}$$

and is related to the thermodynamic value by

$$K_{SP}^* = K_{SP} / \gamma_T(Ca^{2+}) \gamma_T(CO_3^{2-}) \tag{1.18}$$

Since the dissociation of a number of other acids (H_2O, $B(OH)_3$, H_3PO_4, NH_4^+, H_2S, and $Si(OH)_4$ can be components of natural waters, it also necessary to know their dissociation constants in solution (Millero 1995). Equations needed to calculate the values of these acids in sea water are given in the appendix. In dilute solutions the equations given for sea water may not be valid due to differences in the composition of a given river, lake or estuary (Millero 1981; 1983). Recently we have constructed a basic program that can determine all the dissociation constants needed for any natural water of a known composition from 0 to 50 °C and 0 to 6 m (Millero and Roy 1997).

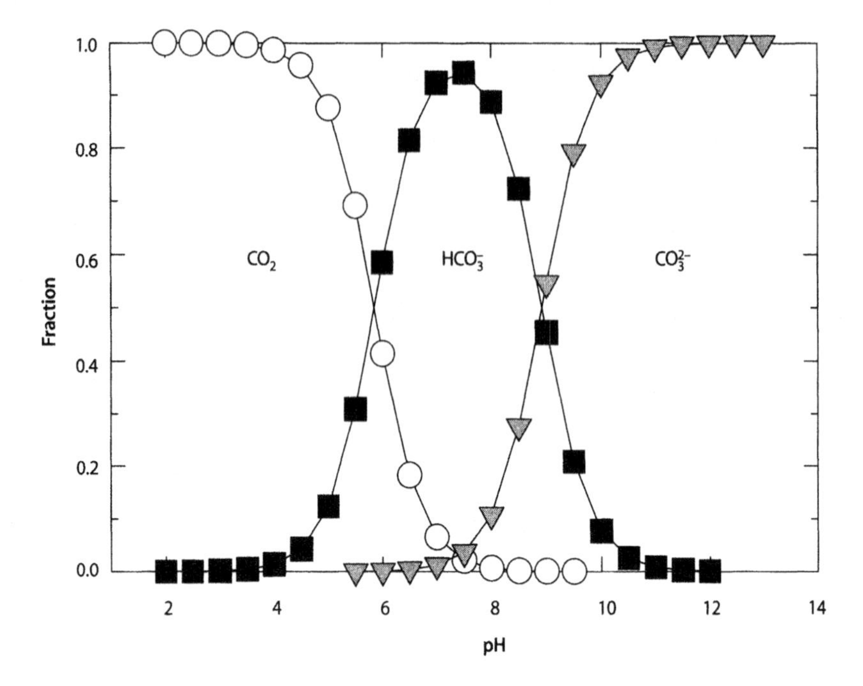

Fig. 1.5. The fractions of carbonic acid in sea water as a function of pH (Millero 1996)

1.3 Parameters of the CO_2 System in Sea Water

To characterize the various components of the carbonate system one must measure at least two of the four measurable parameters

1. pH
2. total alkalinity (TA)
3. total CO_2 (TCO_2)
4. partial pressure of CO_2 (p_{CO_2})

The pH can be measured using electrodes or indicators. If the electrodes are calibrated using seawater buffers (Millero 1981; Millero et al. 1993a), the accuracy is ±0.01 and the precision can be ±0.002 pH units. If the measurements are made at 25 °C and 1 atm, it is necessary to determine the "in situ" values at a given depth in the ocean. The effect of pressure and temperature can be determined from equations given elsewhere (Millero 1995).

The total alkalinity of sea water is defined as the concentration of all the bases that can accept H^+ when a titration is made with HCl to the carbonic acid end point. The value of TA is given by (Dickson 1981)

$$TA = [HCO_3^-] + 2\,[CO_3^{2-}] + [B(OH)_4^-] + [OH^-] - [H^+] + [SiO(OH)_3^-] + [MgOH^+] + [HPO_4^{2-}] + [PO_4^{3-}] \quad (1.19)$$

The percent of *TA* due to the various bases is shown in Table 1.2 for ocean waters with pH = 8, $[SiO(OH)_3^-] = 10^{-5.25}$ and $[HPO_4^{2-}] = 10^{-5.52}$. For most waters, $[HCO_3^-]$, $[CO_3^{2-}]$ and $[B(OH)_4^-]$ are the most important bases. For anoxic waters, HS^- and NH_3 can also contribute to the total alkalinity.

The carbonate alkalinity, *CA*, is defined by

$$CA = [HCO_3^-] + 2\,[CO_3^{2-}] \qquad (1.20)$$

and is calculated from

$$CA = TA - \Sigma B_i \qquad (1.21)$$

where $\Sigma B_i = [B(OH)_4^-] + \ldots$, the sum of all the bases other than HCO_3^- and CO_3^{2-}. The $[B(OH)_4^-]$ concentration, which is the largest source of $[B]_T$, can be calculated from

$$[B(OH)_4^-] = K^*_{HB}[B]_T / (K^*_{HB} + [H^+]_T) \qquad (1.22)$$

where $[B]_T$ is the total concentration of boron. The contribution due to other bases (phosphate, silicate, ammonium, and hydrogen sulfide) can be determined using the dissociation constants for the appropriate acid (Appendix). The methods used to determine the *TA* are given elsewhere (Millero et al. 1993b).

Although it is possible to determine TCO_2 from a titration, more reliable values can be obtained by direct measurements. This is done by stripping the inorganic CO_2 with nitrogen after the addition of phosphoric acid. The CO_2 can be collected in a liquid air trap and analysed by gas chromatography, infrared spectroscopy or by conductivity. By collecting the CO_2 in a DMSO solution with ethyleneamine, the CO_2 can be coulometrically titrated with OH^- produced on a Pt electrode. Routine measurements of TCO_2 on 30 cm^3 sea water can be made to a precision of 1 $\mu mol\ kg^{-1}$ and an accuracy of 2 $\mu mol\ kg^{-1}$ (Johnson et al. 1993).

The partial pressure of CO_2 in sea water is determined by equilibrating the sample with air or nitrogen. The CO_2 in the equilibrated gas is measured using gas chromatography or an IR analyser. By passing sea water through a shower head equilibrator, one can make continuous measurements on surface sea water (Wanninkhof and Thoning 1993). The system can be calibrated using standard CO_2 gas mixtures and yield values of p_{CO_2} to a precision of 1 μatm.

Table 1.2. Contribution of various components to the total alkalinity (*TA*) of sea water

Species	*TA* (%)
HCO_3^-	89.8
CO_3^{2-}	6.7
$B(OH)_4$	2.9
$SiO(OH)_3^-$	0.2
$MgOH^+$	0.1
OH^-	0.1
HPO_4^{2-}	0.1

As mentioned earlier, any two combinations of the four observable parameters can be used to characterize the carbonate system. It is also possible to use three parameters. This gives a total of ten combinations that can be used. The investigator must make a selection based on his or her needs after considering both the desired analytical precision and area of interest. The basic equations for determining the various parameters are given elsewhere (Park 1969). A number of computer codes are available to calculate the components of the carbonate system in sea water (Millero 1995; Lewis and Wallace 1998) from various inputs of the measured parameters. As mentioned earlier, the equations of Millero and Roy (1997) can be used to estimate the dissociation constants of acids in any natural water of known composition. Our computer code (Millero 1995) has been combined with the equations of Millero and Roy (1997), and can be used to determine the components of the carbonate system in any natural water (Gleitz et al. 1995).

To select the best two parameters needed to study the carbonate system, we can examine how the system changes during the formation and breakdown of organic carbon and the dissolution or precipitation of $CaCO_3$. The largest changes in the CO_2 system in deep waters are due to the oxidation of organic carbon. This oxidation can be followed by considering the changes in the apparent oxygen utilization ($AOU = [O_2]_{meas} - [O_2]_{sat}$). The effects of a change in AOU of 0.13 and 0.26 mM are given in Table 1.3 and shown in Fig. 1.6. The largest change occurs in p_{CO_2} followed by TCO_2 and pH. The carbonate alkalinity ($CA = [HCO_3^-] + 2\,[CO_3^{2-}]$), does not change. If one considers the present capabilities of measuring p_{CO_2} (±0.1%), TCO_2 (±0.17%), TA (±0.2%) and pH (±0.04%), the best selection would be p_{CO_2}-TCO_2 followed by pH-TCO_2 and p_{CO_2}-TA.

The changes in the CO_2 system due to the dissolution of $CaCO_3$ in deep waters are are given in Table 1.4 and shown in Fig. 1.7. The greatest change occurs in p_{CO_2}, with CA and pH following. The best combination is CA/TCO_2, followed by pH/CA and CA/TCO_2 obtained by an acid titration; this represents the best approach for studying changes in the carbonate system.

If one combines the AOU and $CaCO_3$ effects (Brewer et al. 1975; Chen 1978), one finds

$$(CH_2O)_{106}(NH_3)_{16}H_3PO_4 + 138\ O_2 + 124\ CO_3^{2-} \longrightarrow 16\ NO_3 + HPO_4^{2-} + 230\ HCO_3^- + 16\ H_2O \tag{1.23}$$

The CO_3^{2-} ions formed from the dissolution of $CaCO_3$ react with the protons formed from the oxidation of plant material. If x μM of $CaCO_3$ and y μM of organics are decomposed, the changes in TA, TCO_2 and NO_3^- are given by (Chen 1978)

$$\Delta TA = 2x - 17y \tag{1.24}$$

$$\Delta TCO_2 = x + 106y \tag{1.25}$$

$$\Delta NO_3 = 16y \tag{1.26}$$

The changes in Ca^{2+} are given by

$$\Delta Ca = 0.463\,\Delta TA + 0.074\,\Delta TCO_2 = 0.5\,\Delta TA + 0.53\,\Delta NO_3 \tag{1.27}$$

Table 1.3. Changes in the CO_2 system due to the oxidation of plant material (Millero 1997)

	Initial (mM)	ΔAOU (mM) 0.13	0.26
ΔCO_2	0	0.10	0.20
TCO_2	2.200	2.300	2.400
CA	2.487	2.487	2.487
p_{CO_2}	350	610	1 160
pH	8.200	8.001	7.753
$[CO_2]$	0.012	0.021	0.040
$[HCO_3^-]$	1.889	2.072	2.234
$[CO_3^{2-}]$	0.299	0.208	0.126

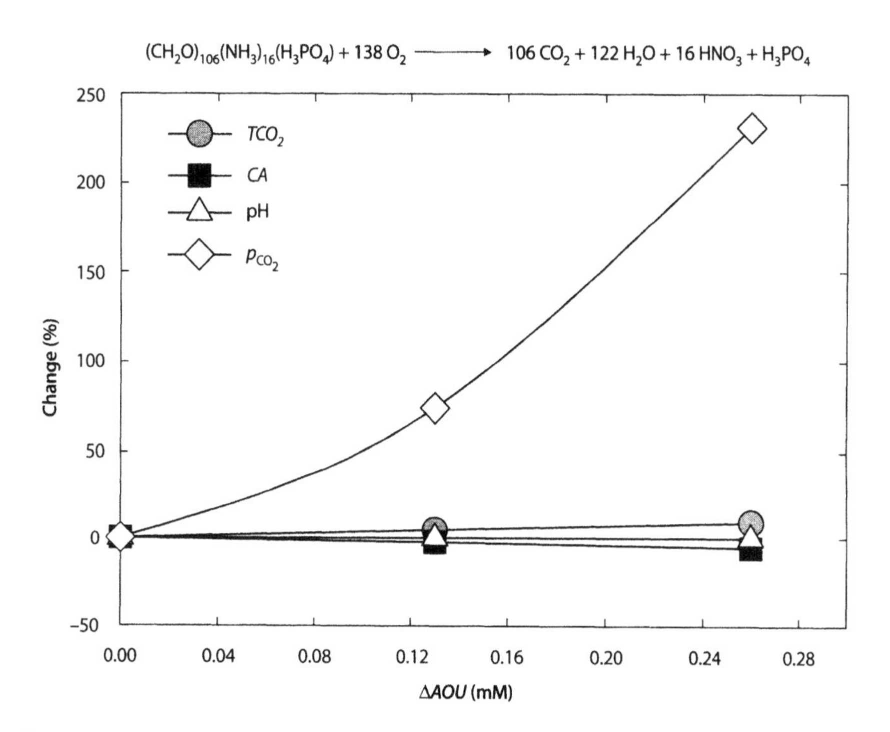

Fig. 1.6. Changes in parameters of the carbonate system due to oxidation of plant material (Millero 1996)

This equation has been used by Chen (1978) to predict changes in Ca^{2+} as a function of depth in the Pacific that agree very well with the measured values. The changes in inorganic carbon to organic carbon are given by

$$\begin{aligned} \text{Inorg C / Org C} &= x / 106y = 16\ \Delta\text{Ca} / 106\ \Delta\text{NO}_3 \\ &= (8\ TA + 8.5\ \Delta\text{NO}_3) / 106\ \Delta\text{NO}_3 \end{aligned} \quad (1.28)$$

Table 1.4. Changes in the CO_2 system due to the dissolution of $CaCO_3$ (Millero 1997)

	Initial (mM)	$\Delta CaCO_3$ (mM)	
		0.05	0.10
ΔCO_2	0	0.05	0.10
TCO_2	2.200	2.250	2.300
CA	2.487	2.587	2.687
CO_2	350	310	290
pH	8.200	8.264	8.321
$[CO_2]$	0.012	0.011	0.010
$[HCO_3^-]$	1.889	1.892	1.844
$[CO_3^{2-}]$	0.299	0.348	0.397

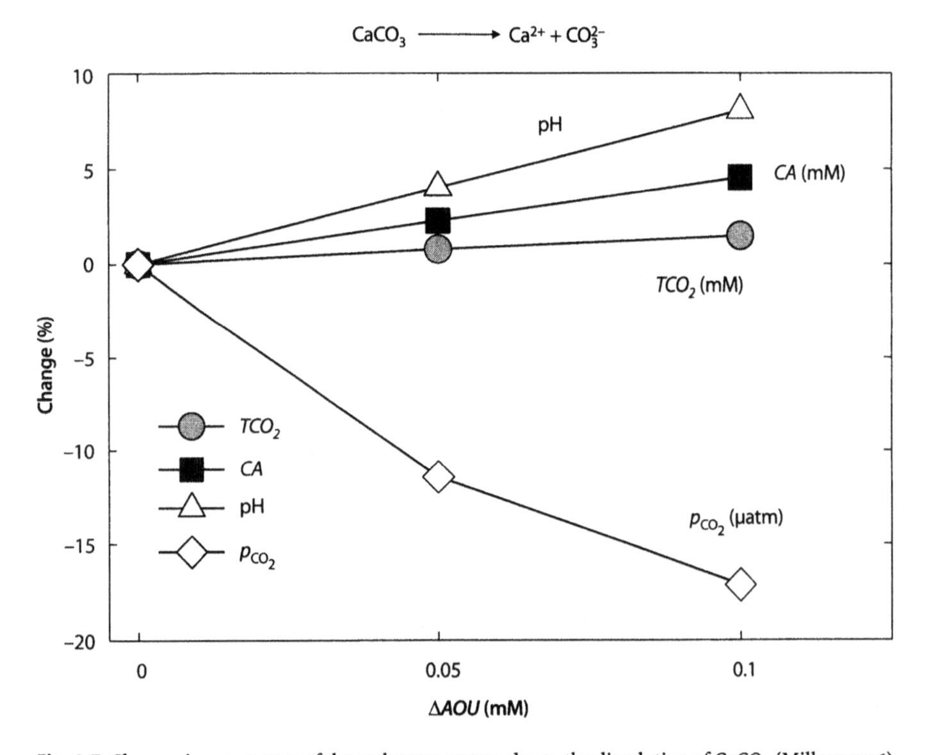

Fig. 1.7. Changes in parameters of the carbonate system due to the dissolution of $CaCO_3$ (Millero 1996)

This equation has also been shown by Chen (1978) to predict reasonable ratios of Inorg C/Org C in ocean waters. Later we will examine how these relationships can be used to determine the excess CO_2 that penetrates the oceans due to the anthropogenic increase of CO_2 in the atmosphere due to the burning of fossil fuels.

1.4
Distribution of Carbonate Species

The distribution of the various components of the CO_2 system in the oceans has been studied by many workers. I will show the results at two stations in the North Atlantic and Pacific Ocean (Millero 1997) to demonstrate the depth dependence in the two major oceans. The current JGOFS CO_2 measurements made as part of the WOCE hydrographic program should yield new CO_2 measurements that will serve as a benchmark for the future. In this section we will discuss the distribution of the CO_2 parameters.

1.4.1
p_{CO_2}

Unfortunately no historical data of sufficient accuracy is available for p_{CO_2} in surface waters of the oceans over time. The limited data presently available indicates that the increase of p_{CO_2} in the North Atlantic has a slope similar (1.5 ppm yr^{-1}) to the atmospheric values. The changes in p_{CO_2} in surface waters can be caused by:

1. removal by photosynthesis
2. removal by dissolution of $CaCO_3$
3. removal by solar heating
4. addition by oxidation of plant material
5. addition by formation of $CaCO_3$
6. addition by increases in CO_2 in the atmosphere from fossil fuel burning

Unraveling all these effects is even more difficult due to the sluggish response of the oceans to changes in the level of CO_2 in the atmosphere. Recent measurements of the p_{CO_2} in the surface waters of the Atlantic Ocean are shown in Fig. 1.8. The higher values of p_{CO_2} near the equator are due to equatorial upwelling, making the region a source of CO_2. The lower values of p_{CO_2} in the polar regions are due to waters being cold, and therefore serving as a sink for CO_2. The complicated structure in the North and South Atlantic oceans may be related to phytoplankton blooms. A north-south profile of p_{CO_2} in the Pacific Ocean is similar to the Atlantic. The cold waters are a sink for CO_2 ($-\Delta p_{CO_2}$), while the upwelling waters are a source of CO_2 ($+\Delta p_{CO_2}$) to the atmosphere.

A number of workers have attempted to separate the causes of the changes in the p_{CO_2} of surface waters due to physical (temperature and salinity) and biological (Chlorophyll) parameters. An example of such a separation is shown in Fig. 1.9 using the Atlantic data. The relative importance, based on the magnitude of the changes, is in the order of temperature, chlorophyll and salinity. This way of examining the importance of physical and biological processes neglects the importance of nutrients and the formation of phytoplankton blooms that can pull down the surface values of p_{CO_2}. The effect of primary production on the levels of p_{CO_2} and TCO_2 have recently been demonstrated (see Fig. 1.10) in the Ironex-II study (Steinberg et al. 1998). The addition of Fe to the waters decreased the p_{CO_2} by 75 μatm and TCO_2 by 26 μmol kg^{-1}. Studies have shown that the pull down of CO_2 by plankton occurs in the North Atlantic dur-

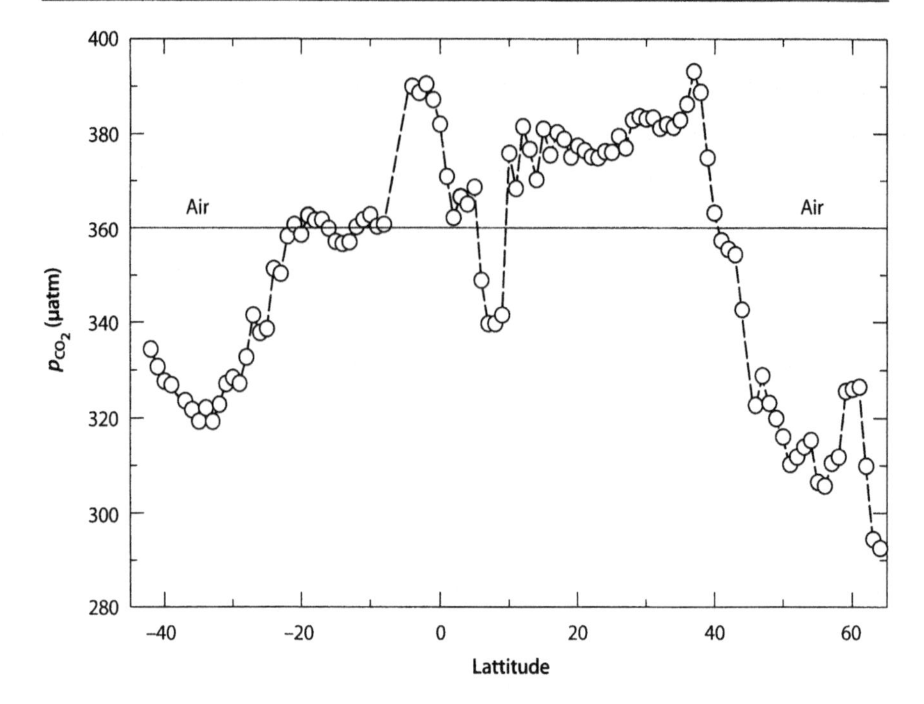

Fig. 1.8. The partial pressure of carbon dioxide in the atmosphere and surface waters of the Atlantic Ocean (25° W) (Millero 1996)

ing the spring bloom. As discussed earlier, for this biological pump to take CO_2 permanently from the atmosphere, the organic carbon must sink below the thermocline and be oxidized back to CO_2 where it can be stored for hundreds of years in deep waters.

The depth profiles of p_{CO_2} for the Atlantic and Pacific oceans are shown in Fig. 1.11. These values were calculated from *TA* and TCO_2, so they are not as accurate as direct measurements. The general trends, however, are real and as expected. The surface values are similar to atmospheric values. The values increase to a maximum (500 µatm in the North Atlantic Ocean and 1200 µatm in the Pacific Ocean) at 1 km due to the oxidation of plant material. The higher values at 1 km in the Pacific Ocean are due to the higher productivity of the surface waters resulting in the production of more organic carbon. The deep water values in the Pacific Ocean are higher than in the Atlantic Ocean due to the fact that the waters are older and have accumulated more CO_2 from the oxidation of organic carbon as the waters make the trip from the N. Atlantic to the N. Pacific (600 years). The deep waters originally formed at the surface have lower p_{CO_2} than the values at the oxygen minimum.

1.4.2 pH

The pH of most surface waters in equilibrium with the atmosphere is 8.2 ±0.1. Recent measurements of pH in the surface waters of the Atlantic and Pacific Ocean are shown

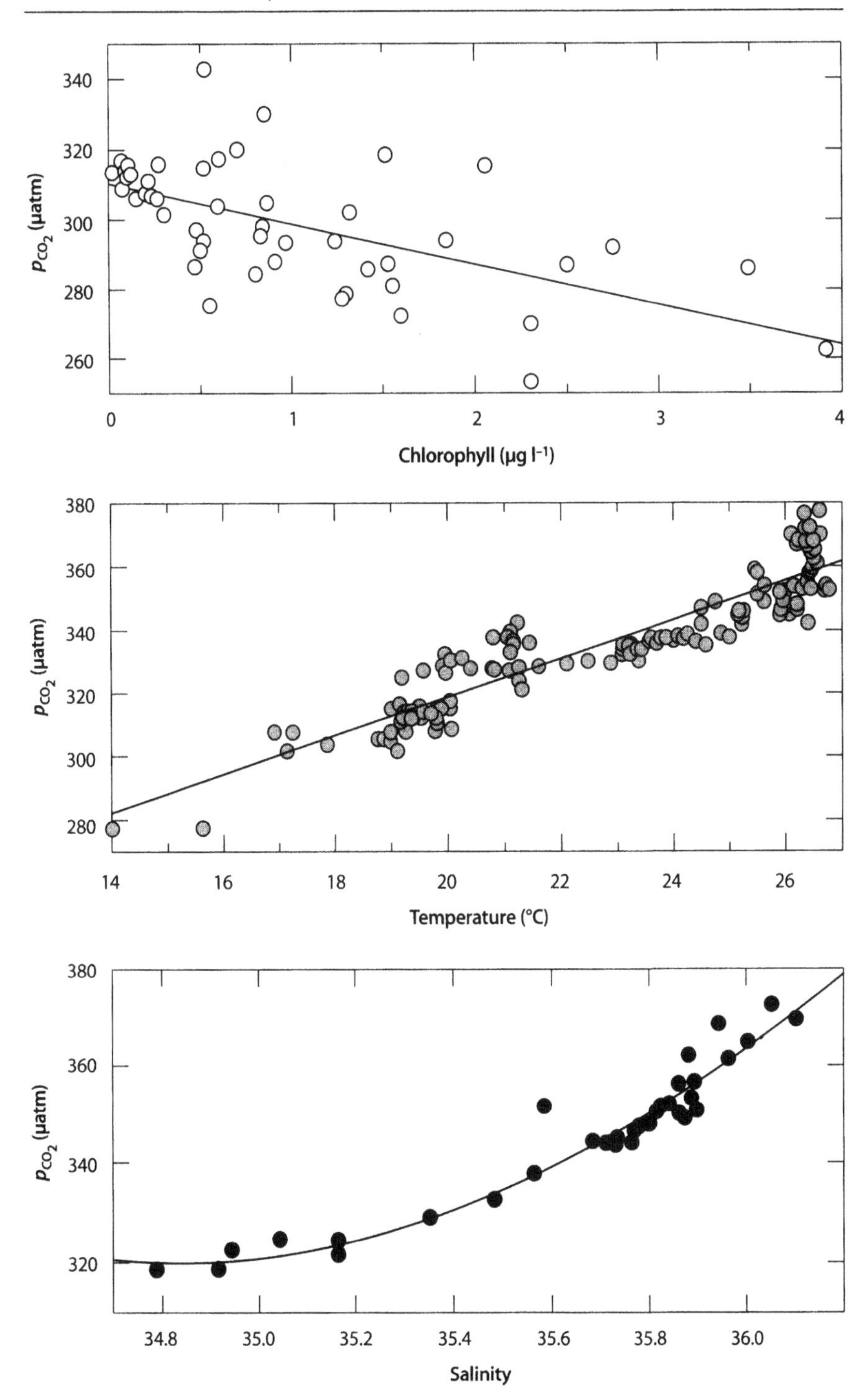

Fig. 1.9. The relationship between the partial pressure of carbon dioxide in Atlantic surface waters to the concentrations of chlorophyll, temperature and salinity (Millero 1996)

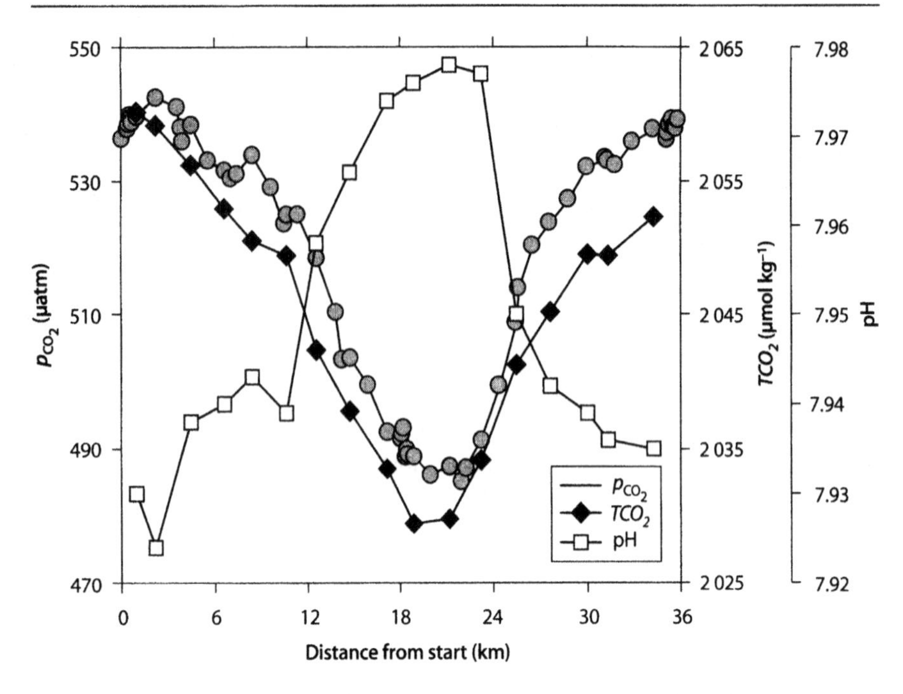

Fig. 1.10. The changes in the p_{CO_2}, and TCO_2 in Pacific waters seeded with Fe during the Ironex-II Study (Steinberg et al. 1998)

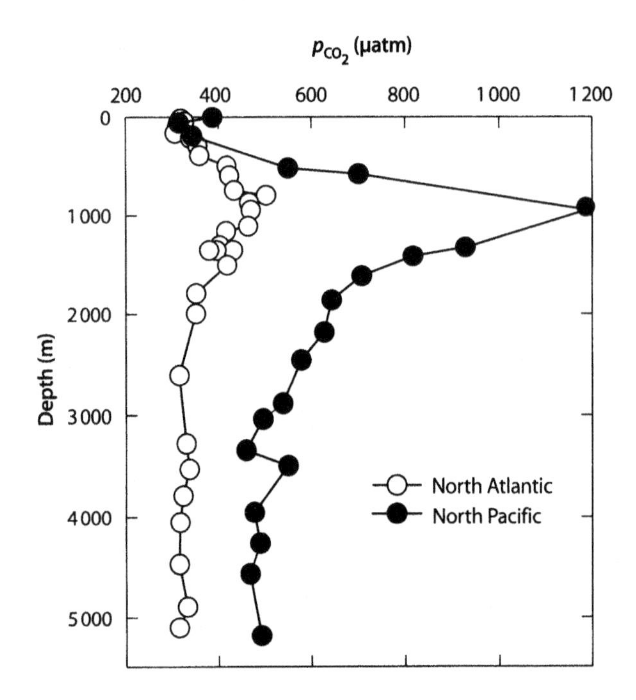

Fig. 1.11. Depth profile for the partial pressure of carbon dioxide in the Atlantic and Pacific Ocean (Millero 1996)

in Fig. 1.12. The gross trends in pH are those expected from the surface p_{CO_2} (the higher the p_{CO_2} the lower the pH and vice versa). The values are lower in upwelling waters in the Equatorial regions and are proportional to temperature. In closed or small bodies of water, the pH can show diurnal variations and cycle between 8.2 and 8.9. The de-

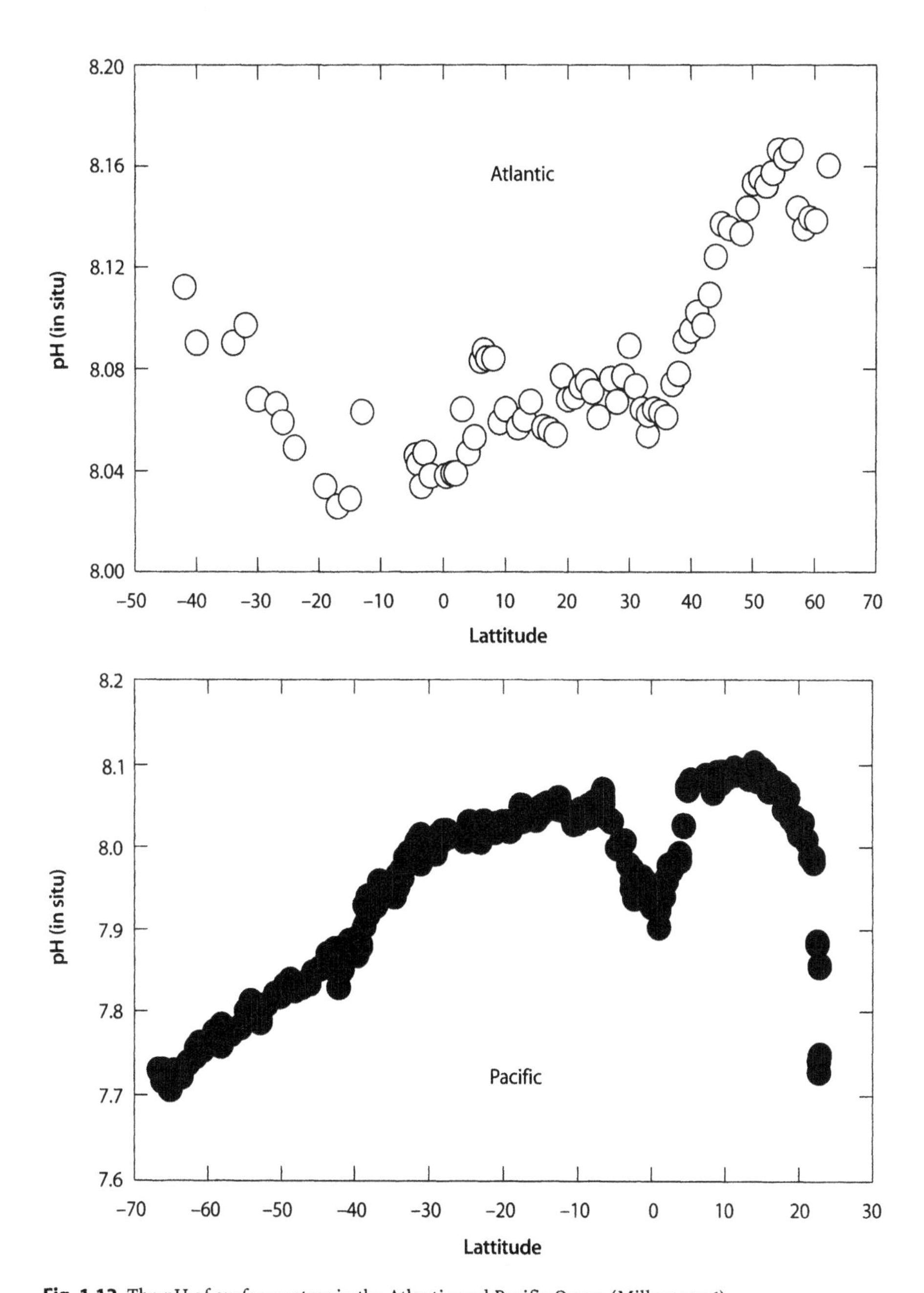

Fig. 1.12. The pH of surface waters in the Atlantic and Pacific Ocean (Millero 1996)

crease occurs in the evening due to the respiration of organisms, and the increase occurs in the afternoon due to photosynthesis. Changes in the pH with depth in the Atlantic and Pacific Ocean are shown in Fig. 1.13. Although it is difficult to see in this figure, the pH goes through a maximum in surface waters due to photosynthesis. The loss of CO_2 increases the pH. The pH then decreases due to the oxidation of plant material and goes through a minimum at about 1 km. This minimum coincides with the O_2 minimum and the maximum in p_{CO_2}

The pH increase in deep waters is due to the dissolution of $CaCO_3$. The pH of deep waters can be as low as 7.5 near 1 000 m. In very deep waters the pH can go through a maximum due to the effect of pressure on the ionization of carbonic acid. Park (NOAA) has used the Redfield model to calculate the pH as a function of depth. He attributed the changes to two factors

$$\Delta pH = \Delta pH_{(a)} + \Delta pH_{(b)} \tag{1.29}$$

where $\Delta pH_{(a)} = -2.0\ AOU$, the decrease due to the oxidation of plant material (AOU is the apparent oxygen utilization) and $\Delta pH_{(b)} = 2.4\ \Delta Ca$ (where ΔCa is the change in Ca^{2+} due to the dissolution of $CaCO_3$).

1.4.3
TA

The total alkalinity (*TA*) of surface and deep waters was measured extensively during the recent global change programs. The normalized total alkalinity ($NTA = TA \times 35 / S$) for surface waters in the Atlantic and Pacific oceans are shown in Fig. 1.14. The normal-

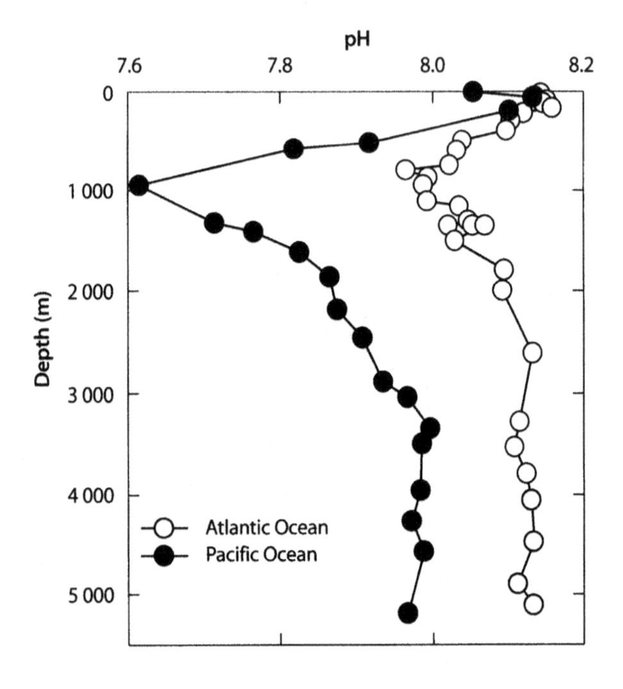

Fig. 1.13. Depth profile of pH in the Atlantic and Pacific oceans (Millero 1996)

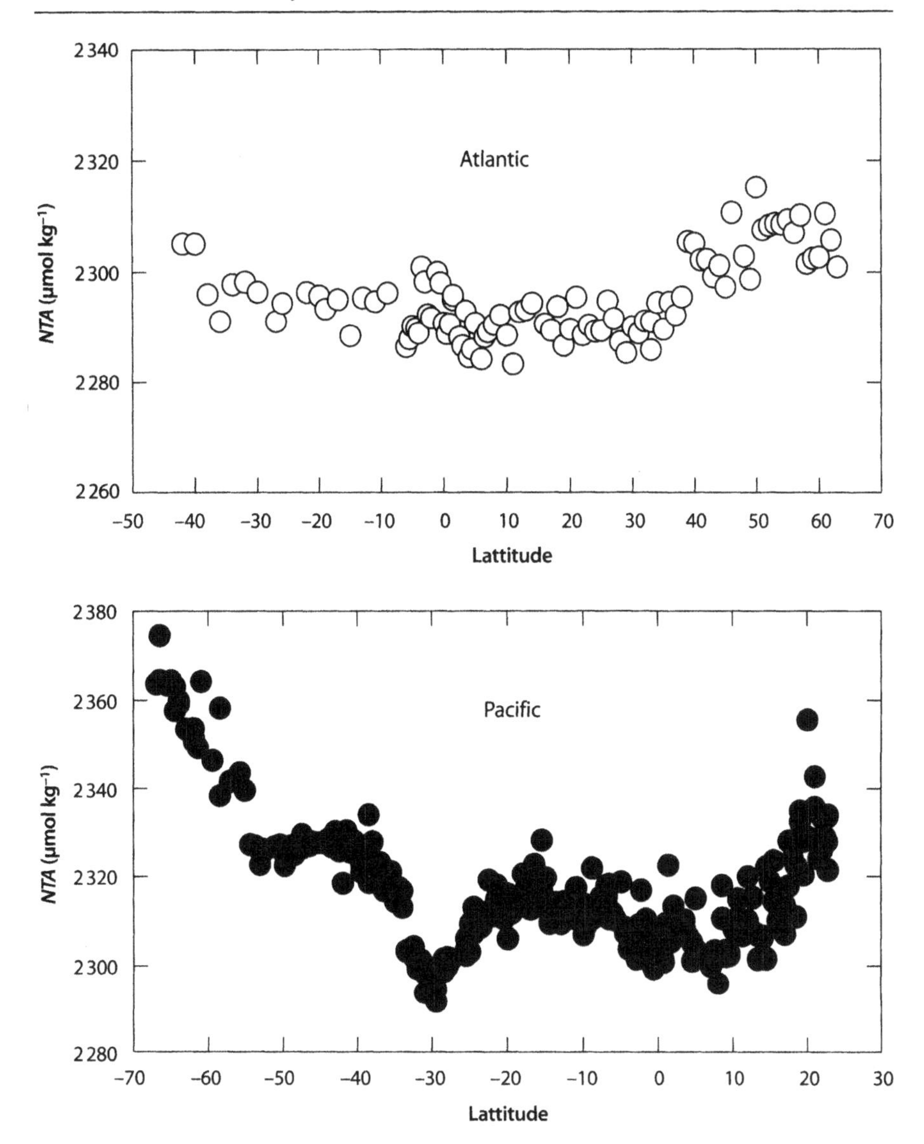

Fig. 1.14. The normalized total alkalinity (*NTA*) of surface waters in the Atlantic and Pacific Ocean (Millero 1996)

ization of *TA* accounts for changes in HCO_3^- (a major component of *TA*) due to changes in salinity. The values of *NTA* are higher in colder waters and nearly the same in Atlantic and Pacific oceans. The *NTA* values for most surface waters is 2300 µmol kg^{-1}. This is related to the near conservative behavior of HCO_3^- in seawater. The values are higher in polar waters ($TA = 2380$ µmol kg^{-1}). The higher values in polar regions are due to the outcropping and upwelling of colder deep waters that have higher *TA*.

A profile of *NTA* values in the North Atlantic and North Pacific oceans are shown in Fig. 1.15. The surface values in both oceans are the same, and the *NTA* in deep wa-

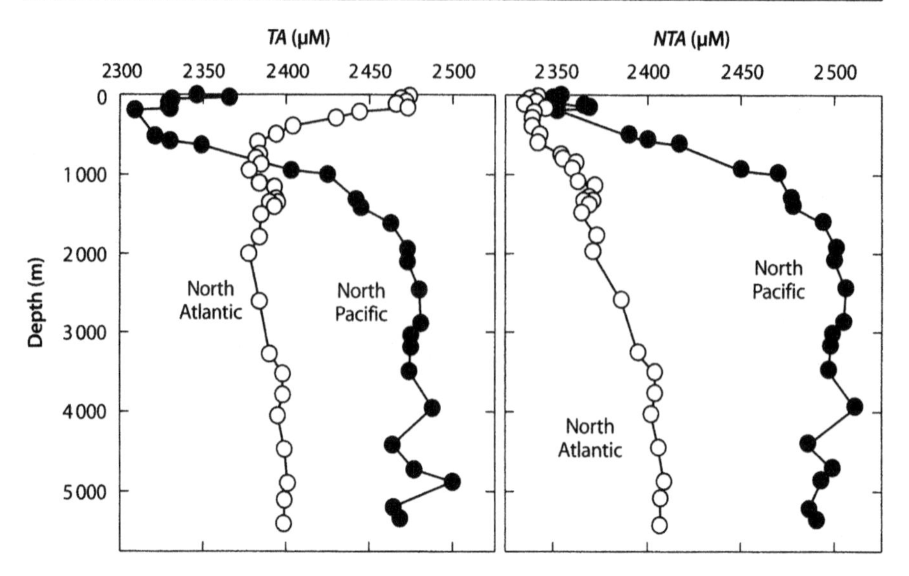

Fig. 1.15. The total alkalinity (*TA*) and normalized total alkalinity (*NTA*) as a function of depth in the Atlantic and Pacific Ocean (Millero 1996)

ters are higher in the Pacific than in the Atlantic. The higher *NTA* in deep waters is related to the dissolution of $CaCO_3$. The deep Pacific alkalinity values are higher than those in the Atlantic, because they are older and have accumulated more CO_3^{2-} from the dissolution of $CaCO_3$.

1.4.4 TCO_2

The total normalized dissolved inorganic carbon dioxide ($NTCO_2 = TCO_2 \times 35 / S$) in the Atlantic and Pacific oceans for surface waters is shown in Fig. 1.16. Unlike the alkalinity, the total CO_2 in the equatorial waters shows a large increase, due to equatorial upwelling. The TCO_2 shows little latitudinal change. For rapid exchange the p_{CO_2} in the water and air are similar, and the TCO_2 is higher in polar regions. The depth profiles of $NTCO_2$ in the Atlantic and Pacific are shown in Fig. 1.17. The values decrease to a minimum in surface waters due to photosynthesis. In deeper waters the $NTCO_2$ increases due to the oxidation of plant material. The values of $NTCO_2$ for deep Pacific waters are higher than those for the Atlantic because the waters are older and have had more time to accumulate CO_2 due to microbial oxidation. The values of $NTCO_2$ and *NTA* correlate very well with each other, and can be used to characterize various water masses. The deep waters increase by 20 μmol kg^{-1} from the North to South Atlantic. A section of TCO_2 in the Atlantic Ocean is shown in Fig. 1.18. The values clearly show the differences due to the major water masses (North Atlantic Deep Water, Antarctic Intermediate Water and Antarctic Bottom Water).

Due to the buffering effect of sea water, only a small amount of CO_2 needs to be transferred to the oceans to restore the equilibrium between the atmosphere and surface. This buffering is called the Revelle factor (*R*). It is the ratio of the fractional rise

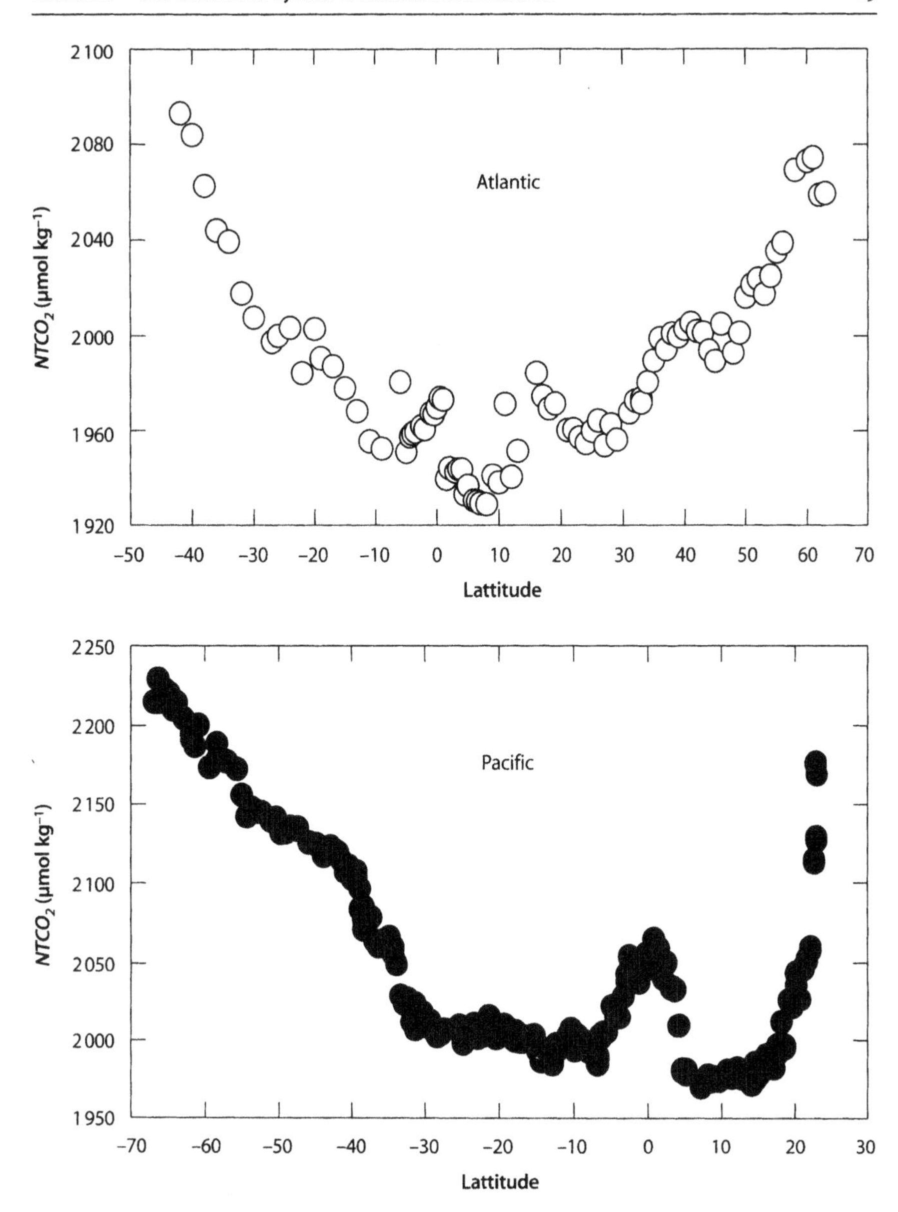

Fig. 1.16. The normalized total carbon dioxide ($NTCO_2$) of surface waters in the Atlantic and Pacific Ocean (Millero 1996)

in the partial pressure of carbon dioxide in the atmosphere to the fractional increase of the total carbon dioxide in the ocean. The increase in the total CO_2 near the equator is related to the Revelle factor

$$R = (\Delta p_{CO_2} / p_{CO_2}) \, / \, (\Delta TCO_2 / TCO_2) \tag{1.30}$$

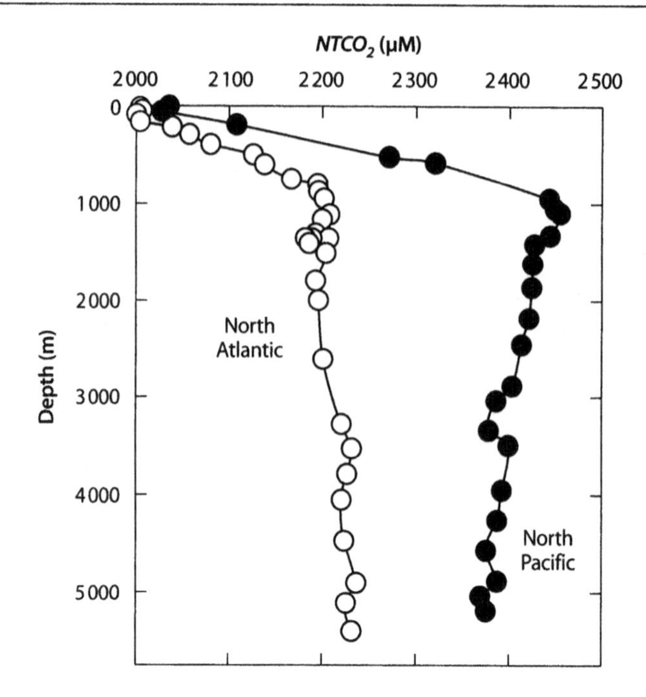

Fig. 1.17. Depth profile of the normalized total carbon dioxide ($NTCO_2$) in the Atlantic and Pacific oceans (Millero 1996)

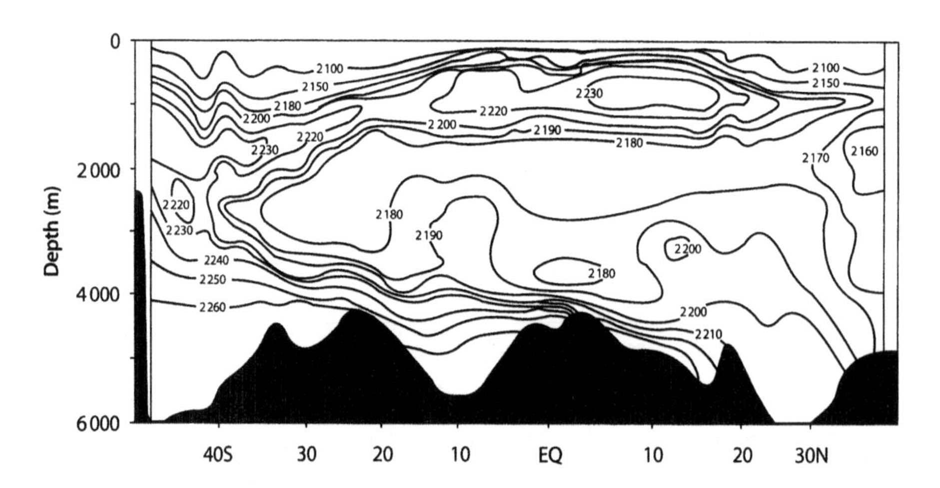

Fig. 1.18. A section of $NTCO_2$ in the Atlantic Ocean (Millero 1996)

This value is about 14 for cold waters and 8 for warmer waters (the average is about 10). Thus, a 10% change in p_{CO_2} results in only a 1% change in TCO_2. This factor is important when considering the effect that increases of CO_2 in the atmosphere have on the carbonate system.

1.5 $CaCO_3$ Dissolution in Sea Water

The precipitation or formation of solid $CaCO_3$ in surface waters and the dissolution of solid $CaCO_3$ in deep waters is very important in transferring CO_2 from surface waters to deep waters. $CaCO_{3(s)}$ is also present in pelagic sediments of the world oceans. The saturation state of sea water with respect to $CaCO_3$ is determined from

$$\Omega = [Ca^{2+}][CO_3^{2-}] / K^*_{SP} \quad (1.31)$$

where $[Ca^{2+}][CO_3^{2-}]$ is the ion product of the concentration of Ca^{2+} and CO_3^{2-} and K^*_{SP} is the solubility product, at the "in situ" conditions, of S, t and P. Since Ca^{2+} is a major constituent of sea water (within 1%), its concentration can be estimated from the salinity. The solubility product for calcite formed by formanifera and aragonite formed by pteropods can be determined from equations given in the appendix. The values of $[CO_3^{2-}]$ can be determined from the measured carbonate parameters (pH and TA or TA and TCO_2).

Values of Ω for calcite and aragonite for Atlantic and Pacific waters are shown in Fig. 1.19. The surface values of Ω for calcite are near 5.0 and decrease below 1.0 in deep water. The surface water value of Ω is 3.0 for aragonite. Aragonite is 1.5 times more soluble than calcite at a given t, P and salinity. The waters of the Pacific become undersaturated ($\Omega < 1.0$) at shallower depths than in the Atlantic. Approximate saturation levels in North Atlantic and North Pacific waters are given in Table 1.5.

The greater solubility of these minerals in deep waters is related to the effect of pressure on the solubility of $CaCO_{3(s)}$. Since two divalent ions are formed during the dissolution, the volume change is large and negative due to electrostriction. The Pa-

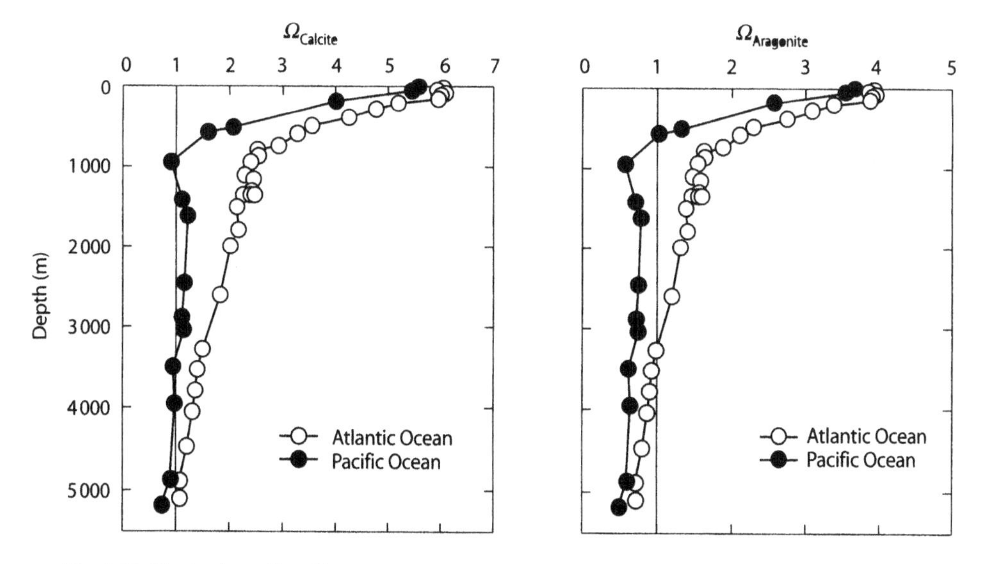

Fig. 1.19. The depth profile of the aragonite and calcite saturation state for the Atlantic and Pacific oceans (Millero 1996)

Table 1.5. Approximate saturation levels (m) of calcite and aragonite in North Atlantic and North Pacific waters

	North Atlantic	North Pacific
Calcite	4300	750
Aragonite	1500	500

cific deep waters become undersaturated at shallower depths due to the lower pH or higher CO_2 formed by the oxidation of plant material. This decreases the concentration of CO_3^{2-} due to the shift in the equilibrium

$$CO_3^{2-} + H^+ \longrightarrow HCO_3^- \tag{1.32}$$

The differences in Ω between the two oceans become smaller in the deep oceans due to the effect of pressure controlling solubility.

Although much of the deep oceans is undersaturated with respect to $CaCO_{3(s)}$, large amounts of calcite are present in ocean sediments. Geologists call the layer where $CaCO_{3(s)}$ is above 5% of the sediments, the carbonate compensation depth (*CCD*). The $CaCO_3$ compensation depth in the Atlantic Ocean is about 2 km below the saturation depth. These results indicate that the solubility of $CaCO_{3(s)}$ in sea water is not controlled by equilibrium but by kinetic constraints.

At a depth of about 4 000 m, the solution rate dramatically increases. This depth of rapid increase in the rate of dissolution is called the lysocline. The aragonite lysocline was found to be higher than the depth for calcite. The depth of the lysocline found by suspending $CaCO_{3(s)}$ agrees very well with the decrease in the mineral found in surface sediments at various depths in the same area. These results indicate that the lysocline and $CaCO_3$ compensation depths in sediments are frequently the same. Thus, the causes of the compensation depth being deeper than the saturation depth are the variable rates of dissolution of various forms of $CaCO_3$. If the sedimentation rates are high, it is possible that $CaCO_{3(s)}$ could be preserved before it dissolves. This would cause the calcium carbonate compensation depth to be below the lysocline.

The lysocline is higher in the Pacific because of the greater undersaturation at lower depths. A comparison of the saturation horizon with the lysocline and calcium carbonate compensation depths (*CCD*) is shown in Fig. 1.20. The values of the lysocline and the *CCD* are not affected by the saturation states. The *CCD* is close to the lysocline except in the equatorial region. This is due to the higher productivity of these waters. The higher the supply rate of $CaCO_{3(s)}$, the deeper the *CCD* will be.

Laboratory studies indicate that when the saturation of waters gets to a critical value, $CaCO_{3(s)}$ starts to dissolve. This critical value is about 30% undersaturation or at a $\Delta[CO_3^{2-} - CO_3^{2-}{}_{(sat)}] = -10\ \mu mol\ kg^{-1}$ (i.e. the solution could absorb another 10 $\mu mol\ kg^{-1}$ of $CaCO_3$). It should be pointed out that this so-called critical value is strongly dependent upon the value selected for the solubility product of the $CaCO_{3(s)}$. Studies of the solubility dissolution rates using minerals and waters collected in the oceans have led to dissolution rates of aragonitic $CaCO_{3(s)}$ measured at sea, could be described by

$$R = 130\ \{1 - [Ca^{2+}][CO_3^{2-}] / K_{SP(aragonite)}\}^{3.1} \tag{1.33}$$

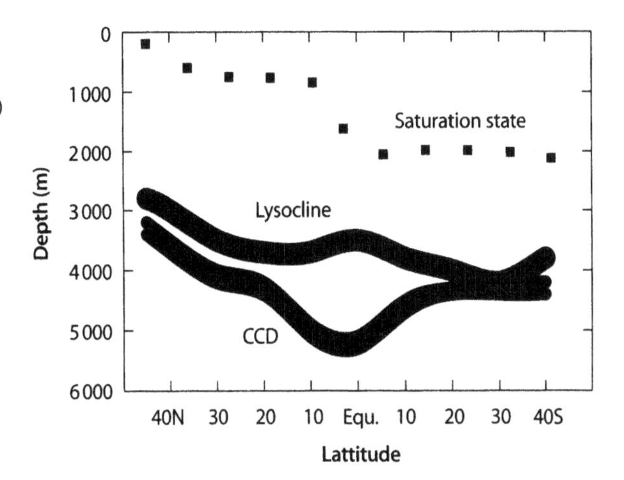

Fig. 1.20. Comparisons of the thermodynamic saturation state, lysocline, and calcium carbonate compensation (CCD) depths in the Atlantic Ocean (Millero 1996)

where $K_{SP(aragonite)} = 1.78\ K_{SP(calcite)}$. The factor of $\rho = 1.78$ is slightly higher than the theoretical value of 1.5. This is related to the changes in $K_{SP(aragonite)}$ and Ω_A for aragonite solubility as a function of time. Since aragonite production appears to be high in the surface waters of the Pacific Ocean, and the deeper waters are undersaturated, the transport of the aragonite and dissolution could result in transporting carbon to deep waters. Approximately 90% of this aragonite flux is thought to be dissolved in the upper 2.2 km of the water column.

1.6 Calculating the Penetration of Anthropogenic CO_2 into the Oceans

The estimation of the penetration of anthropogenically produced CO_2 into the oceans has been made using two methods:

1. The time series method, which examines the changes in CO_2 as a function of time (Brewer et al. 1995; Wallace 1995);
2. Attempting to correct for the addition of CO_2 due to the dissolution of $CaCO_3$ and oxidation of plant material (Brewer 1978; Chen and Millero 1979; Gruber et al. 1996).

1.6.1 Time Series Method

To use the time series method one needs reliable measurement of TCO_2 as a function of temperature, salinity, oxygen and *TA* (or silicate). The values of TCO_2 are fit to equations of the form

$$TCO_2 = a + bS + c\theta + dTA + eAOU \tag{1.34}$$

where a, b, etc. are empirical constants, *S* is salinity, θ is the adiabatic temperature, *TA* is the total alkalinity, and *AOU* is the apparent oxygen utilization. Sabine et al. (1997, 1999) have fit the GEOSECS data (1977–78) in the Indian Ocean to Eq. 1.34 where

$a = 706.5, b = 7.7, c = -6.68, d = 0.513$, and $e = 0.7255$ with $\sigma = 5.2\ \mu mol\ kg^{-1}$. By comparing the GEOSECS fit with the more recent measurements 18 years later during the WOCE/JGOFS studies, Sabine et al. (1999) have been able to examine the changes in the TCO_2 in the deep waters (>200 db). The penetration of CO_2 into the Indian Ocean calculated in this manner is shown in Fig. 1.21. The penetration of CO_2, since the GEOSECS measurements, has increased by 29%, which is similar to the atmospheric increase (31%) over the past 18 years. Models (Sarmiento et al. 1995) predict that as the CO_2 continues to rise, this trend may change. This is due to the fact that the buffering capacity of the ocean will decrease. Since the surface waters have a wide seasonal variability, the excess values were calculated using the annual means. Although this method gives reasonable estimates of the penetration of CO_2, it is quite sensitive to the errors in the individual measurements. The excess CO_2 for the Indian Ocean determined by this method has a maximum of 20 $\mu mol\ kg^{-1}$ over the 18 years. This is reasonable agreement with the expected value 18 $\mu mol\ kg^{-1}$ due to an increase of 30 μatm (assuming $TA = 2\,300\ \mu mol\ kg^{-1}$, $S = 35$, and $t = 20$ °C).

1.6.2 Calculation by Correcting for Dissolution of $CaCO_3$ and Oxidation of Plant Material

Attempting to calculate the anthropogenic CO_2 by correcting for the oxidation of plant material and dissolution of $CaCO_3$ was independently examined by Brewer (1978) and Chen and Millero (1979). The changes in the TCO_2 in a given water sample can be attributed to

$$\Delta TCO_2 = \Delta[TCO_2]_{CaCO_3} + \Delta[TCO_2]_{Organic} + \Delta[TCO_2]_{Anthro} \tag{1.35}$$

The excess TCO_2 can be calculated from the rearrangement of this equation

$$\Delta[TCO_2]_{Anthro} = \Delta TCO_2 - \Delta[TCO_2]_{CaCO_3} - \Delta[TCO_2]_{Organic} \tag{1.36}$$

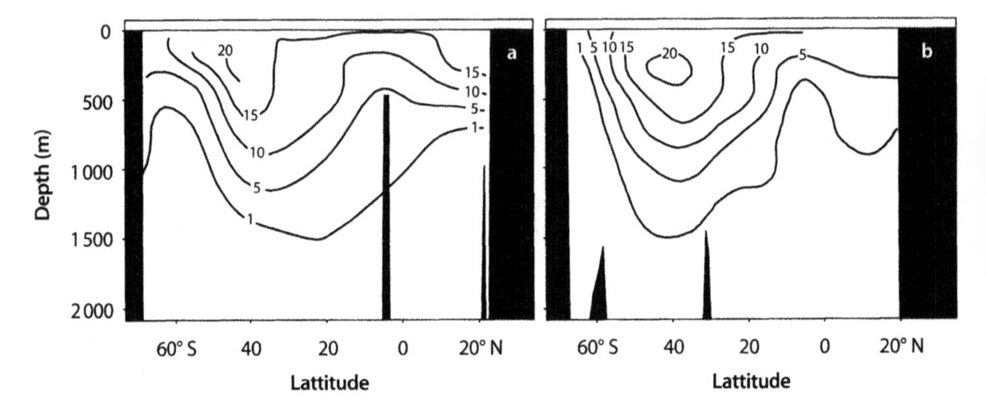

Fig. 1.21. Sections of excess CO_2 in the Indian Ocean determined by the time method; **a** along 57° E; **b** along 92° E (Sabine et al. 1999)

As discussed earlier one can make an estimate of the changes in the TCO_2 due to the oxidation of plant material and dissolution of $CaCO_3$ using equations discussed earlier.

If one uses the equations of Brewer (1978) and Chen and Millero (1979) to examine the changes in TCO_2 in old deep waters and younger waters, it is possible to make an estimate of the changes due to the increase of CO_2 in the atmosphere due to the burning of fossil fuels. The estimated increase (260 ±20 ppm for preindustrial values) is in reasonable agreement with the values obtained from ice cores (Fig. 1.1). As discussed elsewhere (Gruber et al. 1996), these methods are subject to large uncertainties. The method recently introduced by Gruber et al. (1996) has some improvements. The technique is based on the assumption that the natural carbon cycle is in steady state, meaning the carbon to oxygen and nitrate to oxygen ratios (rC/O_2 and rN/O_2) are constant. They use the conservative tracer (ΔC^*) defined by

$$\Delta C^* = TCO_2 - (TCO_2)_{eq}(S, t, TA^0)]_{f_{CO_2}=280} - rC/O_2([O_2] - [O_2]_{sat}) - \tfrac{1}{2}[TA - TA^0 + rN/O_2([O_2] - [O_2]_{sat})] \tag{1.37}$$

where TCO_2 is the measured total inorganic CO_2, $(TCO_2)_{eq}$ is the TCO_2 for the preindustrial value at a $f_{CO_2} = 280$ µatm, at the temperature (t), salinity (S) and performed total alkalinity (TA^0) and $[O_2] - [O_2]_{sat}$ is the difference between the measured and calculated oxygen (in situ t and S). The value of TA^0 is estimated from a multiple regression of S and $PO = [O_2] - rO_2/PO_4[PO_4]$ as independent variables. For the Indian Ocean, Sabine et al. (1999) have determined the values of TA^0 from the equation

$$TA^0\,(\mu\text{mol kg}^{-1}) = 357.6 + 55.98\,S + 0.054\,PO - 1.345\,\theta \tag{1.38}$$

where PO (µmol kg^{-1}) $= [O_2] + 170\,[PO_4]$ and θ is the potential temperature (°C).

The ratios used for the organic carbon are $P / N / C / O_2 = 1 / 16 / 117 / -170$ (Anderson and Sarmiento 1994). The value of ΔC^* reflects the uptake of anthropogenic CO_2 (ΔC_{Ant}) and the disequilibrium at the time the water left the surface and any uncertainties in the choice of the end members, etc. (ΔC_{diseq}). This disequilibrium term is related to

$$\Delta C_{diseq} = \sum_{i=1}^{n} f^i \Delta C^i_{diseq} \tag{1.39}$$

where f^i represents the relative contribution of the different end members ($\Sigma f^i = 1$). They assume that the water is transported along constant density surfaces (isopycnals) and the disequilibrium has stayed constant within an outcrop region of the isopycnal. The value of ΔC^* should, thus, reveal the history of the CO_2 uptake. The values of ΔC_{diseq} are determined on a number of isopycnal surfaces.

The CO_2 disequilibrium is determined by two methods. In the first, the variability of ΔC^* in deep waters far away from the outcrop is assumed to contain no anthropogenic CO_2. This method is used for all deep waters (>2000 m), except for the water masses in the North Atlantic near Greenland and the Norwegian Seas. For shallow waters this method does not work because of the penetration of anthropogenic CO_2. For these density surfaces, the $(TCO_2)_{eq}$ in Eq. 1.37 is replaced by the value $(TCO_2)_{eq(t)}$, which is the equilibrium concentration at the time the water left the surface. This requires the age of the water parcel. The modified ΔC_t^* is given by

$$\Delta C_t^* = TCO_2 - (TCO_2)_{eq}(S, t, TA^0, f_{CO_2(t_{sample} - \tau)} - rC/O_2(O_2 - O_{2(sat)}) \\ - ½[TA - TA^0 + rN/O_2([O_2] - [O_2]_{sat})] \quad (1.40)$$

where $f_{CO_2}(t_{sample} - \tau)$ is the value of the fugacity of CO_2 at the time the water parcel was in contact with the atmosphere. The age is estimated from the ratio of CFC-11 and CFC-12 in the water (Weiss et al. 1985) and the fugacity is estimated from a fit of the results shown in Fig. 1.1 (Keeling and Whorf 1994; Neftel et al. 1994).

Recently these methods have been used to examine the penetration of fossil fuel CO_2 into the Indian (Sabine et al. 1999) and Atlantic (Gruber 1998) oceans. The calculations for the Indian Ocean determined by the correction method are shown in Fig. 1.22 and tabulated in Table 1.6. Since part of the Indian Ocean (e.g. the Arabian Sea) has areas of denitrification, Sabine et al. (1999) had to correct for these effects. They made these corrections using the equations developed by Gruber and Sarmiento (1997). The patterns of the penetration of CO_2 in the Indian Ocean determined by both methods are quite similar (Fig. 1.21 and 1.22). The calculated penetration of CO_2 into the oceans can be compared to model estimates. The model used by Sabine et al. (1999) and Gruber (1998) is the Princeton Ocean Biogeochemistry Model (OBM), which is based on the circulation of Toggweiler et al. (1989) and the biological and physical pumps of Sarmiento et al. (1995). Although the total inventory of CO_2 in the Indian Ocean (Table 1.6) is in good agreement with the model calculations, the spatial distributions do not agree. This is thought to be due to limitations in the model.

The Gruber (1998) results in the Atlantic are tabulated in Table 1.6. The results calculated from the data are in reasonable agreement with model calculations (Sarmiento et al. 1995). The total inventory for the Atlantic calculated from the measurements (22 ±5 Gt C) in the North and South Atlantic oceans (18 ±4 Gt C) are comparable to the model calculations (20 and 17.7 Gt C, respectively). It should be pointed out that the 1973 inventory for the North Atlantic (25 ±5 Gt C) and South Atlantic (16 ±3 Gt C) of

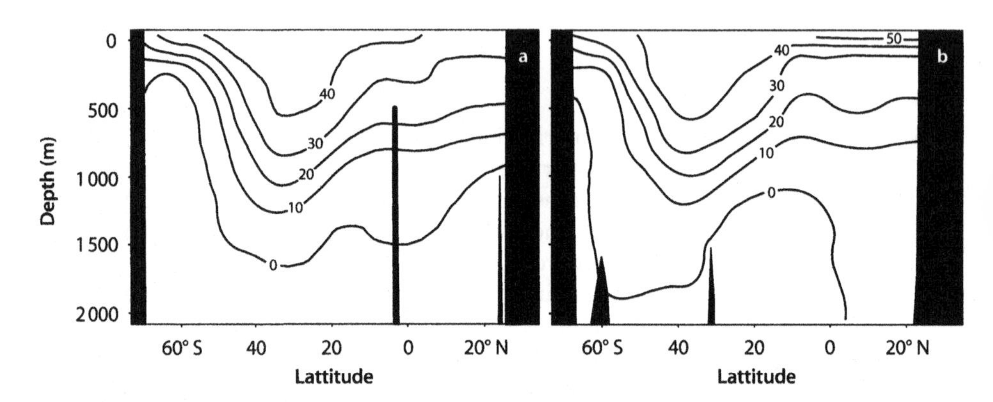

Fig. 1.22. Sections of excess CO_2 in the Indian Ocean determined by the correction method; **a** along 57° E; **b** along 92° E (Sabine et al. 1999)

Table 1.6. Comparison of the calculated inventory of CO_2 from data and model calculations (Sabine et al. 1999; Gruber 1998)

Ocean	Location	Experimental (Gt C)	Model (Gt C)
Indian Ocean	20° E to 120° E	21.0 ±2	26.7
	Lat. >35° S	4.0	2.5
	Lat. <35° S	14.0	9.3
Atlantic Ocean	10° S to 80° N	40.0 ±5	37.7
North Atlantic	Equator to 80° N	22.0 ±5	20.0 (1982)
South Atlantic	Equator to 60° S	18.0 ±4	17.7 (1989)

Chen (1993) adjusted to 1982 (Gruber 1998) is also in good agreement with the Gruber (1998) estimates. Gruber (1998) also showed that the calculated penetration of CO_2 in the Atlantic was in reasonable agreement with the estimates made by Tans et al. (1990), calculated from the flux of CO_2 into the oceans (Eq. 1.6).

Although these estimates have some deficiencies, they provide a framework that can be used to examine the present and future penetration of fossil fuel CO_2 into the oceans and limits for the models. With improvements in the model, one may be able to make better predictions of the future levels of CO_2 in the atmosphere and its effect on global warming.

Acknowledgements

The author wishes to acknowledge the support of the Oceanographic Section of the National Science Foundation and the National Oceanic and Atmospheric Admistration for supporting my studies of the CO_2 system.

References

Anderson LA, Sarmiento JL (1994) Redfield ratios of remineralization determined by nutrient data analysis. Global Geochem Cycles 8:65–80
Brewer PG (1978) Direct observation of the oceanic CO_2 increase. Geophys Res Lett 5:997–1000
Brewer PG, Wong GTF, Bacon MP, Spencer DW (1975) The calcium problem. Earth Planet Sci Lett 26:81–87
Brewer PG, Glover DM, Goyet C, Shaver DK (1995) The pH of the North Atlantic Ocean, Improvement to the global model for sound absorption in seawater. J Geophys Res 100:8761–8776
Chen C-T (1978) Decomposition of calcium carbonate and organic carbon in the deep oceans. Science 201:735–736
Chen C-T (1993) The oceanic anthropogenic CO_2 sink. Chemosphere 27:1041–1064
Chen C-T, Millero FJ (1979) Gradual increase of oceanic carbon dioxide. Nature 277:205–206
Clayton T, Byrne RH (1993) Calibration of m-cresol purple on the total hydrogen ion concentration scale and its application to the CO_2-system characteristics in seawater. Deep-Sea Res 28:609–623
Dickson AG (1981) An exact definition of total alkalinity and a procedure for the estimation of alkalinity and total CO_2 from titration data. Deep-Sea Res 28:609–623
Dickson AG (1984) pH scales and proton-transfer reactions in saline media such as seawater. Geochim Cosmochim Acta 48:2299–2308
Dickson AG (1993) pH buffers for sea water media based on the total hydrogen ion concentration scale. Deep-Sea Res 40:107–118
Gleitz M, Rutgers v.d. Loeff M, Thomas DN, Dieckmann GS, Millero FJ (1995) Seasonal changes of inorganic carbon, oxygen and nutrient concentrations in Antarctic sea ice brines. Mar Chem 51:81–91

Gruber N (1998) Anthropogenic CO_2 in the Atlantic Ocean. Global Biogeochemical Cycles 12:165–191
Gruber N, Sarmiento JL (1997) Global patterns of marine nitrogen fixation and denitrification. Global Biogeochemical Cycles 11:235–266
Gruber N, Sarmiento JL, Stocker TF (1996) An improved method for detecting anthropogenic CO_2 in the oceans. Global Biogeochemical Cycles 10:809–837
Johnson KM, Wills KD, Butler DB, Johnson WK, Wong CS (1993) Coulometric total carbon dioxide analysis for marine studies: maximizing the performance of an automated gas extraction system and coulometric detector. Mar Chem 44:167–187
Keeling CD, Whorf TP (1994) Atmospheric CO_2 records from sites in the SIO air sampling network. In: Boden T et al. (eds) Trends '93: a compendium of data on global change. Rep. ORNL/CDIAC-65, Carbon Dioxide Inf. Anal. Cent., Oak Ridge Natl. Lab., Oak Ridge, Tenn., pp 16–26
Lewis E, Wallace DWR (1998) Program developed for the CO_2 system calculations. Oak Ridge, Oak Ridge National Laboratory, ORNL/CDIAC-105
Millero FJ (1981) The ionization of acids in estuarine waters. Geochim Cosmochim Acta 45:2085–2089
Millero FJ (1983) The estimation of the pK^*_{HA} of acids in seawater using the Pitzer equations. Geochim Cosmochim Acta 47:2121–2129
Millero FJ (1986) The pH of estuarine waters. Limnol Oceanogr 31:839–847
Millero FJ (1995) Thermodynamics of the carbon dioxide system in the oceans. Geochim Cosmochim Acta 59:661–677
Millero FJ (1996) Chemical oceanography. CRC Press, Boca Raton
Millero FJ (1997) The influence of iron on CO_2 in the oceans. Science Progress 80:147–168
Millero FJ, Roy RN (1997) A chemical equilibrium model for the carbonate system in natural waters. Croatia Chemica Acta 70:1–38
Millero FJ, Zhang J-Z, Fiol S, Sotolongo S, Roy RN, Lee K, Mane S (1993a) The use of buffers to measure the pH of seawater. Mar Chem 44:143–152
Millero F J, Zhang JZ, Lee K, Campbell DM (1993b) Titration alkalinity of seawater. Mar Chem 44: 153–165
Neftel A, Friedli H, Moor E, Lotscher H, Oeschger H, Siegenthaler U, Stauffer B (1994) Historical CO_2 record from the Siple station ice core. In: Boden T et al. (eds) Trends '93: a compendium of data on global change. Rep. ORNL/CDIAC-65, Carbon Dioxide Inf. Anal. Cent., Oak Ridge Natl. Lab., Oak Ridge, Tenn., pp 11–14
Park K (1969) Oceanic CO_2 system: an evaluation of ten methods of investigation. Limnol Oceanogr 14:179–186
Quay PD, Tilbrook B, Wong CS (1992) Oceanic uptake of fossil fuel CO_2: Carbon -13 evidence. Science 256:74–79
Redfield AC, Ketchum BH, Richards FA (1963) The influence of organisms on the composition of seawater. In: Hill MN (ed) The sea, vol II. Interscience, New York, pp 26–77
Sabine CL, Wallace DWR, Millero FJ (1997) Survey of CO_2 in the oceans reveals clues about global carbon cycle. Eos TransAGU 78, 49:54–55
Sabine CL, Key RM, Goyet C, Johnson KM, Millero FJ, Sarmiento JL, Wallace DWR, Winn CD (1999) Anthropogenic CO_2 inventory of the Indian Ocean. Global Biogeochemical Cycles 13:179–198
Sarmiento JL (1993) Ocean carbon cycle. Chem. & Eng. News 71:30–43
Sarmiento JL, Murnane R, LeQuere C (1995) Air-sea CO_2 transfer and the carbon budget of the North Alantic. Philos Trans R Soc London Ser. B 348:211–219
Steinberg PA, Millero FJ, Zhu X (1998) Carbonate system response to iron enrichment. Mar Chem, 62:31–43
Takahashi T, Broecker WS, Langer S (1985) Redfield ratio based on chemical data from isopycnal surfaces. J Geophys Res 90:6907–6924
Tans PP, Fung IY, Takahashi T (1990) Observational constraints on the global atmospheric CO_2 budget. Science 247:1431–1438
Toggweiler JR, Dickson K, Bryan K (1989) Simulations of radiocarbon in a coarse-evolution world ocean model, 1 steady state pre-bomb distributions. J Geophys Res 94:8217–8242
Wallace DWR (1995) Monitoring Global Ocean Carbon Inventories. Ocean Observ Dev Panel Background Report 5, Ocean Observing System Development Panel, Tex. A&M Univ., College Stations
Wanninkhof R (1992) Relationship between wind speed and gas exchange over the ocean. J Geophys Res 97:7373–7382
Wanninkhof R, Thoning K (1993) Measurement of fugacity of CO_2 in surface water using continuous and discrete sampling methods. Mar Chem 44:189–204
Weiss R (1974) Carbon dioxide in water and seawater: The solubility of a non-ideal gas. Mar Chem 8:347–359
Weiss R, Bullister J, Gammon R, Warner M (1985) Atmospheric chlorofluormethanes in the deep equatorial Atlantic. Nature 314: 608–610

Appendix – Thermodynamic Equations for the Carbonic Acid System in Sea Water

Solubility of CO_2 (Weiss 1974)

$$\ln K_0 = -60.2409 + 93.4517(100 / T) + 23.3585 \ln(T / 100) + S[0.023517 - 0.023656 (T / 100) + 0.0047036 (T / 100)^2] \quad (A.1)$$

Carbonic Acid (Millero 1995)

$$\ln K_1^* = \ln K_1 + A\, S^{0.5} + B\, S + C\, S^{1.5} + D\, S^2 + \ln(1 - 0.001005\, S) \quad (A.2)$$

$$\ln K_1 = 290.9097 - 14\,554.21 / T - 45.0575 \ln T$$
$$A_1 = -228.39774 + 9\,714.36839 / T + 34.485796 \ln T$$
$$B_1 = 54.20871 - 2\,310.48919 / T - 8.19516 \ln T$$
$$C_1 = -3.969101 + 170.22169 / T + 0.603627 \ln T$$
$$D_1 = -0.00258768$$

$$\ln K_2^* = \ln K_2 + A\, S^{0.5} + B\, S + C\, S^{1.5} + D\, S^2 + \ln(1 - 0.001005\, S) \quad (A.3)$$

$$\ln K_2 = 207.6548 - 11\,843.79 / T - 33.6485 \ln T$$
$$A_2 = -167.69908 + 6\,551.35253 / T + 25.928788 \ln T$$
$$B_2 = 39.75854 - 1\,566.13883 / T - 6.171951 \ln T$$
$$C_2 = -2.892532 + 116.270079 / T + 0.45788501 \ln T$$
$$D_2 = -0.00613142$$

Boric Acid (Dickson 1990b)

$$\ln K_{HB}^* = \ln K_{HB} + A\, S^{0.5} + B\, S + C\, S^{1.5} + D\, S^2 + \ln(1 - 0.001005\, S) \quad (A.4)$$

$$\ln K_{HB} = 148.0248 - 8\,966.90 / T - 24.4344 \ln T$$
$$A_{HB} = 137.194 - 2\,890.51 / T + 25.085 \ln T + 0.053105\, T$$
$$B_{HB} = 1.62247 - 77.942 / T - 0.2474 \ln T$$
$$C_{HB} = 1.726 / T$$
$$D_{HB} = -0.0993 / T$$

Water (Millero 1995)

$$\ln K_W^* = \ln K_W + A\, S^{0.5} + B\, S \quad (A.5)$$

$$\ln K_W = 148.9802 - 13\,847.26 / T - 23.6521 \ln T$$
$$A = -5.977 + 118.67 / T + 1.0495 \ln T$$
$$B = -1.615 \times 10^{-2}$$

HSO_4 (Dickson 1990a)

$$\ln K_{HSO_4} = -4\,276.1 + 141.328 - 23.093 \ln T + A\, I^{0.5} + B\, I + C\, I^{1.5} + D\, I^2 \quad (A.6)$$

$$A = -324.57 + 13\,856 / T + 47.986 \ln T$$
$$B = 771.54 - 35\,474 / T - 114.723 \ln T$$
$$C = 2\,698 / T$$
$$D = -1\,776 / T$$

HF (Dickson and Riley 1979)

$$\log K_{HF} = -1\,590.2 / T + 12.641 - 1.525\, I^{0.5} \quad (A.7)$$

Solubility of $CaCO_3$ (Mucci 1983)

$$\log K^*_{SP} = \ln K_{SP}(i) + A\, S^{0.5} + B\, S + C\, S^{1.5} \quad (A.8)$$

$$\log K_{SP(calcite)} = -171.9065 - 0.077993T + 2\,839.319 / T + 71.595 \log T$$
$$A_{cal} = -0.77712 + 0.0028426\, T + 178.34 / T$$
$$B_{cal} = -0.07711$$
$$C_{cal} = 0.0041249$$

$$\log K_{SP(aragonite)} = -171.945 - 0.077993\, T + 2\,903.293 / T + 71.595 \log T$$
$$A_{arag} = -0.068393 + 0.0017276\, T + 88.135 / T$$
$$B_{arag} = -0.10018$$
$$C_{arag} = 0.0059415$$

Relationships for the Various pH Scales (Millero 1995)

$$10^{-pHNBS} = f_H[H^+]_T = f_H[H^+]_F(1 + [SO_4^{2-}] / K_{HSO_4} + [F^-] / K_{HF}) \quad (A.9)$$

where f_H is the apparent total proton activity coefficient, K_i values are the dissociation constants (given above)

$$pH_T = pH_F - \log(1 + [SO_4^{2-}] / K^*_{HSO_4}) \quad (A.10)$$

$$pH_{SWS} = pH_T - \log\{(1 + \beta_{HSO_4}[SO_4^{2-}] + \beta_{HF}[F^-]) / (1 + \beta_{HSO_4}[SO_4^{2-}])\} \quad (A.11)$$

$$pH_{SWS} = pH_F - \log\{(1 + \beta_{HSO_4}[SO_4^{2-}] + \beta_{HF}[F^-])\} \quad (A.12)$$

Effect of Temperature and Pressure on the pH of Sea Water (Millero 1979, 1995)

$$pH_t = pH_{25} + A + B\, t + C\, t^2 \quad (A.13)$$

$$A = -2.6492 - 0.0011019\, S + 4.9319 \times 10^{-6}\, S^2 + 5.1872\, X - 2.1586\, X^2$$
$$B = 0.10265 - 0.20322\, X + 0.084431\, X^2 + 3.1618 \times 10^{-5}\, S$$
$$C = 4.4528 \times 10^{-5}$$

where $X = TA / TCO_2$

$$\mathrm{pH}_t^{\mathrm{P}} = \mathrm{pH}_t^0 + A\,P \tag{A.14}$$

$$-10^3 A = 0.424 - 0.0048\,(S - 35) - 0.00282\,t - 0.0816\,(\mathrm{pH}_t^0 - 8)$$

where pH_t^0 is the pH at temperature t (°C) and 1 atmosphere.

References for Appendix

Dickson AG (1990a) Standard potential of the reaction: $AgCl(s) + 1.2H_2(g) = Ag(s) + HCl(aq)$, and the stastandard acidity constant of the ion HSO_4^- in synthetic sea water from 273.15 to 318.15 K. J Chem Thermodyn 22:113–127

Dickson AG (1990b) Thermodynamics of the dissociation of boric acid in synthetic seawater from 273.15 to 318.15 K. Deep-Sea Res 37:755–766

Dickson AG, Riley JP (1979) The estmation of acid dissociation constants in seawater from potentiometric titrations with strong base. I. The ion product of water – K_W. Mar Chem 7:89–99

Millero FJ (1979) The thermodynamics of the carbonic acid system in seawater. Geochim Cosmochim Acta 43:1651–1661

Millero FJ (1995) Thermodynamics of the carbon dioxide system in the oceans. Geochim Cosmochim Acta 59:661–677

Mucci A (1983) The solubility of calcite and aragonite in seawater at various salinities, temperatures and one atmosphere total pressure. Amer J Sci 283:780–799

Weiss R (1974) Carbon dioxide in water and seawater: The solubility of a non-ideal gas. Mar Chem 8:347–359

Chapter 2

Marine Organic Geochemistry: A General Overview

S. Pantoja · S. Wakeham

2.1 Introduction

Organic geochemistry developed traditionally from the petroleum geologists' need to find and extract petroleum, since a thorough understanding of the chemical principles involved in the origin, migration, accumulation and alteration of petroleum would greatly aid in its discovery. Early on it became clear that organic matter that has been transformed into petroleum and gas was largely of marine origin and was deposited in marine sediments. A more fundamental understanding of the processes of petroleum and gas generation required a better understanding of the cycling of organic matter in the ocean. Simultaneously, there has been increasing awareness of the important role of the ocean in mediating global-scale processes, notably global climate change through the ocean's buffering capacity for atmospheric carbon dioxide. Further, marine sediments hold the record of past environments, and realistic interpretations of past earth history hinge on understanding the behavior of organic matter in the ocean. It thus becomes important to better characterize the biogeochemical cycles that influence the production and preservation of organic matter in the sea. The term biogeochemistry emphasizes the close linkage between biology, geology, and chemistry into a cross-disciplinary science that strives to define the relationship between the biosphere and the geosphere, and between living and non-living organic matter.

Dynamic biogeochemical cycles link the distributions of the major chemical elements, C, N, S, P, and O, between living and non-living organic matter and various inorganic reservoirs (reviewed by Summons 1993). The current view of organic biogeochemical cycles in the ocean is illustrated in Fig. 2.1. Primary organic matter is biosynthesized from inorganic nutrients by the photosynthetic plankton, using light as the major energy source. This particulate organic matter (*POM*) is subject to grazing by heterotrophic bacteria and zooplankton. Grazing, excretion, cell lysis and enzymatic hydrolysis of cellular material transfer *POM* to the dissolved organic matter (*DOM*) pool. Much of the *DOM* becomes refractory, but a significant fraction feeds the microbial loop involving bacteria and protozoa. Whereas conventional wisdom was that most *POM* was grazed by zooplankton, it is now becoming clear that organic matter flux though the microbial loop may process half of oceanic primary production (Azam 1998). A very small fraction of primary production becomes part of the sinking flux, by which *POM* is transferred from the upper ocean to deep-sea sediments.

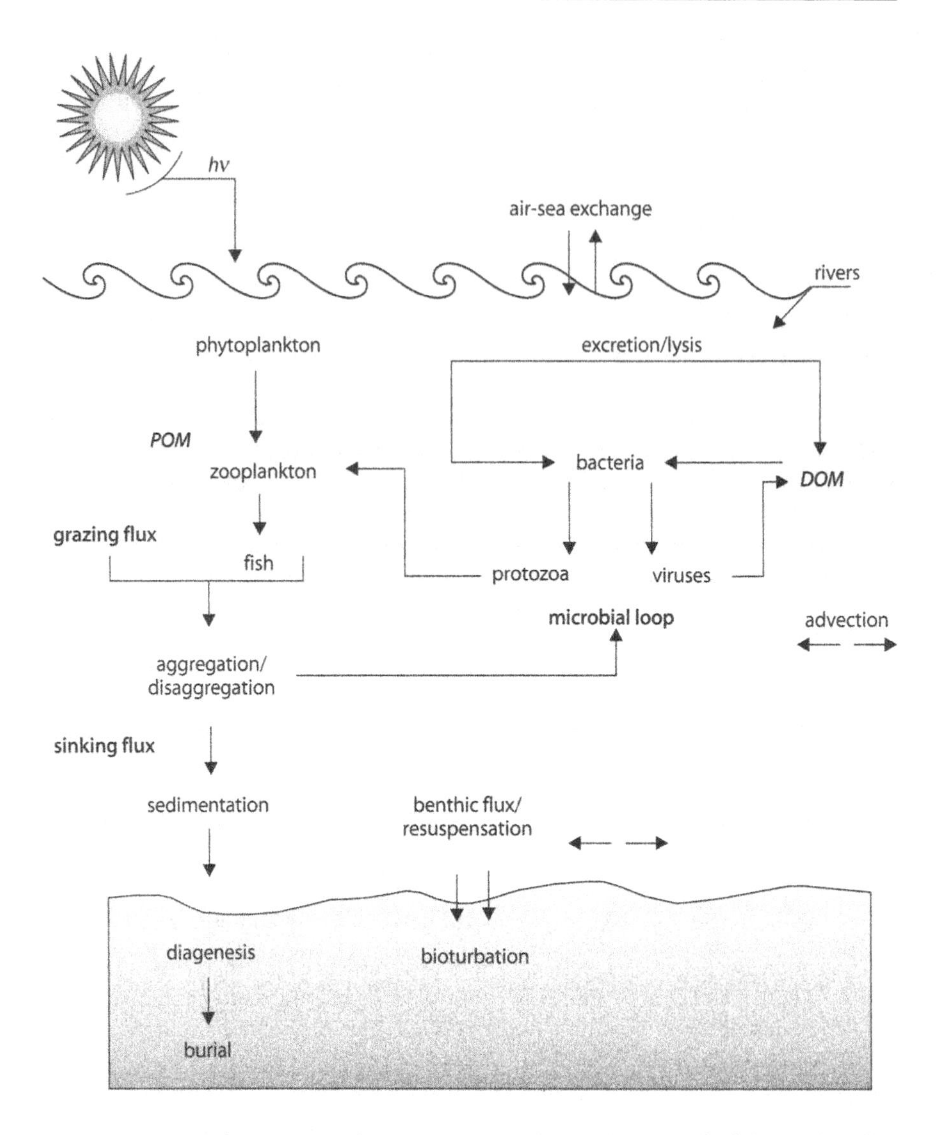

Fig. 2.1. Schematic of biogeochemical cycles in the ocean, showing the grazing food chain, sinking flux, and microbial loop (adapted from Azam 1998)

2.1.1 Organic Carbon Cycle

Carbon is a fundamental element of life and thus is a key component of all organic compounds. Hedges and Oades (1997) have presented a recent compilation of the reservoirs and fluxes within the global carbon cycle (Fig. 2.2). Most carbon on earth is sequestered on geologic time scales in sedimentary rocks, predominately as inorganic carbonates (60 000 000 Gt C [Gt = 10^{15}g]). About 15 000 000 Gt of organic carbon (*OC*)

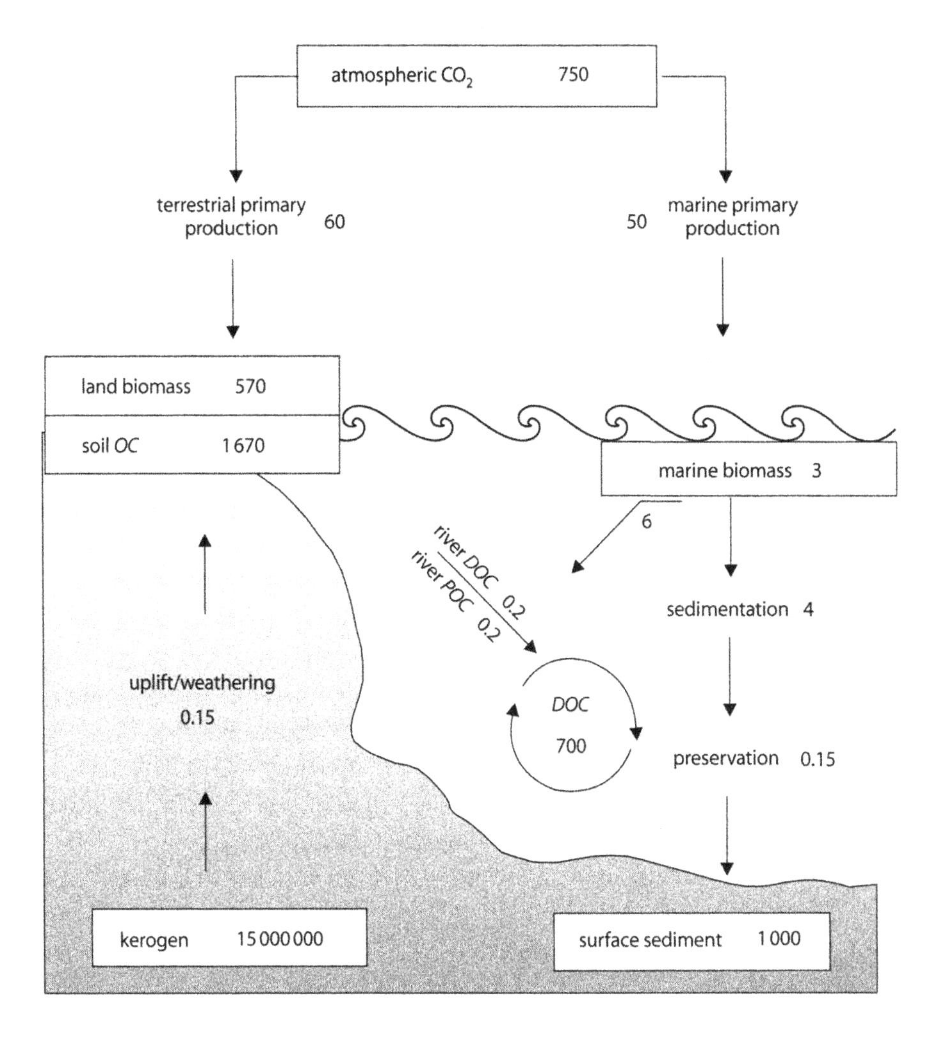

Fig. 2.2. Global reservoirs (Gt) and fluxes (Gt yr^{-1}) of *OC* on land and in the ocean (adapted from Hedges and Oades 1997)

is stored in sediments as kerogen at concentrations of <1% by weight. Actively cycling carbon is dominated by dissolved inorganic carbon, largely in the form of bicarbonate, in seawater (39 000 Gt). Organic carbon in the contemporary terrigenous environment includes 570 Gt *OC* in plant biomass and 1 670 Gt *OC* in soil litter and humus. In the ocean, the dominant *OC* reservoir is 700 Gt of dissolved organic carbon (*DOC*), 95% of which is in the deep-sea. Marine biomass adds 3 Gt *OC*, and 1 000 Gt *OC* is buried in surface sediments. Progressive alteration of sedimentary organic carbon by diagenesis, catagenesis, and metagenesis produces the kerogen that ultimately generates petroleum, coal and gas. Fossil fuels may represent ≈ 0.05% of sedimentary organic carbon.

Photosynthetic fixation is the dominant process by which inorganic carbon is transformed to organic carbon. Net global photosynthesis is estimated at 110 Gt yr^{-1}, roughly equally distributed between terrigenous and marine systems (55% and 45%, respectively). On land, most organic carbon becomes incorporated into soils, while about 0.2 Gt yr^{-1} each is transported as *POC* and *DOC* to the ocean. Marine photosynthesis, dominated by the nano- and picoplankton, fixes about 50 Gt yr^{-1}. However, since the reservoir of marine biomass is relatively small, most *OC* is rapidly recycled in surface waters. About 6 Gt yr^{-1} is converted to *DOC*, and 4 Gt yr^{-1} sinks as particulate matter out of the upper ocean. The remaining 40 Gt yr^{-1} is efficiently remineralized back to inorganic nutrients and eventually re-enters the cycle via photosynthesis. Sinking *OC* is also efficiently degraded in the interior of the ocean, with the result that 0.15 Gt yr^{-1} (or <1% of primary productivity), is buried in surface sediments and a fraction of this is converted into kerogen. Tectonic uplift and weathering of sedimentary rocks returns *OC* from kerogen to the contemporary *OC* cycle at a rate of 0.15 Gt yr^{-1}.

Carbon undergoes transitions from highly oxidized to reduced states, ranging from +4 for CO_2 and HCO_3^- to −4 for CH_4 as it passes through the carbon cycle (Fig. 2.3). Photosynthetic fixation of inorganic carbon, whether CO_2 or HCO_3^-, involves a 4 electron reduction to produce organic matter with the generic formula CH_2O, in which C is in the zero-oxidation state. Subsequent degradation of organic matter by a sequence of terminal electron acceptors further reduces the oxidation state of C. In aerobic respiration, O_2 is the electron acceptor for the reaction that converts glucose to CO_2 and H_2O, with an energy yield of 686 kcal mol^{-1}. In fermentation, a much less efficient process producing about 57 kcal mol^{-1}, energy is transferred by oxidation of part of the organic molecule and reduction of another part. The low molecular weight organic products are available to be used as substrates by other groups of microbes. Once oxygen has been depleted, bacteria use compounds other than oxygen as terminal electron acceptors, beginning with NO_3^- (nitrate reduction; 650 kcal mol^{-1}), and progressing to SO_4^{2-} (sulfate reduction; ~10 kcal mol^{-1}) and CO_2 (methanogenesis; ~40 kcal mol^{-1}). Low molecular weight substrates, such as lactate, acetate, and H_2 serve as electron donors in sulfate reduction and methanogenesis, and H_2S, CO_2, H_2O and CH_4 are products. Since low molecular weight compounds are used in sulfate reduction and methano-genesis, there is a close coupling of activities of fermenters, sulfate reducers, and methanogens.

2.1.2 Nitrogen Cycle

Most of the nitrogen in earth is present as molecular dinitrogen in the atmosphere (4 000 000 Gt). The oceanic nitrogen reservoir is also dominated by N_2 (20 000 Gt), followed by nitrate (570 Gt), which comprises about 6% of seawater N. The world ocean appears to be experiencing a net loss of nitrogen. The magnitude of the input to the sea from rivers, biological fixation and precipitation accounts for about 70% of the losses by denitrification plus burial (Schlesinger 1997). Sediment burial removes nitrogen from seawater mainly as organic matter. Organic nitrogen trapped in sedimentary rocks is a major reservoir for nitrogen in the earth (2 000 000 Gt, Wada and Hattori 1991). During diagenesis, the C/N ratio of organic matter does not increase with sediment age as fast as between the surface of the ocean and the sediment surface. The result is that 1 atom of nitrogen out of ca. 10 000 produced in the photic zone is preserved in sediments (0.1%), similar to carbon burial efficiency (Hedges 1992).

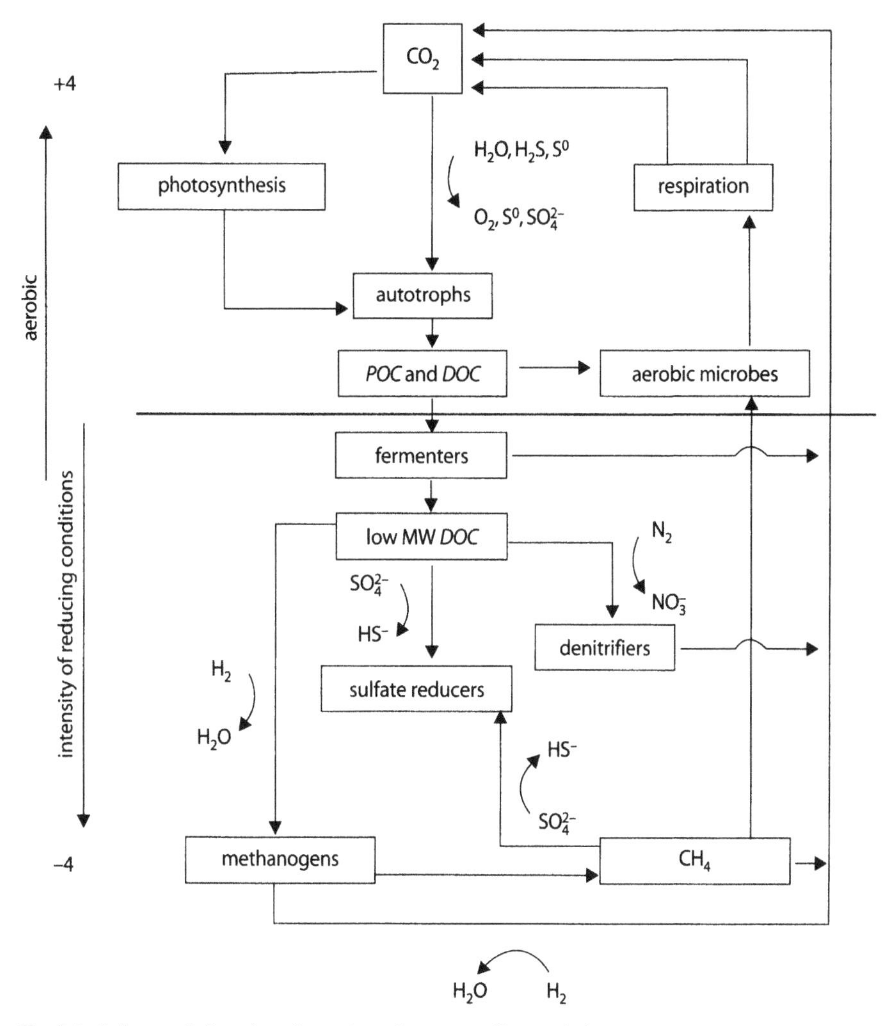

Fig. 2.3. Carbon cycle in oxic and anoxic environments (from Valiela 1995)

The nitrogen cycle involves electron shifts between the most oxidized form (NO_3^-) and the most reduced (NH_4^+) (Fig. 2.4). Uptake by phytoplankton efficiently removes nitrate, nitrite and ammonium during photosynthesis in the photic zone. During this transformation from *DIN* (dissolved inorganic nitrogen) to *PON* (particulate organic nitrogen), the oxidation state of nitrogen changes from +3 (NO_2^-), or +5 (NO_3^-) to −3 (organic nitrogen) by intracellular reduction. Ammonium is preferentially consumed by phytoplankton, presumably because it is already more reduced (−3) than nitrite and nitrate. Similarly, NH_4^+ availability reduces rates of N_2 fixation, probably because reduction of N_2 is an energy-demanding process.

Organic nitrogen thus formed is eventually decomposed during respiration. In oxygenated environments O_2 is used as a terminal electron acceptor. Ammonium produced during decomposition of organic nitrogen can be taken up by bacteria and algae or

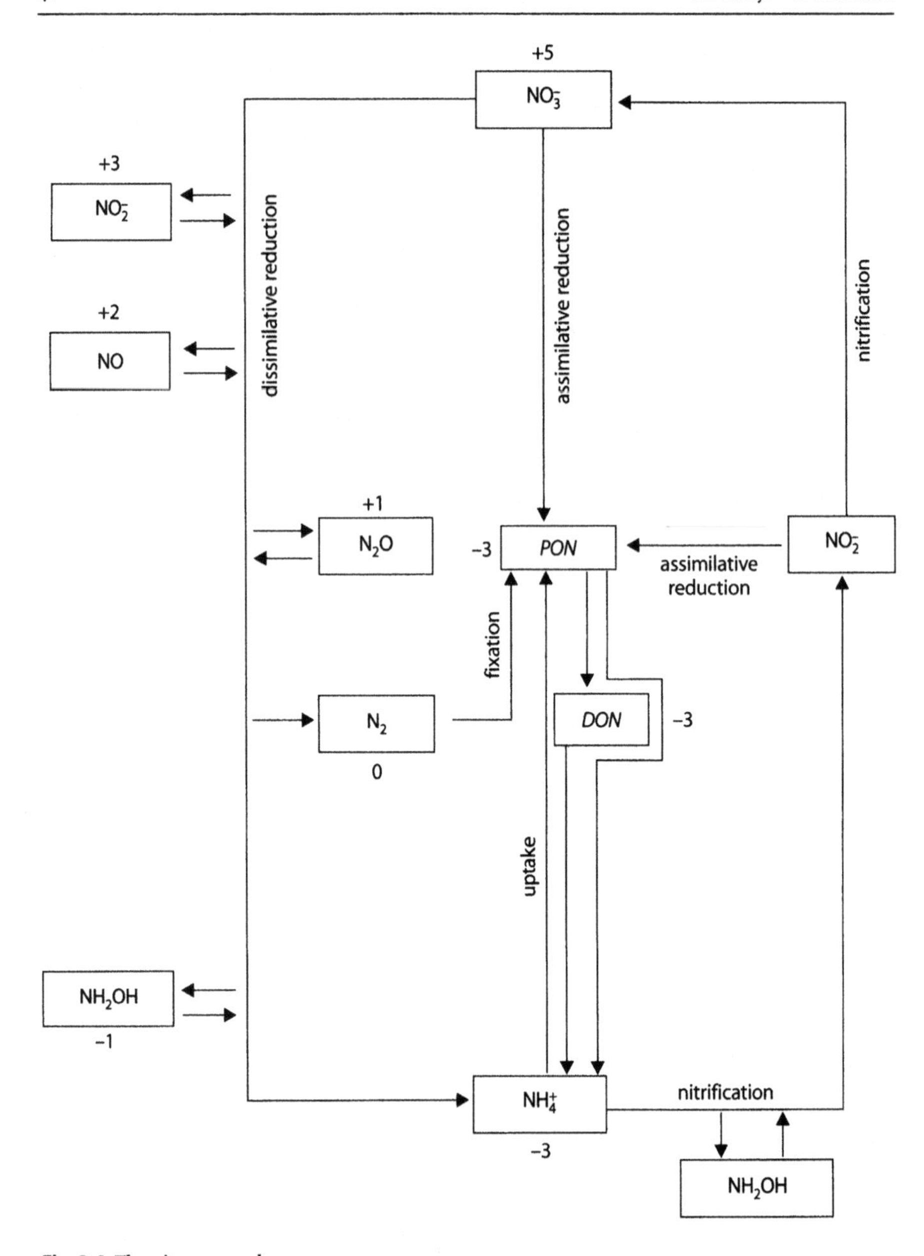

Fig. 2.4. The nitrogen cycle

can be oxidized to nitrate (nitrification). Nitrification is a microbially-mediated process in which ammonium is first oxidized to nitrite with the consequent production of 66 kcal mol^{-1}. Bacteria involved in this step are usually of the genus *Nitrosomonas*. The biochemistry of nitrification is not very well understood, but it is known that intermediates like hydroxylamine (NH_2OH) are produced during the reaction (Fig. 2.4). Those intermediates do not seem to accumulate (Fenchel and Blackburn 1979). Nitrite is further oxidized to nitrate by *Nitrobacter* and other genera. This step also yields energy (−17.5 kcal mol^{-1}). Energy yielded from oxidation of NH_4^+ by these chemosynthetic bacteria is used in reduction of CO_2 to organic carbon.

Another sink of nitrate (besides uptake) is dissimilatory[1] reduction during respiration of organic matter. Nitrate is used as an electron acceptor by heterotrophic bacteria in a series of steps (Fig. 2.4). Nitrate is reduced to nitrite, nitric oxide (NO), and then to the gases nitrous oxide (N_2O) and dinitrogen (N_2). When the products of reduction of nitrate are gaseous, the process is called denitrification. Denitrification only occurs at low concentrations of oxygen, mostly because the synthesis of enzymes is repressed by the presence of O_2 (Fenchel and Blackburn 1979). Denitrifying microbes may reduce nitrate to ammonium with the intermediate production of hydroxylamine. This pathway is thought to be minor compared with ammonium production by degradation of organic nitrogen (Valiela 1995).

2.2 Molecular Constituents of Organic Matter in the Ocean

For years, marine geochemists have attempted to infer the sources and behavior of organic matter in the ocean on the basis of bulk parameters, such as organic carbon concentration, C/N ratios, stable carbon isotopic compositions, etc. While bulk parameters are informative, their use ignores the vast amount of information available at the molecular level. Organisms biosynthesize organic compounds containing a wide range of organic functional groups, many of which are specific to structural or metabolic functions within the organisms. Many molecules have novel structures that provide organic geochemists with a wealth of information related to the compounds' sources and, by inference, the origin of bulk organic matter. Finally, organic chemical reactions affect various functional groups in different ways that are diagnostic of how biogeochemical reactions proceed. In the following section, we discuss aspects of the biogeochemistries of important biochemical classes in the ocean, providing examples of how molecular-level information can be used to characterize the sources and fates of organic matter as a whole.

2.2.1 Lipids

Lipids are operationally defined as substances that are practically insoluble in water but extractable with non-polar organic solvents. After amino acids and carbohydrates, lipids are the next most abundant biochemical in organisms. Plankton generally contain lipids equivalent to 10–60% of organic carbon (Parsons et al. 1984; Sargent and Henderson 1986; Wakeham et al. 1997a,b), while in sediments 1% or less of *OC* is lipid (e.g. Santos et al. 1994; Wakeham et al. 1997a,b). In organisms, lipids are involved in energy storage and mobilization, membrane structure, and hormonal control of metabolic processes (Lehninger 1981). The wide variety of molecular structures of lipids (Fig. 2.5) makes them valuable "biomarkers" for tracing sources of organic matter in

[1] Dissimilatory processes do not involve cellular incorporation, only processes associated to energy production.

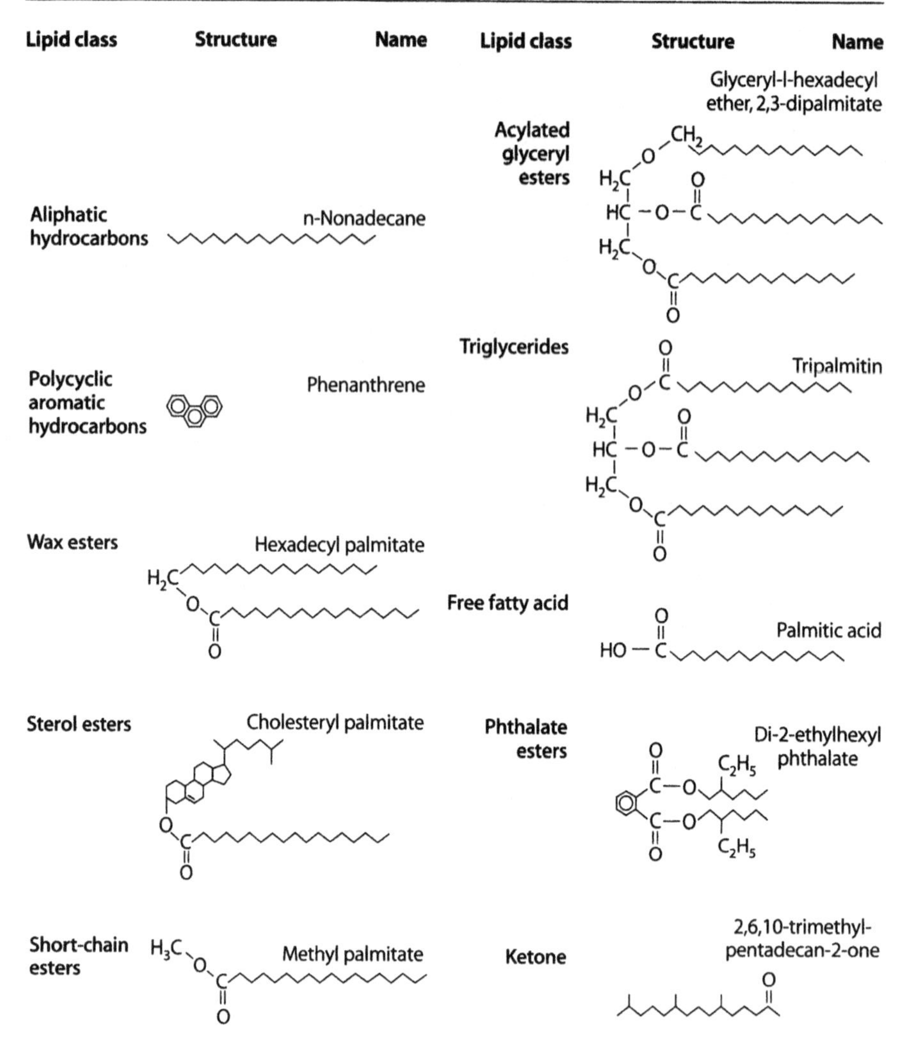

Fig. 2.5 a. Structures of representative lipids

the ocean and for elucidating the biogeochemical processes that affect the fate of organic carbon (Table 2.1) (reviewed by Lee and Wakeham 1988; Wakeham and Lee 1993). The reactive nature of many of the organic functional groups in lipids makes it possible to trace biogeochemical reaction pathways (e.g. Gagosian et al. 1980; Repeta and Gagosian 1984; Sun et al. 1998); yet some structures are stable toward diagenetic reactions and are preserved in sediments, even to ancient sediments and fossil fuels (Mackenzie et al. 1982). This stability provides an unambiguous link between ancient sedimentary organic matter and its contemporary biological analog, allowing geochemists to infer past oceanographic conditions (Brassell 1993).

There have been many compilations of lipid compound distributions in marine organisms (e.g. Sargent 1976; Volkman 1986; Sargent et al. 1987; Volkman et al. 1989;

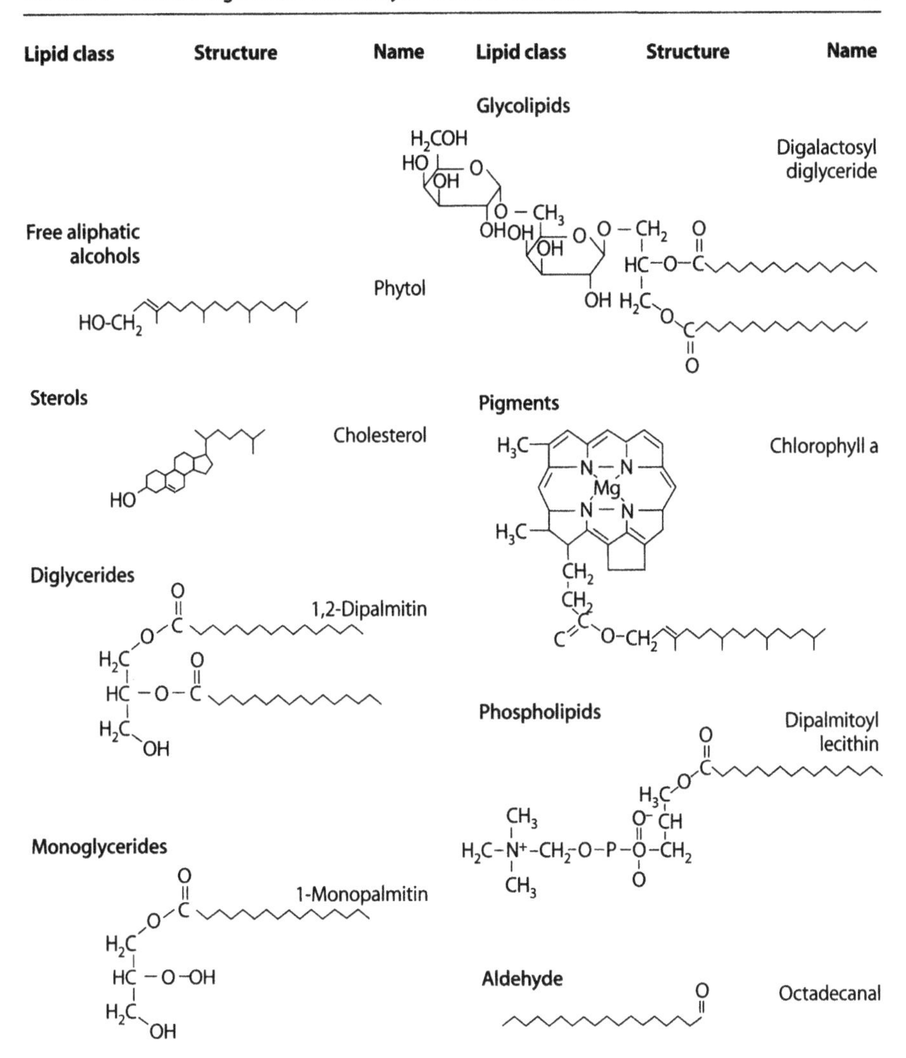

Fig. 2.5 b. Structures of representative lipids

Conte et al. 1994) and of inventories and biogeochemical behaviors of lipids in particulate matter and sediments (e.g. Cranwell 1982; Lee and Wakeham 1988; Wakeham and Lee 1993). Several fundamental facts have emerged from this work. First, lipids are rapidly degraded as particulate material moves from surface waters where it is produced to sediments where a tiny fraction of production is buried and preserved. The behavior of specific lipids is highly dependent on molecular structure, whereby short-chain compounds and highly unsaturated compounds tend to be more reactive than long-chain, unsaturated molecules. The nature of the water column and sediment bed, whether oxic or anoxic, influences the form of organic matter degradation (aerobic vs. anaerobic), and resulting differences in biochemical reaction mechanisms will influence the structure of decomposition products thus formed. Terrigenous, higher

Table 2.1. Lipid biomarkers as source indicators

	Marine	Terrigenous	Bacteria
Fatty acids	C14-C24-alkanoic acids; even C predominance; few branched; highly unsaturated Δ^9, cis-monounsaturated; mid-chain hydroxy acids? – –	C24-C46-alkanoic acids; even C predominance; few branched; predominately saturated; – α-hydroxy acids; – diterpenoid acids	C15-C19-alkanoic acids; mixed odd, even predominance; highly branched (iso/anteiso); cyclopropyl C17 and C19 Δ^{11}, trans monounsaturated; β-, ω-, (ω-)-hydroxy acids; $\alpha\omega$-diacids; hopanoic acids
Hydrocarbons	C15-C19-alkanes; odd C predominance; C17, C20, C25-isoprenoid	C23-C35-alkanes; even C predominance; –	isoalkanes; hopenes/hopanes; –
Fatty alcohols	C14-C24-alkan-1-ols; even C predominance	C24-C36-alkan-1-ols; even C predominance	branched-alkan-1-ols; hopanols
Ketones	C37-C39 alkenones	C24-C38-alkan-2-ones	6,10,14-trimethylpentadecan-2-one
Aldehydes	?	C24-C36-alkanals	–
Sterols	complex mix C26-C30	simple mix C28-C29	5α(H)-stanols

plant-derived compounds tend to be more efficiently preserved in deep-sea sediments than are marine-derived compounds; this preferential preservation is likely related to differences in molecular structure and a particle matrix effect that protects terrigenous compounds from degradation. Furthermore, there may be in-growth of molecules that have been produced in situ by heterotrophs, both zooplanktonic and microbial. The major effect of selective degradation/preservation is that the lipid composition of sediments may be highly altered compared to the source materials. Thus, there is both quantitative and qualitative uncoupling between surface water production and sediment preservation for lipids, and this decoupling needs to be sorted out in order to use the sedimentary record for paleoceanographic reconstructions.

2.2.1.1
Lipid Degradation Rates

Relatively less is known about rates of lipid degradation in the ocean. Data from early sediment trap studies in the equatorial North Atlantic (e.g. de Baar et al. 1983) could be used to estimate degradation rate constants based on changes in vertical flux as a function of depth. Thus, the rate of net loss of fatty acids increased with number of double bonds and decreased with number of carbon atoms. It was estimated that 20–40% of total oxygen consumption in this region might be accounted for by degradation of sinking organic material. It is noteworthy that no similar lipid-based oxygen consumption rates have been estimated since the de Baar et al. report, despite the generation of a large data set for sediment trap lipids.

Considerably more effort has gone toward quantifying lipid degradation rates in sediments. Using laboratory simulations, Harvey and co-workers (Harvey et al. 1995; Harvey and Macko 1997a,b) have compared the kinetics of phytoplankton decay under oxic and anoxic conditions: in addition to lipids, their study examined organic carbon, nitrogen, carbohydrates and amino acids. Among the major biochemical fractions, carbohydrates were most rapidly degraded under oxic conditions, followed by protein and lipid. Degradation rates for all components were significantly lower under anoxic conditions. Similar comparisons of degradation of various organic matter pools in oxic and anoxic regimes have been made in the water column (Lee 1992) and sediments (Sun et al. 1993, 1997). There is still considerable controversy as to the role of oxygen in the degradation of organic matter (Henrichs and Reeburgh 1987; Emerson and Hedges 1988; Canfield 1994; Pedersen and Calvert 1990; Lee 1992). Nonetheless, degradation of organic constituents in sediments behaves in a manner consistent with the "multi-G" model of Westrich and Berner (1984), where the highly reactive material degrades most rapidly, but the more refractory pools degrade more slowly. However, a recent analysis of sedimentary lipid degradation (Canuel and Martens 1996) in a variety of sediments having widely different accumulation rates indicates that degradation rates in fact vary as a function of time since burial (Fig. 2.6). The highest rates and rate constants occur in surface sediments, and the time scale used in calculating degradation rates thus becomes critical.

2.2.1.2 *Lipid Degradation Mechanisms*

Organic functional groups are altered during biogeochemical reactions in the water column and sediments, and it is possible to use changes in molecular structure to follow reaction mechanisms. The diagenetic behavior of steroids has been widely studied because of their biomarker potential in organisms and recent and ancient sediments (Gagosian et al. 1980; Edmunds et al. 1980; Mackenzie et al. 1982; Mermoud et al. 1984; Gaskell et al. 1975; de Leeuw and Baas 1986). Gagosian et al. (1980) summarized a

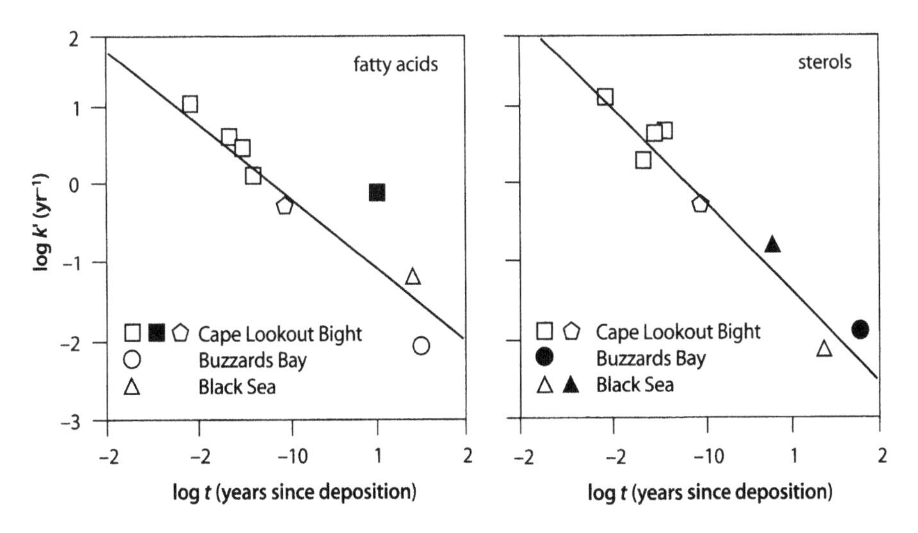

Fig. 2.6. Comparison of rate constants (k') for fatty acids and sterols in surface sediments of various locations (adapted from Canuel and Martens 1996)

mechanistic pathway for biologically- and chemically-mediated degradation of sterols in seawater and sediments. Unsaturated steroidal alcohols (stenols) derived from plankton are sequentially oxidized and dehydrated to form steroidal ketones (stanones and stenones), steroidal diols, saturated steroidal alcohols (stanols) and steroidal hydrocarbons (sterenes). Conversion of stenols to stanols has been shown to be an anaerobic microbial process in the laboratory, and thus ratios of stanol to stenol have been used as indicators of redox conditions in the water column and sediments. Increasing stanol/stenol ratios have been observed in a series of water columns progressing from oxic to suboxic and anoxic (Wakeham 1989) and downcore in reducing sediments (Nishimura 1978); sterenes are abundant in suboxic seawater (Wakeham et al. 1984b). As organic matter is further altered during maturation of ancient sediments, steroidal compounds that have survived burial in recent sediments are transformed to the saturated and aromatized steroidal hydrocarbons (steranes) that are widely used biomarkers in petroleum geochemistry (Mackenzie et al. 1982).

Several other diagenetic pathways of lipids have also been elucidated. The geochemistry of chlorophyll a, the predominant of several photosynthetic chlorin pigments in plants, has been extensively studied (Callot 1991). Grazing by herbivorous zooplankton demetallates chl a to phaeophytins and hydrolyses the phytyl side-chain to phaeophorbides; in fact, ratios of chl a/phaeopigments are often used as an indicator of grazing. Recent work has also shown that chlorophylls are esterified to sterols during the grazing processes (King and Wakeham 1996; Harradine and Maxwell 1998). Phorbin steryl esters may better reflect the sterol distribution of the phytoplankton community that is being grazed than free sterols that are selectively removed during diagenesis (King and Repeta 1994). Hydrolysis of the phytyl side-chain of chl a releases the isoprenoid alcohol, phytol. Phytol decomposition has also been examined, with a complex suite of alkanes, acids, ketones, and alcohols being formed (Volkman and Maxwell 1986). Oxidative degradation of phytol can also produce 6,10,14-trimethylpentadecan-2-one and phytenic acid (Gillan et al. 1983). Hydrogenation of phytol in anaerobic guts of pelagic herbivores or by benthic organisms may be the major source of sedimentary dihydrophytol (Prahl et al. 1984).

While pathways of phytol degradation have been well characterized, rates of the various transformations have not. A recent study (Sun et al. 1998) has attempted to remedy this deficiency by simultaneously measuring phytol and several transformation products in laboratory incubations and coastal sediments (Fig. 2.7). Phytol is delivered to the sediment surface predominately as a free, extractable form that is diagenetically transferred to a non-extractable "bound" pool. At the molecular level, phytol is converted to oxidized (6,10,14-trimethylpentadecan-2-one and 4,8,12-trimethyltridecanoic acid) and reduced (dihydrophytol) isoprenoid products. Sedimentary distributions of precursor and products were modelled after the Westrich and Berner (1984) model, generating pseudo-first order rate constants for degradation of phytol, transfer to the bound pool, formation of 6,10,14-trimethylpentadecan-2-one, 4,8,12-trimethyltridecanoic acid, dihydrophytol, and subsequent decay of dihydrophytol. Anaerobic decomposition of extractable phytol is slow, as is anaerobic decomposition of bound phytol. Under anoxic conditions, therefore, the fate of extractable phytol reaching the sediment is controlled by transfer between the extractable to bound pools, such that the overall apparent loss rate was equivalent to the extractable to bound conversion rate. However, under oxic conditions, extractable phytol is both transformed

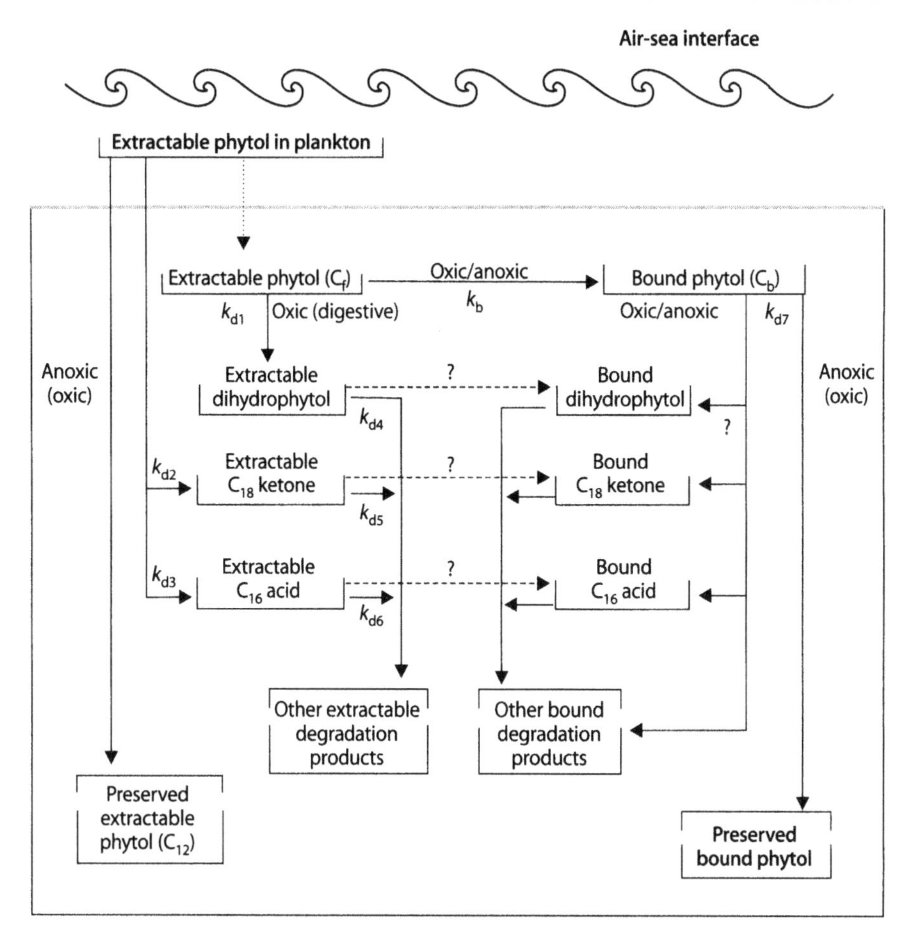

Fig. 2.7. Schematic representation of the degradation pathway of phytol in coastal sediments. C_{18} ketone = 6,10,14-trimethylpentadecan-2-one; C_{16} acid = 4,8,12-trimethyltridecanoic acid. Rate constants k_b^{an}, k_b^{ox}, k_{d1}^{an}, k_{d4}^{ox}, k_{d7}^{an}, and k_{d7}^{ox} can be estimated. Subscripts "b" and "d" indicate conversion of extractable to bound phytol; superscripts "an" and "ox" indicate anoxic and oxic degradation, respectively (adapted from Sun et al. 1998)

to 6,10,14-trimethylpentadecan-2-one and 4,8,12-trimethyltridecanoic acid as well as being incorporated into the bound pool.

2.2.2 Amino Acids and Proteins

Amino acids are the building blocks of protein in living biomass. They are mainly formed with carbon, nitrogen, oxygen and hydrogen. Some of them contain sulfur in their structure, as in the amino acids cysteine and methionine (Fig. 2.8a). Amino acids could be uncharged (e.g. H- in glycine, HO-CH_2- in serine, CH_3- in alanine), negatively charged (e.g. ^-OOC-CH_2- in aspartic acid), or positively charged at pH 6–7 (e.g. H_3N^+-$(CH_2)_4$- in lysine). That diversity in chemical structure determines differ-

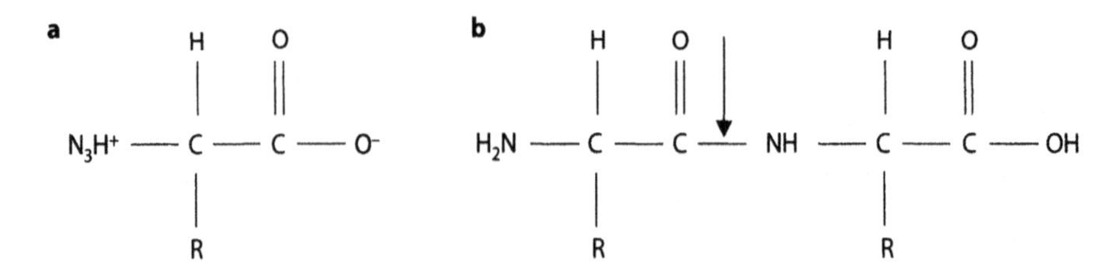

Fig. 2.8. a Structure of amino acids; **b** The peptide bond between two amino acids. R-groups could be neutral, charged or uncharged at pH 7 (see text)

ential acid-base properties of amino acids, and has proven important in understanding chemical properties of proteins. Thus, for example, alanine is a dibasic acid in its fully protonated form at low pH. In the course of titration with a base, it donates two protons to form a dipolar ion ("zwitterion") and a conjugated base (Fig. 2.9). At pH 6.02, there is no net electrical charge on the molecule, therefore it does not move toward either positive or negative pole in an electrical field. This pH is called the isoelectric point (the arithmetic mean of pK_1 and pK_2).

In the marine environment most of the amino acids occur as polymers. For instance, protein nitrogen accounts for >80% of the nitrogen in plankton (Parsons et al. 1984). Proteins are formed by condensation of amino groups and carboxyl groups of α-amino acids to form an amide bond. The resulting bond is called a peptide bond (Fig. 2.8b). Depending upon the number of residues per molecule, the final product is known as a dipeptide, oligopeptide, or polypeptide. Peptides larger than about 10 000 Daltons are called proteins.

The primary structure of proteins corresponds to the chain of amino acids (number of amino acids and distribution). The formation of helical coils by the interaction of H and O atoms of different amino acids in the chain is called the secondary structure. The tertiary structure is given by the association of several α-helixes (subunits). A quaternary structure results from the interaction among subunits of proteins (for example, the enzyme ribulose-1,5-carboxylase-oxygenase).

2.2.2.1.
Racemization of Amino Acids and Geochemical Implications

Amino acids commonly found in living organisms consist almost entirely of the L-enantiomers[2] (L stands for levorotatory, D for dextrorotatory). Some D-amino acids are present in bacterial cell walls and in some antibiotics, but in low quantity compared to the bulk of organic matter. The consequence of this disproportion is that natural amino acids are optically active (except for glycine) and under polarized light rotate the plane to the left. A racemic mixture would be optically inactive since it contains equal amounts of D- and L-amino acids.

[2] Enantiomers are a particular type of stereoisomers in which the spatial orientation of their atoms results in one being the mirror image of the other. Stereoisomers that are not mirror images of each other are called diastereomers.

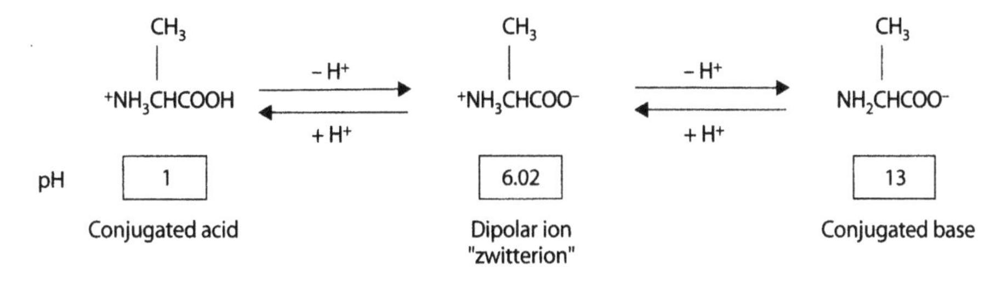

Fig. 2.9. Acid-base behavior of alanine

Racemization of amino acids has been detected in fossil material, and the proportion of D- to L-amino acids increases with the age of the fossil (Bada and Schroeder 1975). Interconversion of amino acids involves abstraction of the α-proton by hydroxide ion to form a planar carbanion, which has an equal chance of being attacked by a proton from either side (Fig. 2.10a). Some amino acids like threonine and isoleucine contain more than one chiral center. Inversion occurring only at the α-carbon results in the formation of epimers or diastereomers[3] (Fig. 2.10b).

As with any chemical process, the extent of the reaction in a fossil (free or polymeric amino acids) depends on time and temperature. If the temperature history of an environment is known, the extent of racemization of amino acids can be used to determine the age of a fossil (e.g. Bada et al. 1970; Wehmiller and Hare 1971). The advantage over other techniques, for example ^{14}C dating, is that rate constants for racemization of several amino acids are much slower than decay rates of radiocarbon, and therefore racemization can be used to date material older than 40 000 yr.

Thus, for the epimerization of L-isoleucine to D-alloisoleucine

$$\text{L} - \text{isoleucine} \underset{k_\text{A}}{\overset{k_\text{I}}{\rightleftharpoons}} \text{D} - \text{alloisoleucine}$$

the following equation can be written, assuming that the reaction follows reversible first-order kinetics:

$$\frac{-d[\text{I}]}{dt} = k_\text{I}[\text{I}] - k_\text{A}[\text{A}] \qquad (2.1)$$

If the initial concentration of D-alloisoleucine is negligible, i.e. $[\text{I}]_{t=0} \sim [\text{I}] + [\text{A}]$, integration of Eq. 2.1 results:

$$\ln\left(\frac{1+[\text{A}/\text{I}]}{1-K'[\text{A}/\text{I}]}\right)_t - \ln\left(\frac{1+[\text{A}/\text{I}]}{1-K'[\text{A}/\text{I}]}\right)_{t=0} = (1+K')k_\text{I}t \qquad (2.2)$$

where K' is the reciprocal of the equilibrium constant for the reaction and [A/I] is the ratio of D-alloisoleucine and L-isoleucine (Bada and Schroeder 1972). Rate con-

[3] Diastereomers have different chemical and physical properties.

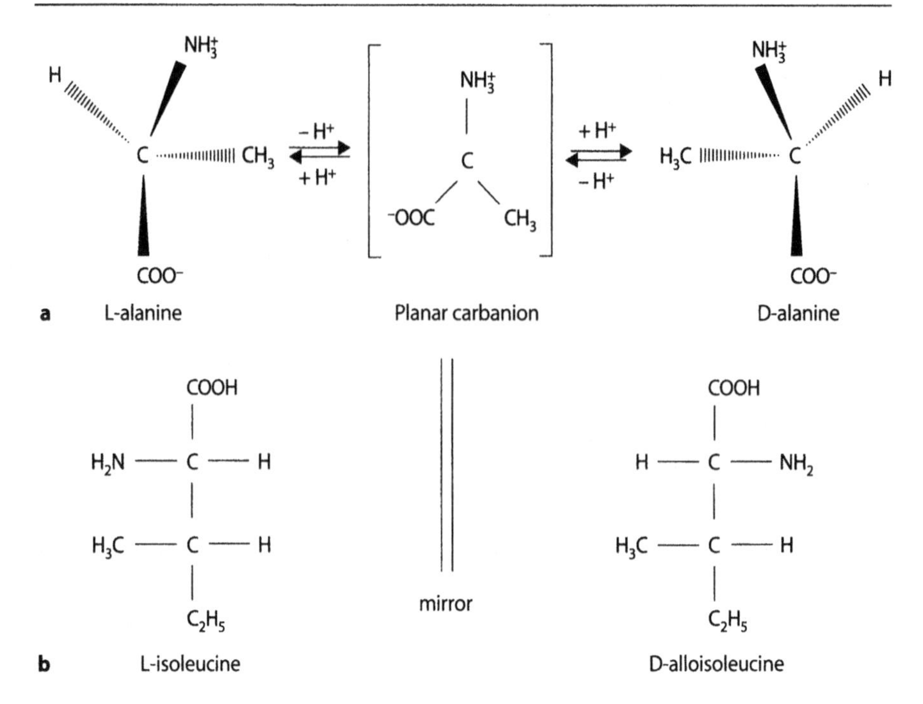

Fig. 2.10. a L- and D-isomers of alanine (enantiomers); **b** L- and D-isomers of isoleucine (diastereomers)

stants for racemization have to be measured, as well as their temperature dependency (Arrhenius relation). Thus, k_I at the lower temperature at which fossils are found in nature can be estimated. From that, Eq. 2.2 can be solved for time. Alternatively, if the age of the fossil can be determined independently, the amino acid racemization reaction can also be used to estimate the average temperature of the environment at the time of deposition of the fossil (e.g. Bada et al. 1973; Schroeder and Bada 1973).

Several considerations must be taken into account for the use of this technique as either paleothermometer or for dating fossil material. The issues discussed in the literature are microbial decomposition of tissues remaining in the bones, percolation through the bone of ground water containing amino acids, deviations from first-order rate kinetics of racemization (or epimerization) by other concurrent reactions in the fossil, and the effects of free or polymeric amino acids on the rate constants (Bada 1972; Steinberg and Bada 1983; Mitterer 1993).

2.2.2.2
Microbial Degradation of Amino Acids and Proteins

Mineralization is the main fate of amino acids in the ocean. They provide carbon skeleton and nitrogen for biosynthesis and are involved in catabolic processes.

It is thought that one of the first steps in microbial degradation of proteinaceous material in the ocean is the extracellular breakdown of polymers into smaller pieces. Small molecules like amino acids and possibly dipeptides may thus enter the cell

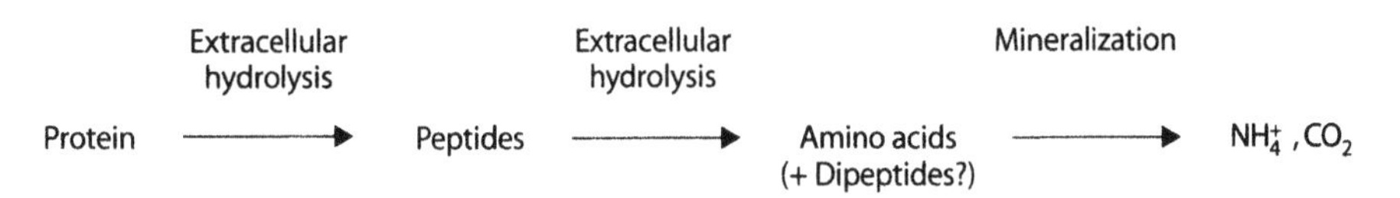

Fig. 2.11. Conceptual model of degradation of proteins by extracellular hydrolysis, and production of smaller transportable molecules

membrane of microorganisms where they are respired (Fig. 2.11). Chemically, this reaction corresponds to hydrolysis because it involves incorporation of water. Hydrolysis of peptides is discussed in Chapter 6.

Uptake[4] of amino acids (and other small molecules) across the cell membrane of microorganisms has been modelled as Michaelis-Menten enzyme kinetics, in which the rate of the reaction is dependent on the concentration of the substrate:

$$\frac{dS}{dt} = \frac{V_m S}{K_S + S} \tag{2.3}$$

S is the concentration of the substrate, V_m is the maximum velocity of the reaction and K_S is the half-saturation constant. K_S is the concentration of the substrate necessary to reach 50% of the maximum velocity (Fig. 2.12).

At low concentration of substrates (lower than K_S), $K_S + S - K_S$, then Eq. 2.3 becomes:

$$\frac{dS}{dt} = \left(\frac{V_m}{K_S}\right) S \tag{2.4}$$

If we define a new constant, $k = V_m / K_S$, then Eq. 2.4 is equivalent to a first-order kinetic reaction, where k is the rate constant in units time^{-1}. The rate of uptake depends on the concentration of the substrate (Fig. 2.12). In the ocean, because of the low concentration of dissolved monomers in sea water, we assume that most of the heterotrophic uptake occurs under substrate limitation conditions.

At very high concentrations of the substrate, i.e. $S >> K_S$, $K_S + S - S$, Eq. 2.3 becomes:

$$\frac{dS}{dt} = \left(\frac{V_m}{K_S}\right) = \text{constant} \tag{2.5}$$

In this case, the rate of reaction is independent of the substrate. We may have a case in which K_S is very large (for example $K_S = 10$, Fig. 2.12); then, first order reaction kinetics may occur over a wide range in concentration.

Ambient substrate concentrations $[S_n]$ can be estimated using a substrate-addition technique and a plot of turnover time (τ) vs. concentration of the added substrate in an incubation experiment (Wright and Hobbie 1966). The x-intercept of such a plot is equal to $-K_S - S_n$ (Fig. 2.13). Jørgensen and Søndergaard (1984) compared with ambi-

[4] Uptake involves incorporation across the cell membrane plus respiration.

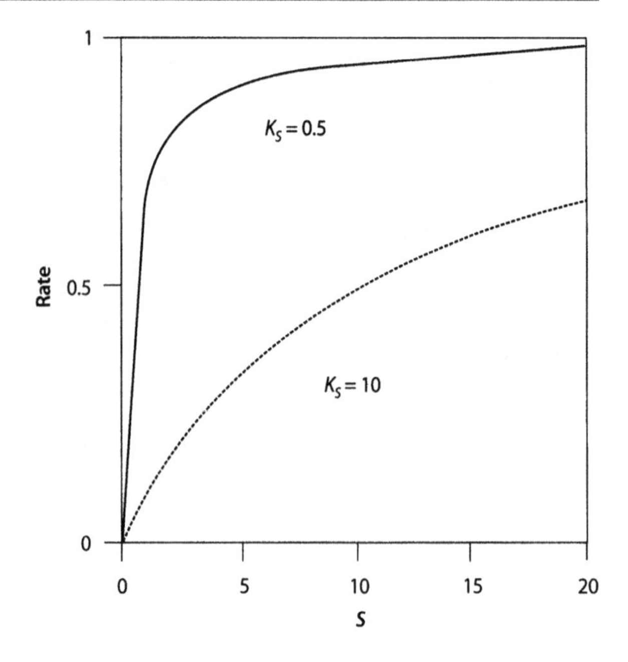

Fig. 2.12. Michaelis-Menten enzyme kinetic model. The hyperbolic function is defined by K_S, the half-saturation constant, and V_m, the maximum velocity of the reaction

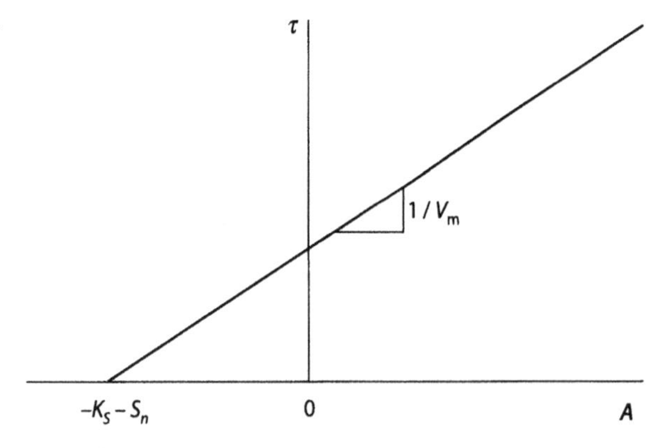

Fig. 2.13. Graphical method for estimating ambient substrate concentration (S_n) using data from incubation experiments adding variable amounts of substrate (A). τ, turnover time of the substrate. K_S and V_m are defined in Fig. 2.12 The linear equation is:
$\tau = (K_S + S_n) / V_m + A / V_m$
(Wright and Hobbie 1966)

ent dissolved free amino acids (DFAA) using this approach with values determined by HPLC in several environments. Estimates of $K_S + S_n$ were always larger than S_n, suggesting that the measured DFAA were actually free, dissolved molecules in sea water.

Half-saturation constants (K_S) have been interpreted as an indication of affinity of an enzymatic system for a substrate (e.g. Dixon and Webb 1964) or microorganism assemblage (e.g. Titman 1976; Billen 1991). Therefore, low K_S values indicate greater affinity. More recently, Button (1993) defined a new "specific affinity" as V_m / K_S. He pointed out that K_S is directly dependent on V_m, but is insensitive to changes in concentration of enzymes per unit surface area. In practice, K_S depends on the shape of the hyperbolic saturation curve. At low substrate concentration (Eq. 2.4), specific affinity approximately equals the rate constant (k) and is thus a better measure of the likelihood that microorganisms will use a specific substrate.

Amino acids undergo several microbiologically-mediated reactions (Meister 1965), some of which are extracellular (see Chapter 6). While some reactions result in the depletion of amino acids (e.g. fermentation, deamination, respiration), others lead to the formation of new amino acids (Lee and Cronin 1982). For example, decarboxylation of aspartic acid and glutamic acid produces β-alanine and γ-aminobutyric acid, respectively. The relative abundances of β-alanine and γ-aminobutyric acid have been used to assess the extent of degradation of natural organic mixtures, since they are consistently enriched in diagenetic sequences (Cowie and Hedges 1994).

2.2.3 Carbohydrates

Carbohydrates are organic compounds that contain only carbon, hydrogen, and oxygen, in which hydrogen and oxygen occur in the same proportion as they do in water. They are polyhydroxy aldehydes and ketones whose general formula is $C_n(H_2O)_n$. They exist as monosaccharides, disaccharides, trisaccharides, etc. A polysaccharide contains many monosaccharide molecules. Monosaccharides can be classified as aldoses or ketoses, depending on the occurrence of an aldehyde or a keto group. According to the number of carbon, they are trioses, tetroses, pentoses, hexoses, etc. Thus, a C_6 monosaccharide can be an aldohexose or a ketohexose. Naturally occurring monosaccharides are usually of D-configuration at C-5. The most common are hexoses, pentoses and deoxy sugars (Fig. 2.14). Modified polysaccharides, liked chitin, are also common. In chitin, the hydroxyl group in C_2 of the monosaccharide (amino sugar) has been replaced by an amino group.

Carbohydrates are important components of marine organisms. For example, they usually comprise 20 to 40% of dry biomass of phytoplankton (Parsons et al. 1961). Carbohydrates are contained in cell walls of plants, bacteria, and fungi, providing structural support. Chitin is the marine equivalent of cellulose. Chitin (a polymer of N-acetyl-D-glucosamine) is present as a structural component in arthropods, mollusks, most fungi and some algae. In bacteria, peptidoglycans in cell walls can account for up to 75% of dry biomass. Peptidoglycans are polysaccharide chains of N-acetyl-D-glucosamine and N-acetylmuramic acid cross-linked by peptides.

Carbohydrates are also involved in energy reserve. D-glucose is stored in the form of polysaccharides: starch in plants and glycogen in animals.

A very high range in reactivity occurs in carbohydrates. Glucose, for instance, is readily metabolized by most organisms, but cellulose (a polymer of glucose) is not. Moreover, starch, a cellular storage material, is water soluble and degradable, whereas the structural component cellulose is insoluble and very resistant to degradation. They are both made of many units of D-glucose and the only difference is the position of the glycoside bond among them. In cellulose, D-glucose units are linked via an β-glycoside linkage whereas in starch all glycoside linkages are α (axial).

Bacteria must initially hydrolyze polysaccharides outside the cell in order to provide substrates small enough to fit bacterial porins and thus enter the cell where glycolysis and respiration occur. Hydrolysis is carried out by bacterial cell surface and extracellular enzymes. This step, previous to complete mineralization of carbohydrates, appears to be dependent on chemical structure of substrates. For instance, the polysaccharide pullulan (α(1,6-linked maltotriose[α(1,4) linkages] units) is preferentially

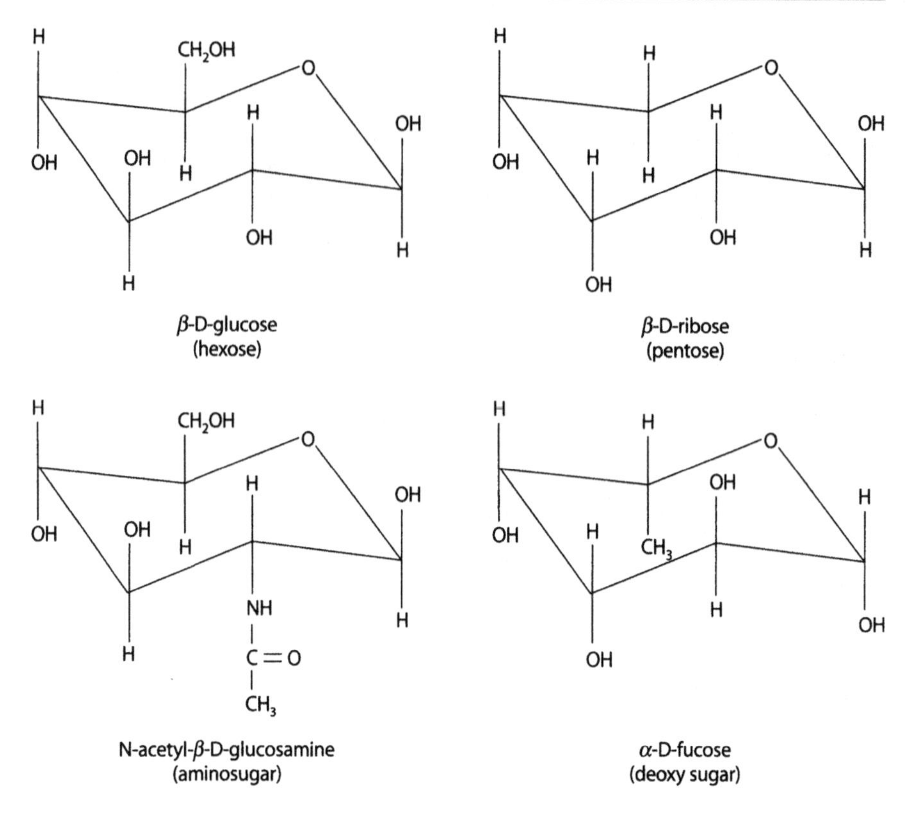

Fig. 2.14. Structure of some common monosaccharides

degraded at the 1,6-linkage by anaerobic marine bacteria (Arnosti and Repeta 1994). Other evidence suggests that enzymatic hydrolysis of macromolecular carbohydrates is, in general, a fast reaction with some substrates, and rates of hydrolysis do not necessarily correlate with the molecular weight of substrates (Arnosti et al. 1994).

2.2.4 Lignin and Other CuO-Oxidation Products

2.2.4.1 *Lignin*

Lignin is a phenolic macromolecule that is biosynthesized by vascular land plants (Sarkanen and Ludwig 1971). Accounting for up to 30 wt% of woody tissues, its functions are both structural and protective. The lignin-hemicellulose complex associated with cellulose microfibrils imposes a structural rigidity to wood, and lignin protects cell wall polysaccharides from microbial attack (Goodwin and Mercer 1972; Kirk and Farrell 1987). Because photosynthesis by vascular plants represents over half of the global primary productivity (Fig. 2.2), lignin is a major component of the global biogeochemical cycle and may be a significant source of organic matter to marine environments. Further, the absence of lignin in non-vascular plants, such as marine algae, makes it an ideal indicator of terrigenous, vascular plant-derived organic matter.

The polymeric structure of lignin precludes direct analysis by conventional techniques. Alkaline hydrolysis with CuO (Hedges and Parker 1976; Hedges and Ertel 1982) has become the widely used method for cleaving the various carbon-carbon and carbon-oxygen bonds of the lignin macromolecule, thus yielding small phenolic products amenable to gas chromatography. Six monomeric vanillyl and syringyl phenols in the forms of aldehydes, ketones, and carboxylic acids (Fig. 2.15) are produced (Hedges and Parker 1976). The parameter lambda (Λ), the sum of the eight lignin phenols (mg / 100 mg *OC*), was developed by Hedges and Mann (1979a) as a measure of total lignin, and hence vascular plant tissue, in sedimentary organic matter. In a study of lignin geochemistry of surface sediments along a transect off the coast of Washington State (USA), Hedges and Mann (1979b) found a distinct maximum in Λ in the mid-shelf silt deposit. Their conclusion was that a major proportion of sedimentary lignin, and hence vascular plant debris, accumulates on the mid-shelf. Mixing of marine and terrestrial end-members is supported by a strong correlation between Λ and ^{13}C of bulk organic matter. In this and subsequent studies (e.g. Haddad and Martens 1987; Hedges et al. 1988a; Prahl et al. 1994) lignin monomers constitute up to about 5% of sedimentary organic carbon. A recent study by Keil et al. (1998) examined lignin distribution in size-fractionated Washington shelf sediment and found significant

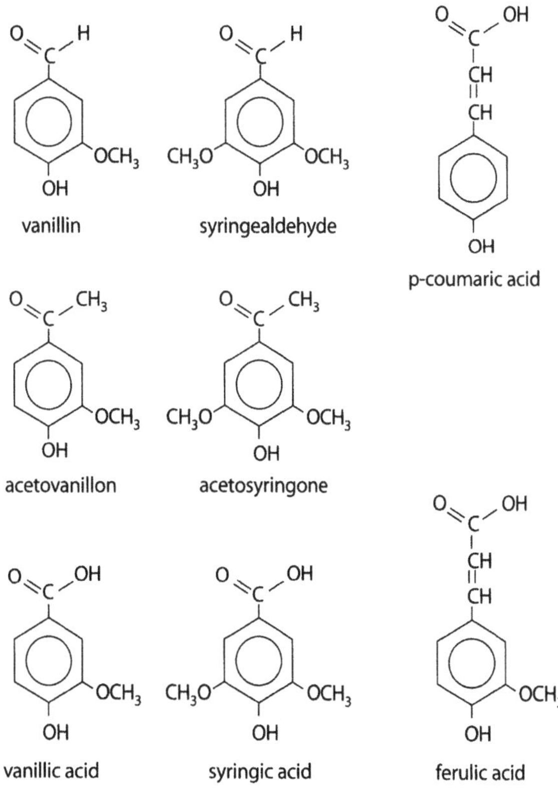

Fig. 2.15. Structures of major monomeric phenols produced by CuO oxidation of lignin

compositional differences related to grain size, with coarse material containing the highest Λ values (Fig. 2.16). This result helps to explain the hydrodynamic sorting of particulate material that occurs in continental shelf sediments, whereby vascular plant material is selectively deposited in mid-shelf sediments, but little is transported to the deep sea on fine particles.

Distributions of lignin-derived CuO-generated monomers vary because of differences in the lignin macromolecular structure biosynthesized by different vascular plant types (Hedges and Parker 1976; Hedges and Mann 1979a,b). Hedges and Mann (1979a) developed a series of lignin parameters to differentiate plant types. For example, CuO oxidation of gymnosperms produces only vanillyl phenols but angiosperms yield syringyl and vanillyl phenols. Thus, the ratio S/V, defined as the total yield of

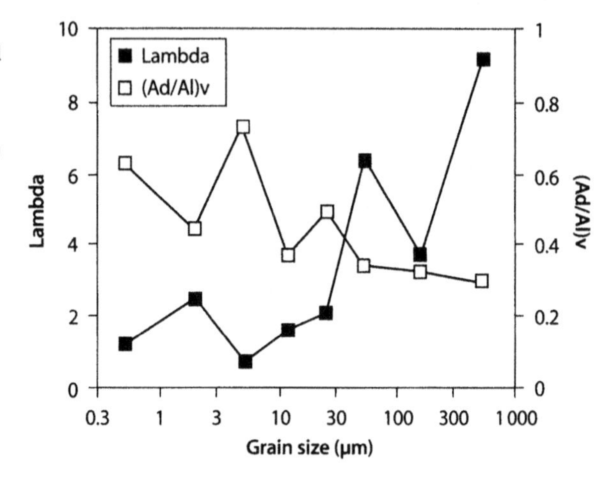

Fig. 2.16. Yield of total lignin phenols, Λ = mg 100 g OC^{-1}, and vanillic acid/vanillin ratios, (Ad/Al)v, for size fractio-nated sediments off the Washington Coast (data from Keil et al. 1998)

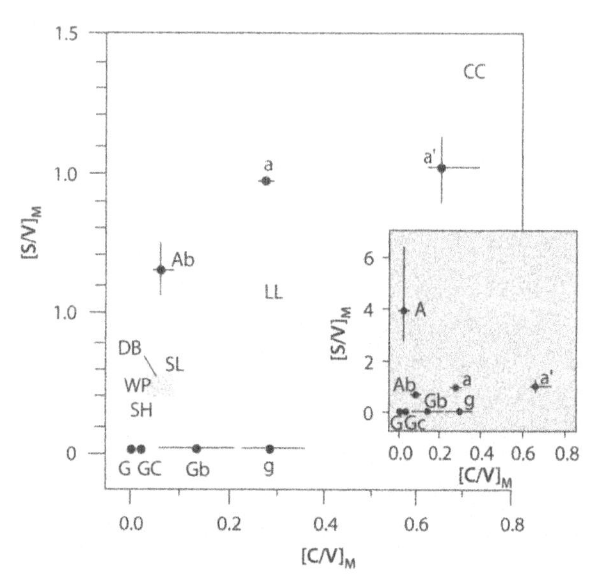

Fig. 2.17. Cross-plot of monomeric cinnamyl/vanillyl, $[C/V]_M$, and syringyl/vanillyl, $[S/V]_M$, in various plant tissue and sediment samples. *G*: gymnosperm wood; *g*: gymnosperm needles; *Gb*: gymnosperm barks; *Gc*: gymnosperm cones; *A*: angiosperm woods; *a*: angiosperm leaves; *Ab*: angiosperm barks; *a'*: angiosperm monocotyls; *CC*: Crab Creek; *LL*: Long Lake; *WP*: Willapa Bay; *SH*: Washington continental shelf; *SL*: Washington continental slope (adapted from Goñi and Hedges 1992)

syringealdehyde, acetosyringone and syringic acid divided by the total of vanillin, acetovanillinone and vanillic acid, is zero for gymnosperm wood (Fig. 2.17) but >0 for angiosperm wood. Likewise, a C/V ratio (p-coumaric + ferulic acids / the three vanillyl phenols) discriminates woody and non-woody vascular plant tissues because only non-woody tissues produce the cinnamyl phenols. These and related parameters have gained wide use in two-dimensional plots to characterize mixtures of vascular plant sources for sedimentary organic matter. For example, for a suite of plant tissues and fresh water and marine sediments (Fig. 2.17; Goñi and Hedges 1992), a plot of monomeric $[S/V]_M$, against $[C/V]_M$ shows wide discrimination between plant tissues and sediment types. The lignin character of sediments indicates that gymnosperm tissues are a major contributor to Dabob Bay and Washington shelf sediments, while a significant non-woody angiosperm tissue source is required to explain the ratios observed for Long Lake and Crab Creek.

2.2.4.2
Lignin Dimers

Gas chromatographic analysis of CuO oxidation products reveals, in addition to the commonly used eight lignin monomers, the presence of additional monomers and a complex series of phenolic dimers (Fig. 2.18) (Goñi and Hedges 1992). These diphenolic products reflect the ultrastructure of the lignin polymer and have no other origin, in contrast to phenolic acids linked to polysaccharides via ester links that are hydroly-

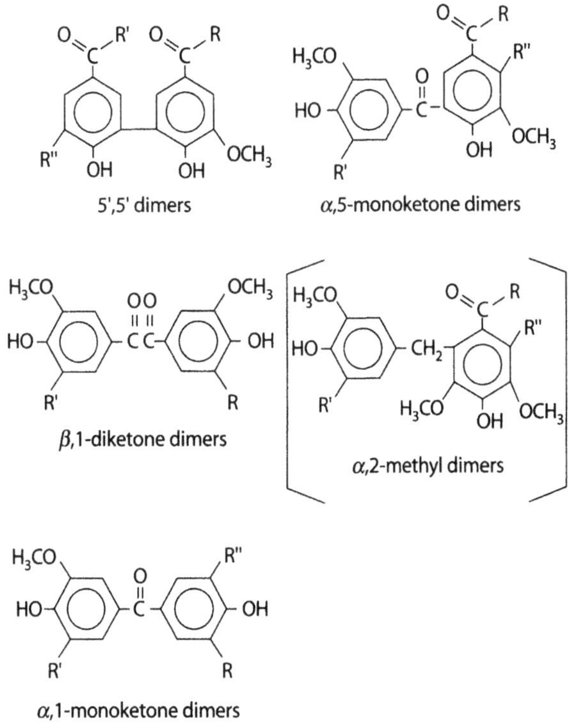

Fig. 2.18. Structures of major dimeric phenols produced by CuO oxidation of lignin

5',5' dimers

α,5-monoketone dimers

β,1-diketone dimers

α,2-methyl dimers

α,1-monoketone dimers

sed during the CuO reaction. Like monomeric phenols, the dimers may be used to develop dimer parameters that appear to be more source specific and less diagenetically sensitive than the conventional monomer-based ratios.

2.2.4.3
Lignin Phenols as Diagenetic Indicators

Lignin phenol distributions aid in assessing diagenetic status of organic material. Degradation by white-rot fungi, the major lignin decomposers, results in pronounced mass loss, decrease in carbon-normalized lignin concentration, and increase in acid/aldehyde ratios, for example vanillic acid to vanillin [(Ad/Al)v] (Hedges et al. 1988b; Goñi et al. 1993). Brown-rot fungus, while less efficient at degrading lignin, produced some new monomers and dimers. An indication of the extent of fungal degradation in sediments can thus be obtained. For example, in the size-fractionated sediments described above (Fig. 2.17), increasing [(Ad/Al)v] from 0.2 in sands to 0.8 in clays suggests that lignin in the sand-sized fraction is only slightly modified vascular plant material but lignin in clays is highly degraded (Hedges and Oades 1997; Keil et al. 1998). The combination of phenol monomers and dimers enhances the information yield for CuO oxidation (Goñi and Hedges 1992). Ratios of dimer to monomer reflect the fraction of lignin phenolic units linked by carbon-carbon bonds rather than ether bonds that are easily hydrolysed during CuO oxidation. Distributions of dimers give information on the type of C-C bonding present in the lignin macromolecule. Angiosperm woods, for example, contain lignin with C bonding through side chain linkages at the C_1 ring carbon, as indicated by elevated ratio of side-chain dimers to ring-ring dimers; gymnosperm tissues appear rich in 5-5'-ring-ring bonds. Variations in ratios of selected dimers may also indicate effects of selective alteration of linkages. Finally, dimeric acids/aldehyde ratios support inferences regarding lignin alteration made from monomeric acid/aldehyde ratios.

2.2.4.4
Cutin Acids

CuO oxidation of plant tissues and sediments also yields a series of non-lignin products, notably long-chain C_{16}-C_{18} hydroxy acids whose structural characteristics suggest cutin as their vascular plant source (Goñi and Hedges 1990a,b,c). Cutin is a polyester-like biopolymer that comprises part of the cuticle of vascular plants and provides a protective covering for aerial parts. Unlike waxy epicuticular coatings that are soluble in non-polar organic solvents, and hence are "lipids", cutin is insoluble and requires alkaline hydrolysis to release the cutin acids. Like lignin phenols, cutin acids are diagnostic for different cutin-bearing plant. Gymnosperms, for instance, yield hydroxy acids dominated by C_{16} acids, predominately 9,16- and 10,16-dihydroxy-hexadecanoic acids. Angiosperms were rich in C_{18} acids, notably 9,10,18-trihydroxy-octadecanoic acid in dicotyledons. Also like lignin phenols, a series of cutin acid parameters have been developed. Ratios of 9,10-ω-C_{18}/total cutin acids in excess of 0.2 are typical of gymnosperms whereas ratios for monocotyledons average 0.02, and a plot of x,ω-C_{16}/total cutin acids vs 9,10,ω-C_{18}/total cutin acids provides a good means of distinguishing non-woody tissues from gymnosperms vs. those of angiosperms.

2.2.5
Uncharacterized Organic Material

Although amino acids, carbohydrates and lipids constitute the major biochemicals in living organisms, they are usually present in macromolecular matrices that are not fully amenable to molecular-level analyses by currently-available analytical techniques. Analyses of cultured plankton provide wide ranges of the proportions of organic carbon present as these biochemical classes (Fig. 2.19), although in most cases all three compound classes were not analysed simultaneously in the same samples. Comprehensive molecular-level analyses of amino acids, carbohydrates, pigments and lipids in particulate material in the equatorial Pacific (Wakeham et al. 1997a) determined that about 84% of *OC* in plankton could be accounted for (Fig. 2.19), mainly as amino acids. Pigments, while important light-harvesting biochemicals in photosynthetic organisms, were relatively minor constituents of *OC*. The remaining 16% of *OC* was not characterized by the molecular techniques used, but might include nucleic acids, uronic acids, and amino sugars. The fraction of characterizable *OC* decreases, and as a consequence the fraction of uncharacterizable material increases, with the extent of diagenesis in the ocean. In the interior of the ocean, only about 30% of *OC* could be characterized, while in sediments the fraction of characterized *OC* dropped still further to less than 20%.

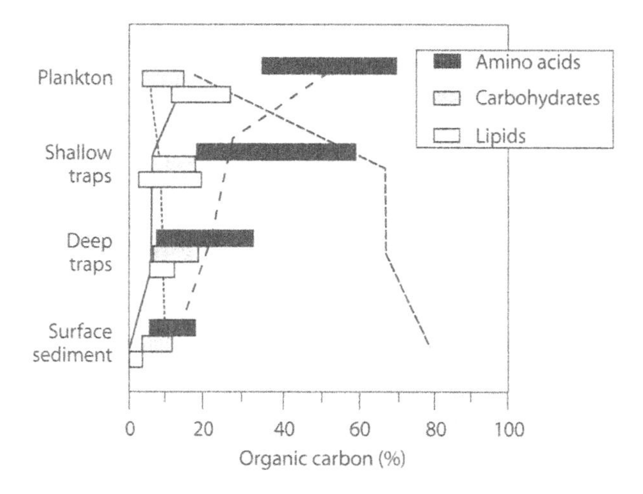

Fig. 2.19. Ranges of amino acid, carbohydrate and lipid contributions to *OC* in plankton, shallow sediment traps deployed near the base of the euphotic zone, deep traps in the mesopelagic and bathypelagic zones, and deep-sea surface sediments. Lines correspond to average fractions of organic carbon represented by amino acid, carbohydrate, lipid, and uncharacterized material in the equatorial Pacific as reported by Wakeham et al. (1997b). References – plankton: Degens and Mopper (1976); Smetacek and Hendrikson (1979); Tanoue et al. (1982); Wefer et al. (1982); Cowie and Hedges (1984, 1996); Parsons et al. (1984); Montani and Okaichi (1985); Sargent and Henderson (1986); Tanoue and Handa (1987); Cowie and Hedges (1992); shallow traps: Wakeham et al. (1993, 1984a,b); Lee and Cronin (1982, 1984); Hedges et al. (1988); Cowie and Hedges (1992); deep traps: Lee and Cronin (1984); Wefer et al. (1982); Ittekkot et al. (1984a,b); Wakeham et al. 1984b; Müller et al. (1986); Wakeham and Lee (1989); Haake et al. (1992); sediments: Whelan (1977); Farrington and Tripp (1977); Tanoue and Handa (1987); Venkatesan et al. (1987); Steinberg et al. (1987)

The progressive alterations of organic matter that occur during diagenesis in the oceanic water column and sediments efficiently remove the readily identifiable organic constituents and leave behind a large fraction that cannot be characterized. This chemical recalcitrance, however, does not imply a corresponding biological recalcitrance, since the rain of organic matter to the sea floor fuels benthic metabolism and indeed the benthic boundary layer is a "hotspot" of organic matter degradation (Mayer 1993; Deming and Baross 1993). Nonetheless, the challenge to organic geochemists of extracting molecular information from this as yet uncharacterized material will require application of novel analytical tools. But how is this uncharacterized material generated, and what is its role in preservation of organic matter?

The uncharacterized organic matter is largely macromolecular in nature. Tissot and Welte (1984) described the classical view that macromolecular material results from sequential and random polymerization and polycondensation of biomonomers that have been released by enzymatic hydrolysis of biopolymers (Fig. 2.20). Free, low molecular weight compounds are cross-linked into macromolecules via incorporation of heteroatoms (e.g. Sinninghe Damste and de Leeuw 1993) or condensation of oxygen-containing functionalities (Richnow et al. 1993). This model fits observations that labile organic compounds become less abundant during diagenesis and poorly characterized, higher molecular weight material increases in proportion. An alternate model is preferential preservation of some fraction of the original biopolymeric material (Philp and Calvin 1976; Tegelaar et al. 1989). In this case, macromolecular material represents the remains of biopolymers after the more labile components have been degraded. Evidence for this pathway comes from the finding of insoluble, non-hydrolysable highly aliphatic material (e.g. algenans, bacterans, suberans, and cutans) in living organisms (de Leeuw and Largeau 1993) and their detritus. Unfortunately, the potential artifactual nature of some of this insoluble, highly-aliphatic material has been noted (Allard et al. 1998).

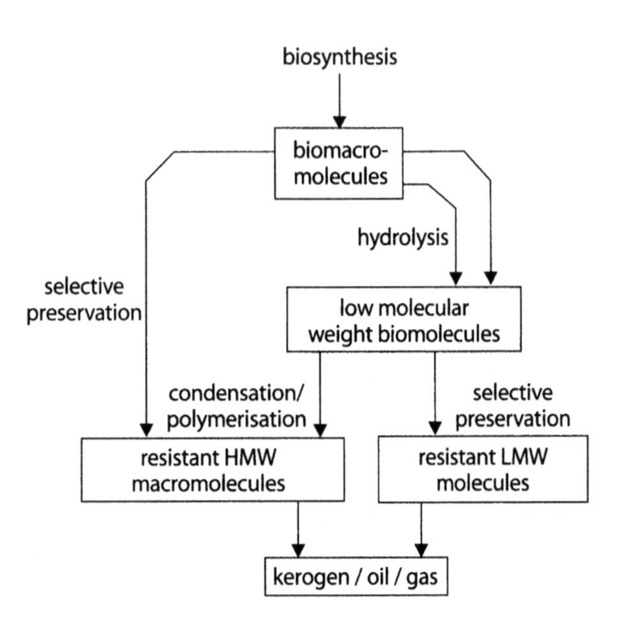

Fig. 2.20. Schematic of selective preservation vs. condensation pathways for preservation of organic matter (adapted from Tegelaar et al. 1989)

2.2.6 Dissolved Organic Matter

Dissolved organic matter (*DOM*, <0.2 μm) is one of the largest reservoirs of organic nitrogen and carbon in the ocean. On average, this reservoir contains 600 Gt C (Williams and Druffel 1987). Dissolved organic carbon (*DOC*) is typically enriched in the surface mixed layers of all oceans compared to the deep sea. This enrichment appears to be related to production of soluble material by phytoplankton. Chemical analysis of *DOM* has resulted in a low proportion of small, rapidly cycling organic compounds (amino acids, amines, sugars, etc.). The majority of *DOM* is made of resistant components of unknown chemical structure.

Between 25–30% of seawater *DOM* is of molecular weight >1 000 D (*UDOM*[5]). Proton NMR of *UDOM* produces similar spectra regardless of the sampling site. Thus, carbohydrates account for 55% of the total carbon in *UDOM*, whereas acetate and lipids account for 7 and 6% of the carbon, respectively (Aluwihare et al. 1997). Polysaccharide linkage analysis of samples from the North Atlantic and Pacific Ocean resulted in 40% of terminal sugars, 40% of sugars linked without branching, and 20% of the sugars cross-linked with one branch point. That observation, plus the relative fixed ratio of biochemicals found in *UDOM*, indicated that the pool is a mixture of acyl-oligosaccharides in which carbohydrates, acetate, and lipids are linked in a common macromolecular structure. The source of *UDOM* appears to be phytoplankton release (Aluwihare et al. 1997).

^{15}N CPMAS[6] NMR spectra of *UDOM* from the Pacific Ocean showed a single peak at 260 ppm in surface and deeper samples (100, 375 and 4 000 m), corresponding to the resonance of amide nitrogen (McCarthy et al. 1997). Although the amide fraction makes about 80% of the nitrogen, <10% of the nitrogen was identified as amino acids. Other nitrogen forms may contribute to *UDOM*. Some of the potential can-didates are "pyrrol-like" nitrogen such as purines, pyrimidines, porphyrins, as well as melanoidins (complex condensates of sugars and amino acids) (McCarthy et al. 1997).

Acknowledgements

We would like to thank A. Gianguzza and the organizing committee of the International School on Marine Chemistry for the invitation to participate in this event. S. Pantoja was supported by the Postdoctoral Scholar Program at the Woods Hole Oceanographic Institution, with funding provided by the Johnson Foundation.

References

Allard B, Templier J, de Leeuw JW (1998) Artifactual origin of mycobacterial bacteran. Formation of melanoidin-like artifactual macromolecular material during the usual isolation process. Org. Geochem 26:691–703

Aluwihare LI, Repeta DJ, Chen RF (1997) A major biopolymeric component to dissolved organic carbon in surface seawater. Nature 387:166–169

[5] *UDOM*, ultrafiltered dissolved organic matter. The fraction of *DOM* isolated with ultra-filters of nominal pore size of ~1 nm (~1 000 daltons).

[6] CPMAS, cross-polarization magic-angle-spinning.

Arnosti C, Repeta DJ (1994) Extracellular enzyme activity in anaerobic bacterial cultures: Evidence of pullulanase activity among mesophilic marine bacteria. Appl Environ Microbiol 60: 840–846
Arnosti C, Repeta DJ, Blough NV (1994) Rapid bacterial degradation of polysaccharides in anoxic marine systems. Geochim Cosmochim Acta 58:2639–2652
Azam F (1998) Microbial control of oceanic carbon flux: The plot thickens. Science 280:694–606
Bada JL (1972) The dating of fossil bones using the racemization of isoleucine. Earth Plan Sci Lett 15:223–231
Bada JL, Schroeder RA (1972) Racemization of isoleucine in calcareous marine sediments: Kinetics and mechanisms. Earth Plan Sci Lett 15:1–11
Bada JL, Schroeder RA (1975) Amino acid racemization reactions and their geochemical implications. Naturwissenschaften 62:71–79
Bada JL, Luykendyk BP, Maynard JB (1970) Marine sediments: Dating by the racemization of amino acids. Science 170:730–732
Bada JL, Protsch R, Schroder RA (1973) The racemization reaction of isoleucine used as a paleotemperature indicator. Nature 241:394–395
Billen G (1991) Protein degradation in aquatic environments. In: Chróst RJ (ed) Microbial enzymes in aquatic environments. Springer-Verlag, New York, pp 123–143
Brassell SC (1993) Application of biomarkers for delineating marine paleoclimatic fluctuations during the pleistocene. In: Engel MH, Macko SA (eds) Organic geochemistry. Principles and applications. Plenum Press, New York, pp 699–638
Button DK (1993) Nutrient-limited microbial growth kinetics: Overview and recent advances. Antonie van Leeuwenhoek 63:225–235
Callot HJ (1991) Geochemistry of chlorophylls. In: Sheer H (ed) Chlorophylls. CRC Press, Boca Raton, pp 339–364
Canfield DE (1994) Factors influencing organic carbon preservation in marine sediments. Chem Geol 114:315–329
Canuel EA, Martens CS (1996) Reactivity of recently deposited organic matter: Degradation of lipid compounds near the sediment-water interface. Geochim Cosmochim Acta 60:1793–1805
Conte MH, Volkman JK, Eglinton G (1994) Lipid biomarkers of the prymnesiophyceae. In: Green JC, Leadbetter BSC (eds) The haptophyte algae. Clarendon Press, Oxford, pp 351–377
Cowie GL, Hedges JI (1984) Carbohydrate sources in a coastal marine environment. Geochim Cosmochim Acta 48:2075–2087
Cowie GL, Hedges JI (1992) Sources and reactivities of amino acids in a coastal marine environment. Limnol Oceanogr 37:703–724
Cowie GL, Hedges JI (1994) Biochemical indicators of diagenetic alteration in natural organic matter mixtures. Nature 369:304–307
Cowie GL, Hedges JI (1996) Degestion and alteration of the biochemical constituents of a diatom (*Thalassiosira weissflogii*) ingested by a herbivorous zooplankton (*Calanus pacificus*). Limnol Oceanogr 41:581–594
Cranwell PA (1982) Lipids of aquatic sediments and sedimenting particles. Prog Lipid Res 21:271–308
De Baar HJW, Farrington JW, Wakeham SG (1983) Vertical flux of fatty acids in the North Atlantic Ocean. J Mar Res 41:19-41
Degens ET, Mopper K (1976) Factors controlling the distribution and diagenesis of organic materials in marine sediments. In: Riley JP, Chester R (eds) Chemical oceanography. Academic Press, London
Deming JW, Baross JA (1993) The early diagenesis of organic matter: Bacterial activity. In: Engel MH, Macko SA (eds) Organic geochemistry. Plenum Press, New York, pp 119–144
Dixon M, Webb EC (1964) Enzymes. Academic Press, London
Edmunds KLH, Brassell SC, Eglinton G (1980) The short-term diagenetic fate of 5α-cholestan-3β-ol: In situ radiolabelled incubations in algal mats. In: Douglas, AG, Maxwell JR (eds) Advances in organic geochemistry 1979. Pergamon Press, Oxford, pp 427–434
Emerson S, Hedges JI (1988) Processes controlling the organic carbon content of open ocean sediments. Paleoceanography 3:621–634
Farrington JW, Tripp BW (1977) Hydrocarbons in western North Atlantic sediments. Geochim Cosmochim Acta 41:1627–1641
Fenchel T, Blackburn TH (1979) Bacteria and mineral cycling. Academic Press, London
Gagosian RB, Smith SO, Lee C, Farrington JW, Frew NM (1980) Steroid transformations in recent marine sediments. In: Douglas AG, Maxwell JR (eds) Advances in organic geochemistry 1979. Pergamon Press, Oxford, pp 407–419
Gaskell SJ, Morris RJ, Eglinton G, Calvert SE (1975) The geochemistry of a recent marine sediment off northwest Africa. An assessment of source of input and early diagenesis. Deep-Sea Res 22:777–789

Gillan FT, Nichols PD, Johns RB, Bavor HJ (1983) Phytol degradation by marine bacteria. Appl Environ Microbiol 41:1423–1428
Goñi MA, Hedges JI (1990a) Cutin-derived CuO reaction products from purified cuticles and tree leaves. Geochim Cosmochim Acta 54:3065–3072
Goñi MA, Hedges JI (1990b) Potential applications of cutin-derived CUO reaction products for discriminating vascular plant sources in natural environments. Geochim Cosmochim Acta 54:3073–3081
Goñi MA, Hedges JI (1990c) The diagenetic behavior of cutin acids in buried conifer needles and sediments from a coastal marine environment. Geochim Cosmochim Acta 54:3083–3093
Goñi MA, Hedges JI (1992) Lignin dimers: Structures, distribution, and potential geochemical applications. Geochim Cosmochim Acta 56:4025–4043
Goñi MA, Nelson B, Blanchette RA, Hedges JI (1993) Fungal degradation of wood lignins: Geochemical perspectives from CuO-derived phenolic dimers and monomers. Geochim Cosmochim Acta 57:3985–4002
Goodwin TW, Mercer EI (1972) Introduction to plant biochemistry, Pergamon Press, Oxford
Haake B, Ittekkot V, Ramaswamy V, Nair RR, Honjo S (1992) Fluxes of amino acids and hexosamines to the deep Arabian Sea. Mar Chem 40:291–314
Haddad RI, Martens CS (1987) Biogeochemical cycling in an organic-rich coastal marine basin: 9. Sources and accumulation rates of vascular plant-derived organic material. Geochim Cosmochim Acta 51:2991–3001
Harradine PJ, Maxwell JR (1998) Pyrophaeoporphyrins c1 and c2: Grazing products of chlorophyll c in aquatic environments. Org Geochem 28:111–117
Harvey HR, Macko SA (1997a) Catalysts or contributors? Tracking bacterial mediation of early diagenesis in the marine water column. Org Geochem 26:531–544
Harvey HR, Macko SA (1997b) Kinetics of phytoplankton decay during simulated sedimentation: Changes in lipids under oxic and anoxic conditions. Org Geochem 27:129–140
Harvey HR, Tuttle JH, Bell JT (1995) Kinetics of phytoplankton decay during simulated sedimentation: changes in biochemical composition and microbial activity under oxic and anoxic conditions. Geochim Cosmochim Acta 59:3367–3377
Hedges JI (1992) Global biogeochemical cycles: Progress and problems. Mar Chem 39:67–93
Hedges JI, Ertel JR (1982) Characterization of lignin by capillary gas chromatography of cupric oxide oxidation products. Anal Chem 54:174–178
Hedges JI, Mann DC (1979a) The characterization of plant tissues by their lignin oxidation products. Geochim Cosmochim Acta 43:1803–1807
Hedges JI, Mann DC (1979b) The lignin geochemistry of marine sediments from the southern Washington coast. Geochim Cosmochim Acta 43:1809–1818
Hedges JI, Oades JM (1997) Comparative organic geochemistries of soils and marine sediments. Org Geochem 27:319–361
Hedges JI, Parker PL (1976) Land-derived organic matter in surface sediments from the Gulf of Mexico. Geochim Cosmochim Acta 40:1019–1029
Hedges JI, Blanchette RA, Weliky K, Devol AH (1988a) Effects of fungal degradation on the CuO oxidation products of lignin: A controlled laboratory study. Geochim Cosmochim Acta 52:2717–2726
Hedges JI, Clark WA, Cowie GL (1988b) Organic matter sources to the water column and surficial sediments of a marine bay. Limnol Oceanogr 33:1116–1136
Henrichs SM, Reeburgh WS (1987) Anaerobic mineralization of marine sediment organic matter: Rates and the role of anaerobic processes in the ocean carbon economy. Geomicrobiol J 5:191–237
Ittekkot V, Deuser W, Degens ET (1984a) Seasonality in the fluxes of sugars, amino acids, and amino sugars to the deep ocean: Sargasso Sea. Deep-Sea Res 31:1057–1069
Ittekkot V, Deuser W, Degens ET (1984b) Seasonality in the fluxes of sugars, amino acids, and amino sugars to the deep ocean: Panama Basin. Deep-Sea Res 31:1071–1083
Jørgensen NOG, Søndergaard M (1984) Are dissolved free amino acids free? Microb Ecol 10:301–316
Keil RG, Tsamakis E, Giddings JC, Hedges JI (1998) Biochemical distributions (amino acids, neutral sugars, and lignin phenols) among size-classes of modern marine sediments from the Washington coast. Geochim Cosmochim Acta 62:1347–1364
King LL, Repeta DJ (1994) Phorbin steryl esters in Black Sea sediment traps and sediments: A preliminary evaluation of their paleoceanographic potential. Geochim Cosmochim Acta 58:4389–4399
King LL, Wakeham SG (1996) Phorbin steryl ester formation by macrozooplankton in the Sargasso Sea. Org Geochem 24:581–585
Kirk TK, Farrell RL (1987) Enzymatic combustion: The microbial degradation of lignin. Annu Rev Microbiol 41:465–505
Lee C (1992) Controls on organic carbon preservation: The use of stratified water bodies to compare intrinsic rates of decomposition in oxic and anoxic systems. Geochim Cosmochim Acta 56: 3323–3335

Lee C, Cronin C (1982) The vertical flux of particulate organic nitrogen in the sea: Decomposition of amino acids in the Peru upwelling area and the equatorial Atlantic. J Mar Res 40:227–251
Lee C, Cronin C (1984) Particulate amino acids in the sea: Effects of primary productivity and biological decomposition. J Mar Res 42:1075–1097
Lee C, Wakeham SG (1988) Organic matter in seawater: Biogeochemical processes. In: Riley JP (ed) Chemical oceanography, vol IX. Academic Press, New York, pp 1–51
Leeuw JW de, Baas M (1986) Early-stage diagenesis of steroids. In: Johns, RB (ed) Biological markers in the sedimentary record. Elsevier, New York, pp 101–123
Leeuw JW de, Largeau C (1993) A review of macromolecular organic compounds that comprise living organisms and their role in kerogen, coal, and petroleum formation. In: Engel MH, Macko SA (eds) Organic geochemistry. Principles and applications. Plenum Press, New York, pp 23–72
Lehninger AL (1981) Biochemistry. Worth Publishers, New York
Mackenzie AS, Brassell SC, Eglinton G, Maxwell JR (1982) Chemical fossils: The geological fate of steroids. Science 217:491–504
Mayer LM (1993) Organic matter at the sediment-water interface. In: Engel MH, Macko SA (eds) Organic geochemistry. Plenum Press, New York, pp 171–184
McCarthy M, Pratum T, Hedges J, Benner R (1997) Chemical composition of dissolved organic nitrogen in the ocean. Nature 390:150–154
Meister A (1965) Biochemistry of the amino acids. Academic Press, London
Mermoud F, Wünsche L, Clerc O, Gülacar FO, Buchs A (1984) Steroidal ketones in the early diagenetic transformations of ?^{5} sterols in different types of sediments. Org Geochem 6:25–29
Mitterer RM (1993) The diagenesis of proteins and amino acids in fossil shells. In: Engel MH, Macko SA (eds) Organic geochemistry. Plenum Press, New York, pp 739–753
Montani S, Okaichi T (1985) Amino acid variations in marine particles during sinking and sedimentation in Harima-Nada, the Seto Inland Sea. In: Sigleo AC, Hattori A (eds) Marine and estuarine geochemistry. Lewis Publishers, New York, pp 15–27
Müller PJ, Suess E, Ungerer CA (1986) Amino acids and amino sugars of surface particulate and sediment trap material from waters of the Scotia Sea. Deep-Sea Res 33:819–838
Nishimura M (1978) Geochemical characteristics of the high reduction zone of stenols in Suwa sediments and the environmental factors controlling the conversion of stenols into stanols. Geochim Cosmochim Acta 42:349–357
Parsons TR, Stephens K, Strickland JDH (1961) On the chemical composition of eleven species of marine phytoplankton. J Fish Res Bd Canada 18:1001–1016
Parsons TR, Takahashi M, Hargrave B (1984) Biological oceanographic processes. Pergamon Press, Oxford
Pedersen TF, Calvert SE (1990) Anoxia vs. productivity: What controls the formation of organic-carbon-rich sediments and sedimentary rocks? Amer Assoc Petrol Geol Bull 74:454–466
Prahl FG, Eglinton G, Corner EDS, O'Hara SCM (1984) Copepod fecal pellets as a source of dihydrophytol in marine sediments. Science 224:1235–1237
Prahl FG, Ertel JR, Goñi MA, Sparrow MA, Eversmeyer B (1994) Terrestrial organic carbon contributions to sediments on the Washington margin. Geochim Cosmochim Acta 58:035–3048
Philp RP, Calvin M (1976) Possible origin for insoluble organic (kerogen) debris in sediment from insoluble cell-wall materials of algae and bacteria. Nature 262:134–136
Repeta DJ, Gagosian RB (1984) Transformation reactions and recycling of carotenoids and chlorins in the Peru upwelling region (15° S, 175° W). Geochim Cosmochim Acta 48:1265–1277
Richnow HH, Jenisch A, Michaelis W (1993) The chemical structure of macromolecular fractions of a sulfur-rich oil. Geochim Cosmochim Acta 57:2767–2780
Santos V, Billett DSM, Rice AL, Wolff GA (1994) Organic matter in deep-sea sediments from the Porcupine Abyssal Plain in the north-east Atlantic Ocean. I. Lipids. Deep-Sea Res 41:787–819
Sargent JR (1976) The structure, function, and metabolism of lipids in marine organisms. In: Malins DC, Sargent JR (eds) Biochemical and biophysical perspectives in marine biology, vol III. Academic Press, New York, pp 149–212
Sargent JR, Henderson RJ (1986) Lipids. In: Corner EDS, O'Hara SCM (eds) The biological chemistry of marine copepods. Clarendon Press, Oxford, pp 59–108
Sargent JR, Parkes RJ, Mueller-Harvey I, Henderson RJ (1987) Lipid biomarkers in marine ecology. In: Sleigh MA (ed) Microbes in the sea. Ellis Horwood, New York, pp 119–138
Sarkanen K, Ludwig CH (eds) (1971) Lignins. Wiley Interscience, London
Schlesinger WH (1997) Biogeochemistry. An analysis of global change. Academic Press, San Diego
Schroeder RA, Bada JL (1973) Glacial-postglacial temperature difference deduced from aspartic acid racemization in fossil bones. Science 182:479–482

Sinninghe Damsté JS, de Leeuw JW (1993) Analysis, structure and geochemical significance of organically-bound sulfur in the geosphere: State of the art and future research. Org Geochem 16:1077–1101
Smetacek V, Hendrikson P (1979) Composition of particulate organic matter in Kiel Bight in relation to phytoplankton succession. Oceanol Acta 2:287–298
Steinberg SM, Bada JL (1983) Peptide decomposition in the neutral pH region via the formation of diketopiperazines. J Org Chem 48:2295–2298
Steinberg SM, Venkatesan MI, Kaplan IR (1987) Organic geochemistry of sediments from the continental margin off southern New England, U.S.A. – Part I. Amino acids, carbohydrates and lignin. Mar Chem 21:249–265
Summons RE (1993) Biogeochemical cycles. A review of fundamental aspects of organic matter formation, preservation and composition. In: Engel MH, Macko SA (eds) Organic geochemistry. Plenum Press, New York, pp 3–21
Sun M-Y, Lee C, Aller RC (1993) Laboratory studies of oxic and anoxic degradation of chlorophyll-a in Long Island sediments. Geochim Cosmochim Acta 57:147–157
Sun M-Y, Wakeham SG, Lee C (1997) Rates and mechanisms of fatty acid degradation in oxic and anoxic coastal marine sediments of Long Island Sound. Geochim Cosmochim Acta 61:341–355
Sun M-Y, Wakeham SG, Aller RC, Lee C (1998) Impact of seasonal hypoxia on diagenesis of phytol and its derivatives in Long Island Sound. Mar Chem 62:157–173
Tanoue E, Handa N (1987) Monosaccharide composition of marine particles and sediments from the Bering Sea and northern North Pacific. Oceanol Acta 10:91–99
Tanoue E, Handa N, Sakugawa H (1982): Difference in chemical composition of organic matter between fecal pellet of *Euphausia superba* and its feed, *Duniella tertiolecta*. Trans Tokyo Univ Fish 5:189–196
Tegelaar EW, de Leeuw JW, Derenne S, Largeau C (1989) A reappraisal of kerogen formation Geochim Cosmochim Acta 53:3103–3106
Tissot B, Welte D (1984) Petroleum formation. Springer Verlag, Berlin
Titman D (1976) Ecological competition between algae: Experimental confirmation of resource-based competition theory. Science 192:463–465
Valiela I (1995) Marine ecological processes. Springer-Verlag, New York
Venkatesan MI, Ruth E, Steinberg S, Kaplan IR (1987) Organic geochemistry of sediments from the continental margin off southern New England, U.S.A. – Part II. Lipids. Mar Chem 21:267–299
Volkman JK (1986) A review of sterol markers for marine and terrigenous organic matter. Org Geochem 9:83–99
Volkman JK, Maxwell JR (1986) Acyclic isoprenoids as biological markers. In: Johns RB (ed) Biological markers in the sedimentary record. Elsevier, New York, pp 1–42
Volkman JK, Jeffrey SW, Nichols PD, Rogers GI, Garland CD (1989) Fatty acid and lipid composition of 10 species of microalgae used in mariculture. J Exp Mar Biol Ecol 128:219–240
Wada E, Hattori A (1991) Nitrogen in the Sea: Forms, abundances, and rate processes. CRC Press, Boca Raton
Wakeham SG (1989) Reduction of stenols to stanols in particulate organic matter at oxic-anoxic boundaries in seawater. Nature 342:787–790
Wakeham SG, Lee C (1989) Organic geochemistry of particulate matter in the ocean: The role of particles in oceanic sedimentary cycles. Org Geochem 14:83–96
Wakeham S, Lee C (1993) Production, transport, and alteration of particulate organic matter in the marine water column. In: Engel M, Macko S (eds) Organic geochemistry. Plenum Press, New York, pp 145–169
Wakeham SG, Farrington JW, Gagosian RB (1984a) Variability in lipid flux and composition of particulate matter in the Peru upwelling region. Org Geochem 6:203–215
Wakeham SG, Gagosian RB, Farrington JW, Canuel EA (1984b) Sterenes in suspended particulate matter in the eastern tropical North Pacific. Nature 308:840–843
Wakeham SG, Hedges JI, Lee C, Pease TK (1993) Effects of poisons and preservatives on the composition of organic matter in a sediment trap experiment. J Mar Res 51:669–696
Wakeham SG, Hedges JI, Lee C, Hernes PJ, Peterson ML (1997a) Molecular indicators of diagenetic status in marine organic matter. Geochim Cosmochim Acta 61:5363–5369
Wakeham SG, Hedges JI, Lee C, Peterson ML, Hernes PJ (1997b) Compositions and fluxes of lipids through the water column and surficial sediments of the equatorial Pacific Ocean. Deep-Sea Res II 44:2131–2162
Wefer G, Suess E, Balzer B, Leibezeit G, Müller PJ, Ungerer CA, Zenk W (1982) Fluxes of biogenic components from sediment trap deployments in circumpolar waters of the Drake Passage. Nature 299:145–147
Westrich JT, Berner RA (1984) The role of sedimentary organic matter in bacterial sulfate reduction: The G model tested. Limnol Oceanogr 29:236–249
Wehmiller JF, Hare PE (1971) Racemization of amino acids in marine sediments. Science 173:907–911

Whelan JK (1977) Amino acids in a surface sediment core of the Atlantic abyssal plain. Geochim Cosmochim Acta 41:803–810
Williams PM, Druffel ERM (1987) Radiocarbon in dissolved organic matter in the central North Pacific Ocean. Nature 330:246–248
Wright RT, Hobbie JE (1966) Use of glucose and acetate by bacteria and algae in aquatic ecosystems. Ecology 47:447–453

Photooxidation of Dissolved Organic Matter: Role for Carbon Bioavailability and for the Penetration Depth of Solar UV-Radiation

B. Sulzberger

3.1 Introduction

Most of the solar radiation that reaches land or water is converted into thermal energy, but a significant part, especially that in the ultraviolet and visible region, is diverted into photochemical and photobiological processes that affect the global carbon cycle. The most prominent photobiological process on the earth's surface is biological photosynthesis. Terrestrial vegetation and marine algae use the solar energy to convert annually approximately 100 Gt (gigatons) of carbon in the form of atmospheric carbon dioxide (CO_2) into organic matter (Zepp 1994). When plants and algae die, the resulting non-living matter is transformed by various biological and chemical processes that either convert it back to CO_2 (and other trace carbon gases) and water or to biologically refractory organic substances. The refractory organic matter is a mixture of substances, including litter and more refractory compounds, a large portion of which consists of humic substances (Thurman 1985). The term "humic substances" is usually used to refer to the organic matter that has been isolated from natural waters or from soils using well-defined techniques (Frimmel and Christman 1988; Huber and Frimmel 1994). Humic substances make up the largest single class of dissolved organic matter (*DOM*), accounting for 30 to 60% of the *DOM* in most natural waters (Thurman 1985). [The term "dissolved organic matter, *DOM*" is here used as synonym of "dissolved organic carbon, *DOC*."] The term "colored dissolved organic matter (*CDOM*)" is used for the fraction of *DOM* that is colored (Blough and Green 1995) and includes humic substances. Based on Orinoco River data, Blough et al. (1993) estimated that only about 65% of the total *DOM* absorbs solar radiation and is subject to direct photochemical reactions (see Table 3.1).

Oceanic *DOM* is mainly of marine origin (Hedges et al. 1992), although about 1 Gt of carbon in the form of organic carbon reaches the ocean from terrigenous sources each year (Meybeck 1982; Sarmiento and Sundquist 1992) (Fig. 3.1). On the time scale of estuarine flushing, riverine *DOM* appears to be resistant to biological oxidation and to pass through estuarine systems with little net loss or chemical alteration (Mantoura and Woodward 1983). This requires a non-biological sink of terrestrial *DOM* or a mechanism of converting it to a biologically labile form. There is growing evidence that photooxidation is such a mechanism and, perhaps, the rate limiting step of terrestrial *DOM* removal from the ocean (Kieber et al. 1990; Mopper et al. 1991; Miller and Zepp 1995).

Table 3.1. Various pathways of $HO_2\cdot/O_2^-\cdot$ formation via photochemical reactions of *CDOM*

Intermolecular electron transfer from *CDOM* in its excited triplet state to O_2 (Haag and Mill 1990):

$$^3CDOM^* + O_2 \longrightarrow CDOM^+ + O_2^-\cdot$$

Intramolecular electron transfer within *CDOM* in its excited singlet state and subsequent reaction of the thereby formed biradical with O_2 (Blough and Zepp 1995):

$$CDOM^* \longrightarrow CDOM^{\pm}$$

$$CDOM^{\pm} + O_2^-\cdot \longrightarrow CDOM^+ + O_2^-\cdot$$

H-atom abstraction or homolytic bond cleavage of singlet excited *CDOM*, followed by reaction of the radical with O_2 under formation of peroxyl radicals, and subsequent $HO_2\cdot$ abstraction (Blough and Zepp 1995):

$$^1CDOM^* \longrightarrow CDOM\cdot + H\cdot\ (R\cdot)$$

$$CDOM\cdot + O_2 \longrightarrow CDOMOO\cdot$$

$$CDOMOO\cdot \longrightarrow HO_2\cdot + CDOM_{ox}$$

Formation of solvated electrons (Zepp et al. 1987; Bruccoleri et al. 1993) and subsequent reaction with O_2:

$$CDOM \xrightarrow{h\nu} CDOM^+ + e^-_{aq}$$

$$e^-_{aq} + O_2 \longrightarrow O_2^-\cdot$$

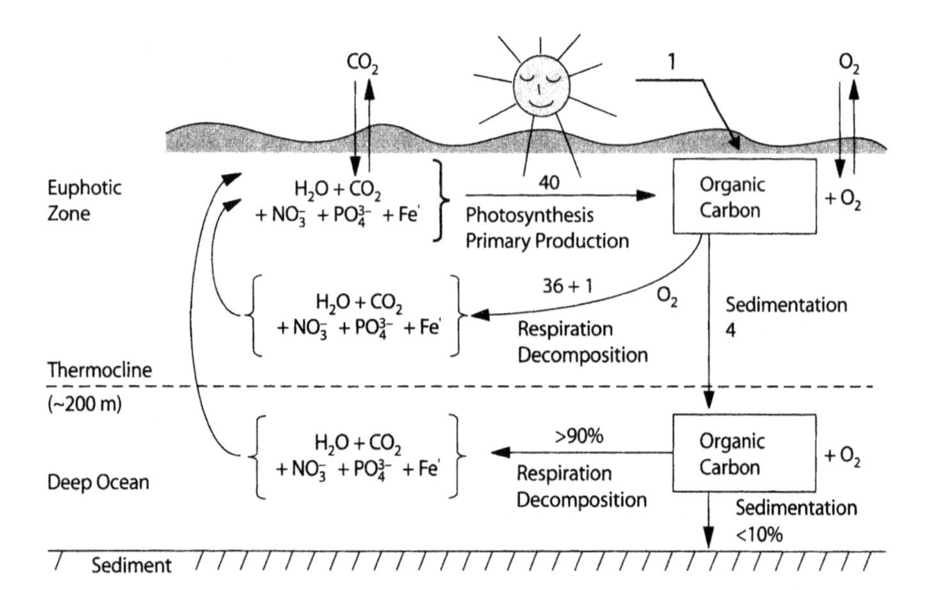

Fig. 3.1. Fate of carbon in the oceans. Numbers are carbon fluxes in Gt yr^{-1} (modified from Hohmann et al. 1990)

3.2 Absorption of Sunlight by *CDOM*

In many surface waters, *CDOM* is the predominant light absorbing component. These materials exhibit broad, featureless absorption spectra, decreasing approximately exponentially throughout the ultraviolet and visible wavelength regimes (Fig. 3.2). *CDOM* spectra have typically been fitted to the expression (Blough and Green 1995),

$$\alpha(\lambda) = \alpha(\lambda_0)e^{-S(\lambda-\lambda_0)} \tag{3.1}$$

where $\alpha(\lambda)$ and $\alpha(\lambda_0)$ are the decadic absorption coefficients at wavelength λ and reference wavelength λ_0, respectively, and S is a parameter that characterizes how rapidly the absorption decreases with increasing wavelength. The larger S is, the less colored *CDOM* is. Studies of marine waters in the eastern Caribbean (Blough et al. 1993) and off the coast of south Florida (Blough and Green 1995) have found that S is generally larger for oligotrophic "blue" waters (= 0.02 nm^{-1}) than for coastal "brown" waters (0.013–0.018 nm^{-1}). These trends of the S values may indicate that *CDOM* that is formed in-situ in open ocean waters (or in fresh waters) has a lower fraction of conjugated functional groups, e.g. of aromatic groups, than *CDOM* of terrestrial origin. Values of S for terrestrial sources of *CDOM* range from 0.01 to 0.02 nm^{-1} (Zepp and Schlotzhauer 1981; Davies-Colley and Vant 1987). Values of α_{300} range from <0.1 m^{-1} for "blue" seawaters to >50 m^{-1} for some coastal waters and fresh waters (Haag and Hoigné 1986; Blough et al. 1993; Green and Blough 1994).

The penetration depth of sunlight into a water body can be calculated with the help of the Beer-Lambert law. For a natural water body, the Beer-Lambert law is given by:

$$\log\frac{W(\lambda)}{W_z(\lambda)} = \alpha(\lambda)z \tag{3.2}$$

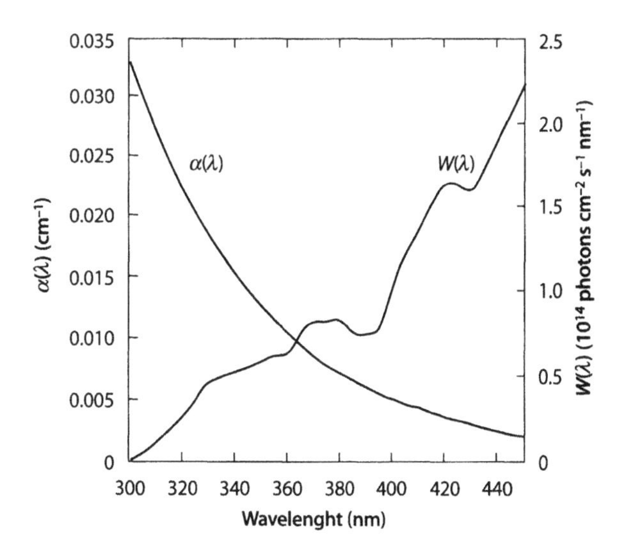

Fig. 3.2. Comparison of the spectral photon flux with the wavelength-dependence of the absorption coefficient of sea water offshore from Delaware (from Blough 1997)

$W(\lambda)$ is the spectral photon flux that hits the surface of a water body (in millieinstein $cm^{-2} s^{-1} nm^{-1}$) (also called incident light intensity), $W_z(\lambda)$ is the light intensity at depth z, and $\alpha(\lambda)$ is the decadic absorption coefficient (in cm^{-1} if the unit of z is cm). Equation 3.2 is only valid under the assumption that sunlight enters a water body perpendicular to the surface, that reflection, backscattering and internal scattering can be neglected, and that the water body is well mixed. Neglecting absorption of light by suspended matter (Miller and Zepp 1979) and by all the other light-absorbing constituents except *CDOM*, the decadic absorption coefficient, $\alpha(\lambda)$, can then be approximated as,

$$\alpha(\lambda) \sim [CDOM]``\varepsilon(\lambda)" \tag{3.3}$$

where "$\varepsilon(\lambda)$" is the decadic extinction coefficient of *CDOM* at wavelength λ (in $l\, mg^{-1} cm^{-1}$ if the unit of $[CDOM]$ is $mg\, l^{-1}$). [Note, that *CDOM* is not a well-defined, light-absorbing chemical compound but includes more than one chromophore.] The depth of a water body at which the incident solar light intensity at a given wavelength is attenuated by a factor of two, $z_{W(\lambda)/2}$, can then be calculated:

$$z_{W(\lambda)/2} = \frac{\log 2}{[CDOM]"\varepsilon(\lambda)"} \tag{3.4}$$

Hence, the higher the *CDOM* concentration is in a water body, the smaller the penetration depth of the sunlight is at a given wavelength.

UV absorption by *CDOM* (see Fig. 3.2) is the primary mechanism shielding aquatic organisms from the harmful effects, especially of UV-B radiation (280–315 nm). However, UV radiation also destroys *CDOM* through photooxidation, resulting in a decrease in the *CDOM* concentration and thus in an increase in the penetration depth of UV light. Increases in UV radiation, due to stratospheric ozone depletion, could therefore cause more damage to aquatic ecosystems than would be expected from direct effects alone.

3.3 Photooxidation of *CDOM*: Role for Carbon Bioavailability

Photooxidation of *CDOM* leads to carbon gases including carbon monoxide (CO) (Valentine and Zepp 1993; Tarr et al. 1995), carbon dioxide (CO_2) (Miller and Zepp 1995), carbonyl sulfide (COS) (Andreae and Ferek 1992; Zepp and Andreae 1994), and to low molecular-weight compounds that are bioavailable, such as formaldehyde, acetaldehyde, glyoxylate and pyruvate (Kieber et al. 1989; Mopper et al. 1991). Kieber et al. (1989) have demonstrated that biological uptake of pyruvate was highly correlated with its rate of photochemical production in sea water. Furthermore, Bushaw et al. (1996) have shown that ammonium is among the nitrogenous compounds produced upon *CDOM* photooxidation. Most of the products are thought to result from the net oxidation of *CDOM* by O_2 (Blough and Zepp 1995), the most abundant marine oxidant, where photooxidation of *CDOM* by O_2 can occur through different pathways (Table 3.1). Zafiriou and collaborators (1994 pers. comm.) calculated the global marine O_2 photochemical uptake to be about 1×10^{14} moles yr^{-1}. As a comparison, terrestrial organic carbon enters the ocean at about 3.4×10^{13} moles yr^{-1} (Meybeck 1982; Sarmiento and Sundquist 1992). As there is more than sufficient photooxidation taking place in the ocean to oxidize all the terrestrial *CDOM*, photooxidation is probably the dominant removal mechanism for this *CDOM* as well as a removal mechanism for some marine-derived *CDOM* (Zafiriou 1994 pers. comm.).

In open ocean waters, organic matter from terrigenous sources represents only a small fraction of the total organic matter. This fraction may be significantly different in coastal waters and in fresh-water rivers and lakes. Analysis of lignin-derived phenols, unique biomarkers of terrestrial plant carbon, indicated that terrestrially-derived carbon accounts for a major source of fresh-water *DOC* (Meyers-Shulte and Hedges 1986). Furthermore, Herndl and co-workers (1997) have shown that in the oligotrophic northern Adriatic Sea, humic substances contribute between 10–15% to the total *DOC* pool, while their contribution to the *DOC* pool range between 15–47% in the shallow Lake Neusiedl, a lake surrounded by a reed belt of *Phragmites australis* (Fig. 3.3a). These authors have also shown that mainly the humic fraction of the total *DOC* pool was subject to photooxidation (Fig. 3.3b), and that this fraction is efficiently altered upon exposure to solar radiation, resulting in an increase in the 250/365 nm absorbance ratio. An increase in the 250/365 nm absorbance ratio means a transformation to less colored compounds (see Eq. 3.1), i.e. to compounds with a lower fraction of aromatic functional groups, which tend to be better bioavailable. In this study, the humic and non-humic fraction of *DOC* was separated by XAD-extraction.

Hence, in coastal waters and in fresh-water rivers and lakes, photooxidation of non-bioavailable humic substances may significantly increase the abundance of bioavailable organic carbon. Furthermore, coastal waters and lakes often exhibit higher concentrations of iron due to larger riverine inputs (De Vitre et al. 1988; Davison 1993; Sigg et al. 1991). Thus, questions arise about the roles of iron in the photooxidation of *CDOM*, in particular the following questions:

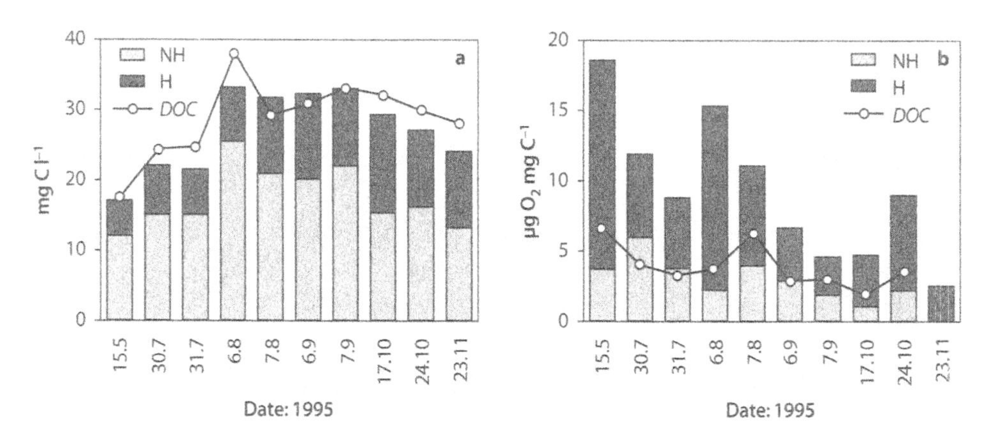

Fig. 3.3. a Contribution of the non-humic (*NH*) and the humic (*H*) fraction to the *DOC* pool in Lake Neusiedl; **b** Photooxidation of the different *DOC* fractions from Lake Neusiedl as measured by O_2 consumption (from Herndl et al. 1997)

i. What are the various pathways of *CDOM* photooxidation in iron-rich, sunlit aquatic systems?
ii. Which are the major reactive oxygen species formed concommitantly with *CDOM* photooxidation in the presence of iron?
iii. Does the iron-related oxidation of humic substances result in the formation of bioavailable carbon compounds, as does direct photooxidation of humic substances, or does it lead to further polymerization of the biologically refractory humic substances?

In the experimental study described in the next section, these questions are addressed.

3.4 Experimental Study on the Roles of Iron in *CDOM* Oxidation in Irradiated Aqueous Systems (Voelker et al. 1997)

In this experimental study, lepidocrocite (γ-FeOOH) was chosen as a model solid Fe(III) phase for crystalline Fe(III)-(hydr)oxides present in surface waters, and Suwannee River Fulvic Acid (SRFA) as a model compound for humic substances. In illuminated, aerated systems containing lepidocrocite and SRFA, four pathways or fulvic acid oxidation are linked to the redox cycling of iron (Fig. 3.4). These pathways are:

i. Photooxidation of SRFA that is adsorbed on the surface of lepidocrocite
ii. Photooxidation of SRFA via photolysis of dissolved Fe(III)-fulvate complexes
iii. Thermal oxidation of SRFA by dissolved Fe(III)
iv. Oxidation of SRFA by HO· that is formed in the Fenton reaction

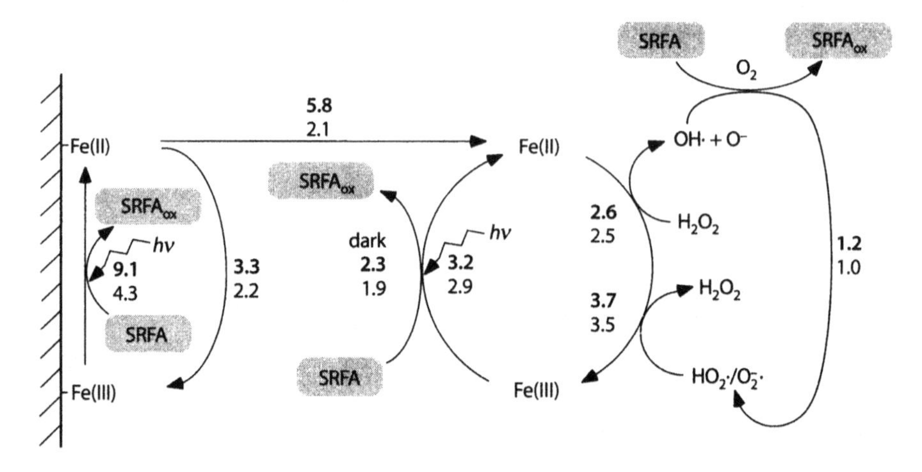

Fig. 3.4. Pathways of oxidation of SRFA in illuminated, aerated aqueous systems, containing 40 μM Fe in the form of initially γ-FeOOH and 10 mg l^{-1} SRFA. The numbers in italics indicate the amount of iron (in μM) oxidized or reduced by each of the depicted processes after 300 min of illumination at pH 3 (*upper numbers*) and at pH 5 (*lower numbers*). The rate of photoreduction of surface Fe(III) was assumed to be equal to the rate of photoreductive dissolution in deaerated suspensions. The amount of surface Fe(II) that reoxidized instead of detaching was calculated as the difference in the amount of total dissolved iron (normalized to total Fe(III)-hydroxide concentration) formed in the deaerated and aerated suspensions (modified from Voelker et al. 1997)

An additional, presumably iron-independent pathway is direct photooxidation resulting in reduction of O_2 to $O_2^{\bar{}}\cdot$, (see Table 3.1). Of the redox processes, where iron is directly involved, photooxidation of adsorbed SRFA under reduction of surface Fe(III) is the most important process in systems containing initially lepidocrocite, both at pH 3 and 5 (Fig. 3.4). With regard to the oxidation of SRFA by dissolved Fe(III), photochemical and thermal oxidation play almost equally significant roles at both pH values.

The relative rates of the individual processes shown in Fig. 3.4 were assessed by measuring net formation of dissolved Fe(II), total dissolved Fe, and H_2O_2 and by simulating the data with kinetic modelling (Fig. 3.5) (for more information on the kinetic modelling, see Voelker and Sulzberger 1996, and Voelker et al. 1997).

3.4.1 Photooxidation of SRFA by Fe(III)

Voelker and collaborators (1997) hypothesize that photooxidation of SRFA with Fe(III) as oxidant occurs through photolysis of Fe(III)-SRFA surface and solution complexes involving carboxyl-groups of SRFA (Leenheer et al. 1995a, b), in a similar way as photolysis of Fe(III)-oxalate or other simple Fe(III)-carboxylate complexes (in the following shown for photolysis of the Fe(III)-fulvate surface complex, where the symbol *i* stands for the surface lattice of lepidocrocite):

1. Ligand-to-metal charge-transfer transition creating a charge-transfer state, which is in part thermally deactivated,

$$i\mathrm{Fe^{III}OOCR} \xleftrightarrow{h\nu} \mathrm{R} + i\mathrm{Fe^{II}O{\cdot}OCR^{*}} \tag{3.5}$$

and in part by formation of primary photoproducts which are surface Fe(II) and the fulvate radical, RCOO·:

$$i\mathrm{Fe^{II}O{\cdot}OCR^{*}} \longrightarrow i\mathrm{Fe(II)} + \mathrm{RCOO}\cdot \tag{3.6}$$

2. Detachment of surface Fe(II) from the crystal lattice of lepidocrocite and reconstitution of the surface site:

$$i\mathrm{Fe(II)} + 2\,\mathrm{H^+} \longrightarrow \mathrm{Fe^{2+}} + i_{\mathrm{rec}} \tag{3.7}$$

3. Decarboxylation of the fulvate radical to form CO_2 and the radical R·:

$$\mathrm{RCOO}\cdot \longrightarrow \mathrm{R}\cdot + \mathrm{CO_2} \tag{3.8}$$

If photooxidation of humic substances by Fe(III) occurs indeed in a similar way as photooxidation of simple carboxylic acids, then one would expect formation of CO_2 as the sole carbon gas. Carbon monoxide production (Valentine and Zepp 1993; Tarr et al. 1995) and COS production (Andreae and Ferek 1992; Zepp and Andreae 1994) have also been observed upon *DOM* photooxidation, presumably without the involvement of iron. Carbon monoxide formation is thought to occur from reactions involving ketones, aldehydes, or other organic carbonyls that make up natural organic matter (Averett et al. 1990; Zafiriou 1994 pers. comm.; Zepp 1994). These functional groups

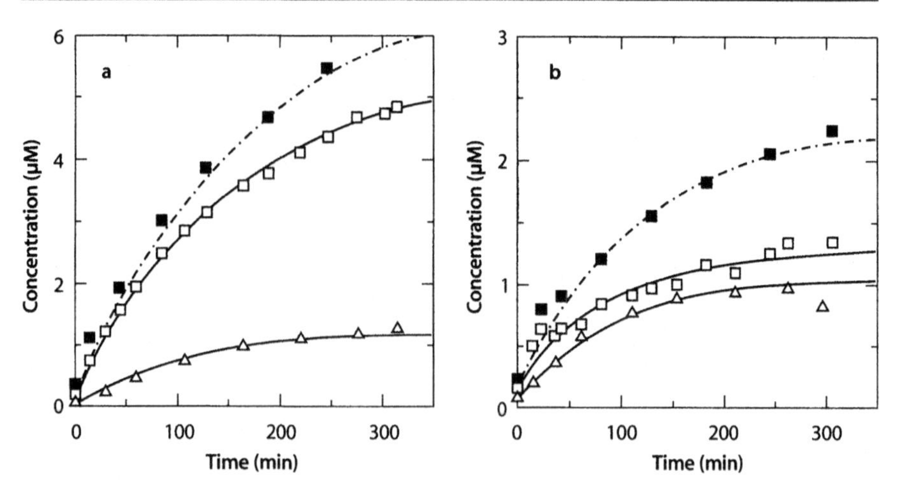

Fig. 3.5. a Total dissolved Fe (*black squares*), Fe(II) (*white squares*), and hydrogen peroxide (*triangles*) in the illuminated, aerated lepidocrocite suspension containing 10 mg l^{-1} SRFA at pH 3 (Fe_T 40 µM). Dashed-dotted lines represent the modelled rates of photoreductive dissolution of lepidocrocite. Solid lines represent best model fit of the Fe(II) and hydrogen peroxide data; **b** Same as Fig. 3.5a except at pH 5 (Fe_T 50 µM) (from Voelker et al. 1997)

are probably not involved in ligand-to-metal charge-transfer reactions of Fe(III)-humate or Fe(III)-fulvate complexes.

The question arises about the fate of the radical R· formed in Reaction 3.8. Formation of organic peroxyl radicals and subsequent reactions are likely to take place if the carbon-centered radical is an alkyl radical (Blough and Zepp 1995):

$$R\cdot + O_2 \longrightarrow ROO\cdot \tag{3.9}$$

Radicals R· may also react with each other:

$$R\cdot + R\cdot \longrightarrow \text{non-radical products} \tag{3.10}$$

Peroxyl radicals can further react by different pathways (Blough and Zepp 1995): *(i)* termination reactions to form non-radical and non-peroxidic products, e.g. polymerization products, *(ii)* abstraction of $HO_2\cdot$, and *(iii)* H-atom abstractions to generate organic peroxides and secondary radicals (Eq. 3.11–3.13, respectively):

$$2\,ROO\cdot \longrightarrow \text{non-radical products} \tag{3.11}$$

$$ROO\cdot \longrightarrow R_{ox} + HO_2\cdot \tag{3.12}$$

$$ROO\cdot + R'H \longrightarrow RO_2H + R\cdot \tag{3.13}$$

Table 3.2 compares the observed rate of $HO_2\cdot/O_2^-\cdot$ photoformation with the rates of $HO_2\cdot/O_2^-\cdot$ photoformation as expected from iron-independent and from iron-dependent pathways. The difference between the observed rate of $HO_2\cdot/O_2^-\cdot$ photoformation and

Table 3.2. Comparison of potential $HO_2\cdot/O_2^-\cdot$ sources in the experiments described in Fig. 3.5 (μM $HO_2\cdot/O_2^-\cdot$ formed during 300 min irradiation)

Sources	pH 3	pH 5
Total observed photoformation	*2.5*[a]	*2.5*[a]
Expected from:		
▪ iron-independent photooxidation of SRFA	0.4–1.8[b]	0.4–1.8[b]
▪ surface ligand-to-metal charge-transfer reaction	9.1[c]	4.3[c]
▪ solution ligand-to-metal charge-transfer reaction	3.2[c]	2.9[c]

[a] Numbers were obtained from kinetic model (see Voelker et al. 1997). Errors estimated are <15%.
[b] Estimated from observed H_2O_2 photoformation rates in the absence of iron, corrected for the different light conditions [note, that $HO_2\cdot/O_2^-\cdot$ formation via reaction of HO· with SRFA (see Fig. 3.4) is not included in this table].
[c] Numbers were obtained from kinetic model (see Fig. 3.4). Errors estimated are <15%.

the rate expected from the iron-independent photooxidation of fulvic acid suggests that an additional source of $HO_2\cdot/O_2^-\cdot$ is likely. However, the difference is much smaller than the expected amount of $HO_2\cdot/O_2^-\cdot$ resulting from ligand-to-metal charge-transfer reactions of surface and solution Fe(III)-fulvate complexes, if one mol $HO_2\cdot/O_2^-\cdot$ is formed per mole Fe(III) reduced by the Reactions 3.5–3.9 and 3.12. This suggests that Reaction 3.12 is not dominant. Further experimental studies need to be done in order to assess the products formed upon photooxidation of humic acids via ligand-to-metal charge-transfer reactions. A pertinent question is whether thereby low molecular-weight, bioavailable carbon compounds are formed as in direct photooxidation of humic substances (Kieber et al. 1989).

3.5 Roles of Ligands in the Redox-Cycling of Iron in the Euphotic Zone of Surface Waters

With regard to the effect of iron on the photooxidation of biologically refractory humic substances, an essential question is whether these materials are the main complexing agents of iron in natural surface waters. Information on this question can be gained by investigating the roles of ligands in the redox-cycling of iron in natural waters, as described in this section.

In the euphotic zone of circumneutral surface waters, ligands are expected to play a key role as reductants of Fe(III), both in photoreductive dissolution of amorphous Fe(III)-hydroxide phases (Rich and Morel 1990; Johnson et al. 1994), and in the photochemical and thermal reduction of dissolved Fe(III) (King et al. 1993; Deng and Stumm 1994; Voelker and Sulzberger 1996). Quantum yields of photolysis of inorganic Fe(III) complexes present in surface waters, i.e. of Fe(III)-hydroxo and -carbonato complexes, are generally smaller than those of photolysis of organic Fe(III) complexes (Faust and Hoigné 1987; King et al. 1993; Faust and Zepp 1993). Hence, an important question is whether Fe(III) is bound to organic or inorganic ligands in the euphotic zone of surface waters.

Several studies have provided increasing evidence that in the oceans dissolved iron is strongly bound by organic ligands (Gledhill and Van den Berg 1995; Rue and Bruland

1995; Van den Berg 1995; Wu and Luther 1995; Kuma et al. 1996; Luther and Wu 1997; Rue and Bruland 1997). These studies have used ligand-exchange methods in combination with voltammetric detection, which give information about the conditional stability constants and the concentrations of the organic ligands involved. These organic ligands, for which no structures are yet known, are likely to play an essential role in regulating the dissolved iron concentration in the oceans. The occurrance of organic complexes of iron in fresh waters has so far not been systematically investigated. Measured "dissolved" iron concentrations (determined by filtration) are usually in excess of the theoretical solubility of Fe(III)-hydroxides. The discrepancy between measured "dissolved" iron concentrations and the solubility is generally attributed to the presence of colloidal Fe(III)-hydroxides (Davison 1993; Perret et al. 1990). The presence of organic Fe(III) complexes may however also enhance the solubility of Fe(III) (Kuma et al. 1996; Millero 1997).

Hitherto, it is not known whether the major organic ligands of Fe(III) in marine and fresh waters are humic acids or low molecular-weight organic compounds formed, e.g. from microbial degradation of organic matter. This may not matter regarding Fe(II) production, since the rate coefficients of photolysis of Fe(III)-humate or -fulvate complexes may be similar to those of photolysis of Fe(III) bound to low molecular-weight ligands, e.g. simple carboxylic acids (Faust and Zepp 1993; King et al. 1993; Voelker et al. 1997). From the point of view of the effect of iron on the photooxidation of biologically refractory humic substances, however, the type of organic Fe(III) ligands present in the eupotic zone of surface waters is essential.

An important class of low molecular-weight organic molecules which specifically chelate Fe(III) are siderophores. Siderophores are excreted by bacteria, e.g. by blue-green algae, to take up iron. They are highly selective for Fe(III). Their chelating structures include typically hydroxamates and catecholates that bind Fe(III) in octahedral coordination. The stability constants of these chelates are very high (Winkelmann et al. 1987; Crumbliss 1991; Reid et al. 1993). Employing voltammetric methods, Lewis and co-workers (1995) have estimated thermodynamic stability constants of approximately 10^{40} M^{-1} for the complexation of Fe^{3+} with catecholate-type siderophores isolated from the marine bacterium *Alteromonas luteoviolacea* and from the marine cyanobacterium *Synechoccus* sp. PCC 7 002. Production of siderophores under iron-limiting conditions is considered to be a frequent strategy among both marine and fresh-water cyanobacteria and eubacteria (Reid and Butler 1991; Wilhelm and Trick 1994). Because siderophores are excreted by bacteria to take up iron, they may be present at low levels in natural waters. Whether Fe(III)-siderophore complexes are subject to photolysis is a still unanswered question.

3.5.1
Roles of Ligands in the Oxidation of Iron(II) in the Euphotic Zone of Surface Waters

The kinetics of Fe(II) oxidation largely depend on Fe(II) speciation (Wehrli 1990). For example, the net rate of Fe(II) oxidation by O_2 can be enhanced (Liang et al. 1993) or retarded (Theis and Singer 1974) by complexation with organic ligands, depending on the relative stability of the Fe(II)- and Fe(III)-organic complexes and on the reducing power of the ligands. Emmenegger and co-workers (1998) have investigated the Fe(II) oxidation kinetics in samples from Lake Greifen, a eutrophic lake in Switzerland, in the pH range between 6.8 and 8.3. The pH was varied by bubbling with CO_2 and synthetic air. Fe(II) concentrations were followed using an automated flow injection analysis system employing luminol-based chemiluminescence detection of Fe(II) (King et al. 1995). Above pH 7.4, apparent rate constants were consistent with the rate law determined in pure carbonate systems (Stumm and Lee 1961; Singer and Stumm 1969; Millero et al. 1987; Davison and Seed 1983; Millero 1988; King 1998). Between pH 6.8 and 7.3, however, the apparent rate constant was independent of pH (Fig. 3.6). Emmenegger and co-workers hypothesize that some naturally occurring (organic, colloidal, or surface) ligand(s), L, may accelerate the oxidation of Fe(II).

The apparent rate constant, k_{app}, is the sum of the pseudo first-order rate constants of Fe(II) oxidation by O_2 and by H_2O_2:

$$k_{app} = k_{O_2}[O_2] + k_{H_2O_2}[H_2O_2] \tag{3.14}$$

Emmenegger and co-workers (1998) have also determined the second-order rate constants k_{O_2} and $k_{H_2O_2}$. The latter was assessed by adding excess H_2O_2 (100–1 000 μM) to water samples from Lake Greifen. With help of the experimentally determined rate constants k_{app} and $k_{H_2O_2}$ (see Fig. 3.6 and 3.7), k_{O_2} was calculated using Eq. 3.14. While $\log k_{H_2O_2}$ exhibited a pH dependence as expected for pure carbonate systems (Millero and Sotolongo 1989) in the whole measured pH range, $\log k_{O_2}$ became independent of pH below pH 7.3 (Fig. 3.7).

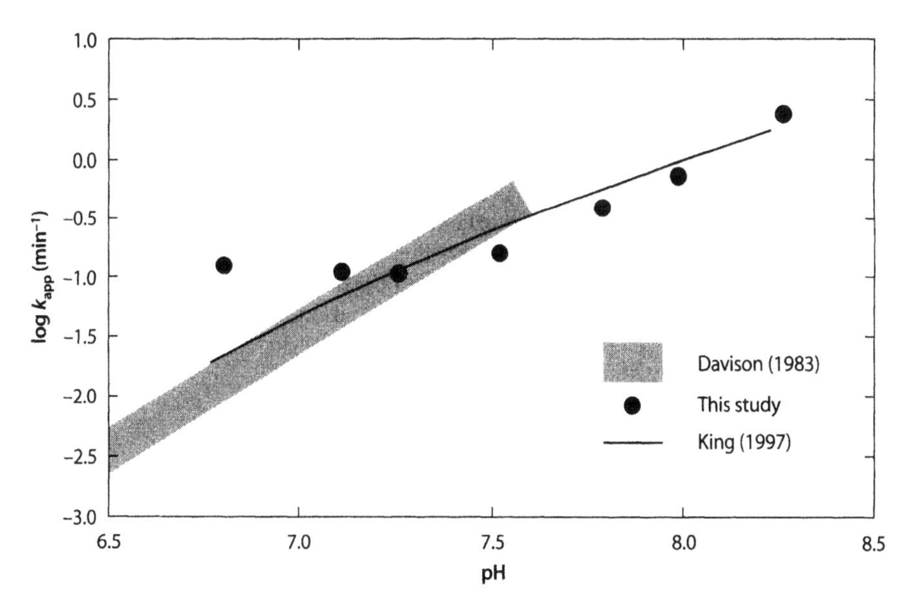

Fig. 3.6. Apparent rate constants of Fe(II) oxidation in unfiltered lake water as a function of pH. $T = 25$ °C, $p_{O_2} = 0.209$ atm, alkalinity = 3.88 mM, $[H_2O_2]$ = 20–50 nM, initial [Fe(II)] = 30 nM. Also plotted, assuming steady-state concentrations of $[H_2O_2]$, are the range for $\log k_{app}$ given by Davison and Seed (1983), and $\log k_{app}$ calculated using the model by King (1998) (from Emmenegger et al. 1998)

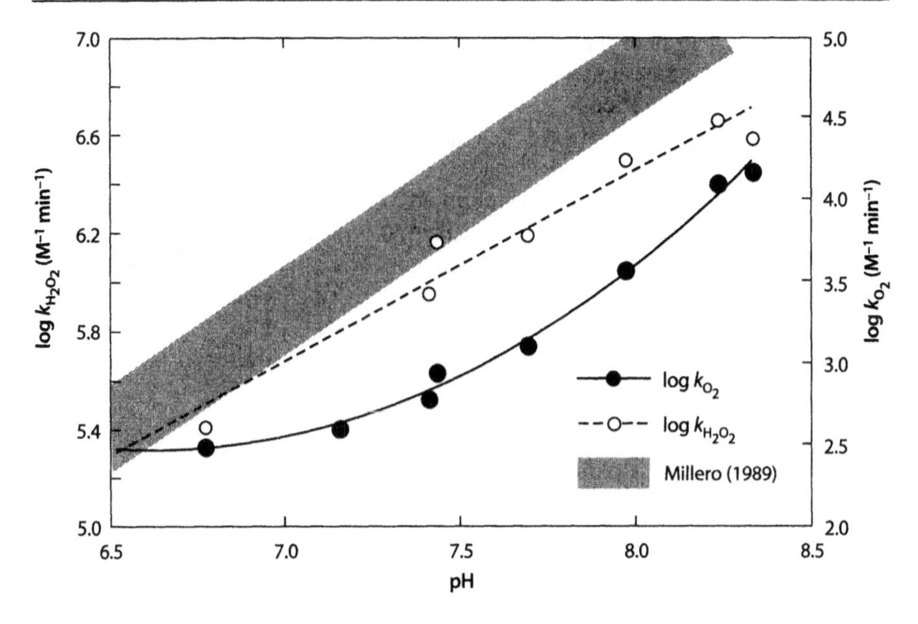

Fig. 3.7. Log $k_{H_2O_2}$ and log k_{O_2} as a function of pH. The data were interpolated by a linear fit for log $k_{H_2O_2}$ and a second-order polynom for log k_{O_2}. In gray are the values obtained for log $k_{H_2O_2}$ by Millero and Sotolongo (1989) ($I = 7$ mM, $T = 25$ °C, $HCO_3^- = 3.8$ mM) multiplied by 0.5 to account for the scavenging of HO·. The range shown was determined by the standard deviation of the bicarbonate correction for salinity = 0 ‰ (from Emmenegger et al. 1998)

These experimental results led Emmenegger et al. (1998) to the following conclusions:

1. The enhancement of the overall apparent rate constant below pH 7.3 can be attributed to the reaction of Fe(II) with O_2.
2. Complexation of Fe(II) by (organic, colloidal or surface) ligand(s) affects Fe(II) oxidation by O_2 much stronger than Fe(II) oxidation by H_2O_2.
3. Fe(II)-L is only a trace species in Lake Greifen [less than about 10% of total Fe(II)].
4. Exchange rates for L are fast compared to the oxidation rates.

3.6 Conclusions

Dissolved organic carbon (*DOC*) has been recognized as one of the largest reactive reservoirs of organic carbon on earth (Hedges 1992) and a major reservoir of chemical energy in most aquatic systems (Wetzel 1992). The bulk of the organic carbon in natural waters is distributed throughout the water column as particulate and dissolved organic matter. Most of the organic carbon in fresh waters is dissolved and not associated with living organisms. In coastal waters and in fresh waters with a high input of *DOC* from terrigeneous sources, photooxidation of biologically refractory *DOC* can result in a significant increase in the abundance of biologically available organic carbon. This may result in a decrease in biodiversity and thus in the biotic stability (Wetzel 1992).

Iron catalyzes the light-induced oxidation of humic substances where both Fe(III) and HO· (formed in the Fenton reaction) act as oxidants. A redox cycle compared to the one depicted in Fig. 3.4 should be a common phenomenon in iron-rich sunlit surface waters, since both Fe(II) and hydrogen peroxide are rapidly produced by photochemical reactions involving humic substances. Questions which remain to be answered are:

1. Does photooxidation of humic substances via ligand-to-metal charge-transfer reactions of Fe(III)-humate complexes lead to the formation of bioavailable carbon compounds, as does the iron-independent photooxidation of humic substances?
2. Are Fe(III)-humate complexes major Fe(III) species in the euphotic zone of surface waters?

Regarding the catalytic effect of iron on the light-induced oxidation of colored dissolved organic matter (*CDOM*), a relevant question is the following: What are the combined effects of increased solar UV radiation, due to stratospheric ozone depletion, and of iron on *CDOM* photooxidation? Increased UV radiation is expected to accelerate the redox-cycling of iron and in turn the oxidation of *CDOM*. If thereby biologically available carbon compounds are formed, these will be back-transformed to CO_2 and water through respiration. As a consequence, UV light will penetrate deeper into surface waters (see Eq. 3.4), and hence impact aquatic organisms, e.g. bacterio- and phytoplankton. It has been demonstrated (Karentz et al. 1994) that fresh water and marine bacteria are impacted by changes in solar UV radiation, and action spectra indicate that UV-B radiation (280–315 nm) is mainly involved (Calkins and Barcelo 1982). This has been confirmed by field studies which indicate that current levels of solar UV radiation reduce bacterioplankton growth in the upper ocean (Herndl et al. 1993).

References

Andreae MO, Ferek RJ (1992) Photochemical production of carbonyl sulfide in seawater and its emission to the atmosphere. Global Biogeochem Cycles 6:175–183
Averett RC, Leenheer JA, McKnight DM, Thorn KA (1990) Humic substances in the Suwannee River, Georgia: Interactions, properties, and proposed structures. U.S. Geological Survey Open File Report No. 87–557, Denver, Colorado
Blough NV (1997) Photochemistry in the sea-surface microlayer. In: Liss PS, Duce R (eds) The sea surface and global change. Cambridge University Press, Cambridge, pp 383–424
Blough NV, Green SA (1995) Spectroscopic characterization and remote sensing of nonliving organic matter. In: Zepp RG, Sonntag C (eds) Role of nonliving organic matter in the earth's carbon cycle. Wiley-Interscience, New York
Blough NV, Zepp RG (1995) Reactive oxygen species in natural waters. In: Foote CS, Valentine JS (eds) Reactive oxygen species in chemistry. Chapman and Hall, London, pp 280–333
Blough NV, Zafiriou OC, Bonilla J (1993) Optical absorption spectra of waters from the Orinoco River outflow: Terrestrial input of colored organic matter to the Caribbean. J Geophys Res 98:2271–2278
Bruccoleri A, Pant BC, Sharma DK, Langford CH (1993) Evaluation of primary photoproduct quantum yields in fulvic acid. Environ Sci Technol 27:889–894
Bushaw KL, Zepp RG, Tarr MA, Schulz-Jander D, Bourbonniere RA, Hodson RE, Miller WL, Bronk DA, Moran MA (1996) Photochemical release of biologically available nitrogen from aquatic dissolved organic matter. Nature 381:404–407
Calkins J, Barcelo J (1982) Action spectra. In: Calkins J (ed) The role of solar ultraviolet radiation in marine ecosystems. Plenum Press, New York, pp 143–150
Crumbliss AL (1991) Aqueous solution equilibrium and kinetic studies of iron siderophore and model siderophore complexes. In: Winkelmann G (ed) CRC Handbook of microbial iron chelates. CRC Press, Boca Raton, pp 177–233

Davies-Colley RJ, Vant WN (1987) Absorption of light by yellow substance in fresh-water lakes. Limnol Oceanogr 32:416–425
Davison W (1993) Iron and manganese in lakes. Earth-Science Rev 34:119–163
Davison W, Seed G (1983) The kinetics of the oxidation of ferrous iron in synthetic and natural waters. Geochim Cosmochim Acta 47:67–79
Deng Y, Stumm W (1994) Reactivity of aquatic iron(III) oxyhydroxides – implications for redox cycling of iron in natural waters. Appl Geochem 9:23–36
DeVitre RR, Buffle J, Perret D, Baudat R (1988) A study of iron and manganese tranformations at the O_2/S(–II) transition layer in a eutrophic lake (lake Bret, Switzerland): A multimethod approach. Geochim Cosmochim Acta 52:1601–1613
Emmenegger L, King DW, Sigg L, Sulzberger B (1998) Oxidation kinetics of Fe(II) in a eutrophic Swiss lake. Environ Sci Technol 32:2990–2996
Faust BC, Hoigné J (1987) Sensitized photooxidation of phenols by fulvic acid and in natural waters. Environ Sci Technol 21:957–964
Faust BC, Zepp RG (1993) Photochemistry of aqueous iron(III)-polycarboxylate complexes: Roles in the chemistry of atmospheric and surface waters. Environ Sci Technol 27:2517–2522
Frimmel FH, Christman RC (1988) Humic substances and their role in the environment. Wiley Interscience, New York
Gledhill M, Van den Berg CMG (1995) Measurement of the redox speciation of iron in seawater by catalytic cathodic stripping voltammetry. Mar Chem 50:51– 61
Green SA, Blough NV (1994) Optical absorption and fluorescence properties of chromophoric dissolved organic matter in natural waters. Limnol Oceanogr 39:7337–7346
Haag WR, Hoigné J (1986) Singlet oxygen in surface waters. 3. Photochemical formation and steady-state concentrations in various types of waters. Environ Sci Technol 20:341–348
Haag WR, Mill T (1990) Survey of sunlight-produced transient reactants in surface waters. In: Blough NV, Zepp RG (eds) Effects of solar ultraviolet radiation on biogeochemical dynamics in aquatic environments. Woods Hole Oceanographic Insitution Technical Report, WHOI-90-09, pp 82–88
Hedges JI (1992) Global biogeochemical cycles: Progress and problems. Mar Chem 39:67–93
Hedges J, Hatcher, Ertel PJ, Meyers-Schulte KA (1992) Comparison of dissolved humic substances from seawater with Amazon River counterparts by ^{13}C-NMR spectrometry. Geochim Cosmochim Acta 56:1753–1757
Herndl, GJ, Mueller-Niklas G, Frick J (1993) Major role of ultraviolet-B radiation in controlling bacterioplankton growth in the surface layer of the ocean. Nature 361:717–719
Herndl GJ, Brugger A, Hager S, Kaiser E, Obernosterer I, Reitner B, Slezak D (1997) Role of ultraviolet-B radiation on bacterioplankton and the availability of dissolved organic matter. Plant Ecology 128:42–51
Hohmann R, Zumbrunn S, Staudenmann J, Schellenberg T, Imboden D, Thierstein H, Eckert V (1990) Der C-Transport von der Atmosphäre in die Tiefsee: Schlüssel zum Verständnis des globalen C-Kreislaufes. Umwelttag 1990 der ETH: Probleme der anthropogenen Klimaänderung, S 93–94
Huber SA, Frimmel FH (1994) Direct gel chromatographic characterization and quantification of marine dissolved organic carbon using high-sensitivity *DOC* detection. Environ Sci Technol 28:1194–1197
Johnson KS, Coale KH, Elrod VA, Tindale NW (1994) Iron photochemistry in seawater from the equatorial Pacific. Mar Chem 46: 319–334
Karentz D, Bothwell ML, Coffin RB, Hanson A, Herndl GJ, Kilham SS, Lesser MP, Lindell M, Moeller RE, Morris DP, Neale PJ, Sanders RW, Weiler CS, Wetzel RG (1994) Impact of UV-B radiation on pelagic fresh-water ecosystems: Report of working group on bacteria and phytoplankton. Arch Hydrobiol, Beih Ergebn Limnol 43:31–69
Kieber DJ, McDaniel J, Mopper K (1989) Photochemical source of biological substrates in sea water: Implications for carbon cycling. Nature 341:637–639
Kieber RJ, Zhou X, Mopper K (1990) Formation of carbonyl compounds from UV-induced photodegradation of humic substances in natural waters: Fate of riverine carbon in the sea. Limnol Oceanogr 35:1053–1515
King DW (1998) Role of carbonate species on the oxidation rate of Fe(II) in aquatic systems. Environ Sci Technol 32:2997–3003
King DW, Aldrich RA, Charnecki SE (1993) Photochemical redox cycling of iron in NaCl solutions. Mar Chem 44:105–120
King DW, Lounsbury HA, Millero FJ (1995) Rates and mechanism of Fe(II) oxidation at nanomolar total iron concentrations. Environ Sci Technol 29:818–824
Kuma K, Nishioka N, Matsunaga K (1996) Controls on iron(III) hydroxide solubility in seawater: The influence of pH and organic chelators. Limnol Oceanogr 41:396–407
Leenheer JA, Wershaw RL, Reddy MM (1995a) Stong-acid, carboxyl-group structures in fulvic acid from the Suwannee River, Georgia. 1. Minor structures. Environ Sci Technol 29:393–398

Leenheer JA, Wershaw RL, Reddy MM (1995b) Stong-acid, carboxyl-group structures in fulvic acid from the Suwannee River, Georgia. 2. Major structures. Environ Sci Technol 29:399–405
Lewis BL, Holt PD, Taylor SW, Wilhelm SW, Trick CG, Butler A, Luther III GW (1995) Voltammetric estimation of iron(III) thermodynamic stability constants for catecholate siderophores isolated from marine bacteria and cyanobacteria. Mar Chem 50:179–188
Liang L, McNabb JA et al. (1993) Kinetics of iron(II) oxygenation at low partial pressure of oxygen in the presence of natural organic matter. Environ Sci Technol 27:1864–1870
Luther III GW, Wu J (1997) What controls dissolved iron concentration in the world ocean – a comment. Mar Chem 57:173–179
Mantoura R, Woodward E (1983) Conservative behaviour of riverine dissolved organic carbon in the Severn estuary. Geochim Cosmochim Acta 47:1293–1309
Meybeck M (1982) Carbon, nitrogen, and phosphorus transport by world rivers. Am J Sci 282:401–450
Meyers-Shulte KJ, Hedges JI (1986) Molecular evidence for a terrestrial component of organic matter dissolved in ocean water. Nature 321:61–63
Miller GC, Zepp RG (1979) Effects of suspended sediments on photolysis rates of dissolved pollutants. Water Res 13:453–459
Miller WL, Zepp RG (1995) Photochemical production of dissolved inorganic carbon from terrestrial organic matter: Significance to the oceanic organic carbon cycle. Geophys Res Lett 22:417–420
Millero FJ (1988) Effect of ionic interactions on the oxidation of Fe(II) and Cu(I) in natural waters. Mar Chem 28:1–18
Millero FJ (1997) The influence of iron on carbon dioxide in surface seawater. In: Gianguzza A, Pelizzetti E, Sammartano S (eds) Marine Chemistry. Kluwer Academic Publishers, Dordrecht, pp 381–398
Millero FJ, Sotolongo S (1989) The oxidation of Fe(II) with H_2O_2 in seawater. Geochim Cosmochim Acta 53:1867–1873
Millero FJ et al. (1987) The oxidation kinetics of Fe(II) in seawater. Geochim Cosmochim Acta 51:793–803
Mopper K, Zhou X, Kieber RJ, Kieber DJ, Sikorski RJ, Jones RD (1991) Photochemical degradation of dissolved organic carbon and its impact on the oceanic carbon cycle. Nature 353:60–62
Perret D, DeVitre RR, Leppard G, Buffle J (1990) Characterizing autochthonous iron particles and colloids – the need for better particle analysis methods. In: Tilzer MM, Serruya C (eds) Large lakes: Ecological structure and function. Springer-Verlag, Heidelberg, pp 224–244
Reid RT, Butler A (1991) Investigation of the mechanism of iron acquisition by the marine bacterium *Alteromonas luteoviolaceus*: Characterization of siderophore production. Limnol Oceanogr 36:1783–1792
Reid RT, Live DH, Faulkner DJ, Butler A (1993) A siderophore from a marine bacterium with an exceptional ferric ion affinity constant. Nature 366:455–458
Rich H, Morel FMM (1990) Availability of well-defined iron colloids to the marine diatom *Thalassiosira weissflogii*. Limnol Oceanogr 35: 652–662
Rue EL, Bruland KW (1995) Complexation of iron(III) by natural organic ligands in the Central North Pacific as determined by a new competitive ligand equilibration/adsorptive cathodic stripping voltammetric method. Mar Chem 50:117–138
Rue EL, Bruland KW (1997) The role of organic complexation on ambient iron chemistry in the equatorial Pacific Ocean and the response of a mesoscale iron addition experiment. Limnol Oceanogr 42:901–910
Sarmiento JL, Sundquist ET (1992) Revised budget for the oceanic uptake of anthropogenic carbon dioxide. Nature 356:589–593
Sigg L, Johnson CA, Kuhn A (1991), Redox conditions and alkalinity generation in a seasonally anoxic lake (Lake Greifen). Mar Chem 36:9–26
Singer PhC, Stumm W (1969) Acidic mine drainage – the rate-determining step. Science 167:1121–1123
Stumm W, Lee GF (1961) Oxygenation of ferrous iron. Industrial and Engin Chem 53:143–146
Tarr MA, Miller WL, Zepp RG (1995) Direct carbon monoxide photoproduction from plant matter. J Geophys Res 100:11403–11413
Theis TL, Singer PC (1974) Complexation of iron(II) by organic matter and its effect on iron(II) oxygenation. Environ Sci Technol 8:569–573
Thurman EM (1985) Organic geochemistry of natural waters. Martinus Nijhoff/DR Junk Publishers, Boston
Valentine RL, Zepp RG (1993) Formation of carbon monoxide from the photodegradation of terrestrial dissolved organic carbon in natural waters. Environ Sci Technol 27:409–412
Van den Berg CMG (1995) Evidence for organic complexation of iron in seawater. Mar Chem 50:139–157
Voelker BM, Sulzberger B (1996) Effects of fulvic acid on Fe(II) oxidation by hydrogen peroxide. Environ Sci Technol 30:1106–1114
Voelker BM, Morel FMM, Sulzberger B (1997) Iron redox cycling in surface waters: effects of humic substances and light. Environ Sci Technol 31:1004–1011

Wehrli B (1990) Redox reactions of metal ions at mineral surfaces. In: Stumm W (ed) Aquatic chemical kinetics. Wiley Interscience, New York, pp 311–336
Wetzel RG (1992) Gradient-dominated ecosystems: Sources and regulatory function of dissolved organic matter in fresh-water ecosystems. Hydrobiol 229:181–198
Wilhelm SW, Trick CG (1994) Iron-limited growth of cyanobacteria: Siderophore production is a common response mechanism. Limnol Oceanogr 39:1979–1984
Winkelmann G, Van der Helm D, Neilands JB (1987) Iron transport in microbes, plants and animals. VCH, Weinheim
Wu J, Luther III GW (1995) Complexation of Fe(III) by natural organic ligands in the Northwest Atlantic Ocean by a competitive ligand equilibration and a kinetic approach. Mar Chem 50:159–177
Zepp RG (1994) Effects of solar radiation on organic matter cycling: Formation of carbon monoxide and carbonyl sulfide. In: Zepp RG (ed) Climate biosphere interaction: Biogenic emissions and environmental effects of climate change. Wiley Interscience, New York, pp 203–221
Zepp RG, Andreae MO (1994) Factors affecting the photochemical production of carbonyl sulfide in seawater. Geophys Res Lett 21:2813–2816
Zepp RG, Schlotzhauer PF (1981) Comparison of the photochemical behavior of various humic substances in water. III. Spectroscopic properties of humic substances. Chemosphere 10:479–486
Zepp RG, Braun A, Hoigné J, Leenheer JA (1987) Photoproduction of hydrated electrons from natural organic solutes in aquatic environments. Environ Sci Technol 21:485–490

Chapter 4

Redox Processes in Anoxic Waters

F.J. Millero

4.1 Introduction

Anoxic waters are defined as those waters that have no dissolved oxygen (Richards 1965). This condition can occur in natural waters when the rate of consumption of oxygen exceeds the supply. The rate of oxidation of organic matter by bacteria is greater than the supply of oxygen from the atmosphere. The supply of O_2 below the photic zone is dependent upon diffusion and advection. Anoxia normally occurs in enclosed basins where physical barriers (sills) and density stratification limit the advection of O_2 to the deep waters (Grasshoff 1975). There are two types of anoxic basins. The most common occurs because of a strong halocline (salinity gradient) which is the result of a net outflow of low salinity water from a positive estuary. This is shown in Fig. 4.1. The halocline prevents low salinity oxic waters from mixing with the high salinity deep waters. Examples of this type of basin are the Black Sea, the Baltic Sea and many fjords such as the Framvaren in Norway. The second type of basin arises because of a strong thermocline preventing the mixing of surface and deep waters. The Cariaco Trench off the coast of Venezuela is an example of this type of basin. It is a deep trench with a maximum depth of 1400 m. The water is isohaline and isothermal from 600 m to the bottom. It is permanently anoxic below a depth of 350 m. The appearance of H_2S above the thermocline is due to mixing. Both basins have a physical obstacle that prevents horizontal mixing of various water masses. In a fjord type basin, a shallow sill prevents the salty seawater rich in O_2 from entering the basin and sinking to the bottom.

In recent years estuarine systems with deep basins like the Baltic and Chesapeake have experienced periodic anoxic behavior. This has been attributed to higher productivity in the surface waters due to increases of nutrients used as fertilizers and perhaps as acid rain. Some examples of anoxic basins are given in Table 4.1. The basins have sill depths of 2 to 150 m and concentrations of H_2S of 20 to 6000 µM.

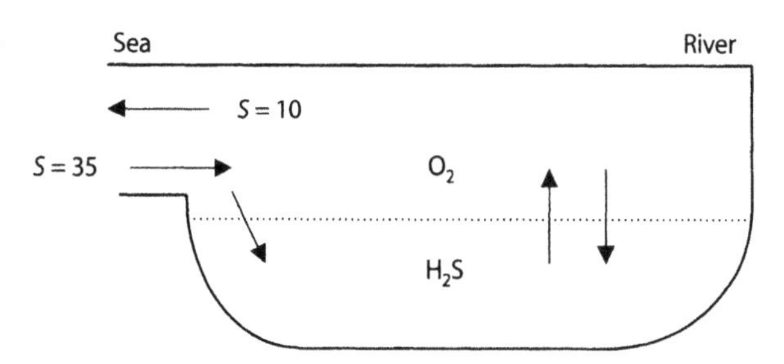

Fig. 4.1. Sketch of a typical anoxic basin

Table 4.1. Typical anoxic basins (Millero 1996)

Name	Location	Bottom (m)	Sill (m)	Maximum H_2S (μM)
Goteland Deep	Baltic Sea	249	60	20
Cariaco Trench	Caribbean Sea	1 390	150	160
Lake Nitinat	British Columbia	250	4	250
Saanich Inlet	British Columbia	236	65	250
Black Sea	Europe	2 243	40	350
Framvaren Fjord	Norway	350	2	6 000

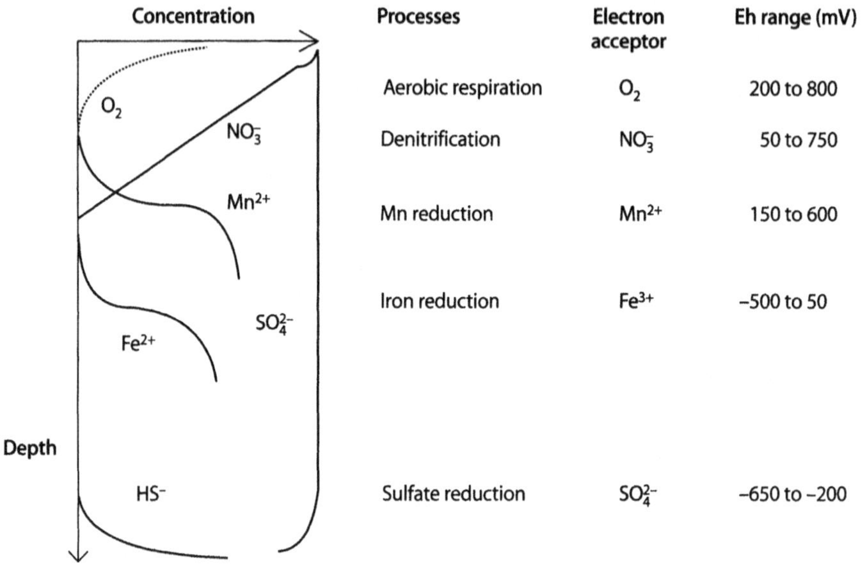
Concentration
Depth
O_2
NO_3^-
Mn^{2+}
SO_4^{2-}
Fe^{2+}
HS^-
Processes
Electron acceptor
Eh range (mV)
Aerobic respiration
O_2
200 to 800
Denitrification
NO_3^-
50 to 750
Mn reduction
Mn^{2+}
150 to 600
Iron reduction
Fe^{3+}
-500 to 50
Sulfate reduction
SO_4^{2-}
-650 to -200

Fig. 4.2. Schematic representation of chemical profiles in a stratified water column

The loss of oxygen in a basin or sediment pore waters causes a series of reactions to occur in a given sequence that is shown in Fig. 4.2. In an aerobic environment, O_2 is the electron acceptor and plant material can be oxidized by

$$\text{-NH}_3 + 0.5\,O_2 \longrightarrow NO_3^- + H_2O \tag{4.1}$$

$$\text{-CH}_2O + O_2 \longrightarrow CO_2 + H_2O \tag{4.2}$$

$$\text{-CH}_2\text{-} + 2\,O_2 \longrightarrow CO_2 + H_2O \tag{4.3}$$

The oxidation of typical plant material (Redfield et al. 1963) gives

$$(CH_2O)_{106}(NH_3)_{16}H_3PO_4 + 138\ O_2 \longrightarrow 106\ CO_2 + 122\ H_2O + 16\ HNO_3 + H_3PO_4 \quad (4.4)$$

After the oxygen is depleted, denitrification occurs with NO_3^- being the electron acceptor. The oxidation of plant material by nitrate to nitrogen is given by the following reactions:

$$\text{-}NH_3 + 0.6\ HNO_3 \longrightarrow 0.8\ N_2 + 0.8H_2O \quad (4.5)$$

$$\text{-}CH_2O + 0.8\ HNO_3 \longrightarrow CO_2 + 0.4\ N_2 + 1.4\ H_2O \quad (4.6)$$

$$\text{-}CH_2\text{-} + 12.8\ HNO_3 \longrightarrow CO_2 + 1.4\ N_2 + 1.4\ H_2O \quad (4.7)$$

The denitrification of typical plant material gives

$$(CH_2O)_{106}(NH_3)_{16}H_3PO_4 + 84.8\ HNO_3$$
$$\longrightarrow 106\ CO_2 + 148.4\ H_2O + 42.4\ N_2 + 16\ NH_3 + H_3PO_4 \quad (4.8)$$

and

$$(CH_2O)_{106}(NH_3)_{16}H_3PO_4 + 94.4\ HNO_3$$
$$\longrightarrow 106\ CO_2 + 177.2\ H_2O + 55.2\ N_2 + H_3PO_4 \quad (4.9)$$

In Reaction 4.8 the NH_3 produced from the plant material is not oxidized, while in Reaction 4.9 it is oxidized all the way to N_2 or N_2O.

MnO_2 reduction occurs next with Mn^{4+} being the electron acceptor, followed by NO_3^- reduction to NH_4^+ and reduction of Fe(III) to Fe(II). Since the concentrations of these electron acceptors are not very high, these processes do not normally dominate the oxidation of organic material. The next electron acceptor is SO_4^{2-} and sulfate reduction occurs according to

$$(CH_2O)_{106}(NH_3)_{16}H_3PO_4 + 53\ SO_4^{2-}$$
$$\longrightarrow 53\ CO_2 + 53\ HCO_3^- + 53\ HS^- + 16\ NH_3 + 53\ H_2O + H_3PO_4 \quad (4.10)$$

It should be noted that no oxidation of NH_3 occurs when the plant material is oxidized by SO_4^{2-} and the ratio of changes in H_2S to NH_3 is 3.3.

The decomposition can also occur with CO_2 being the electron acceptor and the resultant formation of CH_4:

$$(CH_2O)_{106}(NH_3)_{16}H_3PO_4 \longrightarrow 53\ CO_2 + 53\ CH_4 + 16\ NH_3 + H_3PO_4 \quad (4.11)$$

These equations can be used to predict the steady-state concentrations of C, N, P and S in anoxic systems. There are some deficiencies in this approach, because it does not account for mixing or the rates at which the system is driven to a steady state. According to the model, the carbon, nitrogen and phosphate are released in the atomic ratio of 106/16/1. There are a number of processes that can alter this ratio. Under aerobic conditions, the regenerated phosphate can be absorbed on iron and manganese (hydr)oxides. Denitrification, occurring at oxic/anoxic interfaces, can alter the ratio of combined nitrogen (nitrate + nitrite + ammonium) to carbon dioxide and phosphate by producing N_2O and N_2. It is also possible that the organic matter has a C/N/P ratio different than 106/16/1 when lipids rather than carbohydrates are formed or degraded. Phytoplankton can also incorporate C/N/P in ratios different than 106/16/1 depending on the availability of these elements. For example, the ratios of C/N/P for the production of diatoms during the IRONEX-II study (Steinberg et al. 1998) lead to a stoichiometry of $[(C_{90}H_{31}O_{90})(NH_3)_{14}(H_3PO_4)]$ for plant material, which gives the ratio for C/N/P of 90/14/1.

The decomposition of organic matter in anoxic basins based on a theoretical organic molecule having the composition $(CH_2O)_{106}(NH_3)_{16}(H_3PO_4)$, given by Reaction 4.10, can be used to examine the stoichiometry in these basins. According to this model during sulfate reduction, sulfide, ammonia, phosphate and total CO_2 should accumulate in a ratio of 53 moles of sulfide to 16 moles of ammonia, to one mole of phosphate and to 106 moles of total CO_2 in the anoxic water column. The changes in waters below the oxic/anoxic interface should be in the ratios

$$\Delta NH_3 / \Delta H_2S = 16 / 53 = 0.33$$

$$\Delta PO_4 / \Delta H_2S = 1 / 53 = 0.019$$

$$\Delta TCO_2 / \Delta H_2S = 106 / 53 = 2.0$$

The changes *TA* relative to H_2S are given by

$$\Delta TA / \Delta H_2S = (106 + 16 - 2) / 53 = 2.3$$

The changes between *TA* and TCO_2 are given by

$$\Delta TA / \Delta TCO_2 = (106 + 16 - 2) / 106 = 1.13$$

These relationships allow one to examine the chemistry of anoxic basins. Below we will examine the typical behavior of two anoxic basins: the Cariaco Trench and the Framvaren Fjord.

4.2 Cariaco Trench

The Cariaco Trench is located in the Caribbean Sea off the continental shelf north of Venezuela. It is about 200 km long and 50 km wide with a maximum depth of about 1 400 m. The basin is separated from the rest of the Caribbean Sea by a 146 m sill (Richards 1965). The surface waters of the Trench can exchange freely with the offshore water. A saddle at 900 m divides the Trench into two subbasins, the eastern and western basins. The vertical mixing below the mixed layer in the Cariaco Trench is prevented due to the presence of a strong pycnocline (density gradient). The salinity decreases with depth, and the stability of the basin water is caused by the temperature gradient. Coastal upwelling causes the high productivity in the surface waters and contributes the organic matter to the deep waters of the Cariaco Trench. The limited horizontal and vertical exchange limits the dissolved oxygen needed to oxidize organic carbon, and the deep waters have become anoxic. The sedimentary record suggests that anoxic conditions have prevailed in the deep waters for the last 1 100 years, probably punctuated by deep-water renewal (Richards 1965).

The Cariaco Trench was discovered in 1954 and has been studied over the last four decades. It has become increasingly clear that the anoxic portion of the Cariaco Trench is not in a steady state, and that the values of a number of properties have been changing systematically with time (Zhang and Millero 1993b). In June 1990 we participated in a cruise to the Trench and made measurements on a number of chemical parameters (Zhang and Millero 1993b). Temperature and salinity data obtained for both basins of the Cariaco Trench are shown in Fig. 4.3. The maximum temperature occurs at the surface and decreases smoothly with depth approaching a nearly constant value of about 17.2 °C below 600 m. The temperature below 200 m shows no significant differences between the two basins. The maximum salinities occurred at 60 m in both basins, 36.86 in the western basin and 36.90 in the eastern basin. These maxima disappeared during the dry season (January to May), probably due to mixing brought about by the stronger winds and lower air temperatures typical of that season. Below the maxima, the salinities decrease smoothly with depth, and become almost constant at about 36.21 below 500 m in both basins.

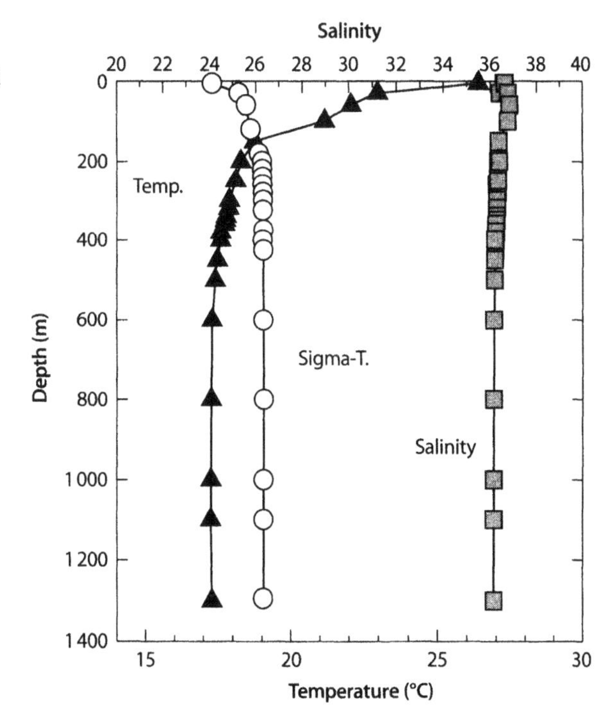

Fig. 4.3. Salinity and temperature and sigma-T, $\sigma_T = 10^3$ (ρ = density) in the eastern and western basins of the Cariaco Trench (Zhang and Millero 1993b)

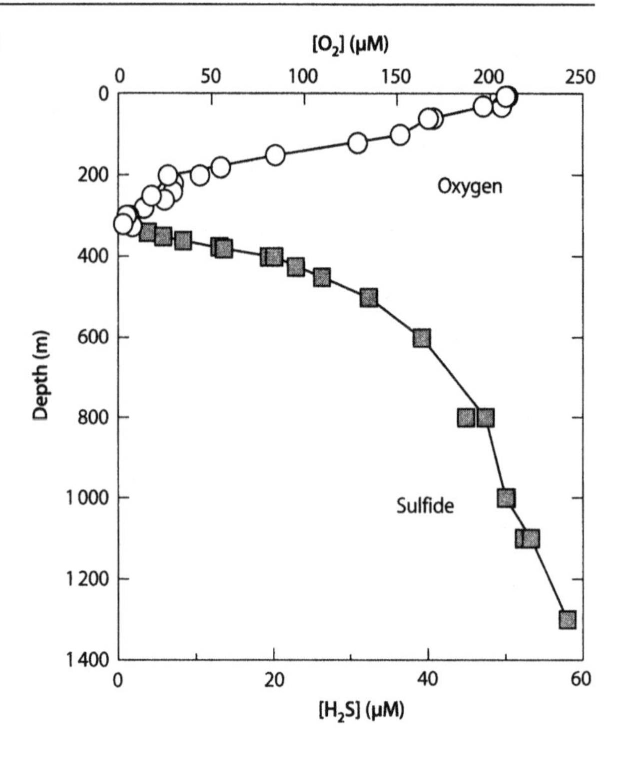

Fig. 4.4. Profiles of oxygen and hydrogen sulfide in the basins of the Cariaco Trench (Zhang and Millero 1993b)

Profiles of oxygen and hydrogen sulfide are shown in Fig. 4.4 for the two basins. The maximum oxygen occurred at the surface in both basins (211 μM in the western basin and 208 μM in the eastern basin). The oxygen concentrations decreased sharply with depth and became undetectable at 330 m where hydrogen sulfide appeared in both basins. The concentrations of hydrogen sulfide increased rapidly below the interface to maximum values of 58 μM in the eastern basin and 55 μM in the western basin at 1300 m. The depth of the oxic/anoxic interface in the Cariaco Trench has fluctuated with time since the first expedition in 1956 between 250 to 375 m. The depth of the interface is controlled by changes in the density or the flux of organic matter, which is produced by photosynthesis in the surface water and sinks into the deep-water column. A shallower oxic/anoxic interface is found during the upwelling season due to the high productivity in the surface water. The concentrations of ammonia, phosphate and silicate in the eastern basin are shown in Fig. 4.5. In surface waters about 1 μM of ammonia was present in the euphotic zone at 30 m and was undetectable below 60 m. The ammonia increased rapidly below the oxic/anoxic interface to 20.2 μM at 1300 m. The concentrations of phosphate in the surface waters were below 0.4 μM, increased rapidly with depth to 2.3 μM at the oxic/anoxic interface, and reached a value of 3.7 μM at 1300 m. The concentrations of silicate in the surface waters were less then 2 μM, increased rapidly with depth to about 42 μM at oxic/anoxic interface, and reached values of 86 μM at 1300 m.

The profiles of sulfite and thiosulfate are presented in Fig. 4.6. The levels of sulfite and thiosulfate were undetectable in the oxic waters, but were found in μM levels in the deep waters (1.8 μM at about 600 m for sulfite and 1.0 μM for thiosulfate). The sulfite

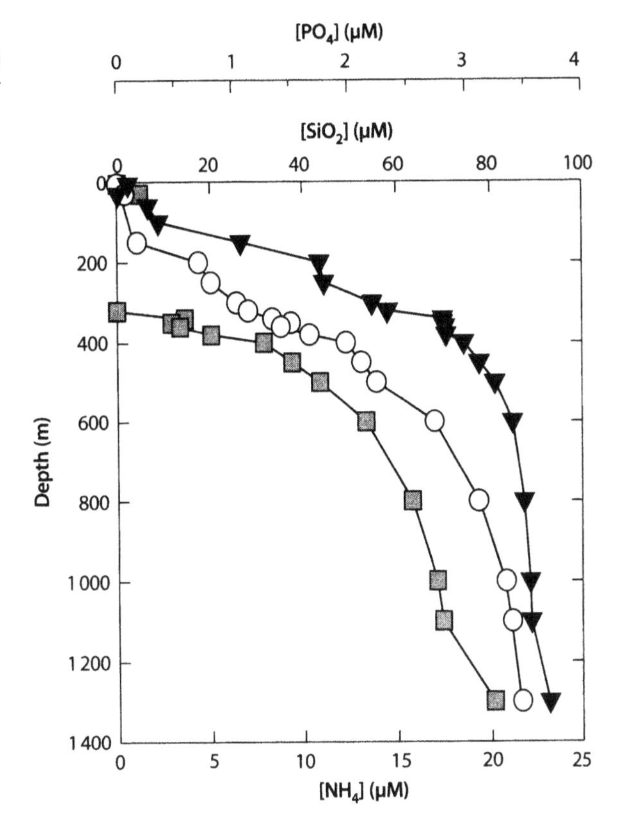

Fig. 4.5. Profiles of ammonia (*squares*), phosphate (*triangles*) and silicate (*circles*) in the Cariaco Trench (Zhang and Millero 1993b)

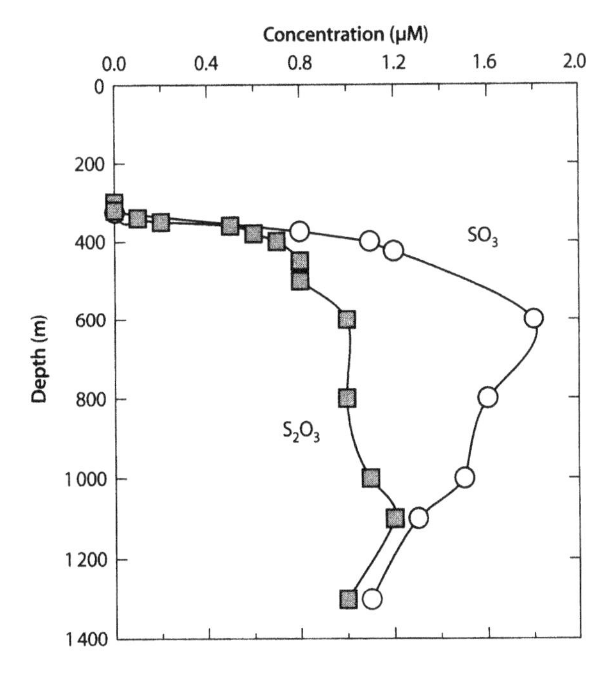

Fig. 4.6. Profiles of sulfite and thiosulfate in basins of the Cariaco Trench (Zhang and Millero 1993b)

and thiosulfate are due to the oxidation of H_2S from oxygenated waters that periodically sink below the oxic/anoxic interface. Tritium ($t_{1/2} = 12$ yr) found in the bottom waters supports the idea that the deep waters have been renewed. The thiosulfate increased gradually to the bottom and showed no maximum near 600 m. The higher values in the deep waters suggest a sediment source for the thiosulfate. A concentration of 178 μM thiosulfate was found in the pore waters of the sediments in the Cariaco Trench, but this may be the result of oxidation of the sediments during collection. The formation of thiosulfate in the sediments could result from the reaction of MnO_2 and FeOOH minerals with H_2S or possibly due the reduction of sulfate.

The pH, *TA* and TCO_2 measurements made during the cruise are shown in Fig. 4.7. The maximum value of pH (8.26–8.30) occurred between 5 to 30 m depth in both basins. The pH decreased to 7.9 below the oxic/anoxic interface. The minimum value of 2.05 mM appeared at the surface and the value became constant below 1 000 m (2.40 mM). The total alkalinity was calculated based on the known pH and total CO_2. The *TA* maximum in the surface water at a depth of 60 m is consistent with the salinity maximum. The salinity corrected *TA* ($NTA = TA \times 35 / S$) does not show this effect. The total alkalinity increased with depth below the oxic/anoxic interface due to the bacterial anaerobic respiration of organic matter to bicarbonate and simultaneously reduction of sulfate to sulfide.

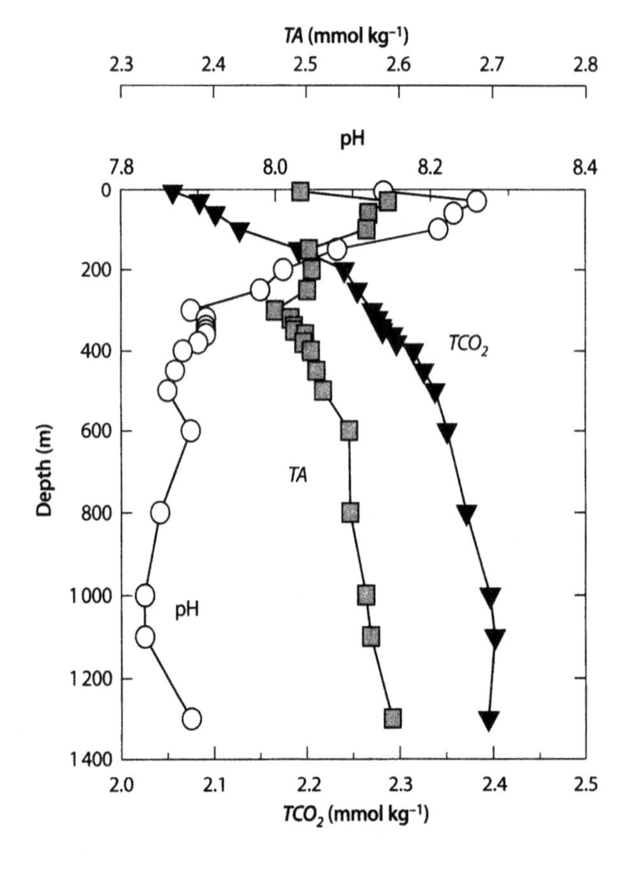

Fig. 4.7. Profiles of the pH, TCO_2, and *TA* in the Cariaco Trench (Zhang and Millero 1993b)

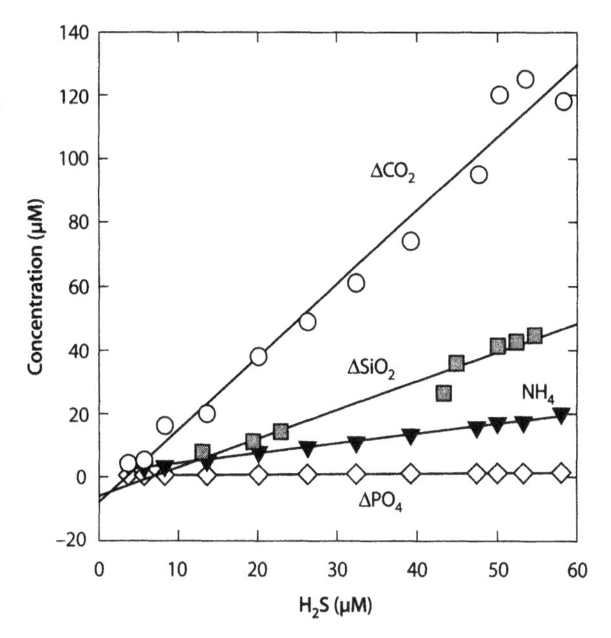

Fig. 4.8. Plots of the changes in concentration of ammonia, phosphate, silicate, and total carbon dioxide in anoxic waters vs. the concentration of H_2S (Zhang and Millero 1993b)

Plots of the concentration of NH_4^+, PO_4^{3-} and ΔTCO_2 vs. H_2S in the anoxic waters of the Cariaco Trench are shown in Fig. 4.8. Good linear correlations were found with slopes of 0.33 ±0.07 for NH_4^+, 0.0191 ±0.0014 for PO_4^{3-}, and 2.25 ±0.10 for TCO_2. These slopes are in good agreement with the theoretical values (0.30 for NH_4^+, 0.0189 for PO_4^{3-}, and 2.00 for TCO_2). Corrections for total sulfur to include sulfite and thiosulfate slightly improve the values of the ratios (0.31 ±0.06 for NH_4^+, 0.0182 ±0.0014 and 2.01 ±0.07 for total CO_2). Since decreases in the concentration of sulfide could also occur due to reactions with MnO_2 and FeOOH (Yao and Millero 1993, 1995a, 1996), we cannot rule out that this effect may be responsible for the slightly larger slopes found without any the corrections.

Scranton et al. (1987) have suggested that the physical and chemical properties of the deep waters of the Cariaco Trench are increasing. For example, there is a significant increase in the temperature of the waters (>1000 m) over the 35 years since the first expedition in 1955 (see Fig. 4.9). Between 1955 and 1982 an average increase of 0.006 °C yr^{-1} occurred, while a much faster rate of increase, 0.028 °C yr^{-1}, occurred in last eight years. Further measurements are needed to verify the recent increase in temperature of the deep basin. The concentrations of hydrogen sulfide in deep waters of the Cariaco Trench have increased with time as suggested by Richards (1965) and Scranton et al. (1987). The increase (Fig. 4.10) in the concentration at 1300 m between 1982 and 1990 (1.58 μM yr^{-1}) is consistent with the increases observed between 1965 and 1982 (1.42 μM yr^{-1}). The average rate of increase over last the 40 years is 1.11 μM yr^{-1}. Salinity variations in the deep portion of the Cariaco Trench are harder to trace than the temperature variations. Scranton et al. (1987) found that the salinity increased 0.008 over a period of 9 years. This variation is small, relative to the analytical error, and it is hard to confirm the increasing pattern. The continuous decrease of the salinity with depth in our data suggests that the upward salt flux from the sediments is insignificant.

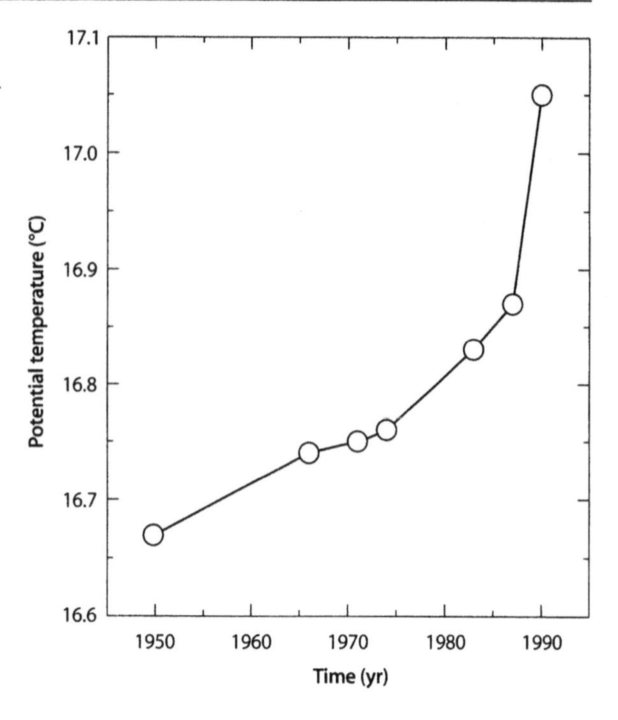

Fig. 4.9. The temporal changes of potential temperature of water below 1200 m in the Cariaco Trench (Zhang and Millero 1993b)

The concentrations of ammonia, phosphate and silicate in the deep waters of the Cariaco Trench have also increased with time since the first measurements were made in the 1950s (see Fig. 4.10). The rates of increase are 0.282 $\mu M\ yr^{-1}$ for ammonia, 0.0372 $\mu M\ yr^{-1}$ for phosphate and 0.854 $\mu M\ yr^{-1}$ for silicate. These increases are consistent with the increase of H_2S with time in the deep waters of the Cariaco Trench. If these compounds behave conservatively they should have accumulated in the anoxic water column over the period in fixed ratios as predicted. The ratio of the rate increase for ammonia to that of sulfide is equal to 0.25, which is close to the value of 0.30 predicted from the model. The ratio of the rate of increase for phosphate to sulfide is 0.034, while the theoretical value is 0.019. The larger increase for phosphate may be due to another source such as the incorporation of PO_4^{3-} on Fe and Mn oxides that sink below the interface. The dissolution of phosphate minerals in the anoxic water or diffusion from sediment's pore waters might also provide the input of the phosphate in the anoxic waters. The flux of silicate from the pore water of the sediment into the water column has been suggested as the cause of the high concentrations of silicate in the deep waters of the Cariaco Trench (Fanning and Pilson 1972; Scranton et al. 1987).

By extrapolating the concentrations of H_2S and NH_4^+ to zero, it is possible to estimate the last time that the trench was oxic. The earlier rate of increase (1955 to 1969) in the H_2S with time extrapolated to zero gives a date of 1916. A similar extrapolation for NH_4^+ gives a date of 1914. These estimations, from two independent chemical compounds, are in good agreement for the last occurrence of a complete turnover of the trench waters. It is interesting to note that the ^{210}Pb dating of the sediments indicates that some disturbance of the sediment water interface occurred around 1932 and 1897 apparently due to earthquakes around 1900 and 1929. These events could also have

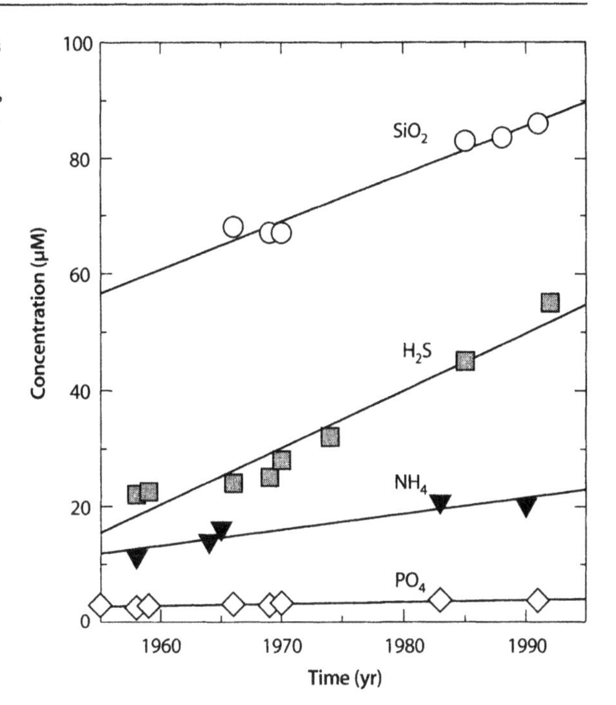

Fig. 4.10. The temporal changes of the maximum concentration of ammonia, phosphate, silicate, and H_2S in water below 1200 m in the Cariaco Trench (Zhang and Millero 1993b)

influenced the turnover of the waters in the trench that we estimate occurred around 1920. The concentrations of PO_4^{3-} and SiO_2 (1 and 2 μM, respectively), at the time of turnover are in good agreement with the concentration of these compounds in the Caribbean Sea (e.g. silica is 2.5 μM in the deep Caribbean Sea).

4.3 Framvaren Fjord

The Framvaren Fjord, located in southern Norway, is a permanent anoxic fjord and has the highest levels of H_2S (6 μM) reported for an open anoxic basin. The fjord has a shallow sill of 2 m that separates the outer water (80 m) from the basin that is 183 m deep. A major river entering the outer fjord causes a flow of fresh water through the canal into the Framvaren. Due to the unusual chemical and microbiological properties, the Framvaren has been widely used as a natural laboratory to study anaerobic processes since the 1930s. Earlier studies on the hydrography, currents, trace metals, isotopes, microbiology, sedimentation and seismology of the fjord have been reviewed (see reference in Yao and Millero 1995b). Unlike most other anoxic basins, the O_2/H_2S interface (~18 m) in the Framvaren is at a depth of significant light penetration. Dense populations of photosynthetic bacteria are present at the redox boundary near the interface. There is also a concentration of particles at the interface as shown by the vertical distribution of transmission of light. This high biological activity can effectively control the biogeochemistry of metals and non-metals. The increase in the particles in the deep waters is related to the formation of metal sulfides and polysulfides. Framboidal pyrite (FeS_2) has been found in the deep waters.

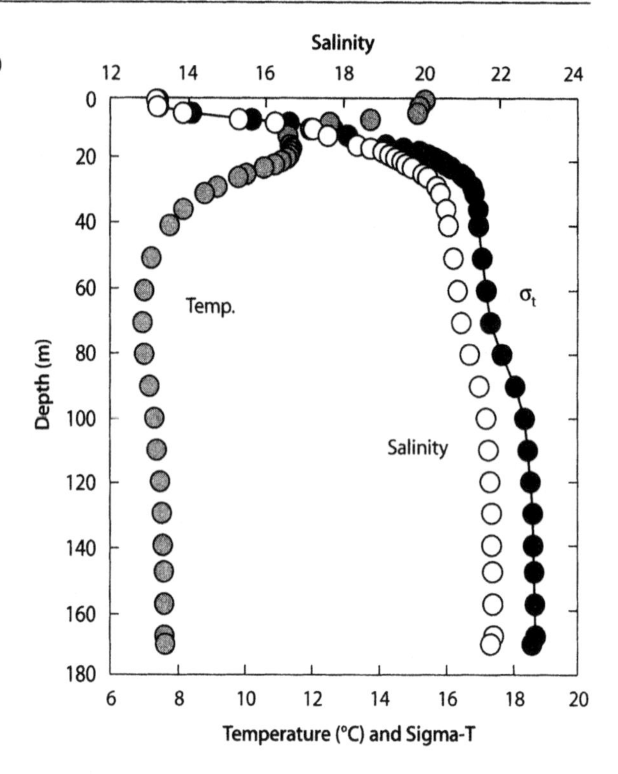

Fig. 4.11. Profile of salinity, temperature, and sigma-T (σ_T) in the Framvaren Fjord (Yao and Millero 1995b)

In 1993 we made some chemical studies of the Framvaren Fjord (Yao and Millero 1995b). The large salinity gradient (Fig. 4.11) accounts for the large density gradient (pycnocline) that separates the surface and deep water. The surface waters have a salinity of 12, while the deep waters have salinities as high as 24. The salinity gradient prevents vertical mixing and the formation of H_2S below a depth of about 18 m. The temperatures of the surface waters (Fig. 4.11) fluctuate from 0 °C in the winter to 19 °C in the summer. The deep waters have a uniform temperature of 7–8 °C. The water masses can be divided into four major layers:

1. the low salinity surface layer (0–2 m) above the sill depth
2. the intermediate oxygenated layer down to ~18 m
3. the deep water where steep gradients in the chemistry occur (18–90 m)
4. the bottom water below 90 m, where changes in salinity and chemistry are small

The deep basin has been anoxic for about 8 000 years. The sill was dredged in 1850, resulting in the formation of a new layer of H_2S. A vertical section of O_2 and H_2S in the central basin is shown in Fig. 4.12. A surface maximum is observed in O_2 due to the photosynthesis of phytoplankton. The O_2 goes to zero at about 18 m and the H_2S increases to concentrations as high as 6 mM or 6 000 μM. These values are the highest levels of H_2S found in any anoxic basin. The dissolved oxygen in the surface water was 283 μM, which is close to the saturated value (285 μM). The maximum value 292 μM in the subsurface (6 m) is due to photosynthesis. The oxygen concentration decreased

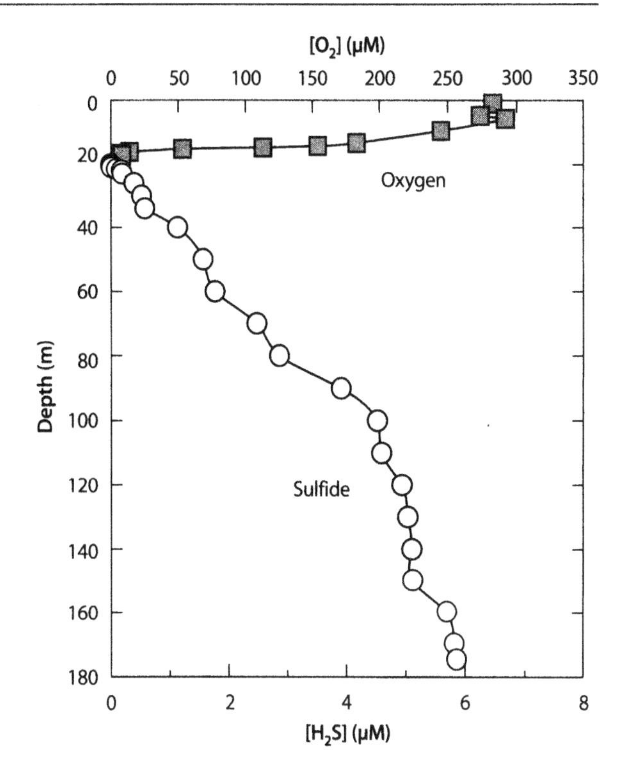

Fig. 4.12. Profiles of oxygen and hydrogen sulfide in the Framvaren Fjord (Yao and Millero 1995b)

sharply from 15 m and became undetectable below 18 m. The oxic/anoxic interface has been relatively stable during the last 20 years. Small fluctuations, however, are expected to occur due to changes in the exchange of water with outside basin and internal waves. The concentrations of H_2S in the bottom water were found to be as high as 5.8 mM. The gradient of H_2S in the bottom water (below 100 m) is much smaller than that in the deep water (20–100 m) (Millero 1991b). This may be due to the separation of the older water and more recent anoxic water. The distributions of nutrients, *TA* and TCO_2 (shown below) show the same pattern.

The concentration of metals in the Framvaren determined by Haraldsson and Westerlund (1988) are shown in Fig. 4.13a,b. The profiles are similar to those found in other anoxic basins. The metals Mn^{2+}, Fe^{2+}, Co^{2+} have the characteristic maximum near the oxic/anoxic interface; Ni^{2+} shows no changes across the interface; and the metals Cu^{2+}, Zn^{2+} and Cd^{2+} show large decreases in the concentration at the interface due to the low solubility of their metal sulfides. The oxidation of H_2S with O_2 in the Framvaren has been shown to be greatly enhanced by the high content of Fe and Mn (Yao and Millero 1995b). Only low levels of the intermediates SO_3^{2-} and $S_2O_3^{2-}$ were found near the interface.

The dissolved Mn and Fe together with O_2 and H_2S near the interface are shown in Fig. 4.14 (Yao and Millero 1995b). The concentration of dissolved Mn increases rapidly below 15 m, corresponding to the rapid decrease of O_2 and reaches a maximum at 21 m where the concentration of H_2S was 12 μM. The maximum dissolved Mn found in this study, 18.0 μM, is higher than previous values of 10.5 μM (Jacobs et al. 1985) and 15.3 μM

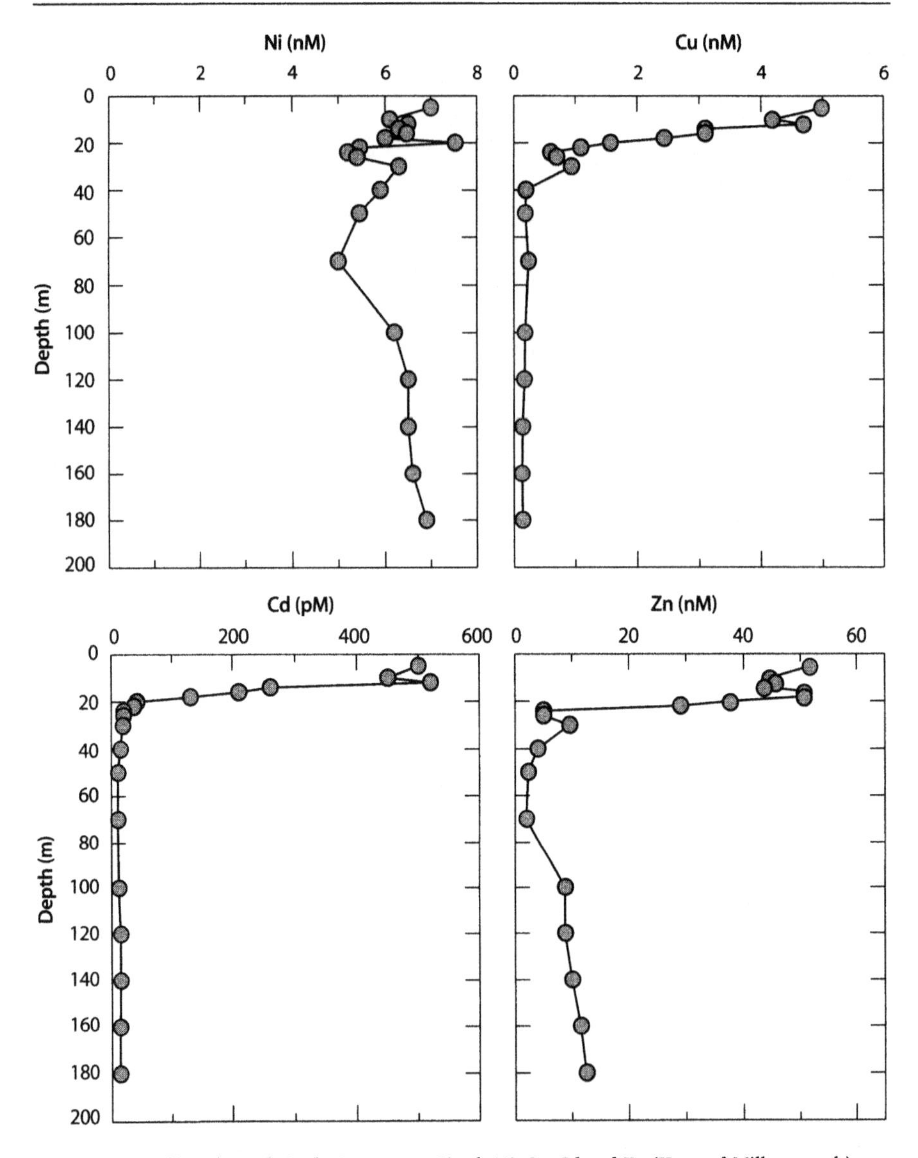

Fig. 4.13 a. Profiles of metals in the Framvaren Fjord: Ni, Cu, Cd and Zn (Yao and Millero 1995b)

(Haraldsson and Westerlund 1988). Our higher value might be due to the better sampling resolution with the submersible pumping system. It might also reflect the continuing accumulation of Mn in the transition zone. The concentrations of dissolved Mn decreased rapidly below the maximum peak at 21 m and reached nearly constant values (less than 1 μM) below 100 m.

The concentrations of dissolved Fe started increasing just below the O_2/H_2S interface (Fig. 4.14) to a sharp maximum at 21 m. The maximum value found in this study, 2.85 μM, is higher than earlier values of 2.0 μM (Haraldsson and Westerlund 1988). Again,

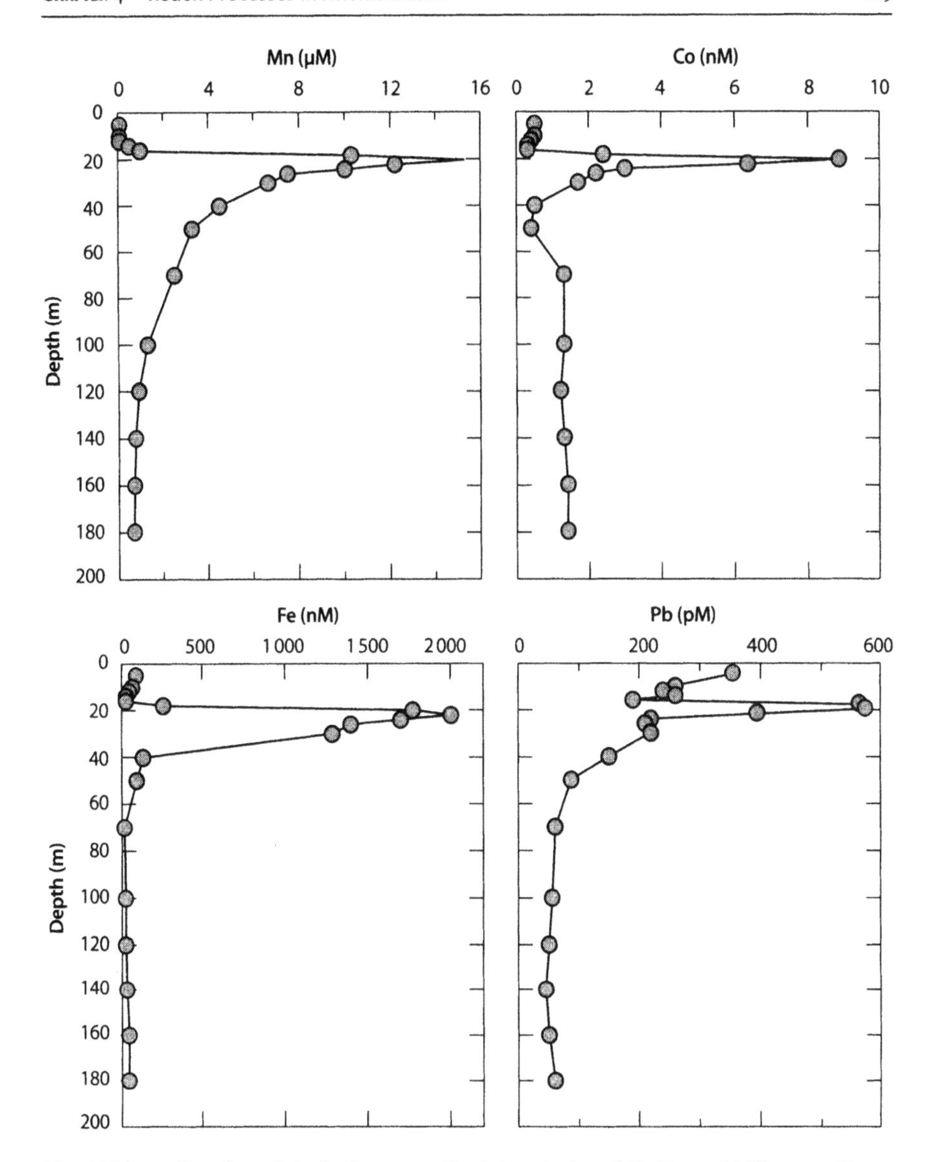

Fig. 4.13 b. Profiles of metals in the Framvaren Fjord: Mn, Co, Fe and Pb (Yao and Millero 1995b)

these differences could be caused by various sampling depth and/or fluctuation of Fe(II) concentration with time. The lower value of 0.89 μM found in the winter might result from the oxidation of Fe(II) near the interface by O_2 due to the enhanced water mixing in this season. The concentrations of dissolved Fe(II) decreased rapidly below the maximum to less than 40 nM in the bottom water, probably due to the formation of iron sulfides.

The concentration of phosphate in the surface water was about 0.4 μM and reached a minimum (0.18 μM) between 16–18 m, which may be due to the high productivity at this region or absorption onto Mn and Fe particles. The maximum values of particu-

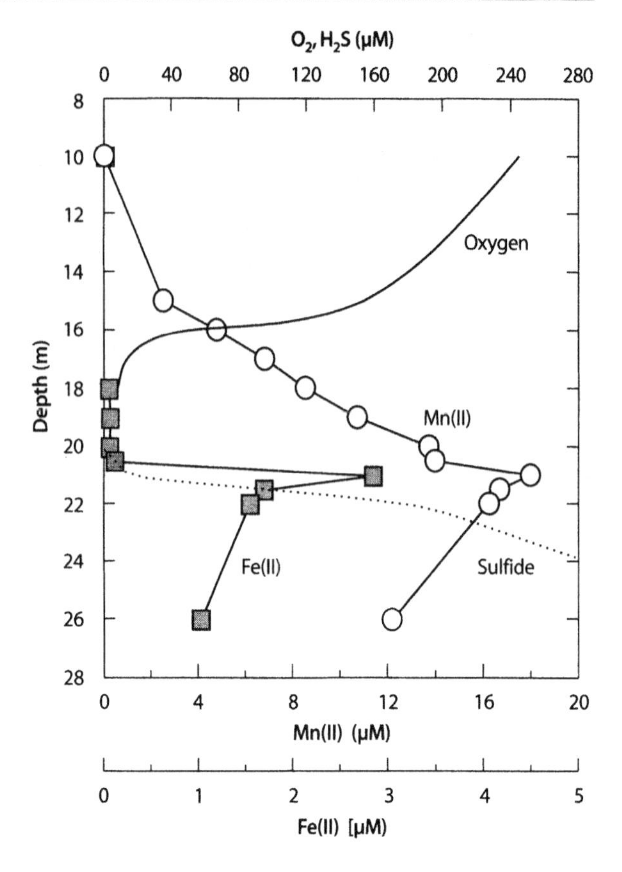

Fig. 4.14. The concentrations of oxygen, hydrogen sulfide, Mn(II) and Fe(II) near the interface in the Framvaren Fjord (Yao and Millero 1995b)

late Mn and Fe were found at or above the O_2/H_2S interface. The PO_4^{3-} increased rapidly below the interface and reached a maximum of 100–102 μM in the bottom water (Fig. 4.15). Ammonia was present in the oxic euphotic zone at about 5.0 μM. It increased rapidly below the interface and reached a maximum of 1.6 mM in the bottom waters (Fig. 4.15). Relatively high (~20 μM) silicate was found in the oxic waters and the concentrations increased rapidly below the interface to a maximum of 640 μM in the bottom water (Fig. 4.15). The distributions of pH, *TA* and TCO_2 are plotted in Fig. 4.16. The pH in the surface waters was about 7.89 and decreased to a minimum (6.98) at 15 m. The low pH at 15 m is difficult to explain, although it agrees with the calculated value using *TA* and TCO_2. The pH decreased below the interface and was constant below 90 m (about 6.90). The *TA* increased with depth from the surface to the interface corresponding with the increase in salinity. The average normalized (to $S = 35$) *TA* was 2.41 mM in the surface water, which is slightly higher than the value in ocean surface water (about 2.35), probably due to the impact of fresh water with high *TA*. Below the interface the *TA* increased rapidly due to the bacteria anaerobic respiration of organic matter to bicarbonate and simultaneously reduction of sulfate to hydrogen sulfide. The *TA* was found to be 19.8 mM in the bottom water. The TCO_2 in the water samples was calculated from the pH and *TA* (Fig. 4.16). The TCO_2 increased rapidly below the interface due to the oxidation of organic matter to inorganic carbon.

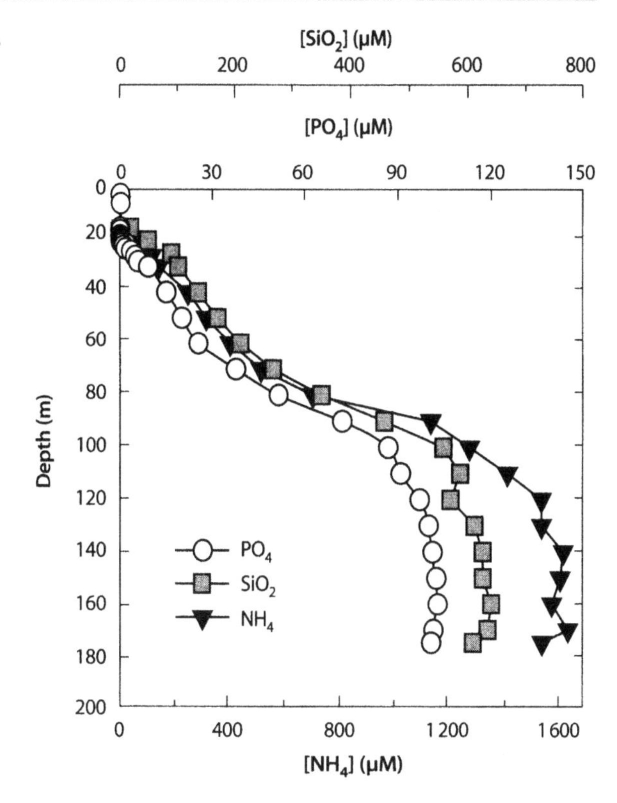

Fig. 4.15. Profiles of ammonia, phosphate, and silicate in the Framvaren Fjord (Yao and Millero 1995b)

At present, it is difficult to balance the sulfur budget in the Framvaren due to the incomplete measurements of all the sulfur species (especially the particulate sulfur, S^0 and FeS_2) and organic sulfur compounds. The ratio of thiols (mercaptants) RSH to H_2S was found to be 0.036 in the Framvaren (Dyrssen 1988). The difference between the initial sulfate concentration, $[SO_4]_{int}$, as calculated from salinity, and the measured sulfate concentration is found to be significantly higher (1–3 mM) than the measured sulfide in bottom water. This discrepancy is explained by the removal of sulfide by oxidation and the formation of iron sulfides (FeS and FeS_2), in addition to the possible diffusion of H_2S. If the major product of oxidation of H_2S is sulfate (Millero 1991b), the discrepancy may be even higher. The removal of sulfide by the formation of iron sulfides is restricted to a continuing supply of iron. At pH = 7.0, about 30% of the total hydrogen sulfide is in $H_2S_{(aq)}$. Diffusion might be a pathway for H_2S to escape into upper oxic water and be oxidized. In addition, there is evidence for the early diagenetic incorporation of inorganic sulfur into organic compounds.

The normalized *TA* ($NTA = TA \times 35 / S$) and normalized TCO_2 ($NTCO_2 = TCO_2 \times 35 / S$) in the bottom waters of the Framvaren are about 12 times more than in the open ocean, suggesting an enormous production of both *TA* and TCO_2 from oxidation of organic matter through the reduction of sulfate. The value of *NTA* in the Framvaren surface water was 2.41 mM, which is only slightly higher than the typical value of the open ocean (2.35 mM), while the $NTCO_2$ is much higher (2.31 mM) than the value of the open ocean (2.05 mM). The lower *NTA* in the Framvaren surface water indicates that

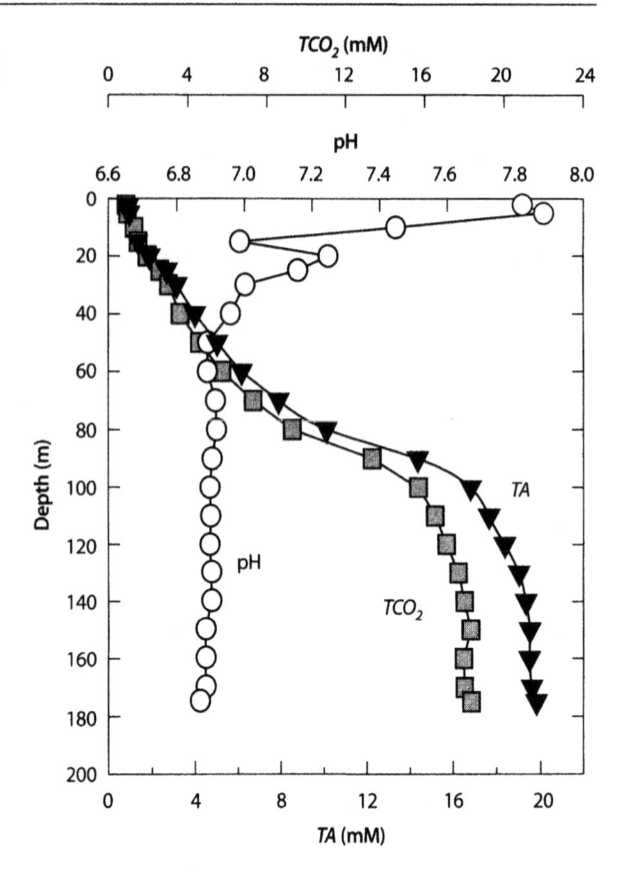

Fig. 4.16. Profiles of pH, TCO_2, and TA in the Framvaren Fjord (Yao and Millero 1995b)

the impact of fresh water is minor. The cause of the higher values of $NTCO_2$ in the surface waters may be due to the oxidation of organic matter in oxic water or diffusion from below the interface.

Plots of ΔTA and ΔTCO_2 vs. H_2S are shown in Fig. 4.17. The linear correlation between these properties demonstrates that hydrogen sulfide is produced proportionately to the formation of both TCO_2 and TA. However, the values of $\Delta TA / \Delta H_2S = 3.42 \pm 0.1$ and $\Delta TCO_2 / \Delta H_2S = 3.00 \pm 0.12$ are both higher than the model values. Based on previous data, a value of $\Delta TCO_2 / \Delta H_2S = 2.54$ was obtained by Dyrssen (1988). The ratio of ΔTA to ΔTCO_2 in anoxic waters of the Framvaren is 1.14 and equal to the model value. Dyrssen (1988) found a similar value of $\Delta TA / \Delta TCO_2$ (1.15). The differences between the observed ratios and the model values may be attributed to a combination of the high values of TA and TCO_2 and the low values of H_2S. The low H_2S concentration could be due the formation of Fe sulfides

$$2\,FeOOH + 3\,H_2S \longrightarrow FeS + FeS_2 + 4\,H_2O \qquad (4.12)$$

without changing TA and TCO_2. The inflow of O_2 containing water through the sill may cause the removal of H_2S in the deep waters of the Framvaren, such as the renewal of the bottom waters when the channel was dug in 1850. In order to keep the same devia-

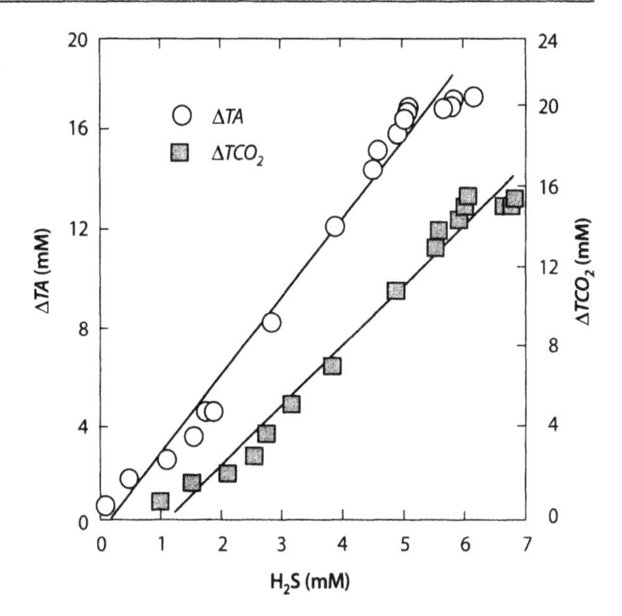

Fig. 4.17. Plots of the changes of TCO_2 and TA vs. the concentration of H_2S in the anoxic waters of the Framvaren Fjord (Yao and Millero 1995b)

tions in the ratios TA/H_2S and TCO_2/H_2S, the removal process of H_2S should not change the TA. In other words, sulfide should be removed from the water column as FeS_2 and elemental sulfur or polysulfides

$$2\,H_2S + O_2 = 2\,S^0 + 2\,H_2O \tag{4.13}$$

$$(x-1)S^0 + HS^- = HS_X^- \tag{4.14}$$

again, without altering both TA and TCO_2. If the product is SO_4^{2-} instead of S^0, the alkalinity will be lowered. Millero et al. (1987a) found that the main reaction product between H_2S and O_2 was SO_4^{2-} when O_2 was in excess. The calculated p_{CO_2} in the surface water in June 1993 is 250 μatm.

Ammonium increased below the interface due to the anaerobic decomposition of organic matter. Based on the model, a plot of NH_4^+ vs. H_2S should give a slope of 0.30. Our experimental result gives the same value 0.30 (±0.01) for all data from 20–175 m (Fig. 4.18). A lower value of $NH_4^+ / H_2S = 0.25$ (±0.02) is found for data from 30–80 m. The ratio of NH_4^+ to H_2S in Cariaco Trench (0.31, Zhang and Millero 1993b) agreed with the model value while it is lower (0.23) in the Black Sea (Table 4.2). Due to the possible removal of H_2S as discussed earlier, the real ratio of NH_4^+ to H_2S must be lower than the model value. One explanation for this is that the C/N ratio in organic matter is higher than 106/16.

The phosphorus transformations unlike nitrogen are simple because of its single oxidation state. Phosphate, however, can absorb onto particles such as FeOOH and MnO_2. The rapid increase of phosphate below the interface also results from the anaerobic respiration of organic matter. The ratio of PO_4^{3-} to H_2S should be 0.019. Our experimental result gives exactly the same value 0.0191 (±0.0004) for all data from 20–175 m (Fig. 4.18). A similar value of 0.0188 (±0.0021) is found for 30–80 m. The ratio of PO_4^{3-}

Fig. 4.18. Plots of the changes in ammonia, phosphate, and silicate vs. the concentration of H_2S in the anoxic waters of the Framvaren Fjord (Yao and Millero 1995b)

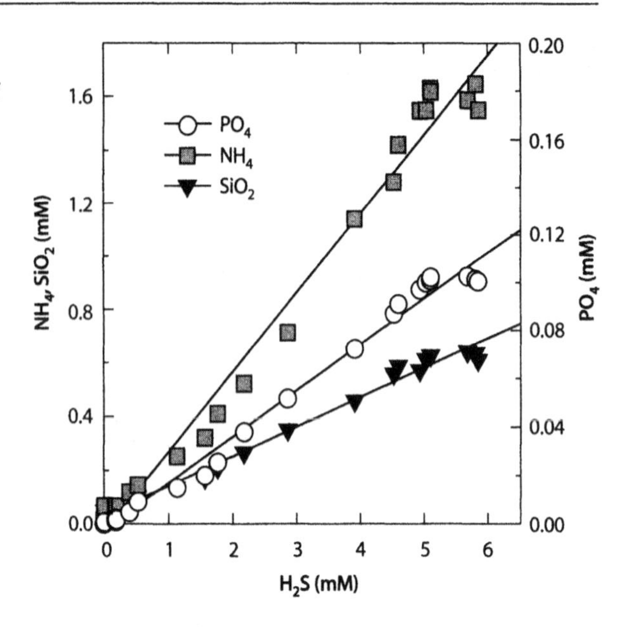

Table 4.2. The stoichiometric ratios in anoxic waters of different basins (Millero 1996)

Basin	TCO_2/H_2S	TA/H_2S	PO_4/H_2S	NH_4/H_2S	Si/H_2S	C/N/P
Framvaren	3.00	3.42	0.019	0.30	0.11	155/16/1
Black Sea	2.30	3.05	0.009	0.23	0.61	255/25/1
Cariaco Trench	2.01	2.43	0.018	0.31	0.91	112/17/1
Model Value	2.00	2.30	0.019	0.30	–	106/16/1

to H_2S in Cariaco Trench (0.018, Zhang and Millero 1993b) agreed fairly well with the predicted value, while it was only half of the value (0.009) in the Black Sea. This low recycling efficiency of PO_4^{3-} in the Black Sea might be due to the removal of PO_4^{3-} by absorption onto particles. Again, the good agreement between the measured ratio of PO_4^{3-} to H_2S in the Framvaren with the model value might be incidental when we consider the possible removal of H_2S produced by sulfate reduction.

The linear correlation between sulfide and silicate in the anoxic waters of the Framvaren gives a slope of 0.11 (±0.01), indicating that silicate is also released proportionately to the production of hydrogen sulfide. There is no theoretical ratio of Si to H_2S from the oxidation of biogenic organic matter, because it depends on the relative abundance of siliceous phytoplankton (e.g. diatoms). The ratios are 0.61 and 0.91 for the Black Sea and Cariaco Trench, respectively. Silicate in the Framvaren is relatively low with respect to the sulfide level. At present little is known about the effect of anoxic environments on the solubilization of silicate.

The linear correlation between concentrations of NH_4^+ and concentrations of PO_4^{3-} for samples collected below the oxic layer give a slope N/P of 16.0 (±0.3), which is the same as the Redfield ratio. A plot of concentrations of <TCO_2 vs. NH_4^+ in the deep anoxic waters gives a slope C/N of 9.6 (±0.2), which is higher than the Redfield value of 6.6. A similar value of C/N = 10.0 (±0.6) is found for 30–80 m (Fig. 4.19). Sediment values give a ratio of C/N of 8. The particulate C/N ratios in the surface waters are higher (10). A plot of concentrations of TCO_2 vs. PO_4^{3-} in the anoxic waters (20–175 m) of the Framvaren shows a slope C/P of 155 (±2) which is much higher than the Redfield ratio of 106, and also higher than the value of 127 for ocean waters. These results indicate that the nutrients (N, P) show Redfield behavior in the Framvaren while ratios of C/N and C/P are higher. Dyrssen (1988) suggested that the higher values could be related to carbohydrates from tree leaves and other terrestrial matter. It has also been proposed that phytoplankton can incorporate C/N/P in ratios different than 106/16/1 depending on the availability of these elements. The high C/N and C/P ratios might result from the excess supply of inorganic carbon compared to N and P.

If we assume the C/N/P ratio of the organic matter in the Framvaren to be 155/16/1 based on the average value for all the data (20–175 m), Reaction 4.10 can be rewritten as

$$(CH_2O)_{155}(NH_3)_{16}(H_3PO_4) + 77.5\ SO_4^{2-} \longrightarrow 155\ HCO_3^- + 77.5\ H_2S + 16\ NH_3 + H_3PO_4 \tag{4.15}$$

The increase of the C/N and C/P ratios does not change the value of 2.0 for $\Delta TCO_2/\Delta H_2S$, while the ratio of $\Delta TA/\Delta H_2S$ decreases slightly from 2.3 to 2.2. The ratios of NH_4^+ and PO_4^{3-} to H_2S both decrease from 0.30 to 0.20 and 0.019 to 0.013, respectively.

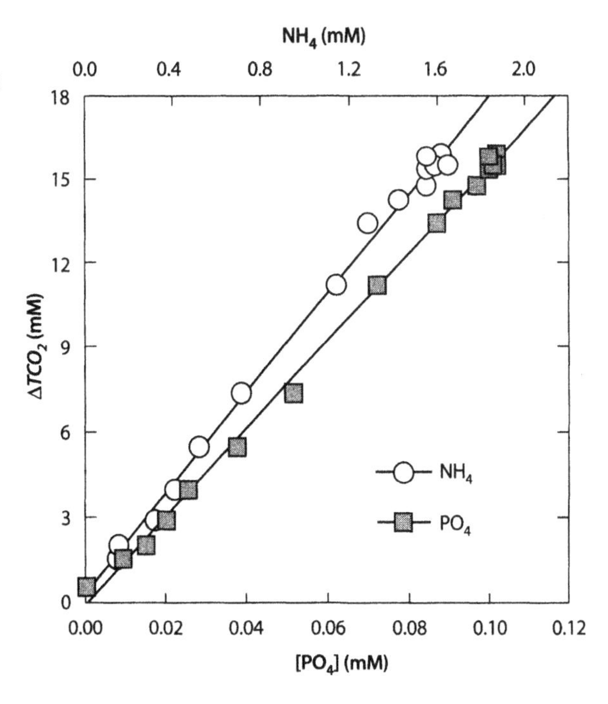

Fig. 4.19. Plots of the changes in TCO_2 vs. the changes in the concentrations of ammonia and phosphate (Yao and Millero 1995b)

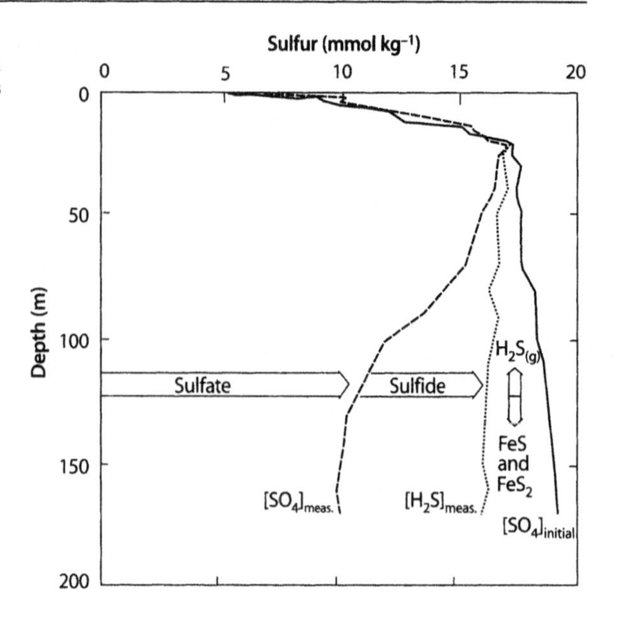

Fig. 4.20. Plots of the various sulfur species in the Framvaren Fjord. The initial concentrations are calculated from the salinity using SO_4 / S = 28.28 / 35 mM (Anderson et al. 1988)

If one further assumes (Yao and Millero 1995b) that total inorganic carbon, ammonia and phosphate in the anoxic waters of the Framvaren are produced only through the decomposition of organic matter based on Reaction 4.9 and there are no significant removals for them. This assumption is supported by the fact that the correlation between C, N and P is better than the correlation between C, N, P with H_2S. In order to rebalance the measured NH_4^+/H_2S and PO_4^{3-}/H_2S ratios (0.30 and 0.019, respectively) to the new model values (0.20 and 0.013, respectively), the initially formed H_2S should equal to $[H_2S]_{initi} = (77.5 / 53) \times [H_2S]_{meas.}$. In other words, about 30% of the H_2S produced from sulfate reduction has been removed, which corresponds to 2.8 mM SO_4^{2-} anomalies in the bottom water. This is equal to the estimate (about 3 mM in the bottom water) made by Dyrssen (1988) by calculating the SO_4^{2-} balance (see Fig. 4.20). As discussed above, the removal of sulfide includes the processes such as oxidation, formation of FeS_2, diffusion and incorporation into organic matter.

If we take the initially formed H_2S ($[H_2S]_{initi}$) instead of the measured values, the ratios of $TCO_2/[H_2S]_{initi.}$ and $TA/[H_2S]_{initi.}$ are 2.04 and 2.33, respectively, which are close to the model values. The C/N ratio is 10 in the Black Sea, which is the same value as in the Framvaren. The extremely high value of C/P ratio may be due to the anomalous behavior of phosphate as discussed earlier. Higher C/N and C/P ratios in the Black Sea may also result from the large influence from terrestrial inputs. The C/N/P ratio in the Cariaco Trench, which is a typical open ocean with little terrestrial input, is found to be 107/15/1 and agrees well with the Redfield value.

4.4 The Kinetics Oxidation of Hydrogen Sulfide in Natural Waters

Molecular oxygen is the most important and abundant oxidant for hydrogen sulfide in natural waters, followed by MnO_2 and Fe_2O_3. This oxidation involves a complex mechanism that results in the formation of several reduced sulfur species (i.e. thiosulfate, sulfite, elemental sulfur and polysulfide) as well as sulfate. In recent years we have studied the oxidation of H_2S (Millero 1986; Millero and Hershey 1989; Zhang and Millero 1994) and H_2SO_3 (Zhang and Millero 1991) with O_2, H_2O_2, MnO_2 and FeOOH in the laboratory (Millero et al. 1987a, 1989; Zhang and Millero 1993a; Yao and Millero 1995a,b, 1996) and in the field (Millero 1991a,b,c; Zhang and Millero 1993b; Yao and Millero 1995b). These results have been used to develop a kinetic model that can characterize the rates and distributions of products in natural waters. The results of these studies are briefly reviewed in this section.

The overall rate equation for the oxidation of sulfide can be represented by

$$-d[H_2S] / dt = k[H_2S][O_2] \tag{4.16}$$

where the brackets represent concentrations. When oxygen is in excess the rate of disappearance of H_2S can be simplified to

$$-d[H_2S] / dt = k'[H_2S] \tag{4.17}$$

where $k' = k[O_2]$. Plots of ln $[H_2S]$ vs. time during the oxidation will give a straight line with a slope of k'. At a pH = 8.0, the rate constant (k, kg H_2O mol^{-1} h^{-1}) is given by (T,K)

$$\log k = 11.78 - (3.0 \times 10^3) / T + 0.44 I^{1/2} \tag{4.18}$$

At 25 °C the half time for the oxidation of H_2S with O_2 was $t_{1/2} = \ln 2/k' = 50 \pm 16$ h in water and 26 ±9 h in Gulf Stream sea water. The effect of pH on the reaction in water can be represented by

$$k = (k_0 + k_1 K_1 / [H^+]) / (1 + K_1 / [H^+]) \tag{4.19}$$

where $k_0 = 80$ kg H_2O mol^{-1} h^{-1} for the oxidation of H_2S and $k_1 = 344$ kg H_2O mol^{-1} h^{-1} for the oxidation of HS^-:

$$H_2S + O_2 \xrightarrow{k_0} \text{products} \tag{4.20}$$

$$HS^- + O_2 \xrightarrow{k_1} \text{products} \tag{4.21}$$

The value of K_1 is the dissociation constant for the ionization of H_2S (Millero et al. 1988). The effect of temperature and ionic strength on the rate constants k_0 and k_1 have been given by

$$\log k_0 = 9.22 - (2.4 \times 10^3) / T \tag{4.22}$$

$$\log k_1 = 10.50 + 0.16\text{pH} - (3.0 \times 10^3) / T + 0.44 I^{1/2} \tag{4.23}$$

These equations are valid from pH = 4 to 8, t = 5 to 65 °C, and I = 0 to 6 M.

Field measurements made on the oxidation of H_2S in the Black Sea (Millero 1991b), the Framvaren Fjord (Millero 1991b; Yao and Millero 1995b), the Chesapeake Bay (Millero 1991c), and the Cariaco Trench (Zhang and Millero 1993b) yielded half times that were much faster than determined in the laboratory on Gulf Stream sea water. The rates of oxidation of H_2S in surface waters (with added NaHS), deep waters and mixtures of surface and deep waters in the Cariaco Trench are shown in Fig. 4.21 along with the half times for these runs. As will be discussed later, the cause of the faster rates of oxidation of H_2S in these natural waters is due to the high concentrations of Fe^{2+} and Mn^{2+}. The intermediates SO_3^{2-} and $S_2O_3^{2-}$ were determined, along with the disappearance of H_2S, during the course of the oxidation of sulfide in the waters of the Cariaco Trench (Zhang and Millero 1993b). The results are shown in Fig. 4.22. The decrease of H_2S and resultant increase of SO_3^{2-}, $S_2O_3^{2-}$ and SO_4^{2-} occurring during the oxidation is similar to laboratory studies and in anoxic basins.

To determine if this increase was due to trace metals, we have measured the rates of oxidation of H_2S in seawater with added transition metals (Vazquez et al. 1989). These studies have shown that at concentrations below 300 nM, the rates are only affected by Fe^{2+}, Cu^{2+} and Pb^{2+}. At higher metal concentrations, the rates of oxidation of H_2S increase for all the metals except Zn^{2+} (Fig. 4.23). The order of the increase in the rates at higher concentrations for these metals is $Fe^{2+} > Pb^{2+} > Cu^{2+} > Fe^{3+} > Cd^{2+} > Ni^{2+} > Co^{2+} > Mn^{2+}$.

Only Fe^{2+} and Mn^{2+} have levels in anoxic basins high enough to affect the oxidation of H_2S. The effect of metals on the oxidation of H_2S with oxygen can be estimated from (Fig. 4.24)

$$\log(k / k_0) = a + b \log [M] \tag{4.24}$$

where

- $a = 6.55$, $b = 0.820$ for Fe(II)
- $a = 5.18$, $b = 0.717$ for Fe(III)
- $a = 1.68$, $b = 0.284$ for Mn(II)

These equations are valid, respectively, from 10^{-8} to $10^{-5.3}$, $10^{-7.2}$ to $10^{-3.3}$ and $10^{-5.9}$ to $10^{-3.3}$ M. At the maximum levels of Fe^{2+} in the Cariaco Trench, one would expect the rates of oxidation to be 17 times faster due to Fe^{2+}. These estimates are the same orders as found in our direct measurements. The calculated half times ($t_{1/2} = 17.2$, 2.7, and 1.5 h) respectively, for the surface, mixed and deep waters are in good agreement with the measured values ($t_{1/2} = 17.2$, 3.0, and 1.6 h).

The increase in the rates by Fe(II) at low concentrations is truly a catalytic effect. This is caused by the oxidation of Fe(II) (Millero et al. 1987b):

$$Fe(OH)_2 + O_2 \longrightarrow Fe(OH)_2^+ + O_2^- \tag{4.25}$$

The oxidation products O_2^- and $Fe(OH)_2^+$ may also oxidize H_2S. The reaction of dissolved or particulate Fe(III) with H_2S in the anoxic can regenerate Fe(II) to complete the catalytic cycle. The overall reaction is given by

$$2\,FeOOH + HS^- + 5\,H^+ \longrightarrow 2\,Fe^{2+} + S_0 + 4\,H_2O \tag{4.26}$$

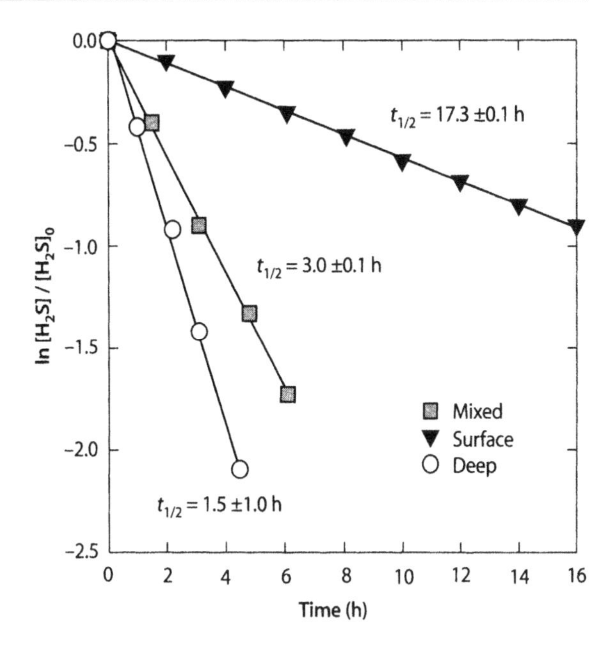

Fig. 4.21. Comparison of the rates of oxidation of H_2S in different waters of the Cariaco Trench (Zhang and Millero 1993b)

The kinetic measurements made on the formation of intermediates during the oxidation of H_2S in Cariaco trench waters give some support to our contention to the source of the SO_2^{2-} and $S_2O_3^{2-}$ found in the anoxic waters of the trench.

The effect of the metals on the rates of oxidation of H_2S (Vazquez et al. 1989) below the observable precipitation of metal sulfides (which may be a slow process) can be attributed to the formation of ion pairs:

$$M^{2+} + HS^- \longrightarrow MHS^+ \tag{4.27}$$

The overall rate constant is given by

$$k[HS^-]_T = k_{HS}[HS^-] + k_{MHS}[MHS^+] \tag{4.28}$$

where k_{HS} and k_{MHS} are the rate constants for the oxidation of HS^- and MHS^+. If k_{MHS} is greater than k_{HS} than the rate can be increased with the addition of the metal.

The presence of Fe and Mn in natural waters not only increases the rate of oxidation of sulfide, but also can have an effect on the oxidation of intermediates such as sulfite. This can change the distribution of the products formed during the oxidation. The final product from the oxidation of sulfide is sulfate, the sulfur compound having the highest oxidation state and the most stable compound in oxic waters. Various intermediates, such as sulfite and thiosulfate, also can be formed during the course of the reaction. The products formed from the oxidation of H_2S in sea water have been studied as a function of pH, temperature, salinity, and reactant concentration (Zhang and Millero 1993a). To examine the mass balance of sulfur compounds during the oxidation the experiments were made in pure water where SO_4^{2-} formed from oxidation could be measured (Fig. 4.25). The major products formed were found to be SO_4^{2-}, SO_3^{2-}

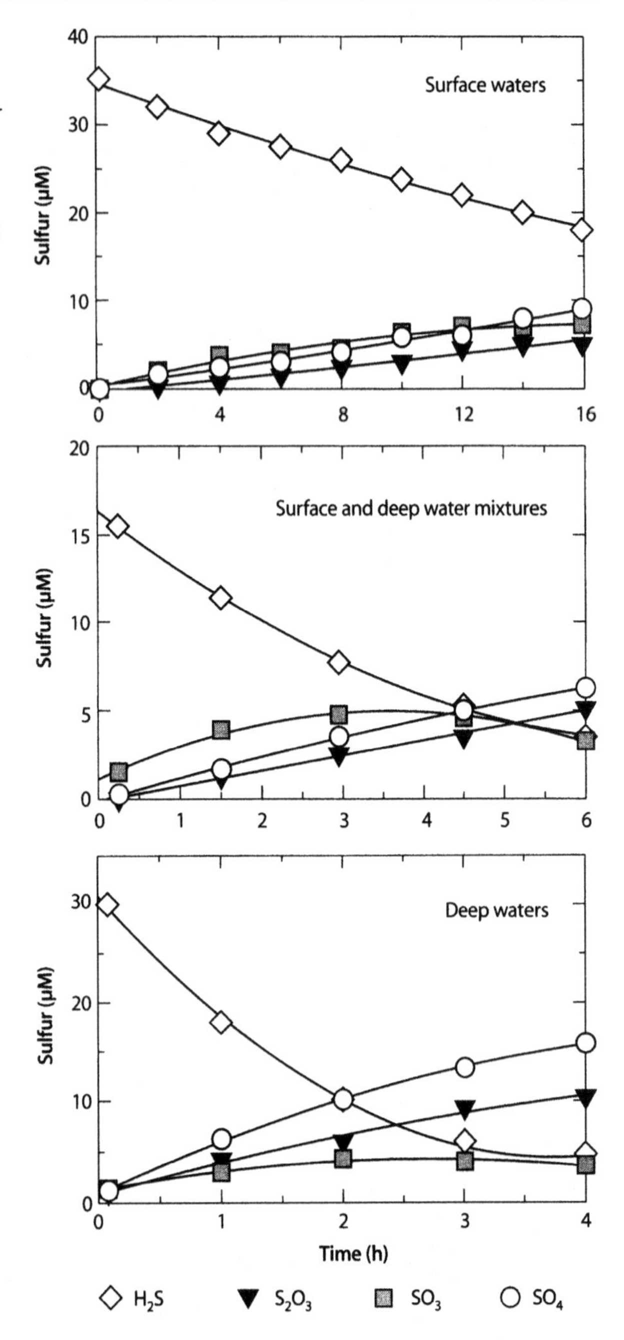

Fig. 4.22. The formation of intermediates during the oxidation of H_2S in different waters of the Cariaco Trench. Top-surface waters, middle-surface and deep-water mixtures, and bottom-deep waters. The smooth curves are calculated from the kinetic model (Zhang and Millero 1993b)

and $S_2O_3^{2-}$. Elemental sulfur or polysulfides were not found by spectroscopic techniques. The total equivalent sulfur of the products and reactants was constant, indicating that SO_4^{2-}, SO_3^{2-} and $S_2O_3^{2-}$ are the main products. The distribution of products from the oxidation of H_2S in sea water is similar to the results in water.

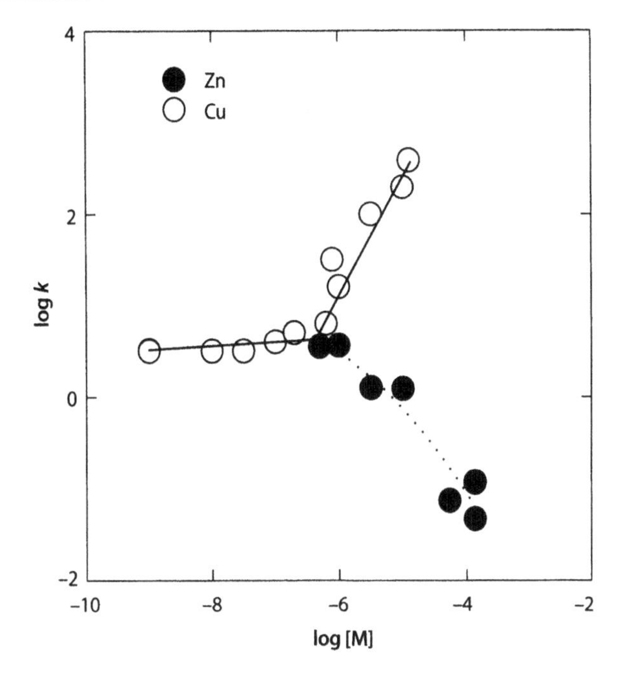

Fig. 4.23. The effect of Zn and Cu on the rate of oxidation of H_2S in sea water at 25 °C and pH = 8.1

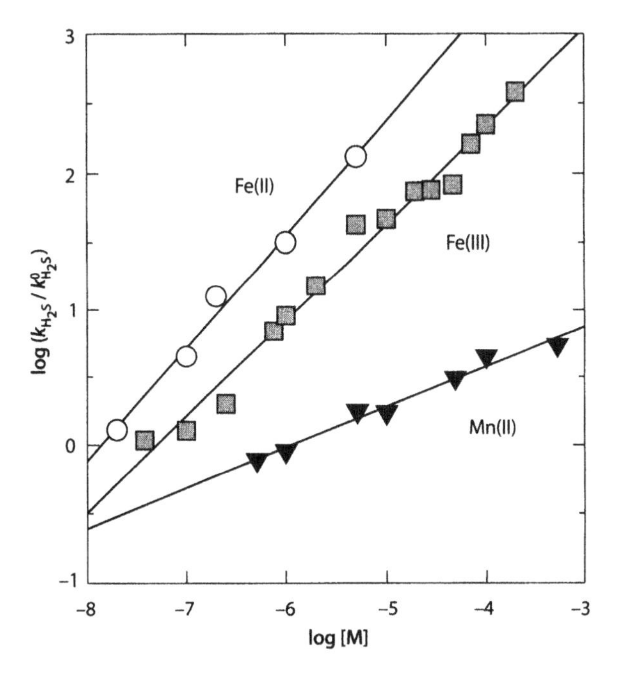

Fig. 4.24. The effect of Fe and Mn on the rate of oxidation of H_2S in sea water at 25 °C and pH = 8.1

The concentration of thiosulfate increases slowly throughout the reaction after an initial lag period. This suggests that thiosulfate is not the initial product of the oxidation. Thiosulfate is a stable product in the absence of bacteria and little oxidation oc-

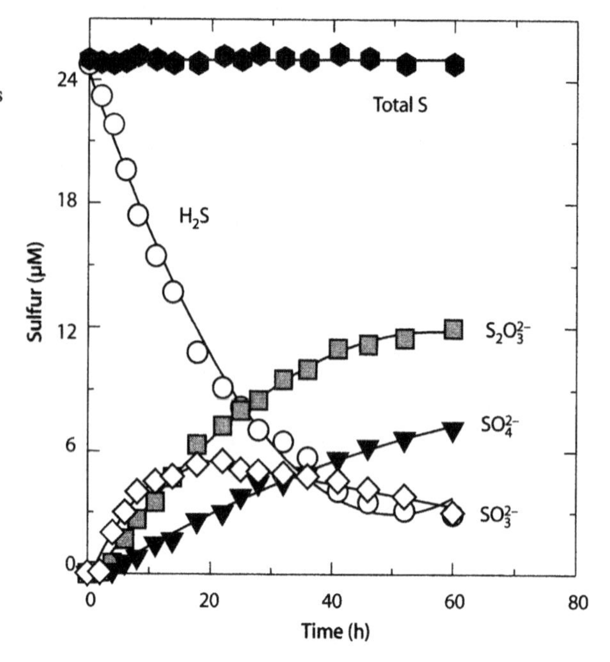

Fig. 4.25. The various sulfur species form during the oxidation of H_2S in water at 50 °C and pH=8.2. The smooth curves are calculated from the kinetic model (Zhang and Millero 1993a)

curs over 80 hours. The effect of pH on the distribution of products has also been examined, and the results have been attributed to changes in the rates of the individual reaction steps.

The effect of metals (Fe^{2+}, Fe^{3+}, Mn^{2+}, Cu^{2+}, Pb^{2+}) and solids (FeOOH and MnO_2) on the distribution of products has also been studied (Zhang and Millero 1993a). The intermediates formed (Fig. 4.26) during the oxidation of Cariaco Trench waters (350 nM Fe^{2+}) clearly show that metals not only increase the rate of oxidation of H_2S, but also change the distribution of products. The oxidation of sulfite with oxygen has also been studied. The values of log k as a function of temperature and ionic strength have been fitted to

$$\log k = 19.54 - 5069.47 / T + 14.74 I^{0.5} - 2.93 I - 2877.0 I^{0.5} / T \quad (4.29)$$

where k is in $M^{-1.5}\,min^{-1}$. This equation should be valid for most estuarine and sea waters.

The effect of pH on the rate of oxidation was found to be significant. The rate increased from pH 4 to a maximum at pH 6.5 and decreased at higher pH. The effect of pH on the rates was attributed to the rate-determining step involving the combination of HSO_3^- and SO_3^{2-}. This yields

$$k = k''\alpha(HSO_3^-)\alpha(SO_3^{2-}) \quad (4.30)$$

where $\alpha(i)$ is the molar fraction of species i. Values of $k'' = 6.66 \pm 0.06$ and 6.17 ± 0.17 were found for NaCl and sea water respectively.

A kinetic model was formulated based on the concentration-time dependence of the reactants (sulfide and oxygen) and products (sulfite, thiosulfate and sulfate). The

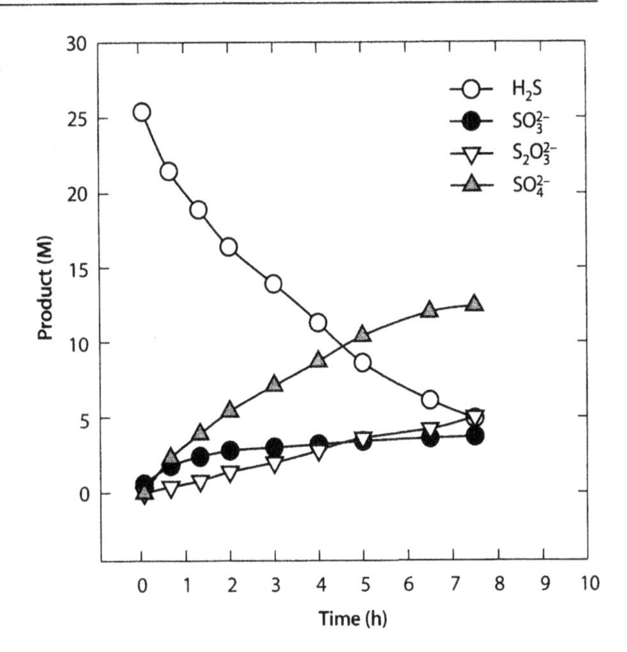

Fig. 4.26. The effect of Mn on the formation of products during the oxidation of H_2S with O_2 in sea water (Zhang and Millero 1993a)

validity of the model was evaluated by comparing the model predictions with the experimental measurements of reactants and products. At low concentrations the overall oxidation of HS^- with O_2 is given by (Zhang and Millero 1993b)

$$H_2S + O_2 \xrightarrow{k_1} \text{products}\ (SO_3) \tag{4.31}$$

$$H_2SO_3 + O_2 \xrightarrow{k_2} \text{products}\ (SO_4) \tag{4.32}$$

$$H_2S + H_2SO_3 + O_2 \xrightarrow{k_3} \text{products}\ (S_2O_3) \tag{4.33}$$

The overall rate equations for H_2S, SO_3^{2-}, $S_2O_3^{2-}$, SO_4^{2-} are given by

$$d[H_2S]\ /\ dt = -k_1[H_2S][O_2] - k_3[H_2S][SO_3^{2-}][O_2] \tag{4.34}$$

$$d[SO_3^{2-}]\ /\ dt = k_1[H_2S][O_2] - k_2[SO_3^{2-}]^2[O_2]^{1/2} - k_3[H_2S][SO_3^{2-}][O_2] \tag{4.35}$$

$$d[S_2O_3^{2-}]\ /\ dt = k_3[H_2S][SO_3^{2-}][O_2] \tag{4.36}$$

$$d[SO_4{}^{2-}]\ /\ dt = k_2[SO_3^{2-}]^2[O_2]^{1/2} \tag{4.37}$$

where $[i]$ is the total concentration of i.

These rate equations have been integrated simultaneously to evaluate the values of k_1, k_2 and k_3 using the experimental time dependence concentrations of all the reactants and products. The experimentally measured concentrations of H_2S, SO_3^{2-}, $S_2O_3^{2-}$, and SO_4^{2-} were found to be in good agreement with the model predictions up to reaction times of 80 hours (curves in Fig. 4.25). The values of k_2 in sea water needed to fit the data were slightly smaller than the values determined in our previous study, especially at higher temperatures. This is probably due to the inhibition of the oxidation of sulfite in the presence of sulfide. This finding is supported by the previous observations that sulfite in the presence of H_2S is more stable in sea water than predicted by its rate of oxidation.

The values of k_1, k_2 and k_3 as a function of salinity (S) and temperature (T in K) have been fitted to the equations (pH = 8.2)

$$\ln k_1 = 26.90 + 0.0322S - 8\,123.21 / T \tag{4.38}$$

$$\ln k_2 = 14.91 + 0.0524S - 1764.68 / T \tag{4.39}$$

$$\ln k_3 = 28.92 + 0.0369S - 8\,032.68 / T \tag{4.40}$$

These equations should be valid for estuarine and sea waters over a wide range of salinity and temperature. This kinetic model can be used to predict the product distribution for the oxidation of sulfide in natural waters with low concentrations of trace metals. The agreement between the model and the observed distribution of reaction products does not provide conclusive proof that the reaction pathways of the overall model actually describe the series of elemental reactions that occur. The detailed mechanisms might involve many elemental reaction steps.

The rates of oxidation of hydrogen sulfide, the effect of metals and the intermediates formed have been examined in a number of natural anoxic basins. A comparison of the measurements made in the Cariaco Trench with laboratory studies on NaHS added to Gulf Stream water are shown in Table 4.3.

The field measurements of the rates of oxidation of H_2S were found (Fig. 4.27) to be in good agreement with those estimated from laboratory studies at the same concentration of Fe^{2+}. The levels of Fe^{2+} are high enough in most anoxic environments to increase the rates of oxidation of H_2S. A kinetic model has been used to analyse the distribution of products (SO_3^{2-}, $S_2O_3^{2-}$, SO_4^{2-}) formed during the oxidation in the Framvaren Fjord and the Cariaco Trench. The rate constants for the production of SO_3^{2-} (k_1), for the production of SO_4^{2-} (k_2) and the production of $S_2O_3^{2-}$ (k_3) estimated for these waters are in reasonable agreement with the predicted values at the same level of Fe^{2+}.

The values of k_2 estimated for the Framvaren Fjord and Cariaco Trench are slightly higher than the predicted values. This could be due to errors in our estimation of the concentration and form of iron in this water. Direct measurements of iron and man-

Table 4.3. Comparison of the rate constants for the oxidation of hydrogen sulfide in different waters

Rate constant	Gulf Stream	Cariaco Trench		
		Surface	Mixed	Deep
k_1	1.7	3.1	18.4	36.3
k_2	48000	48000	72000	240000
k_3	30	15	180	360

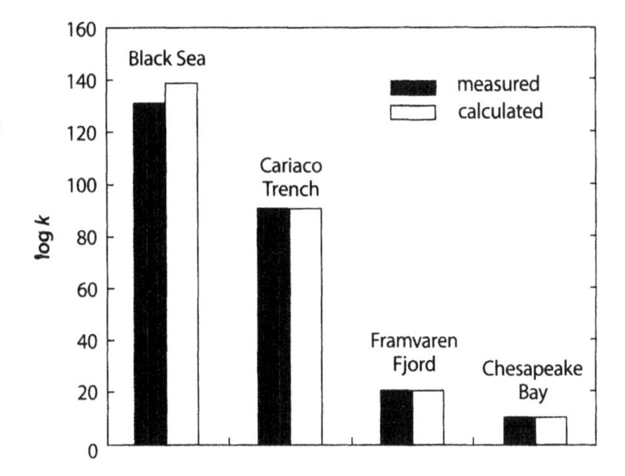

Fig. 4.27. A comparison of the rates of oxidation of H_2S with O_2 measured and calculated in various anoxic basins (The calculated rates are at the same Fe as in the natural water)

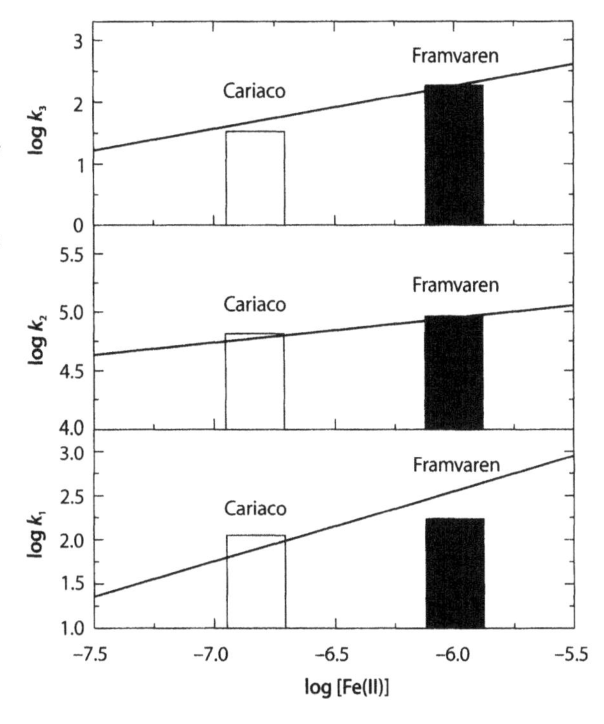

Fig. 4.28. Comparison of the rate constants for the oxidation of H_2S (k_1), H_2SO_3 (k_2), and formation of S_2O_3 (k_3) in the Cariaco Trench and Framvaren Fjord and those produced at the same levels of Fe. The lines are the rate constants predicted from laboratory measurements at various concentrations of Fe(II) (Zhang and Millero 1993a)

ganese in the anoxic waters should be made in all future kinetic studies to avoid this problem. The estimates based upon only the concentration of iron may also lead to some errors. The concentrations of Mn^{2+} below the oxic/anoxic interface reach levels of 0.5 and 15 μM in the Cariaco Trench and Framvaren Fjord, respectively. This Mn^{2+} comes from reduction of MnO_2 sinking from above the oxic/anoxic interface. Reoxidation of Mn^{2+} to MnO_2 by bacteria occurs when Mn^{2+} diffuses up to the oxic layer. This cycling of manganese between Mn^{2+} and MnO_2 is an important feature of

Fig. 4.29. Comparison of the rates of oxidation of H_2S by Mn(IV) and Fe(III) (hydr)oxides in the Framvaren Fjord and laboratory measurements (Yao and Millero 1995a,b, 1996)

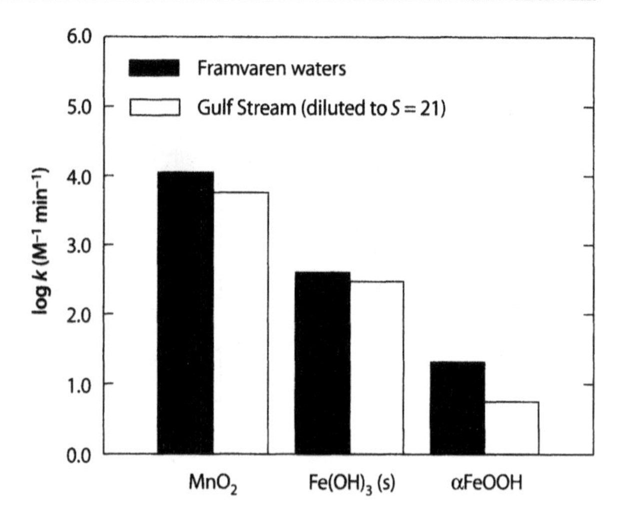

the oxic/anoxic interface and probably affects the distribution of products in the field. The products ($S_2O_3^{2-}$, SO_3^{2-} and SO_4^{2-}) formed during the oxidation in the Framvaren Fjord and Cariaco Trench also have been examined and used to determine k_1, k_2 and k_3. The measured results are compared in Fig. 4.28 with the estimated values at the same level of Fe. The agreement is quite good.

Yao and Millero (1995a,b) have examined how Fe^{3+} (FeOOH) and Mn^{4+} (MnO_2) that occur above the anoxic-oxic interface can oxidize hydrogen sulfide and complete the redox cycle of these metals in anoxic waters. These reactions are much more complicated since the reactions occur on the surfaces of the oxide minerals. The rates of oxidation of H_2S by Mn(IV) and Fe(III) (hydr)oxides in the water near the oxic/anoxic interface were determined and compared with the values obtained in the surface waters of the Gulf Stream (Fig. 4.29) in the laboratory. The agreement is quite good. The slightly higher oxidation rates observed in the field might be related to the microbiological activities near the interface.

References

Andersson LG, Dyrssen D and Hall POJ (1988) On the sulfur chemistry of a super-anoxic fjord, Framvaren, South Norway. Mar Chem 23:283

Dyrssen D (1988) Sulfide complexation in surface seawater. Mar Chem 24:143–153

Fanning KA, Pilson MEQ (1972) A model for the anoxic zone of the Cariaco Trench. Deep-Sea Res 19:847–86

Grasshoff K (1975) The hydrochemistry of landlocked basins and fjords. In: Riley JP, Skirrow G (eds) Chemical oceanography, 2nd edn, vol II. Academic Press, New York, pp 456–597

Haraldsson C, Westerlund S (1988) Trace metals in the water columns of the Black Sea and Framvaren Fjord. Mar Chem 23:417–424

Jacobs L, Emerson S, Skei J (1985) Partitioning and transport of metals across the O_2/H_2S interface in a permanently anoxic basin: Framvaren Fjord, Norway. Geochim. Cosmochim. Acta 49:14331444

Millero FJ (1986) The thermodynamics and kinetics of the hydrogen sulfide system in natural waters. Mar Chem 18:121–147

Millero FJ (1991a) The oxidation of H_2S in Black Sea waters. Deep-Sea Res 38(S2):S1139–S1150

Millero FJ (1991b) The oxidation of H_2S in Framvaren Fjord. Limnol Oceanogr 36(5):1007–1014

Millero FJ (1991c) The oxidation of H_2S in Chesapeake Bay, Estuarine. Coastal Shelf Sci 33:21–527

Millero FJ (1996) Chemical oceanography. CRC press, Boca Raton

Millero FJ, Hershey JP (1989) Thermodynamics and kinetics of hydrogen sulfide in natural waters. In: Saltzman ES, Cooper WJ (eds) Biogenic sulfur in the environment. ACS Press, Washington D.C. (ACS Symp. Ser 393), pp 282–313
Millero FJ, Hubinger S, Fernandez M, Garnett S (1987a) The oxidation of H_2S in seawater as a function of temperature, pH and ionic strength. Env Sci Technol 21:439–443
Millero FJ, Sotolongo S, Izaguirre M (1987b) The oxidation kinetics of Fe(II) in seawater. Geochim Cosmochim Acta 51:793–801
Millero FJ, Plese T, Fernandez M (1988) The dissociation of hydrogen sulfide in seawater. Limnol Oceanogr 33:269–274
Millero FJ, Laferriere AL, Fernandez M, Hubinger S, Hershey JP (1989) Oxidation of H_2S with H_2O_2 in natural waters. Environ Sci Technol 23:209–213
Redfield AC, Ketchum BH, Richards FA (1963) The influence of organisms on the composition of seawater. In: Hill MN (ed) The sea, vol II. Interscience, New York, pp. 26–77
Richards FA (1965) Anoxic basins and fjords. In: Riley JP, Skirrow G (eds) Chemical oceanography, 1st edn, vol I. Academic Press, New York, pp 611–645
Scranton MI, Sayles FL, Bacon MP, Brewer PG (1987) Temporal changes in the hydrography and chemistry of the Cariaco Trench. Deep-Sea Res 34:945–996
Steinberg PA, Millero FJ, Zhu X (1998) Carbonate system response to iron enrichment. Mar Chem, 62:31–43
Vazquez F, Zhang JZ, Millero FJ (1989) Effect of trace metals on the oxidation rates of H_2S in seawater. Geophys Res Lett 16:1363–1366
Yao W, Millero FJ (1993) The rate of sulfide oxidation by δMnO_2 in seawater. Geochim Cosmochim Acta 57:3359–3365
Yao W, Millero FJ (1995a) Oxidation of hydrogen sulfide by Mn(IV) and Fe(III) hydroxides in seawater. In: Vairavamurthy MA, Schooner MAA (eds) Geochemical transformation of sedimentary sulfur. ACS Press, Washington, D.C. (ACS Symp. Ser. 612), pp 260–279
Yao W, Millero FJ (1995b) The chemistry of the anoxic waters in the Framvaren Fjord, Norway. Aquat Geochem 1:53–88
Yao W, Millero FJ (1996) Oxidation of hydrogen sulfide by hydrous Fe(III) oxides in seawater. Mar Chem 52:1–16
Zhang J-Z, Millero FJ (1991) The rate of sulfite oxidation in seawater. Geochim Cosmochim Acta 55:677–685
Zhang J-Z, Millero FJ (1993a) The products from the oxidation of H_2S in seawater. Geochim Cosmochim Acta 57:1705–1718
Zhang J-Z, Millero FJ (1993b) The chemistry of anoxic waters in the Cariaco Trench. Deep-Sea Res 40:1023–1041
Zhang J-Z, Millero FJ (1994) The kinetics of oxidation of hydrogen sulfide in natural waters. In: Alpers CN, Blowers D (eds) Environmental geochemistry of sulfide oxidation. ACS Press, Washington, D.C. (ACS Symp. Ser 550), pp 393–409

Part II

Organic Matter in Marine Environments

Organic Matter Preservation in the Ocean: Lipid Behavior from Plankton to Sediments

S.G. Wakeham

5.1 Introduction

Marine sediments preserve a record of past oceanic environmental conditions. This record, however, may be compromised and difficult to interpret if there is qualitative and quantitative uncoupling between the water column and the sediments. In the case of organic matter, this bias is severe since most organic matter produced in surface waters is degraded in the water column and only a tiny fraction is buried in the sediments. Valid reconstruction of the sediment record requires a firm understanding of the processes that occur in the ocean and how they affect the behavior of organic matter.

The oceanic carbon cycle involves a complex interaction between physical, biological and chemical processes. The bulk of particulate organic carbon (*POC*) in the ocean is biosynthesized in surface waters by photoautotrophic plankton (see review by Wakeham and Lee 1993), with rivers and aeolian sources (rain fall and dry fallout) adding a minor fraction. Once produced, most *POC* is consumed by heterotrophic organisms in the upper ocean, leading to changes in the particle size-spectrum by continuous aggregation/disaggregation ("repackaging") processes, alteration of the chemical composition of particles, and ultimately remineralization of most (>90%) of the *POC* pool. A small fraction, perhaps 5–10%, of *POC* produced in surface waters escapes into the deep ocean. Organic materials associated with fast-sinking, large particles such as marine snow aggregates or faecal pellets may reach the seafloor relatively quickly and with minimal degradation. Compounds associated with slowly sinking fine particles may undergo intense chemical alteration during long transit times. Regardless of the transport mechanism and its time scale, only <1% of the primary production reaches the sea floor, and this in a highly altered state. Further degradation by benthic macro- and microfauna in surface sediments removes most of any remaining organic matter, with the final burial of only ≈ 0.1% of primary production. Interpreting the sedimentary record is thus confounded by the fact that very little of the signal originally produced in surface waters is preserved in the sediments.

Examples from recent Joint Global Ocean Flux Study (JGOFS) experiments in the equatorial Pacific Ocean (EqPac; Murray et al. 1992, 1995) and the Arabian Sea (Lee et al. 1998) are presented below to highlight the processes affecting preservation of organic matter in marine sediments. The goal of these studies was to evaluate the efficiency and dynamics of the carbon pump in these oceanic systems, and to establish how water column biogeochemical processes affect the burial and preservation of biogenic components in sediments.

5.2 Lipids as Tracers of Organic Carbon

Lipids are widely used indicators for inferring organic carbon source and alteration processes. Lipids are generally defined operationally as those compounds that are water-insoluble but extracted with non-polar solvents, and analytical methodologies are often tuned for specific compound classes (Wakeham and Volkman 1991). Along with amino acids and carbohydrates, lipids constitute the major biochemicals in living organisms. Although usually less abundant than amino acids but more abundant than carbohydrates, lipids typically account for 10–60% of organic carbon (*OC*) in marine organisms (Parsons et al. 1984; Sargent and Henderson 1986). In living organisms, lipids play major roles in energy storage and mobilization, membrane structure, and control of metabolic processes.

The utility of lipids as biomarkers lies in the fact that organisms biosynthesize lipids of diverse molecular structures that contain a great variety of organic functional groups (e.g. Cranwell 1982; de Leeuw and Largeau 1993). Organic geochemists are continually searching for novel biomarkers, and in a number of cases biomarkers are quite unique to select organisms. For example, the long-chain C_{37}-C_{39} alkenones that are of such great interest to paleoceanographers (Brassell 1993) are unique to a very limited set of haptophytes (Volkman et al. 1980; Conte et al. 1994). 4-Methylsterols are commonly (but apparently not exclusively) attributed to dinoflagellates (Boon et al. 1979; Robinson et al. 1984). Crustacean zooplankton, primarily calanoid copepods, biosynthesize and store wax (alkyl) esters in the C_{28}-C_{38} carbon number range (Sargent and Henderson 1986) while most other marine organisms use triacylglycerols as energy storage lipids (Sargent 1976). Long-chain hydrocarbons, alcohols, fatty acids ($>C_{20}$) and wax esters (C_{44}-C_{60}) are constituents of epicuticular waxes of terrigenous vascular plants (Kolattukudy 1976) and are thus robust indicators of inputs from higher plants. Bacterial contributions to organic matter are indicated by, for example, branched-chain (e.g. *iso*- and *anteiso*-C_{15}) fatty acids (Kaneda 1991), acyclic isoprenoids (Risatti et al. 1984) and hopanoids (Ourisson et al. 1987).

Lipids are relatively labile toward degradation in the ocean, potentially more reactive than amino acids and carbohydrates (Wakeham et al. 1997a). Rapid degradation of lipid components, either by autolysis or by hydrolytic attack by enzymes from heterotrophic consumers, usually follows the death of the producer organism. Degradation results in qualitative changes in composition between surface water particulate matter and sediment that can be significant as the more labile compounds are lost and more refractory compounds are conserved, further complicating assessments of upper-ocean environmental conditions.

An example of this qualitative decoupling of surface water lipid composition from that of sediments is illustrated in Fig. 5.1, contrasting the molecular distributions of hydrocarbons in a sediment trap sample from a 500 m depth in the Arabian Sea and with that of underlying sediments at a 3500 m depth. Hydrocarbons in the trap are dominated by compounds of phytoplanktonic origin, in keeping with the high primary production in the Arabian Sea. Diatoms are abundant during the monsoon upwelling season and are the probable sources of *n*-C_{17}, 3,6,9,12,15,18-heneicosahexaene (HEH), and a series of C_{25}-isoprenoid alkenes (br25) (Rowland and Robson 1990; Volkman et al. 1994). The sterenes (steroidal hydrocarbons) in the trap sample are thought to be prod-

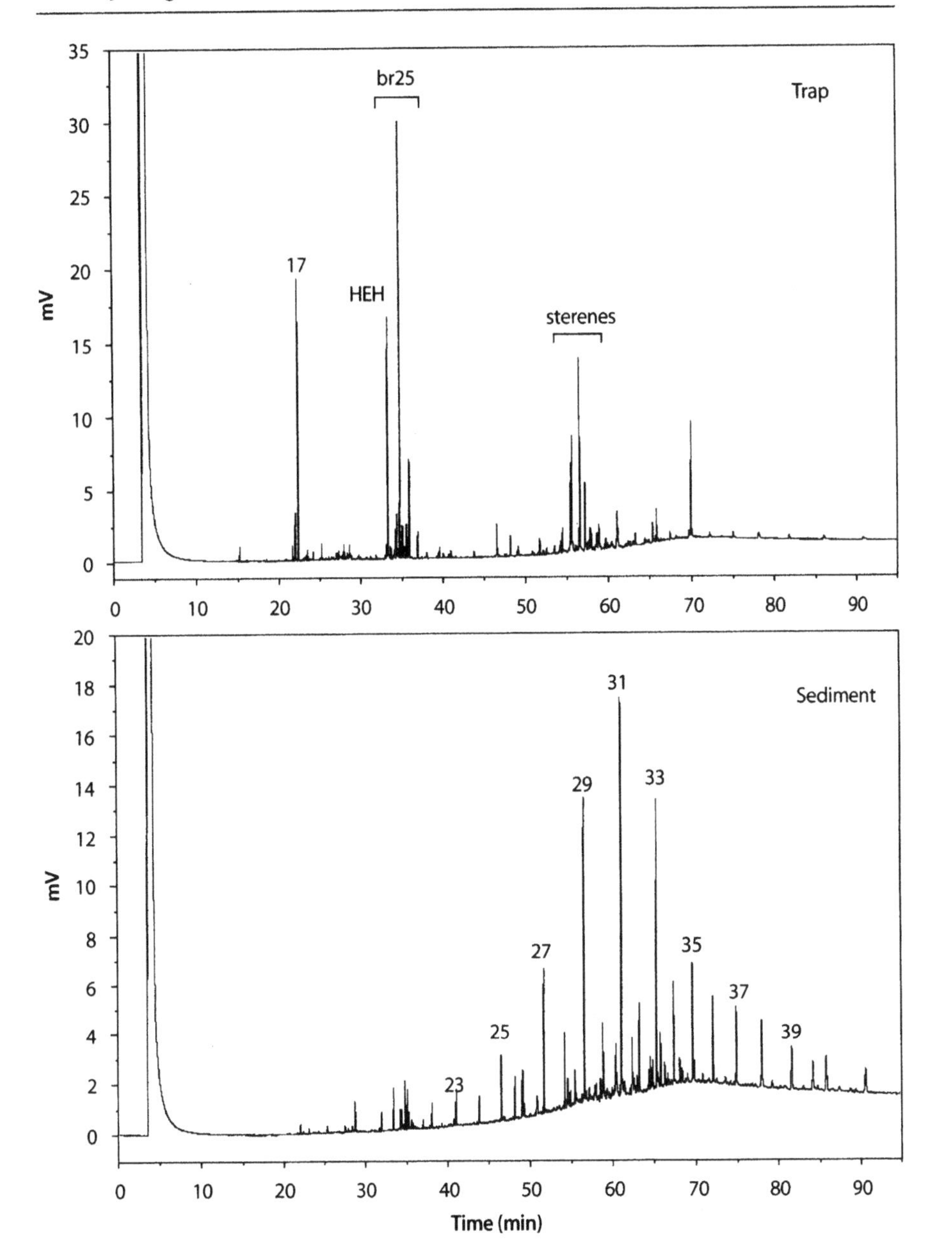

Fig. 5.1. Gas chromatograms of hydrocarbon fractions isolated from Arabian Sea sediment trap (500 m) and 0–1 cm sediment (3500 m) at station M3 (17°20' N, 59°20' E). Numbered peaks are *n*-alkanes; br-25 are C_{25}-isoprenoid alkenes

ucts of microbial dehydration of algal sterols (Wakeham et al. 1984a) in the oxygen minimum zone that spanned 500–1500 m in the water column. On the other hand, the sediments are dominated by long-chain, odd-carbon predominant *n*-alkanes that are constituents of epicuticular waxes of terrigenous vascular plants (Kolattukudy 1976). Vascular plant compounds in deep-sea sediments are quite common (e.g. Prahl et al. 1989; Santos et al. 1994) and are evidence of long-range atmospheric transport of continental material to the open ocean. The algal signal that so dominated sediment trap material was only relatively minor in the sediment, whereas the vascular plant signal that was barely discernable in the trap dominated sedimentary hydrocarbons. Even though the composition of hydrocarbons sinking out of the upper Arabian Sea reflects its phytoplanktonic origin, the planktonic signal has been selectively degraded during transport to the sediments leaving behind a vascular plant component to be preferentially preserved. Thus, not only is sedimentary organic matter quantitatively a small fraction of surface water productivity, the composition of that remaining organic matter may be qualitatively quite different from that produced in overlying waters.

5.3 Lipid Fluxes to the Deep Sea

The vertical export of organic matter between surface waters and sediments reflects quantitative and qualitative balances between production of *POC* in surface waters and its degradation during transit. The export flux of *POC* through the water column generally correlates with primary production over geographically diverse locations, and there is a quasi-exponential decay in *POC* and total-lipid fluxes with depth in the water column (Suess 1980; Martin et al. 1987). *POC* and lipid fluxes in the equatorial Pacific fit this trend (Fig. 5.2) for the three sites (9° N, 5° N and the equator along 140° W) where materials from surface plankton through sediments were collected syn-

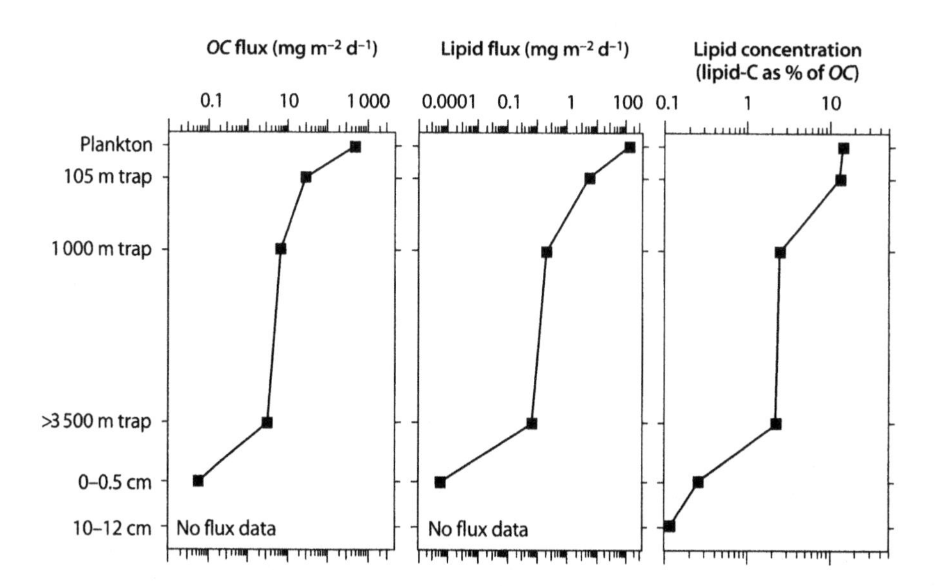

Fig. 5.2. Fluxes of particulate organic carbon (*OC*) and lipids and lipid concentrations for EqPac samples at 9° N, 5° N and equator stations. Plankton were collected by oblique tows (100–0 m) of a 26 μm net; sediment trap samples were collected in floating (105 m) and moored (1000 m below the surface and 1000 m above the seafloor) traps; sediments were collected by multi-coring (see Hernes et al. 1996 for details)

optically as part of the EqPac program. Estimates of lipid production by plankton (see Wakeham et al. 1997b for details) range from 30–290 mg lipid $m^{-2} d^{-1}$ and show a latitudinal dependence corresponding to spatial variations in primary production (300–1 800 mg *OC* $m^{-2} d^{-1}$; Barber et al. 1996). Fluxes of lipids decreased quickly in the water column, resulting in delivery rates of lipids to sediments that were 5–6 orders of magnitude reduced (0.00004–0.0032 mg lipid $m^{-2} d^{-1}$).

Overall, two zones appear most responsible for this substantial decrease in flux. The upper several hundred meters of the water column (the epipelagic zone) is the region where zooplankton grazing and bacterial decomposition are greatest and utilize most organic matter. And, the benthic boundary layer, especially the water-sediment interface, is another biological "hotspot" where benthic macro- and microfauna consume much of that small fraction of organic material that rains down onto the seafloor. The relatively small decrease in *POC* and lipid fluxes between shallow moored traps at 1 000 m below the seasurface and deep moored traps at 1 000 m above the seafloor indicates that decomposition in the ocean's interior is quantitatively relatively minor and particles may transit the deep ocean with minimal degradation.

Simultaneous with the decrease in lipid flux during passage through the water column is a reduction in the proportion of organic carbon that lipids comprise (Fig. 5.2). Lipids represented from up to 25% of *POC* in EqPac net-plankton but only ≈ 0.1% of *POC* in subsurface sediments. This means that the proportion of organic carbon that was lipid decreased by up to 250-fold, in marked contrast to amino acids whose contribution to *POC* decreased by ≈ 4-fold and carbohydrates that actually increased by about 25% (Wakeham at el. 1997a). To a certain extent, the reactivity of the marine lipids that comprise the bulk of the particulate organic matter is due to their high degree of unsaturation, but not all marine lipids are so reactive as will be shown below. Vascular plant lipids that are relatively minor components in particles but that become an increasingly important part of sedimentary organic material often are highly saturated and as a result may be considerably more stable. The relative stability of carbohydrates may be related to that fact that one of their functions is as structural components that may help protect them from degradation (Hernes et al. 1996).

5.4 Seasonal Variations in Lipid Flux

Temporal variations in primary productivity in surface waters, resulting from physical oceanographic forcing functions, can drive strong temporal variability in the flux of material to the deep sea. The Arabian Sea study measured time- and depth-dependent fluxes of organic materials as a function of variations of coastal upwelling in response to the monsoons off the coast of Oman (Lee et al. 1998). A strong seasonal effect on *POC* and lipid flux was observed at three depths in the Arabian Sea (Fig. 5.3). In the shallowest trap, deployed at about 500 m, fluxes varied by 5-fold over the annual cycle. Maximal fluxes were measured from July through September, lagging the south-west monsoon by about a month (Weller et al. 1998) and presumably due to large diatom blooms responding to enhanced upwelling. Secondary flux maxima occurred during December and January following the weaker north-west monsoon. The seasonal variation in flux in the upper water column is significantly greater than corresponding temporal variations in primary productivity (1 100 ±200 mg *POC* $m^{-2} d^{-1}$; Lee

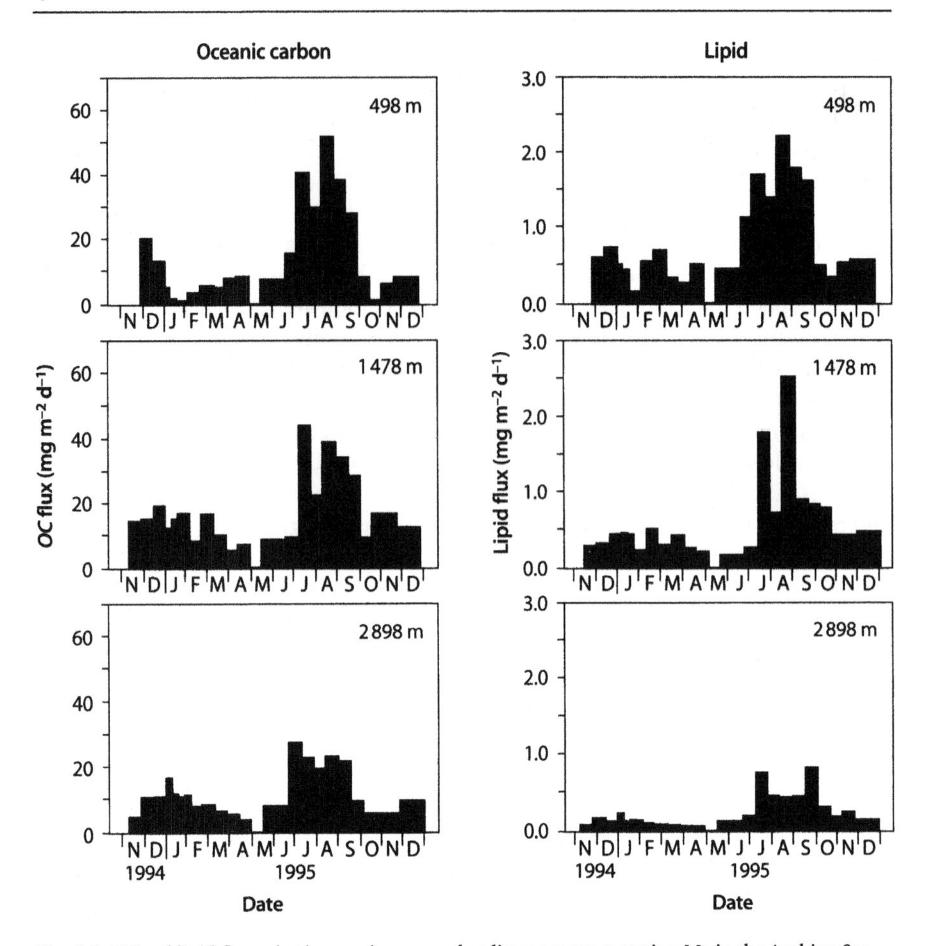

Fig. 5.3. *OC* and lipid fluxes in time-series moored sediment traps at station M3 in the Arabian Sea

et al. 1998). *POC* and lipid fluxes measured in traps at about 3 000 m were, respectively, 3-fold and 5-fold lower than at 500 m. However, it must be stressed that due to the highly dynamic current regime in the western Arabian Sea, notably offshore transport during the monsoons, it is uncertain how directly fluxes measured at 3 000 m are coupled with fluxes at 500 m (see Siegel et al. 1990 for a relevant discussion of particle transport into sediment traps). It is possible that fluxes into mid-depth and deep traps include a significant component of material advected laterally from the west. Nonetheless, it appears that about 0.14% of the surface water primary production in this region of the Arabian Sea reaches the seafloor (Lee et al. 1998), not greatly different from the equatorial Pacific.

Significantly greater seasonal ranges in biomarker fluxes as well as temporal offsets in flux maxima attest to the dynamic nature of the algal community in the Arabian Sea. Diatoms constitute a major phytoplankton during the monsoon seasons. It would appear that the bloom of the several species of diatoms that biosynthesize C_{25}-isoprenoid alkenes (not all diatoms produce isoprenoid alkenes) is intense but short-

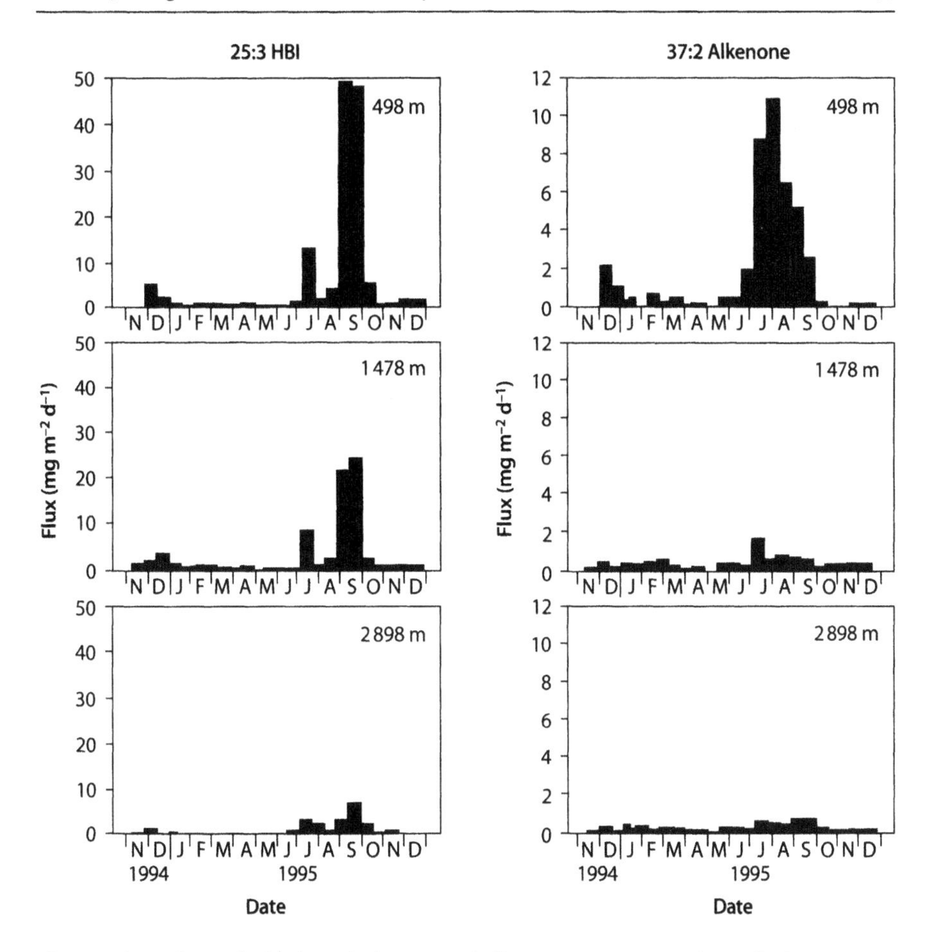

Fig. 5.4. Fluxes for C_{25}-highly branched isoprenoid alkatriene (25:3 HBI) and C_{37}-alkadienone (37:2 alkenone) in time-series traps in the Arabian Sea

lived, producing a strong pulse in flux during September (Fig. 5.4). The flux peak during the south-west monsoon was 50-fold higher than during intermonsoon periods. The sharp peak in flux of the 25:3 isoprenoid alkene was markedly different from the broad flux maximum of *POC* and lipids. On the other hand, the haptophytes that produce alkenones appear to bloom over a longer period coincident with the whole south-west monsoon and more similar to bulk *POC*. For both 25:3 isoprenoid alkene and 37:2 alkenone, fluxes to the deep traps were strongly attenuated, about 10-fold, consistent with their enhanced reactivities compared to *POC* and bulk lipid (2-fold and 4-fold attenuations in flux, respectively).

5.5 Fate of Lipid Biomarkers

Significant changes in the character of lipids occur in the water column in concert with the depth-dependent decrease in lipid flux and preferential loss of lipid relative to *POC*.

To date, the most complete data set available for examining trends from plankton to sediments is the EqPac study. Figure 5.5 illustrates the changing relative compositions of neutral and acidic lipids (see figure caption for the distinction between neutral lipids and acids) at the EqPac equator site. Sterols are the most abundant lipid subclass in the neutral lipid fraction of plankton. Fatty alcohols comprise the only other significant neutral lipid class; hydrocarbons and alkenones were relatively minor components. As diagenesis proceeded down the water column, the proportion of sterols decreased, while hydrocarbons and alkenones increased in relative abundance. In the acid fraction, polyunsaturated (pufa) acids were major components in plankton but degradation made them proportionately less important as particles sank; pufas were barely detectable in sediments. Branched-chain and monounsaturated acids become increasingly more abundant constituents as diagenesis progressed.

The power of molecular-level lipid analysis lies in using structural characteristics of the >80 individual compounds that are typically measured in the neutral and acid to evaluate source and reactivities of lipids in the equatorial Pacific (see Wakeham et al. 1997b for further discussion of lipid biomarker sources). Twelve biomarkers have been selected to represent planktonic (docosahexaenoic acid [22:6ω3 pufa], hexadecanol [16ROH], 28-methylcholesta-5,24(28)-dien-3β-ol [28$\Delta^{5,24(28)}$], cholest-5-en-3β-ol [27Δ^5], 24-ethylcholest-5-en-3β-ol [29Δ^5], 4α,23,24-trimethylcholest-22E-en-3β-ol [30Δ^{22}], and

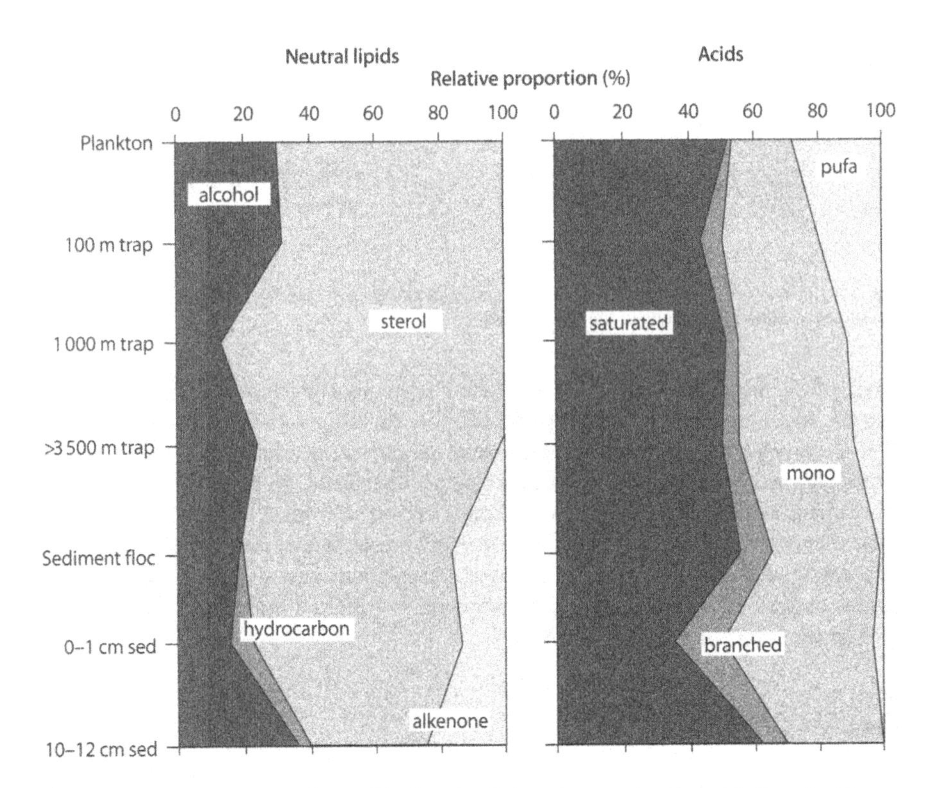

Fig. 5.5. Relative proportions of neutral lipids and acids in EqPac samples. Neutral lipids and acids were obtained following saponification of solvent extracts (see Wakeham et al. 1997b)

C_{37}- + C_{38}-alkenones), bacterial (*cis*-vaccenic acid [18:1ω7], *iso+anteiso*-15:0 acids [br-15 fas], bisnorhopane and squalene), or vascular plant (*n*-nonacosane [C_{29}], hexadecanol [28ROH] and hexacosanoic acid [26fa]) origins. All compounds in this suite, regardless of source, clearly exhibited markedly decreasing fluxes with increasing depth (Fig. 5.6). There were no absolute flux increases below the ocean surface to the extent that would unambiguously indicate significant amounts of net secondary production at depth. Nor were there significant inputs of biomarkers via lateral advection. 24-Ethylcholest-5-en-3β-ol has traditionally been ascribed to vascular plant sources (Huang and Meinschein 1979), but its abundance in many plankton (Volkman 1986) and a stable carbon isotopic composition of −24‰ for these EqPac samples indicates a predominately marine source.

On the other hand, there were significantly different behaviors among these biomarkers when distributions were scaled based on their relative abundances in the sample set (Fig. 5.7). This approach further classifies the biomarkers into subcategories based on their behavior, in addition to source. Three of the seven planktonic biomarkers (22:6ω3 pufa, 16ROH, and $28\Delta^{5,24(28)}$) had maximal abundances in the upper 100 m where planktonic production is highest. Cholest-5-en-3β-ol ($27\Delta^5$) was most abundant in the 1 000 m trap, while the remaining three planktonic compounds ($29\Delta^5$, $30\Delta^{22}$, and alkenones) had enhanced abundances in the sediments. The four bacterial markers (18:1ω7, br-15 fas, bisnorhopane and squalene) and the three vascular plant biomarkers (*n*-C_{29}, 28ROH and 26fa) were most abundant in sediments.

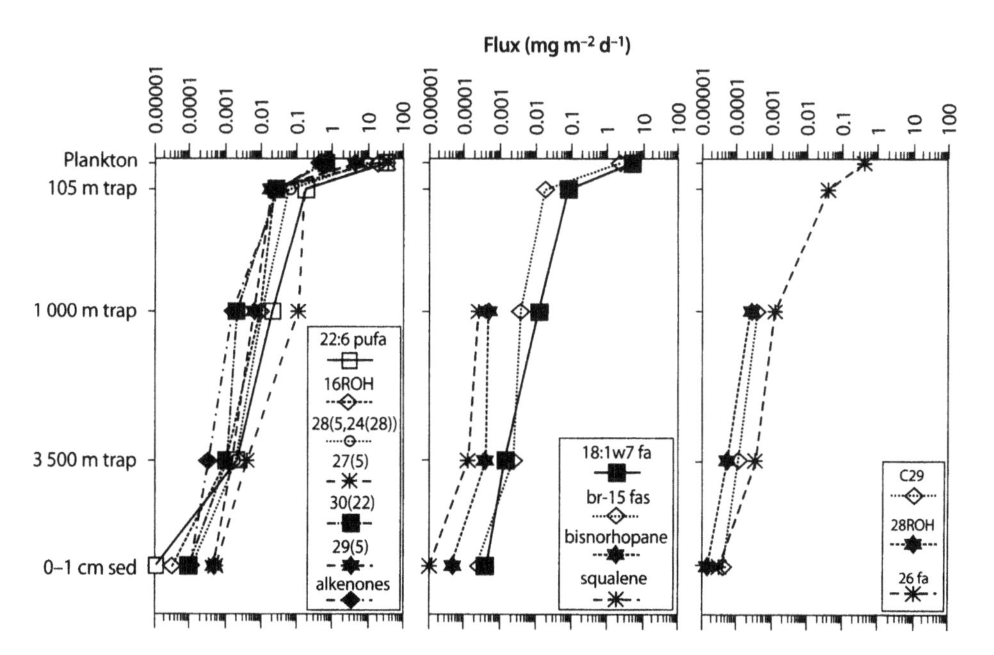

Fig. 5.6. Fluxes for selected biomarkers in EqPac samples from the equator station (docosahexaenoic acid [22:6ω3 pufa], hexadecanol [16ROH], 28-methylcholesta-5,24(28)-dien-3β-ol [$28\Delta^{5,24(28)}$], cholest-5-en-3β-ol [$27\Delta^5$], 24-ethylcholest-5-en-3β-ol [$29\Delta^5$], 4α,23,24-trimethylcholest-22E-en-3β-ol [$30\Delta^{22}$], C_{37}- +C_{38}-alkenones, *cis*-vaccenic acid [18:1ω7], *iso+anteiso*-15:0 acids [br-15 fas], bisnorhopane, squalene, *n*-nonacosane [C_{29}], hexadecanol [28ROH], hexacosanoic acid [26fa])

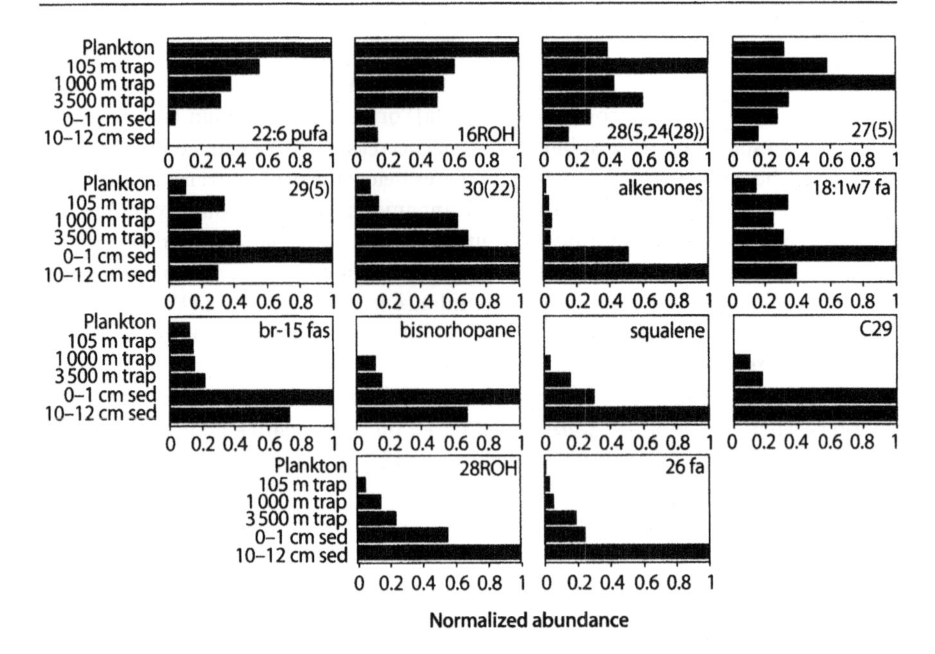

Fig. 5.7. Scaled abundances of biomarkers in EqPac samples

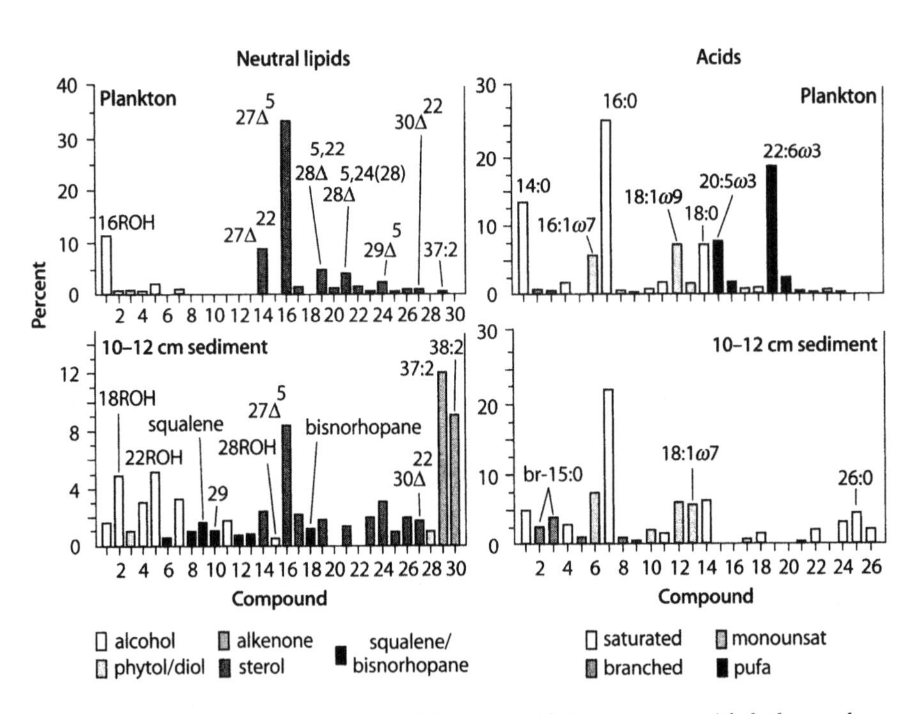

Fig. 5.8. Relative abundances of major neutral lipids and acids in EqPac equatorial plankton and 10–12 cm sediments (see Table 5.1 for compound list and abbreviations)

These scaled profiles of relative abundances demonstrate *(a)* rapid losses of planktonic material in surface waters, *(b)* elevated proportions of heterotrophic biomarkers in the mid-water column and sediments, and *(c)* preservation of selected components of bacteria, plankton, and vascular plants in sediments. This diagenetic selection is largely responsible for the marked compositional differences between plankton and subsurface sediments (Fig. 5.8). Planktonic biomarkers exhibit markedly different types of behavior. Polyunsaturated compounds such as 22:6ω3 and short-chain chain alcohols such as hexadecanol are subject to extensive decomposition early during transit through the water column, 24-methylcholesta-5,24(28)-dien-3β-ol ($28\Delta^{5,24(28)}$) is relatively better preserved, while 24-ethylcholest-5-en-3β-ol, 4α,23,24-trimethylcholest-22E-en-3β-ol and alkenones are remarkably well preserved in sediments. It is uncertain as to why 24-ethylcholest-5-en-3β-ol and 4α,23,24-trimethylcholest-22E-en-3β-ol are so well preserved, but the stability of alkenones has been suggested to result from their unusual double bond configurations (Rechka and Maxwell 1988a,b). The mid-water maximum for cholest-5-en-3β-ol is indicative of a mid-water zooplankton source (Wakeham et al. 1984b). *Cis*-vaccenic acid and *iso*- and *anteiso*-15:0 acids are present in the water column in minor abundances, possibly due to their production by bacteria in zooplankton guts or on particles; their enhanced abundances in sediments could indicate preferential preservation in sediments as well as production in situ. Bisnorhopane and squalene appear to be produced largely in sediments via microfaunal activity. Long-chain hydrocarbons, fatty alcohols and fatty acids derived from vascular plants are relatively refractory constituents of water column particles that become magnified in sediments as most of the surface-derived marine material is respired. The stability of vascular plant biomarkers may be enhanced by relatively high molecular weights and a protective waxy particle matrix (Wakeham et al. 1984b).

5.6 Summary

Interpreting past oceanic environments from the sediment record must be undertaken with caution since the lipid assemblages preserved in sediments may significantly under-represent processes occurring in the overlying water column. Diagenesis in the water column and surface sediments degrades most organic matter. Furthermore, the tiny residue of organic matter that is buried in sediments may present a highly biased view of organic matter inputs. Sedimentary lipid distributions, in general, poorly reflect the activities and abundances of most planktonic organisms, while at the same time overrepresent the importance of other organic matter sources, including some phytoplankton, terrigenous vascular plants, and bacteria. The factors resonsible for these biases require further investigation in oceanic water columns and diagenetically active sediments before sedimentary lipid records may be applied to their full potential.

Acknowledgements

Work described here resulted from collaborative projects, and I thank C. Lee, J. Hedges, P. Hernes, and M. Peterson for their assistance. These projects have been supported by the U. S. National Science Foundation.

Table 5.1. Compound identifications and abbreviations

Bar	Acids	Neutral lipids
1	14:0	C_{16}-alcohol (16 ROH)
2	i-15:0	C_{18}-alcohol (18 ROH)
3	a-15:0	phytol (phytol)
4	15:0	C_{20}-alcohol
5	i-16:0	C_{22}-alcohol (22 ROH)
6	16:1ω7	C_{27}-alkane
7	16:0	C_{24}-alcohol
8	i-17:0	C_{28}-alkane
9	a-17:0	squalene
10	17:1	C_{29}-alkane (n-C_{29})
11	17:0	C_{26}-alcohol
12	18:1ω9	C_{30}-alkane
13	18:1ω7	C_{31}-alkane
14	18:0	cholesta-5,22E-dien-3β-ol ($27\Delta^{5,22}$)
15	20:5ω3	C_{28}-alcohol
16	20:4ω3	cholest-5-en-3β-ol ($27\Delta^{5}$)
17	20:1	5α(H)-cholestan-3β-ol
18	20:0	bisnorhopane
19	22:6ω3	24-methylcholesta-5,22E-dien-3β-ol ($28\Delta^{5,22}$)
20	22:5ω3	24-methylcholest-22E-en-3β-ol
21	22:1	24-methylcholesta-5,24(28)-dien-3β-ol ($28\Delta^{5,24(28)}$)
22	22:0	24-methylcholest-24(28)-en-3β-ol
23	24:1	24-ethylcholesta-5,22E-dien-3β-ol
24	24:0	24-ethylcholest-5-en-3β-ol ($29\Delta^{5}$)
25	26:0	24-ethyl-5α(H)-cholestan-3β-ol
26	28:0	24-ethylcholesta-5,24(28)-dien-3β-ol
27		4α,23,24-trimethylcholest-22E-en-3β-ol ($30\Delta^{22}$)
28		C_{30}-alkanediol (diol)
29		37:2 alkenone (37:2)
30		38:2 alkenone (38:2)

References

Barber RT, Sanderson MP, Lindley ST, Chai F, Newton J, Trees CC, Foley DG, Chavez FP (1996) Primary productivity and its regulation in the equatorial Pacific during and following the 1991–92 El Niño. Deep-Sea Res II 43:933–969

Boon JJ, Rijpstra WIC, De Lang F, de Leeuw JW (1979) Black Sea sterol – a molecular fossil for dinoflagellate blooms. Nature 277:125–127

Brassell SC (1993) Application of biomarkers for delineating marine paleoclimatic fluctuations during the pleistocene. In: Engel MH, Macko SA (eds) Organic geochemistry. Principles and applications. Plenum Press, New York, pp 699–638

Conte MH, Volkman JK, Eglinton G (1994) Lipid biomarkers of the Prymnesiophyceae. In: Green JC, Leadbetter BSC (eds) The haptophyte algae. Clarendon Press, Oxford, pp 351–377

Cranwell PA (1982) Lipids of aquatic sediments and sedimenting particles. Prog Lipid Res 21:271–308

Hernes PJ, Hedges JI, Peterson ML, Wakeham SG, Lee C (1996) Neutral carbohydrate geochemistry of particulate matter in the central Equatorial Pacific. Deep-Sea Res II 43: 1181–1204

Huang W-Y, Meinschein WG (1979) Sterols as ecological indicators. Geochim Cosmochim Acta 43:739–745

Kaneda T (1991) Iso- and anteiso- fatty acids in bacteria: biosynthesis, function, and taxonomic significance. Microbiol Rev 55:288–302

Kolattukudy PE (1976) Chemistry and biochemistry of natural waxes. Elsevier, New York, 459 pp

Lee C, Murray DW, Barber RT, Buesseler KO, Dymond J, Hedges JI, Honjo S, Manganini SJ, Mara J, Moser C, Peterson ML, Prell WL, Wakeham SG (1998) Particulate organic carbon fluxes: Compilation of results from the 1995 U.S. JGOFS Arabian Sea Process Study. Deep-Sea Res (In press)
Leeuw de JW, Largeau C (1993) A review of macromolecular organic compounds that comprise living organisms and their role in kerogen, coal, and petroleum formation. In: Engel MH, Macko SA (eds) Organic geochemistry. Principles and applications. Plenum Press, New York, pp 23–72
Martin J, Knauer GA, Karl DM, Broenkow WW (1987) VERTEX: Carbon cycling in the northeast Pacific. Deep-Sea Res 34:267–285
Murray JW, Leinen MW, Feely RA, Toggweiler JR, Wanninkof R (1992) EqPac: A process study in the central equatorial Pacific. Oceanography 5:134–142
Murray JW, Johnson E, Garside C (1995) A U.S. JGOFS Process Study in the equatorial Pacific (EqPac): Introduction. Deep-Sea Res II 42:275–293
Ourisson G, Rohmer M, Poralla K (1987) Prokaryotic hopanoids and other polyterpenoid sterol surrogates. Annu Rev Microbiol 41:310–333
Parsons TR, Takahashi M, Hargrave B (1984) Biological oceanographic processes. Pergamon Press, Oxford
Prahl FG, Muelhausen LA, Lyle M (1989) An organic geochemical assessment of oceanographic conditions at MANOP Site C over the past 26,000 years. Paleoceanogr 4:495–510
Rechka JA, Maxwell JR (1988a) Unusual long-chain ketones of algal origin. Tetrahedron Lett 29:2599–2600
Rechka JA, Maxwell JR (1988b) Characterization of alkenone temperature indicators in sediments and organisms. In: Mattavelli L, Novelli L (eds) Advances in organic geochemistry 1987. Org Geochem 13:727–734
Risatti JB, Rowland SJ, Yon DA, Maxwell JR (1984) Stereochemical studies of acyclic isoprenoids – XII. Lipids of methanogenic bacteria and possible contributions to sediments. Org Geochem 6:93–104
Robinson N, Eglinton G, Brassell SC, Cranwell PA (1984) Dinoflagellate origin for sedimentary 4α-methylsteroids and 5α(H)-stanols. Nature 308:439–422
Rowland SJ, Robson JN (1990) The widespread occurrence of highly branched acyclic C_{20}, C_{25}, and C_{30} hydrocarbons in recent sediments and biota – a review. Mar Environ Res 30:191–216
Santos V, Billett DSM, Rice AL, Wolff GA (1994) Organic matter in deep-sea sediments from the Porcupine Abyssal Plain in the north-east Atlantic Ocean. I. – Lipids. Deep-Sea Res 41:787–819
Sargent JR (1976) The structure, function, and metabolism of lipids in marine organisms. In: Malins DC, Sargent JR (eds) Biochemical and biophysical perspectives in marine biology, vol III. Academic Press, New York, pp 149–212
Sargent JR, Henderson RJ (1986) Lipids. In: Corner EDS, O'Hara SCM (eds) The biological chemistry of marine copepods. Clarendon Press, Oxford, pp 59–108
Siegel DA, Granata TC, Michaels AF, Dickey TD (1990) Mesoscale eddy diffusion, particle sinking, and the interpretation of sediment trap data. J Geophys Res 95:5305–5311
Suess E (1980) Particulate organic carbon flux in the ocean – surface productivity and oxygen utilization. Nature 288:260–263
Volkman JK (1986) A review of sterol markers for marine and terrigenous organic matter. Org Geochem 9:83–99
Volkman JK, Eglinton G, Corner EDS, Forsberg TEV (1980) Long-chain alkenes and alkenones in the marine coccolithophoprid *Emiliania huxleyi*. Phytochem 19:2619–2622
Volkman JK, Barrett SM, Dunstan GA (1994) C25 and C30 highly branched isoprenoid alkenes in laboratory cultures of two marine diatoms. Org Geochem 21:407–413
Wakeham S, Lee C (1993) Production, transport, and alteration of particulate organic matter in the marine water column. In: Engel MH, Macko SA (eds) Organic geochemistry. Prinicples and applications, Plenum Press, New York, pp 145–169
Wakeham SG, Volkman JK (1991) Sampling and analysis of lipids in marine particulate matter. In: Hurd DC, Spencer DW (eds) Marine particles: Analysis and characterization. American Geophysical Union, Washington, DC, pp 171–179
Wakeham SG, Gagosian RB, Farrington JW, Canuel EA (1984a) Sterenes in suspended particulate matter in the eastern tropical North pacific. Nature 308:840–843
Wakeham SG, Gagosian RB, Farrington JW, Lee C (1984b) Biogeochemistry of particulate organic matter in the oceans – results from sediment trap experiments. Deep-Sea Res 31:509–528
Wakeham SG, Hedges JI, Lee C, Hernes PJ, Peterson ML (1997a) Molecular indicators of diagenetic status in marine organic matter. Geochim Cosmochim Acta 24:5363–5369
Wakeham SG, Hedges JI, Lee C, Peterson ML, Hernes PJ (1997b) Compositions and fluxes of lipids through the water column and surficial sediments of the equatorial Pacific Ocean. Deep-Sea Res II 44:2131–2162
Weller RA, Baumgartner MF, Josey SA, Fischer AS, Kindle J (1998) Atmospheric forcing in the Arabian Sea during 1994–1995: Observations and comparisons with climatology and models. Deep-Sea Res II 45:1961–1999

Chapter 6

Organic Chemical Reaction Rates in the Ocean: Molecular Approaches to Studying Extracellular Biochemical Processes

S. Pantoja

6.1 Introduction

The recycling efficiency of organic matter produced in situ or supplied externally to the ocean is so high that continuous regeneration of recycled nutrients allows levels of production that would not be possible otherwise. In practice, photosynthetic production of organic matter in the ocean (simplified as $H_2O + CO_2 +$ nutrients $\longrightarrow$ organic matter + electron acceptor) is almost balanced by the reverse reaction, respiration (organic matter + electron acceptor $\longrightarrow H_2O + CO_2 +$ nutrients), therefore, more than 99% of the organic matter produced is recycled (Hedges 1992). The remaining organic matter is buried in sedimentary rocks (see Chapter 5). Uplifting and weathering of sedimentary rocks provide a link between biologically and geologically-mediated processes.

A simplified model of organic matter recycling is shown in Fig. 6.1. In this model, organic matter produced in the photic zone, mainly by photosynthetic organisms, is degraded to inorganic compounds, a process called mineralization. In the open ocean, only 1% of the organic matter produced by photosynthesis escapes mineralization and reaches the bottom waters (Henrichs and Reeburgh 1987). In sediments, production and degradation continue, leaving only a small fraction of the organic matter produced by photosynthesis to be preserved in the sedimentary record (ca. 0.1%, Hedges 1992).

Degradation of organic matter has been studied in sea water and sediments in order to understand cycling of chemical compounds in the environment. Furthermore, degradation supplies nutrients for photosynthesis and labile[1] molecules for bacterial consumption. Organic matter is continuously removed from surface waters by the downward sinking of dead organisms and faecal pellets. Degradation provides a mechanism to release nutrients from the organic matrix by transformation into the inorganic form. These nutrients, released at depth due to degradation maybe transported up to the photic zone by mixing, and support new production. Eppley and Peterson (1979) showed that new production approximates the sinking flux of particulate organic matter.

Analysing small concentrations of individual organic compounds in seawater has been a challenge to chemical oceanographers, yet reactions involving these compounds are critical links within biogeochemical processes. For instance, it has been recognized that the chemical structure of organic matter may determine the extent and rates of degradation (e.g. Henrichs and Doyle 1986; Canfield 1994). Understanding how mol-

[1] Labile molecules are defined here as the ones that easily degrade via biochemical reactions.

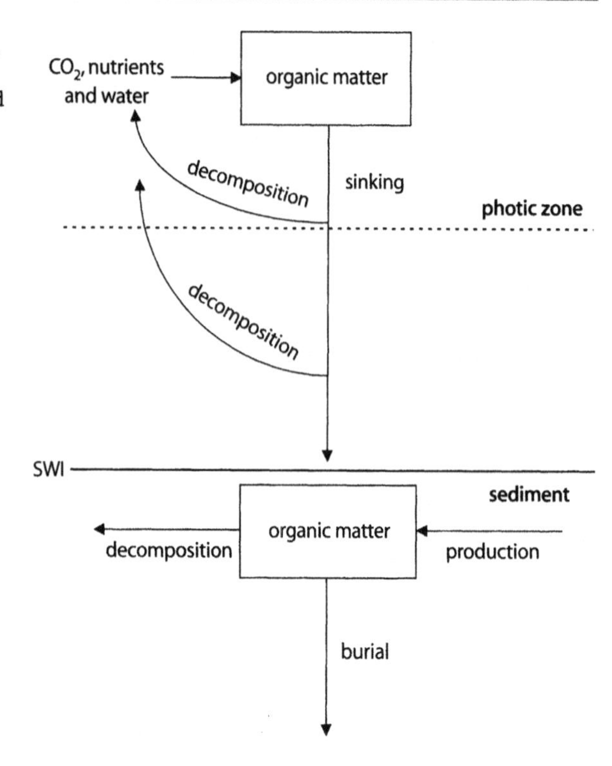

Fig. 6.1. Simplified representation of degradation of organic matter in the water column and sediments (SWI = sediment-water interface; size of compartments not to scale)

ecules are chemically transformed will help understand how those transformations affect:

1. the distribution of organic matter in the ocean,
2. the amount of organic matter that is degraded vs. preserved, and
3. factors that control primary and secondary production.

Moreover, a more detailed study of those processes at the molecular level may reveal new pathways of nutrient production and mechanisms of organic matter degradation.

Several microbially-mediated chemical reactions that use organic matter as a substrate occur outside the cellular membrane of microorganisms (e.g. Palenik and Morel 1990a; Arnosti et al. 1994; Ammerman and Azam 1991; Chróst 1991). Measurements of these reactions at the molecular level are difficult to make, since the concentration of reactants and products is very low, and these compounds participate simultaneously in several reactions in the ocean (sinks and sources).

The core of this chapter will be devoted to reviewing some novel molecular approaches to the study of extracellular biochemical reactions in the ocean. The discussion will center on the two compound classes that pertain to the cycling of organic nitrogen: amino acids and proteins. Thus, Section 6.2 will describe the process of oxidative deamination of amino acids by cell surface deaminases, which results in both removal of amino acids and consequent production of ammonium in sea water. The carbon skeleton and the inorganic nitrogen are presumed to be subsequently used by microorganisms. The second example (Section 6.3) covers degradation of peptides in seawater. Here, we will concentrate on the microbial extracellular hydrolytic step prior to mineralization, because it appears to be a key reaction in degradation of organic matter in the ocean.

6.2 Oxidative Deamination of Amino Acids in Sea Water by Cell Surface Deaminases

Nitrogen is one of the essential nutrients for phytoplankton growth. One of the major steps in the nitrogen cycle in the sea is the uptake of inorganic nitrogen by phytoplankton as a nutrient source for the synthesis of proteins and nucleic acids.

Traditionally, it was thought that phytoplankton only obtain the nitrogen they require from inorganic nitrogen compounds, e.g. nitrate, nitrite and ammonium. Depending on the nitrogen source we differentiate between new and regenerated production, concepts of great ecological value (e.g. Eppley and Peterson 1979). Regenerated production is commonly estimated as ammonium uptake, because ammonium is regenerated by heterotrophic processes in the photic zone. More recently, however, phytoplanktonic uptake of regenerated nitrogen in the form of dissolved organic nitrogen has been identified in some marine systems, but evidence of its uptake by phytoplankton is not straightforward (Antia et al. 1991; Paerl 1991).

A new mechanism of removal of amino acids and amines from seawater was proposed by Palenik and Morel (1990a, 1991). They showed that certain phytoplanktonic species can oxidize organic nitrogen compounds using cell surface deaminases without transferring the complete molecule across the cell membrane (Fig. 6.2). This extracellular process involves the cell surface oxidation of L-amino acids or other primary amines to produce an oxidized organic product (α-keto acid or aldehyde, respec-

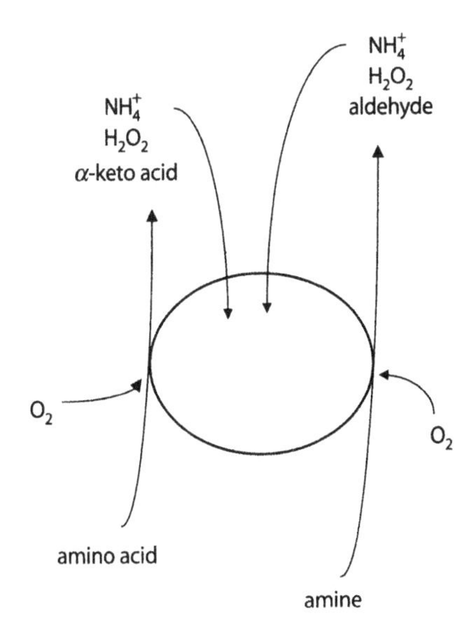

Fig. 6.2. Cell surface oxidative deamination of amino acids and amines (after Palenik et al. 1988)

tively), H_2O_2, and NH_4^+ in equimolar concentrations. They also demonstrated that the ammonium produced was assimilated by the cells, whereas the organic product and hydrogen peroxide remained in solution.

If oxidative deamination is a major pathway, the consequences of this reaction are:

1. Ammonium produced by this mechanism is another source of NH_4^+ for regenerated production.
2. If bacteria consume the α-keto acids produced, whereas the ammonium is taken up by phytoplankton, carbon and nitrogen may be decoupled.
3. Deamination of amino acids to α-keto acids is another pathway for the removal of amino acids, (in addition to heterotrophic consumption by bacteria).
4. Oxidative deamination is another source of hydrogen peroxide in sea water.

However, the quantitative importance of amino acid oxidation in natural waters is difficult to assess. This mechanism (or any involving cell surface processes) may be easily overlooked due to methodological constraints. Amino acid recycling is usually studied using radiolabelled amino acids (^{14}C, ^{3}H) which are added to natural sea water or sediment, and the label is followed into the bacterial biomass, with no information on routes of uptake. Because of the complexity of the sources and sinks of amino acids, α-keto acids, aldehydes, H_2O_2, and NH_4^+ in sea water, measuring the changes in concentration of these compounds due only to cell surface oxidation is not easily accomplished (note that even if radio labelling any molecule in Fig. 6.2, it would be difficult to trace its pathway inside and outside the cell, and more difficult to determine oxidation rates alone).

6.2.1 Synthesis of a Fluorescent Analog of L-lysine as Substrate for Amino Acid Oxidases in the Ocean

Pantoja et al. (1993) synthesized a fluorescent analog of L-lysine by condensing a fluorescent moiety (4-amino-3,6-disulfo-1,8-napththalic anhydride or Lucifer Yellow Anhydride) onto the ε-amino group of the amino acid, leaving the α-amino group free for reaction. The resulting product is a stable water-soluble imide (hereafter LYA-lysine) which exhibits excitation and emission maxima at 424 and 550 nm, respectively (Fig. 6.3).

Oxidative deamination of the fluorescent analog by cell surface deaminases will produce an α-keto acid derivative, which, being unstable in seawater, will further oxidize to its acid (Fig. 6.4). Use of these probes has several advantages:

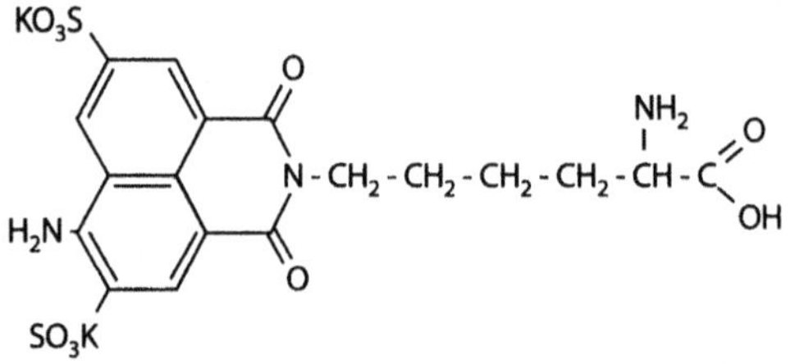

Fig. 6.3. A fluorescent analog of L-lysine: [N(1'-(5'-amino-5'-carboxypentyl))-4-amino-3,6-disulfo-1,8-naphthalimide, dipotassium salt], hereafter LYA-lysine

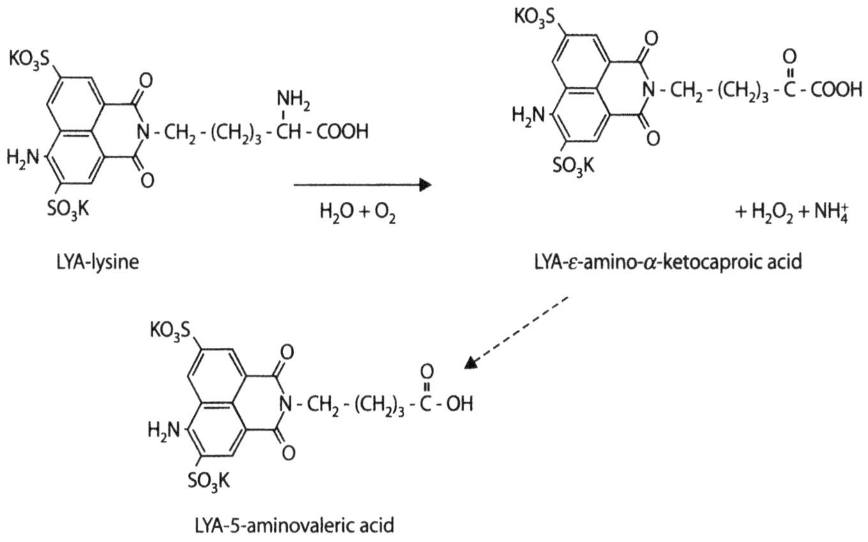

Fig. 6.4. Oxidative deamination of LYA-lysine

1. Fluorescent analogs can be detected by High Pressure Liquid Chromatography (HPLC) with an on-line fluorometer. Therefore, the disappearance of the added substrate from a seawater sample and concurrent production of the oxidation product can be followed with time, allowing the estimation of oxidation rates.
2. The size of the probes precludes incorporation across the cell membrane of microorganisms, therefore only extracellular reactions are detected.
3. The detection limit is about 5 nM, which is in the range of (or lower than) the concentration of an individual amino acid in sea water.
4. They are stable to chemical reactions on the time scale of the biological reactions we are measuring (days to weeks).
5. Experiments determined that natural L-amino acids compete with the fluorescent derivative suggesting similarity between substrates as explained below.

In a phytoplankton culture, competitive inhibition of the oxidation of LYA-lysine was observed in the presence of the mixture of L-alanine and L-isoleucine (0.5 μM each) (Fig. 6.5). The half-saturation constant (K_s) of the oxidation of LYA-lysine increased upon the addition of these amino acids from 0.11 μM (uninhibited) to 0.76 μM (inhibited). The maximum velocity (V_m) of the oxidation remains unaltered (about 0.6 nM min^{-1}) in both cases, as the regression lines share the same y-intercept ($1 / V_m$). Competitive inhibition is demonstrated as an increase in K_s of an enzymatic reaction in the presence of an inhibitor (Dixon and Webb 1958). K_s increases because a larger amount of substrate is needed to convert the enzyme to the enzyme-substrate mixture in the presence of the inhibitor. On the other hand, since the influence of the competitive inhibitor is less at high substrate concentration, V_m is independent of the presence of competitors (Dixon and Webb 1958). In summary, L-amino acid oxidases apparently do not discriminate between natural amino acids and LYA-lysine.

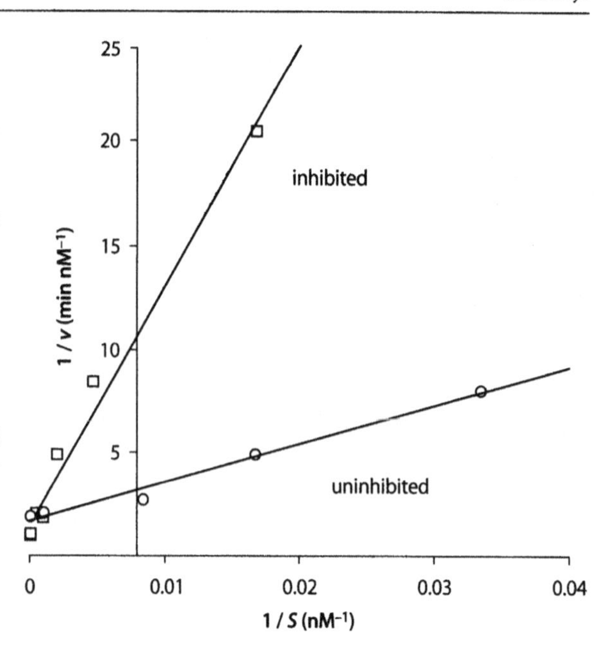

Fig. 6.5. Lineweaver-Burk plot showing the oxidation rate of LYA-lysine by Cocco II (*Pleurochrysis carterae*) before (*circles*) and after (*squares*) addition of a mixture of 0.5 μM L-alanine and 0.5 μM L-isoleucine. $1/v$, reciprocal of rate of oxidation of LYA-lysine (nM min^{-1}) and $1/S$, reciprocal of concentration of LYA-lysine (nM). Coefficient of variation (CV) for each point is 2–10%. Equations for the uninhibited and inhibited cases are $1/v = 191.1\ (1/s) + 1.7$, $r^2 = 0.98$ and $1/v = 1165.9\ (1/s) + 1.5$, $r^2 = 0.98$, respectively (reprinted from Pantoja and Lee 1994, with permission from ASLO)

6.2.2 LYA-lysine as Substrate for Cell Surface Deaminases in Sea Water

In a typical experiment, a water sample is amended with LYA-lysine. Monitoring of the progress of the reaction shows the disappearance of the substrate and the production of the oxidation product (LYA-ε-amino-α-ketocaproic acid) (Fig. 6.6). No oxidation was detected when the probe was incubated in 0.2 μm-filtered culture medium, confirming that oxidation was associated with particles and not with free dissolved enzymes. The absence of cross-membrane transport of the probes was demonstrated by mass balance of the fluorescent analogs in the dissolved fraction during the incubation.

Other reactions that remove amino acids from sea water give rates of similar magnitudes (Table 6.1, Pantoja and Lee 1994). During summer 1993, experiments showed that deamination can account for 20% of the microbial removal of amino acids from sea water. During winter 1992, no deamination was detected.

Pantoja and Lee (1994) explored the potential environmental control of extracellular activity by surveying activity under different temperature regimes. Again, experiments carried out during winter showed no detectable oxidation of LYA-lysine. Oxidation occurred only in the summer in waters above 22 °C (Fig. 6.7). Palenik and Morel (1990b) found that only certain species of cultured phytoplankton possess oxidative capacity. Thus, species succession in the planktonic community due to seasonal and temperature changes are most likely triggering the change in oxidative activity.

The other factor that may affect oxidative deamination could be enzymatic induction, when ammonium is depleted in the medium. Palenik and Morel (1990a) detected higher rates when cultured phytoplankton cells were growing under nitrate, but no activity was detected when they were growing under ammonium. Ammonium is thought to be taken up preferentially over nitrate (presumably because the first intracellular

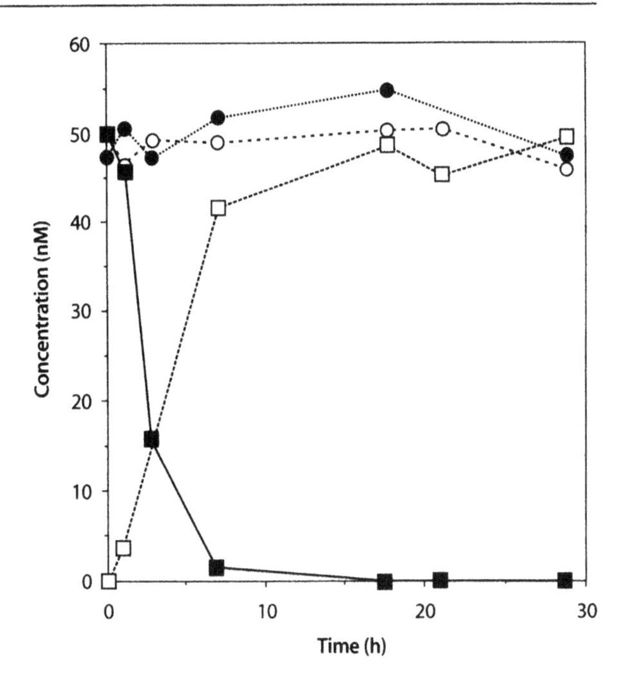

Fig. 6.6. Time course of the oxidation of LYA-lysine (50 nM addition) (*black squares*) and the production of LYA-ε-amino-α-ketocaproic acid plus LYA-5-aminovaleric acid (*white squares*) by the fraction <0.8 μm of a water sample taken in July 1991 from West Neck Bay. Parallel incubations of the probe with killed cells (*black circles*) and with the dissolved (<0.2 μm) fraction (*white circles*) show no significant change from the initial concentration added. CV for each point is 2–10% (reprinted from Pantoja and Lee 1994, with permission from ASLO)

Table 6.1. Rates of uptake and cell surface oxidation amino acids in Flax Pond (NY) sea water; ND, not detected

	February 1992	June 1993
Rate of uptake of ^{14}C-amino acid (nM h^{-1})	6.3	39
Rate of oxidation of LYA-amino acid (nM h^{-1})	ND	10

transformation of nitrate is its reduction to ammonium), therefore uptake of ammonium would be energetically preferable than either uptake of nitrate or oxidation of amino acids. Current attempts to relate the supply of inorganic nutrients and rates of deamination of amino acids are giving hints to the nutrient control of cell surface activity. In a cross-ecosystem comparison of several environments, Mulholland et al. (1998) reported that LYA-lysine oxidation rates may represent up to 12% of ammonium uptake rates, when ammonium concentrations were undetectable in oligotrophic waters (Fig. 6.8).

Production of inorganic nutrients by exo-enzymatic mechanisms may benefit phytoplankton. Thus, ammonium regenerated as a product of oxidative deamination of amino acids could provide regenerated nutrients for phytoplankton. The classical view of marine ecosystems has been that phytoplankton take up inorganic nitrogen, and bacteria take up organic nitrogen compounds, to fulfill their nutrient needs. However, recent evidence shows that bacteria are at least as efficient at consuming both ammonium and phosphate as phytoplankton are (Currie and Kalff 1984; Suttle et al. 1990). The discovery of these new mechanisms reinforces the idea that planktonic popula-

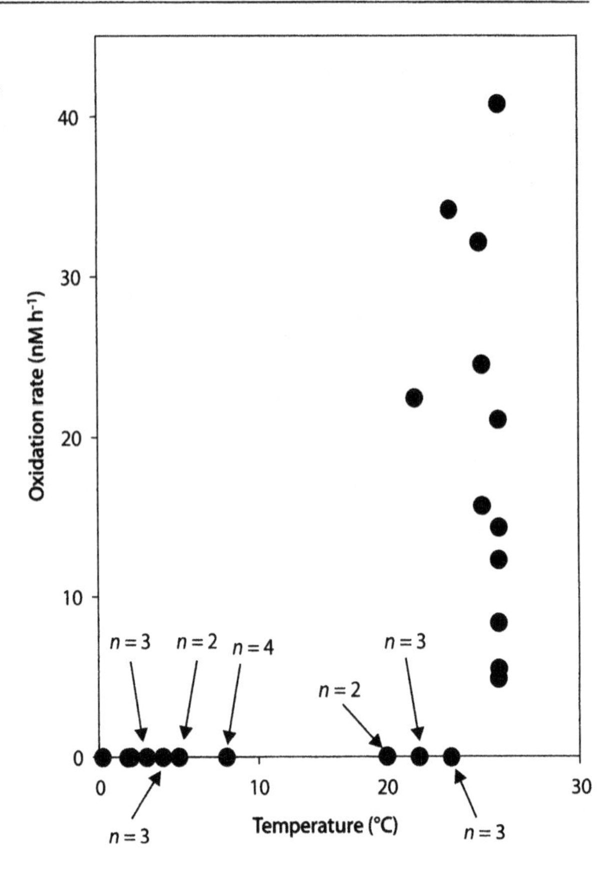

Fig. 6.7. Relationship between oxidation rate of LYA-lysine (nM h^{-1}) and water temperature (°C) at time of sampling. All samples taken are shown; when rates were below detection limits, the number of samples (n) is indicated (reprinted from Pantoja and Lee 1994, with permission from ASLO)

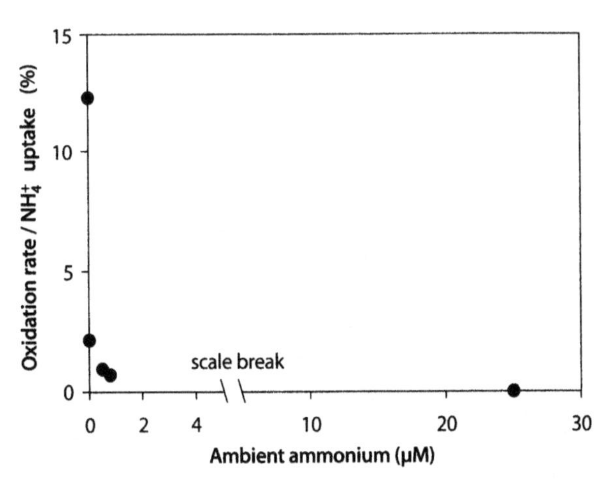

Fig. 6.8. Percentage of oxidation rate of LYA-lysine and ammonium uptake vs. ambient concentration of ammonium at the time of sampling in several environments (data extracted from Mulholland et al. 1998)

tions display more versatile pathways to obtaining nutrients they require than we have traditionally thought.

6.3
Extracellular Hydrolysis of Peptides in Sea Water and Sediments

Peptides and proteins are important components of dissolved organic matter, accounting for 5–20% of dissolved organic nitrogen and 3–4% of dissolved organic carbon (*DOC*) in seawater (Sharp 1983; Thurman 1985). In surface sediments, the protein pool accounts for ca. 10% of organic carbon (Degens 1970; Burdige and Martens 1988) and 30–40% of total nitrogen (Rosenfeld 1979; Henrichs et al. 1984; Burdige and Martens 1988).

In the ocean, photosynthetic organic matter produced in the photic zone involves the cellular synthesis of macromolecules. On average, more than 50% of the dry weight of phytoplankton is in the form of proteins, and more than 30% in polysaccharides (Parsons et al. 1961). In summary, most organic matter (about 80%) in phytoplankton is in the form of macromolecules.[2] As discussed before, most of this planktonic material is degraded in the water column or in surface sediments.

The current model of macromolecular degradation in the ocean (Fig. 6.9) assumes that extracellular hydrolysis of macromolecules yields smaller molecules (e.g. Billen 1984; Chróst 1991; Hoppe 1991). Thus, protein degradation will produce peptides and free amino acids via free enzymes or enzymes associated with microorganisms or other particles. Although the presence of peptides (amino acid sequences of MW < 6 000 D) in marine environments has not been clearly documented, they are believed to be important intermediates in the degradation of organic matter, and most of the carbon and nitrogen in protein may be cycled through this pathway.

Although larger organisms can consume protein and hydrolyze it internally, bacteria must initially hydrolyze proteins and peptides to smaller substrates outside the cell (Payne 1980). When they are small enough (oligopeptides and amino acids), they are incorporated across the cell membrane and mineralized. Molecules larger than approximately 600 Daltons cannot be transported across microbial cell membranes (Nikaido and Vaara 1985).

6.3.1
Experimental Approaches to Studying Degradation of Proteinaceous Material in the Ocean

6.3.1.1
Radiolabelled Substrates

Several radiolabelled substrates have been used to measure rates of degradation of proteins in seawater: methyl-[^{14}C]methemoglobin and [^{125}I]BSA[3] (Hollibaugh and Azam 1983), [^{14}C]-methylated protein mixture (Billen 1991), and [^{14}C]-uniformly-labelled Rubisco[4] and [^{14}C]-glucosylated Rubisco (Kiel and Kirchman 1993). In sediments,

[2] Macromolecules are defined here as compounds of molecular weight larger than about 1000 Daltons. To give a sense of the units: molecular weight of water = 18, average amino acid = 141, a large protein = 600 000 Daltons.]

[3] Bovine serum albumin.

[4] Ribulose-1,5-bisphosphate carboxylase/oxygenase.

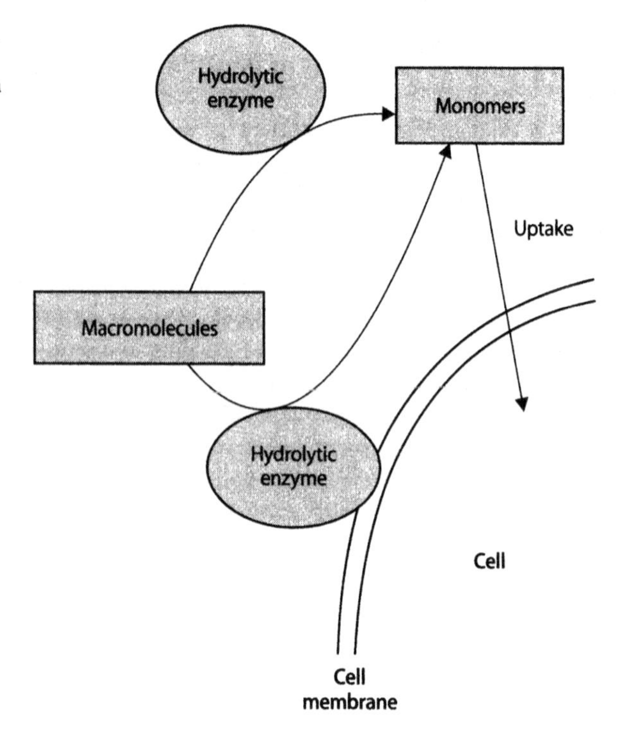

Fig. 6.9. Schematic representation of degradation of macromolecular via the participation of cell surface or extracellular hydrolytic enzymes

Ding and Henrichs (1998) have recently measured adsorption and decomposition of protein in Alaskan sediments using [^{14}C]-uniformly-labelled Rubisco. In these assays, degradation of proteins is estimated by performing some or all the following measurements after incubation of the protein substrate in a sample:

a. Determination of the disappearance of the original protein. Radioactivity of the remaining protein is measured with trichloroacetic acid (TCA);
b. Determination of the appearance of radioactivity in smaller peptides after ultrafiltration;
c. Respiration of incorporated protein measured as $^{14}CO_2$ or ^3H_2O production.

More recently, Luo (1994) elongated the C-terminal end of several oligopeptides with a tritiated amino acid. The resulting ^{3}H-labelled peptides were used as substrates for hydrolysis and mineralization in anoxic marine sediments.

6.3.1.2
Fluorogenic Substrates

These are molecules that contain an amide bond[5] between an amino acid and an aromatic structure. The most common are leucine-4-methyl-7-coumarinyl-amide (Leu-

[5] Amides are compounds in which the -OH of a carboxylic acid has been replaced by $-NH_2$.

MCA, Kanaoka et al. 1977; Hoppe 1983), and leucyl-β-naphthylamide (LLβN, Somville and Billen 1983). The assay is based on the release of a fluorescent moiety (7-amino-4-methylcoumarin and β-naphthylamine, respectively) when the amide bond is cleaved by hydrolytic enzymes. The intensity of the fluorescence is related to proteolytic activity (Fig. 6.10).

6.3.1.3
Fluorescent Substrates (LYA-peptides)

Pantoja et al. (1997) synthesized fluorescent peptide analogs of different amino acid constituents by condensing the amino-nitrogen functional group of commercially available peptides with 4-amino-3,6-disulfo-1,8-napththalic anhydride. The procedure allows us to determine rates of extracellular hydrolysis at specific bonds, since fluorescent peptide substrates and products can be separated by HPLC and quantified with an on-line fluorometer (Fig. 6.11).

6.3.2
Rate Measurements

Most previous measurements of the hydrolysis rates of protein have used fluorogenic dimers such as leucine-methyl-coumarinyl-amide (Section 6.3.1.2). These studies have given us better insight into pathways and rates of degradation of organic matter in the marine environment, pointing out the importance of macromolecular degradation in the recycling of organic matter. The next step is an investigation of the behavior of different molecular classes and individual compounds. The influence of chemi-

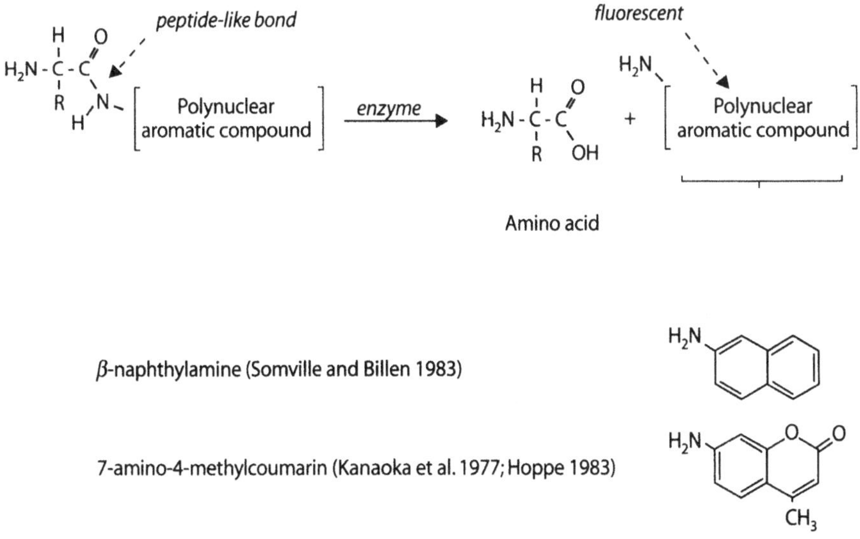

peptide-like bond

Polynuclear aromatic compound

enzyme

fluorescent

Polynuclear aromatic compound

Amino acid

β-naphthylamine (Somville and Billen 1983)

7-amino-4-methylcoumarin (Kanaoka et al. 1977; Hoppe 1983)

Fig. 6.10. General mechanism of reaction of fluorogenic substrates. Fluoropores are β-naphthylamine, methylumbelliferyl, 7-amino-4-methylcoumarin

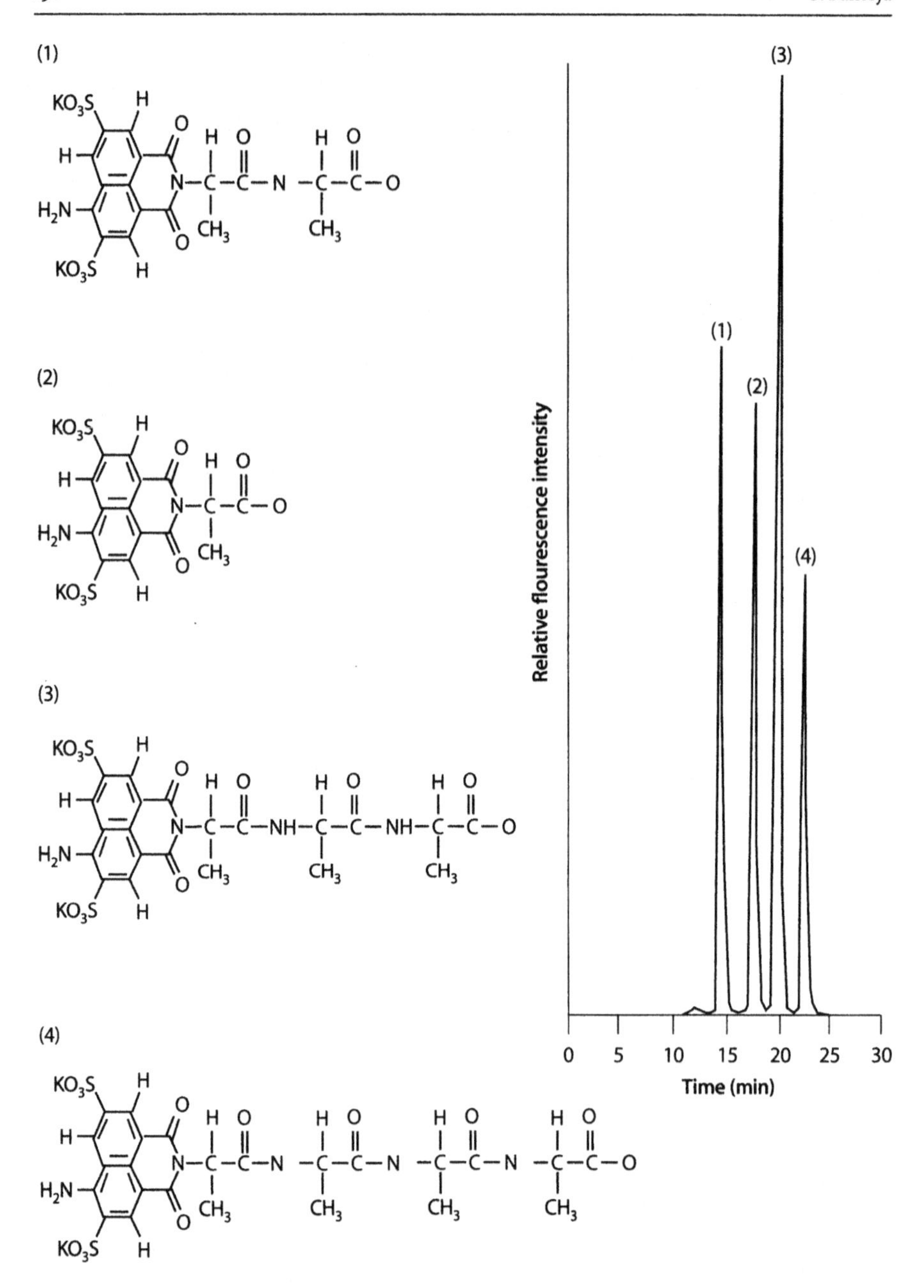

Fig. 6.11. HPLC separation of (*1*) LYA-ala$_2$, (*2*) LYA-ala, (*3*) LYA-ala$_3$ and (*4*) LYA-ala$_4$ (reprinted from Pantoja and Lee 1994, with permission from Elsevier Science)

cal structure on degradation, for example, maybe studied using the fluorescent peptides described in Section 6.3.1.3.

A typical time course of incubation in sea water is shown in Fig. 6.12. Substrates disappear from sea water by hydrolysis, concurrent with the production of smaller pep-

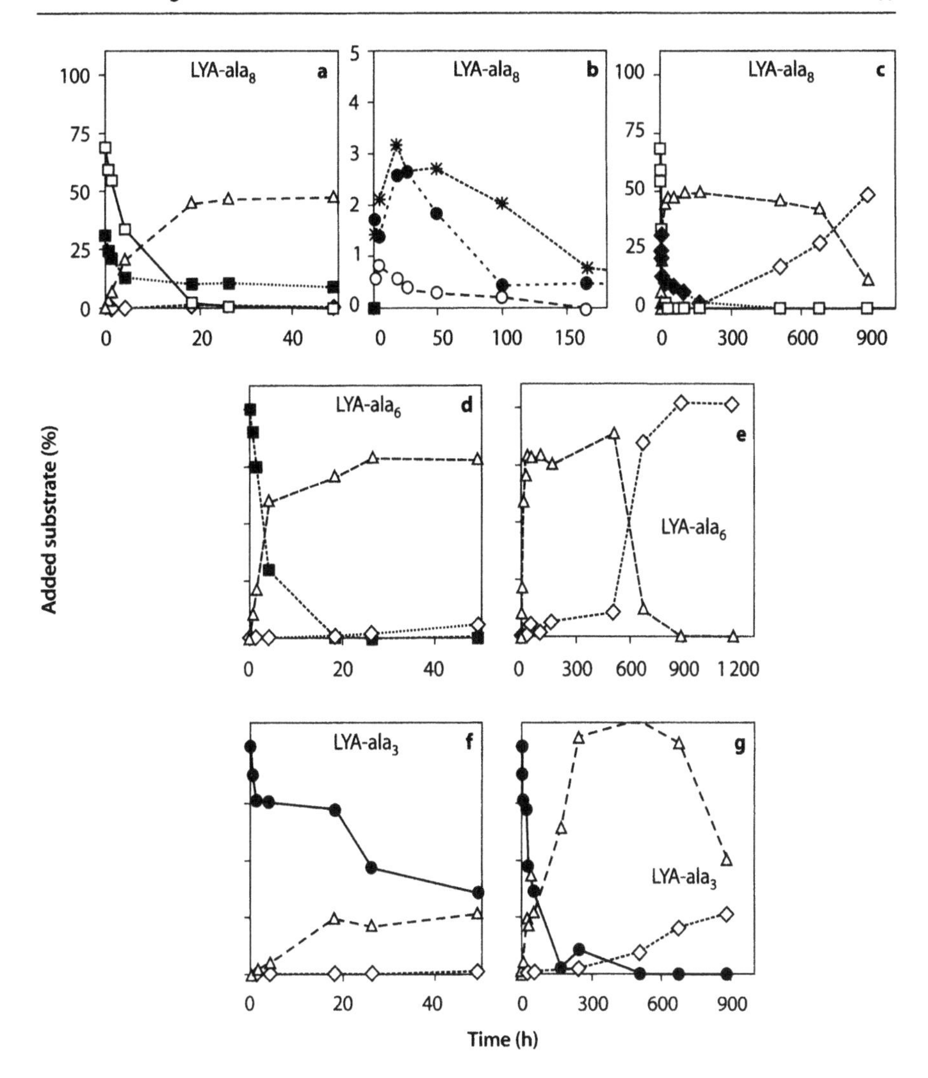

Fig. 6.12. Time course of hydrolysis of; **a–c** LYA-ala$_8$ (*white squares*); **d–e** LYA-ala$_6$ (*black squares*); **f–g** LYA-ala$_3$ (*black circles*) added to Flax Pond sea water. LYA-ala$_5$ (*stars*), LYA-ala (*rhombs*). In panel *b*, y-axis scale was expanded to 5% to show production and consumption of LYA-ala$_6$, LYA-ala$_5$, and LYA-ala$_4$

tides. Incubation of fluorescent derivatives of alanine in seawater resulted in sequential hydrolysis of the larger peptides to smaller peptides and free alanine. For example, LYA-ala$_8$ completely disappeared from sea water in 30 h (Fig. 6.12a). LYA-ala$_2$ was initially the dominant product of the hydrolysis of LYA-ala$_8$ and remained dominant for ca. 600 h (Fig. 6.12c). After 600 h, the LYA-ala$_2$ produced was hydrolyzed to the final product, LYA-ala. LYA-ala produced during the incubation originated principally from hydrolysis of LYA-ala$_2$, but could also be a cleavage product of LYA-ala$_6$, LYA-ala$_5$, LYA-ala$_4$ and LYA- ala$_3$. The intermediate products, LYA-ala$_5$, LYA-ala$_4$, and LYA-ala$_3$, never accounted for more than 7% of the added substrate. Incubation of smaller peptide probes (<8 amino acids, for example LYA-ala$_6$ and LYA-ala$_3$) also produced LYA-ala$_2$

as the dominant product, with subsequent hydrolysis to LYA-ala after 600 h (Fig. 6.12d–g). All end up with the amino acid attached to the tag, which remains in sea water. However, a natural peptide will produce amino acids that are taken up by microorganisms. In this case, we are isolating one reaction because the peptide analogs undergo hydrolysis by ecto- and exo-enzymes because of the size of the fluorescent tag (no incorporation).

Different substrates hydrolyze at different rates. Each of the LYA-peptides containing only the amino acid alanine (LYA-ala_8, LYA-ala_6, LYA-ala_4, LYA-ala_3 and LYA-ala_2) was hydrolyzed at different rates in sea water. Hydrolysis rates decreased in the order LYA-ala_6 > LYA-ala_3 > LYA-ala_8 > LYA-ala_4 > LYA-ala_2 (Fig. 6.13). LYA-ala_2 and LYA-ala_6 hydrolysis differed by a factor of ca. 400. Other dipeptides were also hydrolyzed more slowly in sea water than the longer peptides, although hydrolysis rates among dipeptides of different structure also varied: LYA-dipeptides containing leucine at the C-terminus position were hydrolyzed faster than dipeptides with alanine at that position (Fig. 6.13). The fluorogenic substrate Leu-MCA was hydrolyzed at a rate similar to those of the dipeptides (Fig. 6.13). Most previous estimates of hydrolysis rates of protein and peptides in the marine environment have been based on results from fluorogenic dimers such as Leu-MCA, or LLβN (Section 6.3.1.2). Leu-MCA is hydrolyzed at rates typical of LYA-dipeptides, significantly slower than longer peptides (Fig. 6.13). This observation suggests that rates measured with the fluorogenic substrates may underestimate protein and peptide hydrolysis rates in the marine environment.

The question remains as to whether bond hydrolysis occurred at random or at specific bonds in the peptide. Hydrolysis of LYA-ala_4 in sea water was compared with a model in which all bonds are hydrolyzed at the same rate (Table 6.2). A numerical solution for C_1, C_2, C_3 and C_4 was found using an Euler approximation technique; for the case of random hydrolysis, the boundary conditions were $C_{4(0)} = 55$ nM at $t = 0$ and $C_4 = 0$ at $t = 10$ h (the conditions for the experiment shown in Fig. 6.14a). The model solution can be compared with experimental data from the incubation of LYA-ala_4 in

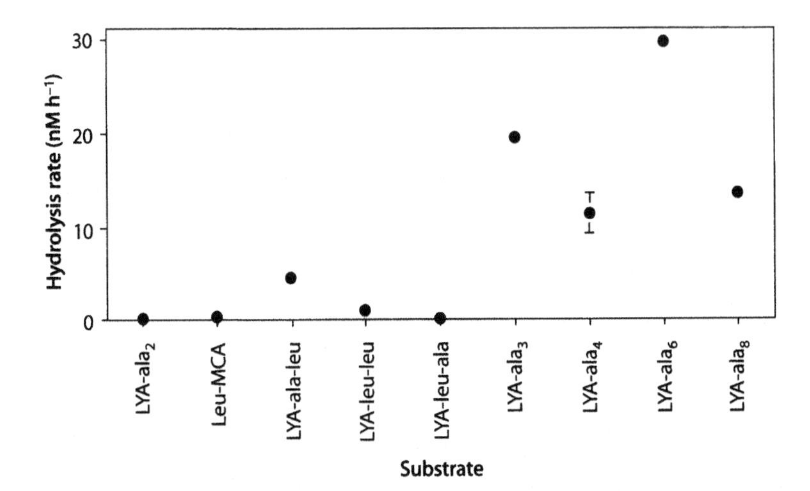

Fig. 6.13. Hydrolysis rates (nM h^{-1}) of alanine and leucine peptides and leucine-methyl-coumarinyl-amide (Leu-MCA) in Flax Pond (NY) sea water

Table 6.2. Reaction equations for peptide hydrolysis. (reprinted from Pantoja and Lee 1997, with permission from Elsevier Science)

Consider a LYA-peptide,

$$\mathrm{LYA}-\mathrm{aa}_1 \underset{k_1^j}{-} \mathrm{aa}_2 \underset{k_2^j}{-} \mathrm{aa}_3 \underset{k_3^j}{-} \mathrm{aa}_4 \cdots \underset{k_{n-1}^j}{-} \mathrm{aa}_n$$

where k_i^j is the rate constant of the *i* bond (numbered from 1 to n–1) of a peptide with *j* amino acids (aa) produced after hydrolysis of the original substrate that contained *n* amino acids.

The following are the reaction equations for the hydrolysis of LYA-ala$_4$:

$$\frac{dC_4}{dt} = -(k_1^4 + k_2^4 + k_3^4)C_4$$

$$\frac{dC_3}{dt} = -(k_1^3 + k_2^3)C_3 + k_3^4 C_4$$

$$\frac{dC_2}{dt} = k_2^3 C_3 + k_2^4 C_4 - k_1^2 C_2$$

$$\frac{dC_1}{dt} = k_1^2 C_2 + k_1^3 C_3 + k_1^4 C_4$$

where C_j is the concentration of peptide containing *j*-amino acids. Numerical solution of these equations allows the calculation of k_i.

For the random case, i.e. $k_i^j = k$, the equations are:

$$\frac{dC_4}{dt} = -3kC_4$$

$$\frac{dC_3}{dt} = k(C_4 - 2C_3)$$

$$\frac{dC_2}{dt} = k(C_3 + C_4 - C_2)$$

$$\frac{dC_1}{dt} = k(C_4 + C_3 + C_2)$$

seawater (Fig. 6.14a). In the experiment (Fig. 6.14a,b) the predominant product is LYA-ala$_2$ with the transient production and subsequent hydrolysis of LYA-ala$_3$ between 2 and 8 h, and the slow production of LYA-ala after 40 h. LYA-ala production continues for 800 h. The random hydrolysis model predicts the simultaneous production of all three peptide products, with subsequent hydrolysis of each to the stable LYA-ala. Thus, the model production of LYA-ala starts at the beginning of the experiment, not after 40 h as we observed experimentally, and reaches the maximum after 15 h instead of 800 h. This exercise illustrates that hydrolysis of peptide bonds is selective rather than random, and that there is preferential production of dipeptides followed by their slow hydrolysis. That could mean that decomposition of peptides in nature may produce dipeptides that could be incorporated across the cell membrane of microorganisms. Another argument for the occurence of this pathway is the previous evidence that aquatic bacterial assemblages are capable of taking up dipeptides (Kirchman and Hodson 1984).

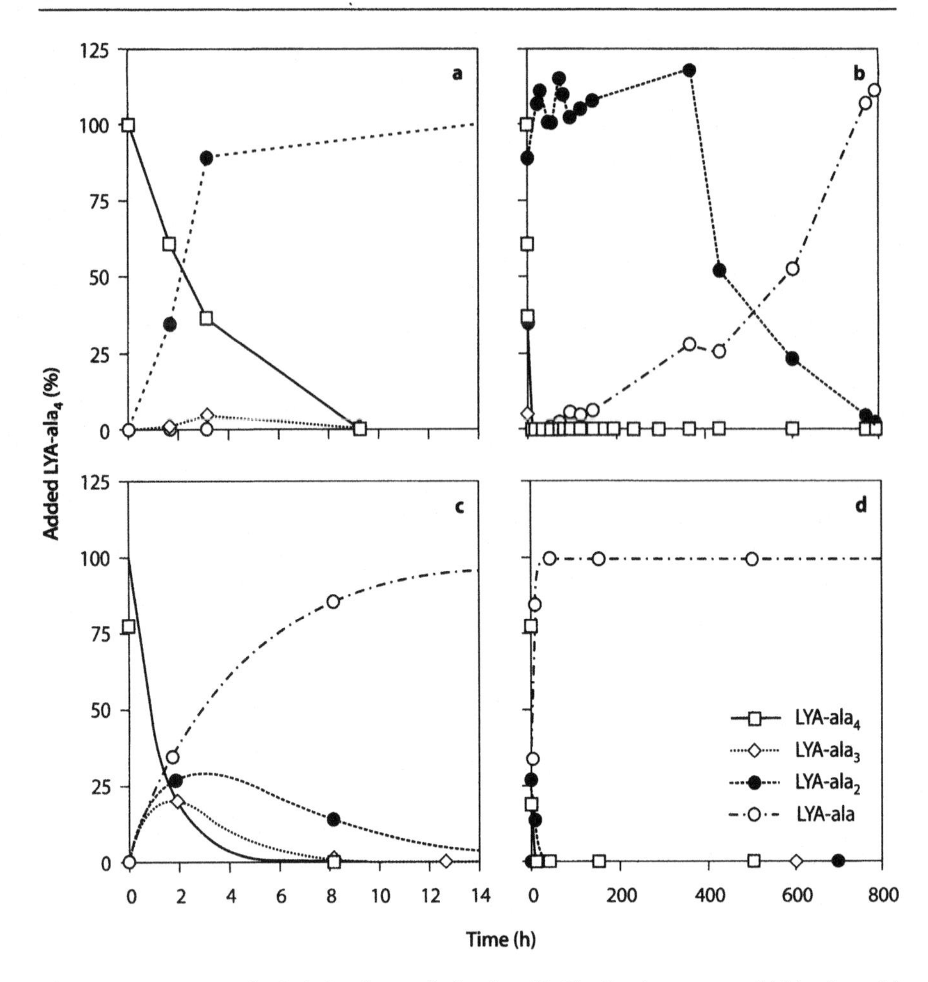

Fig. 6.14. Time course of hydrolysis; **a,b** LYA-ala_4 incubated in Flax Pond sea water; **c,d** A kinetic model in which all bonds are hydrolyzed at the same rate. Expanded scales show the first 14 h of incubation (*a*) and the random model for the same period (*c*), whereas the entire experiment is depicted in b for the incubation and d for the model

Acknowledgements

I would like to thank A. Gianguzza and the organizing committee of the International School on Marine Chemistry for the invitation to participate in this event. Thanks to L. Aluwihare and K. Harper for helpful comments on a previous version of this article. Work with fluorescent molecular probes was developed while the author was conducting his Ph.D. research under the guidance of Dr. C. Lee at Stony Brook (SUNY). S. Pantoja was supported by the Postdoctoral Scholar Program at the Woods Hole Oceanographic Institution, with funding provided by the Johnson Foundation.

References

Ammerman JW, Azam F (1991) Bacterial 5'-nucleotidase activity in estuarine and coastal marine waters: Characterization of enzyme activity. Limnol Oceanogr 36:1427–1436
Antia NJ, Harrison PJ, Oliveira L (1991) The role of dissolved organic nitrogen in phytoplankton nutrition, cell biology and ecology. Phycologia 30:1–89
Arnosti C, Repeta DJ, Blough NV (1994) Rapid bacterial degradation of polysaccharides in anoxic marine systems. Geochim Cosmochim Acta 58:2639–2652
Billen G (1984) Heterotrophic utilization and regeneration of nitrogen. In: Hobbie J and Williams PJleB (ed) Heterotrophic activity in the sea. Plenum, New York pp 313–355
Billen G (1991) Protein degradation in aquatic environments. In: Chróst RJ (ed) Microbial enzymes in aquatic environments. Springer-Verlag, New York pp 123–143
Burdige DJ, Martens CS (1988) Biogeochemical cycling in an organic-rich coastal marine basin: 10. The role of amino acids in sedimentary carbon and nitrogen cycling. Geochim Cosmochim Acta 52:1571–1584
Canfield DE (1994) Factors influencing organic carbon preservation in marine sediments. Chem Geol 114:315–329
Chróst RJ (1991) Environmental control of the synthesis and activity of aquatic microbial ectoenzymes. In: Chróst RJ (ed) Microbial enzymes in aquatic environments. Springer-Verlag, Berlin Heidelberg New York pp 29–59
Currie DJ, Kalff J (1984) A comparison of the abilities of fresh-water algae and bacteria to acquire and retain phosphorus. Limnol Oceanogr 29:196–206
Degens ET (1970) Molecular nature of nitrogenous compounds in seawater and recent marine sediments. In: Hood DW (ed) Organic matter in natural waters. University of Alaska pp 77–106
Ding X, Henrichs SM (1998) Adsorption and decomposition of protein in anoxic sediment. EOS 79:OS4
Dixon M, Webb EC (1958) Enzymes. Academic Press, New York
Eppley RW and Peterson BJ (1979) Particulate organic matter flux and planktonic new production in the deep ocean. Nature 282:677–680
Hedges JI (1992) Global biogeochemical cycles: Progress and problems. Mar Chem 39:67–93
Henrichs SM, Doyle AP (1986) Decomposition of ^{14}C-labelled organic substances in marine sediments. Limnol Oceanogr 31:765–778
Henrichs SM, Reeburgh WS (1987) Anaerobic mineralization of marine sediment organic matter: Rates and the role of anaerobic processes in the oceanic economy. Geomicrobiol J 5:191–237
Henrichs SM, Farrington JW, Lee C (1984) Peru upwelling region sediments near 15° S. 2. Dissolved free and total hydrolyzable amino acids. Limnol Oceanogr 29:20–34
Hollibaugh JT, Azam F (1983) Microbial degradation of dissolved proteins in seawater. Limnol Oceanogr 28:1104–1116
Hoppe H.-G. (1983) Significance of exoenzymatic activities in the ecology of brackish water: Measurements by means of methylumbelliferyl-substrates. Mar Ecol Prog Ser 11:299–308
Hoppe H.-G (1991) Microbial extracellular enzyme activity: A new key parameter in aquatic ecology. In: Chróst RJ (ed) Microbial enzymes in aquatic environments. Springer-Verlag, New York, pp. 60–83
Kanaoka Y, Takahashi T, Nakayama H (1977) A new fluorogenic substrate for amino peptidase. Chem Pharm Bull 25:362–363
Kiel RG, Kirchman DL (1993) Dissolved combined amino acid: Chemical form and utilization by marine bacteria. Limnol Oceanogr 38:1256–1270
Kirchman D, Hodson R (1984) Inhibition by peptides of amino acid uptake by bacterial populations in natural waters: Implications for the regulation of amino acid transport and incorporation. Appl Environ Microbiol 47:624–631
Luo H (1994) Decomposition and adsorption of peptides in Alaskan coastal marine sediments. Ph.D. thesis, University of Atlanta
Mulholland MR, Glibert PM, Berg GM, Van Heukelem L, Pantoja S, Lee C (1998) Extracellular amino acid oxidation by phytoplankton and cyanobacteria: A cross-ecosystem comparison. Aq Microb Ecol 15:141–152
Nikaido H, Vaara M (1985) Molecular basis of bacterial outer membrane permeability. Microbiol Rev 49:1–32
Paerl HW (1991) Ecophysiological and trophic implications of light-stimulated amino acid utilization in marine picoplankton. Appl Environ Microbiol 57:473–479
Palenik B, Morel FMM (1990a) Amino acid utilization by marine phytoplankton: A novel mechanism. Limnol Oceanogr 35:260–269

Palenik B, Morel FMM (1990b) Comparison of cell-surface L-amino acid oxidases from several marine phytoplankton. Mar Ecol Prog Ser 59:195–201
Palenik B, Morel FMM (1991) Amine oxidases of marine phytoplankton. Appl Environ Microbiol 57:2440–2443
Palenik B, Kieber DJ, Morel FMM (1988) Dissolved organic nitrogen use by phytoplankton: The role of cell-surface enzymes. Biol Oceanogr 6:347–354
Pantoja S, Lee C (1994) Cell-surface oxidation of amino acids in seawater. Limnol Oceanogr 39:1718–1726
Pantoja S, Lee C, Marecek JF, Palenik BP (1993) Synthesis and use of fluorescent molecular probes for measuring cell-surface enzymatic oxidation of amino acids and amines in seawater. Anal Biochem 211:210–218
Pantoja S, Lee C, Marecek JF (1997) Hydrolysis of peptides in seawater and sediment. Mar Chem 57:25–40
Parsons TR, Stephens K, Strickland JDH (1961) On the chemical composition of eleven species of marine phytoplankton. J Fish Res Bd Canada 18:1001–1016
Payne JW (1980) Transport and utilization of peptides by bacteria. In: Payne JW (ed). Microorganisms and nitrogen sources. Wiley, New York pp 211–256
Rosenfeld JK (1979) Amino acid diagenesis and adsorption in nearshore anoxic sediments. Limnol Oceanogr 24:1014–1021
Sharp JH (1983) The distribution of inorganic nitrogen and dissolved and particulate organic nitrogen in the sea. In: Carpenter EJ, Capone DG (ed) Nitrogen in the marine environment. Academic Press, New York pp 1–35
Somville M, Billen G (1983) A method for determining exoproteolytic activity in natural waters. Limnol Oceanogr 28:190–193
Suttle CA, Fuhrman JA, Capone DG (1990) Rapid ammonium cycling and concentration-dependent partitioning of ammonium and phosphate: Implications for carbon transfer in planktonic communities. Limnol Oceanogr 35:424–433
Thurman EM (1985) Organic geochemistry of natural waters. Nijhoff/Junk, Boston

Vapour-Particle Phase Interactions of Some Selected Persistent Organic Pollutants in the Marine Atmosphere

M.R. Preston

7.1 Introduction

Unlike the majority of inorganic compounds, many organic chemicals have significant vapour pressures at ambient temperatures. Consequently, organic compounds are frequently transported within the environment at least partially in the vapour phase. The nature of the relationship between this vapour, the solid and liquid aerosol and the dissolved phase is therefore crucial to an understanding of how such chemicals migrate within the environment. In their major review of atmospheric input of trace species to the oceans Duce et al. (1991) highlighted both the complexity of the atmospheric transport processes and the general paucity of data. They compiled a summary of the mean concentrations of chlorinated organic contaminants for the major ocean basins which is reproduced in Table 7.1.

This data represents essentially gas phase material since Duce et al. considered that, on the basis of the available evidence, >95% of these compounds are measured as a gas in typical regions of the marine atmosphere. However, within complex chemical mixtures such as the PCBs, it is clear that there will be a range of gas phase-particle partitioning behaviour dependent on the individual properties of the PCB congeners.

This paper is intended to review two aspects of partitioning behaviour. First, the current status of models of gas-particle interactions and second, the conversion of atmospheric concentration data into air to sea contaminant fluxes. These models derive from early work by Yamasaki et al. (1982) which have been developed to a considerable degree by Terry Bidleman (Atmospheric Environment Service, Canada) and James Pankow (Oregon Graduate Institute, USA) and which have been utilized by other workers including the author (Chen and Preston 1998).

7.2 Theory of Gas-Particle Interactions

7.2.1 Gas-Solid Interactions

The mechanism of association between a gas molecule and a particle may be one of gas-solid adsorption or gas-liquid absorption, so any theory must account for both aspects of this behaviour. In practice most early theory (e.g. Yamasaki et al. 1982; Pankow 1987) concentrated on gas-solid adsorption processes and it is only more recently that the role of liquid phase interactions has been considered (see e.g. Finzio et al. 1997; Pankow 1998). All of the theory below assumes that semivolatile organic

Table 7.1. Mean concentrations (pg m^{-3}) of organic compounds calculated from measured and extrapolated data assigned to 10°×10° grids (Duce et al. 1991)

Compound	Atlantic		Pacific		Indian Ocean
	North	South	North	South	
α-HCH	260	26	420	33	177
γ-HCH	53	3	126	18	71
HCB	126	60	102	60	60
Total PCB	290	33	96	33	117
Chlordane	14	1	9	1	3
Dieldrin	13	1	5	3	3
p,p' DDE	6	4	11	7	25
p,p' DDT	6	4	29	20	52
p,p' DDT (rev)	6	4	25	2	20

carbon (*SOC*)-particle interactions are at equilibrium and that there is free exchange between the two phases. This may not always be the case, but analysis of systems where there is a non-exchangeable proportion of the compound of interest is complex and is not therefore considered further. Work on this aspect has been published by Pankow and Bidleman (1991), which should be consulted if further details are required.

Historically one of the most useful descriptors of the gas-particle partitioning of *SOCs* in the atmosphere was suggested by Yamasaki et al. (1982) and is expressed in Eq. 7.1.

$$K_\mathrm{p} = \frac{\frac{F}{TSP}}{A} = \frac{c_\mathrm{p}}{c_\mathrm{g}} \tag{7.1}$$

where K_p is a partitioning constant, *TSP* (μg m^{-3}) is the concentration of total suspended particulate material, F (ng m^{-3}) and A (ng m^{-3}) are respectively the particle and gas-phase concentrations of the compound of interest. c_p (ng sorbed (μg *TSP*)$^{-1}$) is therefore the concentration in/on the particle phase and $c_\mathrm{g} = A$.

Theory predicts (Yamasaki et al. 1982; Pankow 1987, 1991) that the K_p values will depend on the pure subcooled liquid-vapour pressure p_L^0 according to the equation:

$$\log K_\mathrm{p} = m_\mathrm{r} \log p_\mathrm{L}^0 + b_\mathrm{r} \tag{7.2}$$

where m_r and b_r are constants which depend on the compound class and the nature of the particulate material. Pankow (1994a,b) has suggested that, on the basis of both theoretical and experimental studies, both adsorptive and absorptive mechanisms will produce a value of m_r close to –1. In a recent paper Finzio et al. (1997) have reviewed published literature and indicated ranges of m_r and b_r of –0.61 to –1.04 and –4.26 to 5.95 respectively for PAH and –0.61 to –0.95 and –4.74 to –5.86 respectively for organochlorines (pesticides and PCBs). Chen (1997) has calculated ranges of m_r and b_r of −0.671 to −1.186 and −3.209 to −7.658 respectively for a variety of structurally related azaarene species.

When the partitioning process is controlled by absorption into a primary organic material (om) phase then Pankow (1994a, 1998) has shown that for a neutral compound:

$$K_p = \frac{760RTf_{om}}{10^6 M_{om}\zeta_{om}p_L^0} \tag{7.3}$$

so that

$$\log K_p = -\log p_L^0 + \log\frac{760RTf_{om}}{10^6 M_{om}\zeta_{om}} \tag{7.4}$$

where R is the gas constant, T (K) is the temperature, f_{om} is the weight fraction of the TSP that comprises the absorbing om phase (including any water or other inorganic species in that phase), M_{om} is the number average molecular weight of the species making up the om phase, ξ_{om} is the mole fraction scale activity coefficient of the compound of interest in the om phase at temperature T.

The temperature dependence of K_p follows the same mathematical form as that for the temperature dependence of the Henry's law constant, namely:

$$\log K_p = \frac{m_p}{T} + b_p \tag{7.5}$$

where m_p and b_p also depend on the specific compound and on the nature of the particulate material. Theory of gas-solid partitioning (Pankow 1987) predicts that the value of m_p obtained by linear regression over some ambient temperature range of interest will be given by

$$m_p = \frac{Q_1}{2.303R} - \frac{T_{amb}}{4.606} \tag{7.6}$$

where Q_1 (kJ mol^{-1}) is the enthalpy for desorption from the surface (always positive) and T_{amb} is the centre of the temperature range for the regression. Theory also predicts that

$$b_p = \log\frac{A_{tsp}t_0}{275\sqrt{(MT_{amb})}} + \frac{1}{4.606} \tag{7.7}$$

where A_{tsp} ($cm^2\ \mu g^{-1}$) is a specific surface area for the aerosol, t_0 is a molecular vibration time (s) and M is the relative molecular mass of the compound of interest.

There is evidence that b_p values tend to be reasonably constant within some compound classes (Pankow 1994a). While this is true for PAHs ($b_p \approx 19 \pm 1$; Pankow and Bidleman 1991), Chen and Preston (1998) found much greater variability among azaarenes (b_p ranging from ~283–5 694).

The temperature dependency of p^0_L may also be described by an equation similar to Eq. 7.3 (Falconer and Bidleman 1994) such that

$$\log p^0_L = \frac{m_L}{T} + b_L \tag{7.8}$$

where b_L is a constant and

$$m_L = \frac{-Q_l}{2.303R} \tag{7.9}$$

where Q_l is the heat of vaporizaisation and R the gas constant.

Q_l (and hence m_L) and b_L can be determined from gas chromatographic measurements using the known heats of vaporization of reference materials as calibrants. This method has been used by researchers such as Foreman and Bidleman (1985), Hinkley et al. (1990), and Falconer and Bidleman (1994) to determine p^0_L for PCBs.

The conversion of p^0_L values to 'aerosol associated material' uses the Junge-Pankow formulation (Pankow 1987) which is based effectively on the assumption that adsorption processes follow a linear Langmuir isotherm such that

$$\phi = \frac{c_j\Theta}{p^0_L + c_j\Theta} \tag{7.10}$$

where ϕ is the fraction of a *SOC* adsorbed to particles and θ is the particle surface area available for adsorption. c_j is the constant in Junge's (1977) equation and is described by

$$c_j = 760RTN_s e^{(Q_l - Q_v)/RT} \tag{7.11}$$

where T (K) is the temperature, N_s is the moles of sorption sites per cm^2 of aerosol, Q_l is the enthalpy for desorption directly from the surface and Q_v is the enthalpy of vaporization of the liquid. A value for c_j of 17.2 Pa-cm has been suggested by Falconer and Bidleman in their PCB work. The value of θ depends on the source of the aerosol and Bidleman (1988) suggests 1.1×10^{-5} for urban air, 1.6×10^{-6} for average continental background air and 4.2×10^{-7} for clean continental background air. Calculations by Duce et al. (1991) suggest that when all the components of the marine aerosol are considered (sulfate aerosol, mineral aerosol and sea salt aerosol) the properties are most closely related to urban aerosol values. This conclusion is, however, dependent on the similarity of the adsorption behaviour of aerosols of differing composition and there is doubt about the validity of this assumption. For example, Liang et al. (1997) have calculated both $\log p^0_L$ and $\log K_p$ for several different aerosol types and observed significant variations. Pankow et al. (1994) have made similar measurements relating to environmental tobacco smoke.

Other workers (e.g. Rounds et al. 1993; Liang and Pankow 1996) have used special desorption techniques to calculate K_p more directly, though concentrating largely on hydrocarbons, especially those associated with tobacco smoke.

A compilation of $\log K_p$ and $\log p^0{}_L$ data derived from a variety of sources is given in Table 7.2.

7.2.2
Gas-Liquid Interactions

One significant problem associated with the theory outlined in 7.2.1 above is the difficulty of obtaining the value of the subcooled liquid-vapour pressure (p_L^0). An interesting recent development in theory, which helps to circumvent this problem, derives from the uncertainty over whether the relevant mechanisms are *ab*sorptive or *ad*sorptive. Finzio et al. (1997) have shown that

$$K_p = BK_{OA} \tag{7.12}$$

where K_{OA} is the octanol/air partition coefficient and

$$B = 1.22 \times 10^{-12} f_{OM} \frac{(M_O / M_{OM})}{(\gamma_O / \gamma_{OM})} \tag{7.13}$$

where f_{OM} is the mass fraction of organic matter in the particles that can absorb gaseous *SOCs*, M_O and M_{OM} are the molecular mass of octanol and the mean molecular mass of the organic phase respectively and γ_O and γ_{OM} are the activity coefficients of the chemical in octanol and the organic matter phase respectively.

Note that

$$K_{OA} = \frac{K_{OW}}{K_{AW}} \tag{7.14}$$

where K_{OW} is the dimensionless octanol water partition coefficient and K_{AW} is the dimensionless air-water coefficient. However Harner and Mackay (1995) have noted that for technical reasons a direct measurement of K_{OA} is preferable to the calculated value.

Pankow (1998) has further explored this idea and concluded that log K_{OA} is likely to be a "more universal correlating parameter for log K_p, log $K_{p,OM}$ and log $K_{p,OC}$ than is log p_L^0" [log $K_{p,OM}$ and log $K_{p,OC}$ are organic matter based and organic carbon based partitioning constants respectively]. A compilation of values for these parameters (and the Henry's law constant *H*) is given in Table 7.2.

Values of *B* range between ~0.3 × 10^{-12} and ~6.9 × 10^{-12} for PAH (average = 1.88 × 10^{-12}), ~0.1 × 10^{-12} and ~4 × 10^{-12} for PCBs and ~0.1 and ~4.6 × 10^{-12} for organochlorine pesticides (average all *OCs* = 1.50 × 10^{-12}). Pankow (1998) has suggested that a single value of –11.76 for log *B* might be appropriate.

This new aspect of partitioning theory is in many ways analogous to particle-water partitioning theory and may prove to be an important area for development in the future.

7.3
Practical Applications of Partitioning Theory

The theory outlined above can seem somewhat daunting and it is therefore worth exploring how useful practical information may be extracted. A series of examples are now given and which relate to PCBs.

Table 7.2. Mean values of $\log p_L^0$, $\log K_p$, $\log K_{OW}$, the Henry's law constant H, and calculated and measured $\log K_{OA}$ values for various PAH, PCB, organochlorine pesticides and azaarenes compiled from Pankow et al. (1994), Chen (1997), Harner and Mackay (1995) and Finzio et al. (1997). All values are for 25 °C ($\log p_L^0$ for polychlorinated-p-dioxins and dibenzofurans have been published by Eltzer and Hites 1989)

Compound	$\log p_L^0$ (Torr)	$\log K_p$ ($m^3\ \mu g^{-1}$)	$\log K_{OW}$	H ($Pa\ m^3\ mol^{-1}$)	$\log K_{OA}$ (calculated)	$\log K_{OA}$ (measured)
Alkanes						
C_{16}	-2.91	-4.93				
C_{17}	-3.43	-4.35				
C_{18}	-3.94	-3.95				
C_{19}	-4.45	-3.49				
C_{20}	-4.96	-3.01				
C_{21}	-5.48	-2.52				
C_{22}	-5.99	-2.24				
PAHs						
Fluorene	-0.39	-4.48	4.18	7.87	6.68	
Phenanthrene	-1.15	-4.20	4.57	3.24	7.45	
Anthracene	-1.18	-4.27	4.54	3.96	7.34	
Fluoranthene	-2.16	-3.37	5.22	1.037	8.60	
Pyrene	-2.35	-3.27	5.18	0.92	8.61	
Chrysene	-3.64	-2.16	5.86	0.065	10.44	
Benz[a]anthracene	-3.60	-2.13	5.91	0.581	9.54	
Benzo[a]pyrene	-4.93	-1.08	6.04	0.046	10.77	
Benzo[e]pyrene	-4.89	-1.05	6.04	0.020	11.13	
Benzo[k]fluoranthene	-4.68	-1.32	6.00	0.016	11.19	
Azaarenes						
Quinoline/isoquinoline	-1.169					
ΣMe quinolines	-1.498	-2.187				
ΣDiMe quinolines	-1.679	-1.968				
ΣTriMe quinolines	-2.164	-1.688				
ΣBenzo(f)quinoline, Benzo(h)quinoline, Acridine, Phenanthridine	-3.441	-2.092				
ΣMe 3 ring azaarenes	-2.946	-1.913				
ΣAzapyrene	-4.894	-1.752				
ΣAzachrysene	-5.597	-1.244				
OC pesticides						
α-HCH	-0.66	-4.47	3.81	0.87	7.26	
γ-HCH	-1.20	-4.01	3.80	0.13	8.08	
HCB		-4.43	6.00	7.12	8.54	6.90
p,p'DDD	-3.02	-2.98	6.22	0.64	9.81	
p,p'DDE	-2.49	-3.28	6.96	7.95	9.45	
p,p'DDT	-3.30	-2.70	6.91	2.36	9.93	10.09
Chlorobenzenes						
1,2-	2.292		3.4		4.41	4.36
1,2,3-	1.447		4.1		5.11	5.19
1,2,3,4-	0.715		4.5		5.74	5.64
1,2,3,4,5-	-0.143		4.5		5.81	5.63
Penta-	-0.658		5.0		6.46	6.27
Hexa-	-0.638		5.5		6.78	6.90

Table 7.2. *Continued*

Compound	$\log p_L^0$ (Torr)	$\log K_p$ ($m^3\,\mu g^{-1}$)	$\log K_{OW}$	H ($Pa\,m^3\,mol^{-1}$)	$\log K_{OA}$ (calculated)	$\log K_{OA}$ (measured)
PCB congeners						
3	-0.567		4.5		6.27	6.78
8	-0.81	-4.59	5.1	31.31	7.00	
15	-2.319		5.3		7.46	7.67
29	-0.879		5.6		7.61	7.96
33	-1.58	-4.03	5.8	43.67	7.55	
40	-2.01	-3.72	5.6	21.94	7.65	
44	-1.90	-3.82	6.0	25.43	7.99	
49	-1.78	-3.89	6.10	37.90	7.92	8.40
52	-1.8	-3.80	6.10	47.59	7.82	
61	-2.301		5.9		8.15	8.74
66	-2.21	-3.66	6.31	25.84	8.29	9.02
70	-2.27	-3.61	6.40	19.15	8.50	
87	-2.65	-3.26	6.50	24.81	8.50	
99	-2.53	-3.34	6.60	30.50	8.51	
101	-2.48	-3.39	6.40	35.48	8.24	9.07
110	-2.74	-3.22	6.30	19.15	8.41	
128	-3.47	-2.66	7.00	11.91	9.32	10.36
138	-3.30	-2.78	6.70	10.48	9.06	9.81
153	-3.17	-2.84	6.70	10.84	9.06	9.74
155	-3.319		7.0		8.46	8.99
170	-4.07	-2.22	7.08	19.25	9.19	

7.3.1 PCBs

In a major exercise Falconer and Bidleman (1994) measured and/or calculated the parameters m_L and b_L for approximately 180 of the 209 PCB isomers. This data permits log p_L^0 to be calculated at a given temperature using Eq. 7.8. If it is then required to estimate the fraction of any given PCB congener in the particulate phase at a given temperature, Eq. 7.10 may be used. The parameter c has been estimated to be 17.2 Pa-cm and the characteristic values of the particle surface area parameter Θ relating to the aerosol type (urban ≈ marine, continental background, clean continental background) can be adopted. Some results of such calculations for partitioning as a function of congener number are given in Fig. 7.1 and Fig. 7.2. There is a general increase in particle associ-ation with increasing congener number, as would be expected, but the progression is not smooth and is clearly related to the congener structure. Bidleman and Falconer noted that the partitioning behaviour of PCBs was closely related to the number of ortho-substituted chlorine atoms. Examples of this phenomenon are shown in Fig. 7.3 which shows the partitioning as a function of temperature for a number of congeners calculated using the urban aerosol parameter ($\theta = 1.1 \times 10^{-5}$). This demonstrates that increasing ortho-substitution *reduces* the level of particle association. Within a homolog the degree of adsorption follows the order (number of orthochlorines) $0 > 1 > 2 > 3 > 4$ with an average ϕ at 10 °C for the non-orthotetrachlorobi-phenyls and pentachloro-biphenyls being 3.8 and 2.3 times those for the di-ortho congeners respectively.

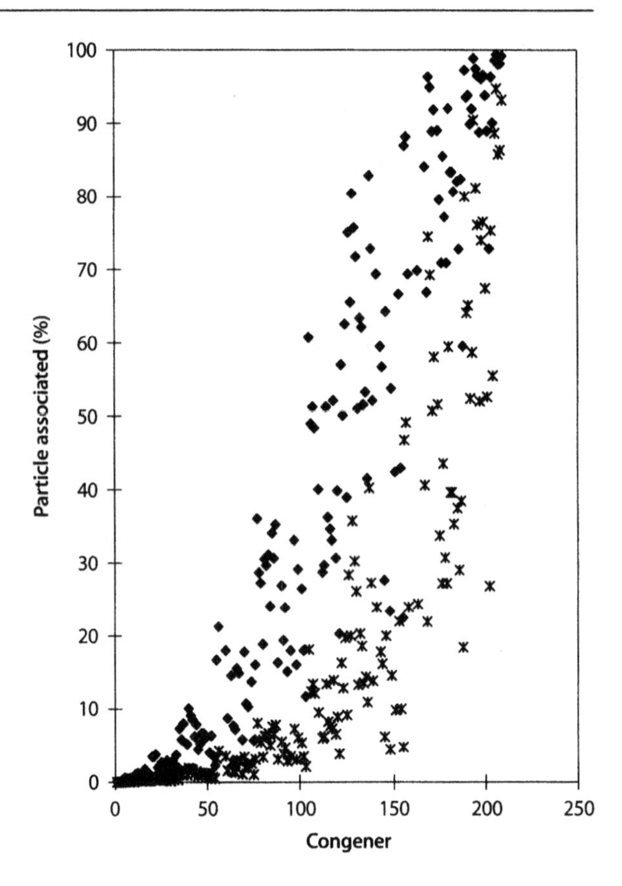

Fig. 7.1. The percentage particle associated PCB at 283 K (◆) and 298 K (x) plotted against congener number

7.3.2 Calculation of Air-Sea Fluxes

Any attempts to calculate air to sea fluxes of *SOCs* must take account of *(i)* dry deposition, *(ii)* wet deposition and *(iii)* gas exchange processes. Various aspects of these phenomena have been reviewed by Duce et al. (1991) and Preston (1992).

7.3.3 Dry Deposition

As Duce et al. (1991) have noted, there is very little data available which can be used to assess dry deposition velocities of organochlorine compounds, and much the same is true for other *SOCs*. From the practical point of view it is very difficult to distinguish between dry deposition and gas phase adsorption. Bidleman and Christensen (1979) and Bidleman (1988) have indicated that deposition velocities lie in the range of 0.05 to 1 cm s^{-1}. Whether such velocities apply to the marine aerosol is still open to question because of the general lack of size distribution data for organic compounds in this material. However, Duce et al. (1991) used an assumed range of 0.05 to 0.5 cm s^{-1} in their work with a "best" estimate of 0.1 cm s^{-1}.

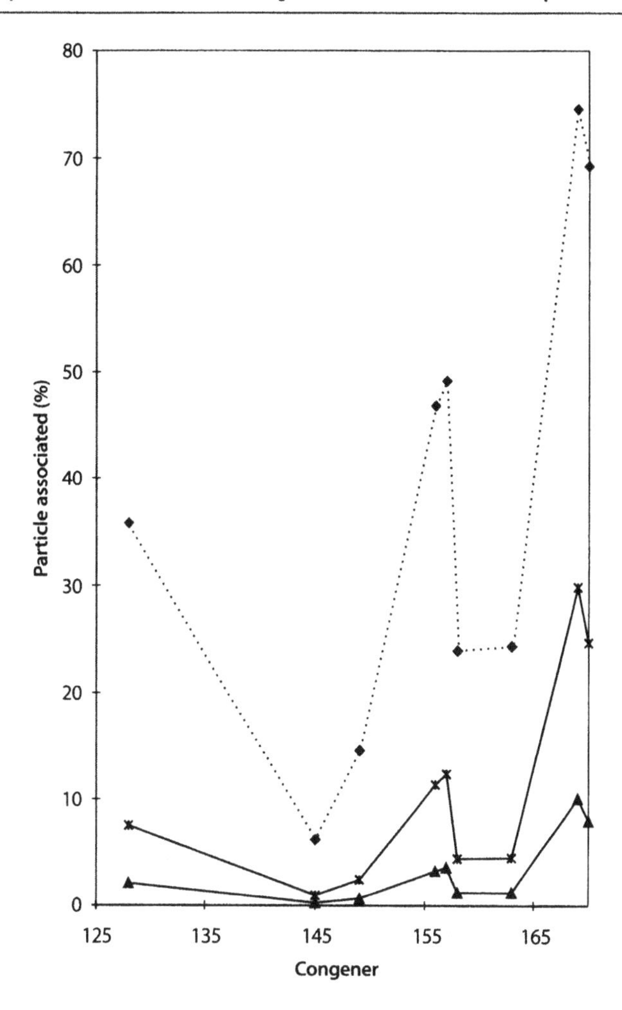

Fig. 7.2. The relationship between partitioning behaviour and aerosol characteristics for urban (◆), average continental (⋇) and clean continental (▲) aerosols for PCB congeners 128, 145, 149, 156 and 169 at 298 K

7.3.4 Wet Deposition

To calculate wet depositional fluxes it is necessary to know the scavenging ratio (i.e. the concentration ratio between a chemical in aerosol and its concentration in rain). There are considerable uncertainties in estimates of the most appropriate scavenging ratios for organic compounds with estimates lying between 13 and 1600. A value of 500 is commonly accepted as reasonable (Duce et al. 1991; Preston 1992).

A further problem in calculating wet depositional fluxes at sea is that rainfall data is very sparse. Anyone with experience of work at sea will know firstly that rain fall is very patchy and secondly that any attempts to measure rain fall from a ship are almost certainly doomed to failure because of the influence of sea spray.

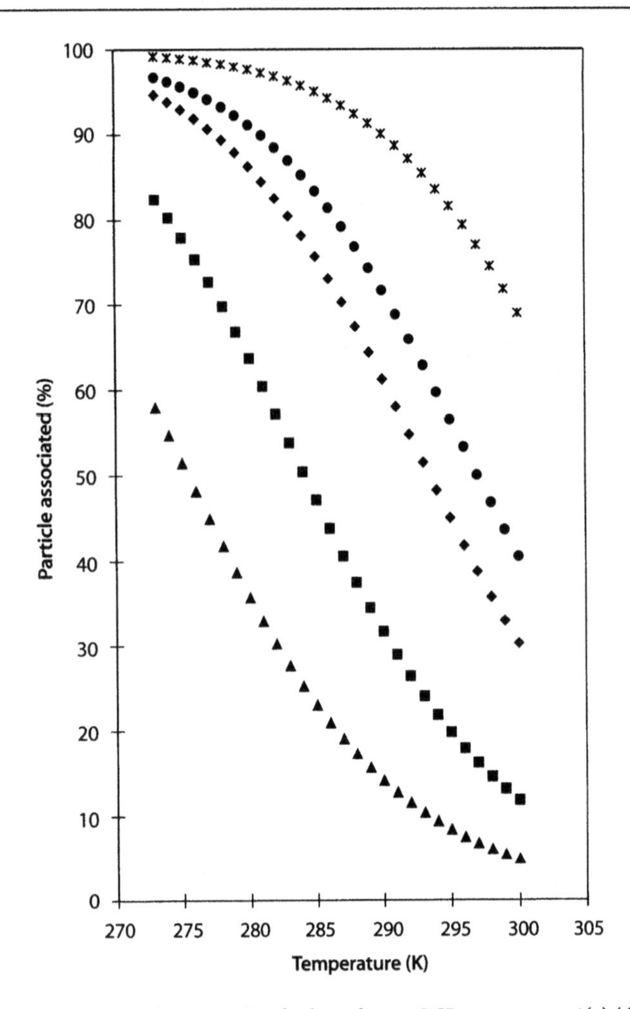

Fig. 7.3. The percentage particle association for homologous PCB congeners 128(2) (◆), 145(4) (▲), 149(3) (■), 156(1) (●) and 169(0) (⁎) plotted against temperature for aerosol with urban characteristics. Each congener has a total of 6 chlorine atoms, and numbers in parentheses are the number of ortho substituted chlorine atoms

7.3.5 Gas Phase Exchange Processes

Gas phase processes are rather more difficult to model than either of the wet and dry deposition processes described above. The flux of a gas (F_g) between air and water (in either direction) is driven by the concentration difference (ΔC) and the rate term K (the piston or transfer velocity) such that

$$F_g = K\Delta C = K_w\left[Cg_a\left(\frac{H}{RT}\right)^{-1} - Cg_w\right] = K_a\left[Cg_a - Cg_w\left(\frac{H}{RT}\right)\right] \quad (7.15)$$

where K_w and K_a refer to whether the calculation uses concentrations expressed on a liquid phase or a gas phase basis. Cg_a and Cg_w are the gas concentrations in air and water respectively, H is Henry's Law constant, R is the gas constant and T the temperature in degrees Kelvin.

It is more common to use the reciprocal of K which is a measure of the resistance R to gas exchange. R may in turn be divided into two components: *(i)* the resistance in air r_a and *(ii)* the resistance in water r_w such that

$$\frac{1}{K_w} = \frac{1}{\alpha k_w} + \frac{RT}{Hk_a} \tag{7.16}$$

$$R_w = r_w + \frac{RT}{H} r_a \tag{7.17}$$

where k_a and k_w are the transfer velocities for chemically unreactive gases in air and water respectively. The parameter α is a correction factor which allows for any enhancement of k_w through chemical reactivity of the gas in water. The low reactivity of most *SOCs* in water means that α can be ignored ($\alpha = 1$). On a gas phase basis the equations become

$$\frac{1}{K_a} = \frac{1}{k_a} + \frac{H}{RT\alpha k_w} \tag{7.18}$$

$$R_a = r_a + \frac{H}{RT} r_w \tag{7.19}$$

7.3.6 Determination of the Air Phase Transfer Velocity (k_a) through Measurement of r_a

The parameter r_a, the air phase transfer resistance, is made up of a turbulent exchange component ($r_{a(turb)}$) and a diffusive exchange component ($r_{a(diff)}$). $r_{a(turb)}$ approximates to

$$r_{a(turb)} = \left(\frac{u}{u^*}\right)^2 = \frac{1}{C_d} \tag{7.20}$$

where u is the wind speed, u^* is the friction velocity and C_d is a drag coefficient. For marine flux calculations C_d can be assumed to have a value of 1.3×10^{-3} (Duce et al. 1991), though the limitations of this assumption must be appreciated.

The calculation of $r_{a(diff)}$ is relatively complicated, but for present purposes it may be assumed to be

$$r_{a(diff)} = \frac{5}{u^*} Sc_a^{2/3} \tag{7.21}$$

where Sc_a, the Schmidt number is equal to

$$Sc_a = \frac{\nu_a}{D_a} \tag{7.22}$$

where ν_a is the air viscosity and D_a is the diffusivity of the substance in air. Conveniently, Sc_a is related to the square root of the relative molecular mass of a trace gas giving

$$r_a = r_{a(turb)} + r_{a(diff)} = \frac{(770 + 45(M)^{1/3})}{u} = \frac{1}{k_a} \tag{7.23}$$

7.3.7 Determination of the Water Phase Transfer Velocity (k_w)

The fundamental work relating to measurement of k_w was performed by Liss and Merlivat (1986). Subsequent confirmation of this work (e.g. Watson et al. 1991) using purposeful trace studies have provided the following equations for k_w:

$$k_w = 0.17u \qquad (\text{for } u \leq 3.6 \text{ m s}^{-1}) \tag{7.24}$$

$$k_w = 2.85u - 9.65 \qquad (\text{for } 3.6 < u \leq 13 \text{ m s}^{-1}) \tag{7.25}$$

$$k_w = 5.9u - 49.3 \qquad (\text{for } u > 13 \text{ ms}^{-1}) \tag{7.26}$$

Values of some gas transfer velocity values (K_w) calculated by Duce et al. (1991) are given in Table 7.3.

Table 7.3. Calculated overall gas transfer velocities (K_w) for some chlorinated contaminants in various oceans. (Duce et al. 1991); K_w (cm h^{-1})

	Atlantic		Pacific		Indian Ocean
Compound	North	South	North	South	
α-HCH	0.32	0.31	0.47	0.35	0.46
γ-HCH	0.17	0.16	0.25	0.18	0.24
HCB	4.1	4.0	4.8	4.3	4.8
PCB (1254)	3.3	3.2	4.0	3.5	4.0
PCB (1242)	3.5	3.4	4.2	3.6	4.2
Chlordane	2.4	2.3	3.1	2.5	3.0
Dieldrin	1.1	1.1	1.5	1.2	1.5
DDE	2.5	2.4	3.1	2.6	3.1
DDT	1.0	0.98	1.4	1.1	1.4

7.4 Conclusions

The estimation of gas-particle phase relationships for chemicals of environmental concern is not straightforward, though considerable advances in both theory and practical measurements have been made over the last 10–15 years. Such estimates are, however, critical for a proper understanding of the transfer of chemicals between the atmosphere and the surface ocean. More widespread measurements, particularly in remote marine atmospheres, will be essential if we are to be able to provide reliable estimates of pollutants entering the oceans.

References

Bidleman TF (1988) Atmospheric processes: Wet and dry deposition of organic compounds are controlled by their vapor-particle partitioning. Environ Sci Technol 22:361–367
Bidleman TF, Christensen EJ (1979) Atmospheric removal processes for high molecular weight organochlorines. J Geophys Res 84:7857–7862
Chen H-Y (1997) Azaarenes as contaminants of the urban atmosphere. Ph.D. thesis, University of Liverpool, UK
Chen H-Y, Preston MR (1998) Gas/particle partitioning behaviour of azaarenes in an urban atmosphere. Environ Poll 97:169–174
Duce RA, Liss PS, Merrill JT, Atlas EL, Buat-Menard P, Hicks BB, Miller JM, Prospero JM, Arimoto R, Church TM, Ellis W, Galloway JN, Hansen L, Jickells TD, Knap AH, Reinhardt KH, Schneider B, Soudine A, Tokos JJ, Tsunogai S, Wollast R, Zhou M (1991) The atmospheric input of trace species to the world ocean. Global Bigeochem Cycles 5:193–259
Eltzer BD, Hites RA (1989) Polychlorinated dibenzo-p-dioxins and dibenzofurans in the ambient atmosphere of Bloomington, Indiana. Environ Sci Technol 23:1389–1395
Falconer RL, Bidleman TF (1994) Vapor pressures and predicted particle/gas distributions of polychlorinated biphenyl congeners as functions of their temperature and orthochlorine substitution. Atmos Environ 28:547–554
Finzio A, Mackay D, Bidleman TF, Harner T (1997) Octanol-air partition coefficient as a predictor of partitioning of semi-volatile organic chemical to aerosols. Atmos Environ 21:2289–2296
Foreman WT, Bidleman TF (1985) Semivolatile organic compounds in the ambient air of Denver, Colorado. Atmos Environ 24A:2405–2416
Harner T, Mackay D (1995) Measurement of octanol-air partition coefficients for chlorobenzenes, PCBs and DDT. Environ Sci Technol 29:1599–1606
Hinkley DA, Bidleman TF, Foreman WT, Tuschall J (1990) Determination of vapor pressures for non-polar and semipolar organic compounds from gas chromatographic retention data. J Chem Eng Data 35:232–237
Junge C (1977) Basic considerations about trace constituents in the atmosphere as related to the fate of global pollutants. In: Suffet HI (ed) Fate of pollutants in the air and water nvironments, Part 1. John Wiley, New York, pp 7–26
Liang C, Pankow JF (1996) Gas/particle partitioning of organic compounds to environmental tobacco smoke: Partition coefficient measurements by desorption and comparison to urban particulate material. Environ Sci Technol 30:2800–2805
Liang C, Pankow JF, Odum JR, Seinfeld JH (1997) Gas/particle partitioning of semivolatile organic compounds to model inorganic, organic and ambient smog aerosols. Environ Sci Technol 31: 3086–3092
Liss PS, Merlivat L (1986) Air-sea exchange: Introduction and synthesis. In: Buat-Menard P (ed) The role of air-sea exchange in geochemical cycling. D Reidel, Norwell, Mass., pp 113–127
Pankow JF (1987) Review and comparative analysis of the theories of partitioning between the gas and aerosol particulate phases in the atmosphere. Atmos Environ 21:2275–2283
Pankow JF (1991) Common y-intercept and single compound regressions of gas-particle partitioning data vs. $1/T$. Atmos Environ 25A:2229–2239
Pankow JF (1994a) An absorption model of gas/particle partitioning of organic compounds in the atmosphere. Atmos Environ 28:185–188
Pankow JF (1994b) An absorption model of the gas/aerosol partitioning involved in the formation of secondary organic aerosol. Atmos Environ 28:189–193

Pankow JF (1998) Further discussions of the octanol/air partition coefficient K_{OA} as a correlating parameter for gas/particle partitioning coefficients. Atmos Environ 32:1493–1497

Pankow JF, Bidleman TF (1991) Effects of temperature, *TSP* and per cent non-exchangeable material in determining the gas-particle partitioning of organic compounds. Atmos Environ 25A:2241–2249

Pankow JF, Lorne IM, Buchholz DA, Luo W, Reeves BD (1994) Gas/particle partitioning of polycyclic aromatic hydrocarbons and alkanes to environmental tobacco smoke. Environ Sci Technol 28:363–365

Preston MR (1992) The interchange of pollutants between the atmosphere and the oceans. Marine Pollution Bulletin 24:477–483

Rounds SA, Tiffany BA, Pankow JF (1993) Description of gas/particle sorption kinetics with an intraparticle diffusion model: Description experiments. Environ Sci Technol 27:366–377

Watson AJ, Upstill-Goddard RC, Liss P (1991) Air-sea exchange in rough and stormy seas measured by a dual tracer technique. Nature 349:145–147

Yamasaki H, Kuwata K, Miyamoto H (1982) Effects of ambient temperature on aspects of airborne polycyclic aromatic hydrocarbons. Environ Sci Technol 16:189–194

Part III

Metals and Organometallic Compounds in Marine Environments

Chapter 8

Determination of Organic Complexation

C.M.G. van den Berg

8.1 Introduction

It has been known for several decades that trace metals can become complexed by organic matter in natural waters. Only relatively recently has it become known that some interesting metals such as copper, zinc, iron, nickel, cobalt and probably more, participate in these reactions. This complexation changes the geochemistry of the metals by preventing them from being scavenged and thus increases their residence time in estuaries and the upper water column. The complexation also changes the availability to organisms (algae, mollusks or fish). It is therefore important to determine the chemical speciation of metals in addition to their dissolved concen tration in sea water. The organic complexation has been determined in the past by several methods, including fluorescence quenching (Berger et al. 1984), solubility increases (Campbell et al. 1977), anodic stripping voltammetry (Chau et al. 1974), competitive adsorption on MnO_2 (van den Berg and Kramer 1979) and preconcentra tion on C18 cartridges (Mills and Quinn 1981). The most important methods to determine the chemical speciation of metals in the marine system now use voltammetry with and without ligand competition. These methods will be explained in this chapter.

8.2 Metal Species in Solution

The metal species in seawater, and their relationships, are shown in Table 8.1. The main species are M' (all inorganic metal ions), ML (the metal complexed by the natural organic matter) and L' (the natural organic ligands not associated with M). The metal-organic speciation in seawater would be known exactly if M' and ML could be determined directly. A problem is that the measurement itself usually removes a fraction of the detected species, thus causing a shift in the equilibrium. The metal speciation can also be calculated from the total metal and ligand concentrations ($[M_t]$ and $[L_t]$) and the conditional stability constant (K'_{ML} or K''_{ML}): in this case the actual species do not need to be determined, but are evaluated from ligand concentrations and conditional stability constants which have to be determined for each sample separately; with this information the concentration of the free metal ion can be calculated. The two methods are discussed below.

Table 8.1. Relationships between metal species in water

Symbols and metal species	
M′	Free inorganic metal (not complexed by organic ligands)
M_t and L_t	Total metal and ligand concentrations
ML	Organic complexes of M
L′	Free ligand (not complexed by M)
Relationships	
$K'_{ML} = [ML] / ([M^{n+}][L'])$	$K''_{ML} = [ML] / [M'][L']$
Relationship between M^{n+} and M′:	Relationship between M^{n+} and M_t:
$[M'] = \alpha_M[M^{n+}]$	$[M^{n+}] = [M_t] / (\alpha_M + \alpha_{ML})$
Relationship between K'_{ML} and K''_{ML}	$\alpha_M = 1 + \Sigma(K_{MXin}[Xi]^n)$ where Xi = inorganic anion
$K'_{ML} = \alpha_M K''_{ML}$	(Cl^-, HCO_3^-, OH^-)
$\alpha_{ML} = K'_{ML}[L']$	$\alpha_{M'L} = K''_{ML}[L']$

8.3 Determination of Organic Complexation Using ASV

In stripping voltammetry metals are first deposited on the electrode, subsequently to be measured by means of a potential scan from their oxidation or reduction current. In anodic stripping voltammetry (ASV) metals are first plated on the electrode: this involves the reduction of the dissolved metal ion to the metallic state by applying a potential more negative than its reduction potential. The electrode usually consists of mercury, either a mercury film or a mercury drop. The metal then remains on the electrode where it dissolves in the mercury as an amalgam. ASV can be used to determine cadmium and lead in natural waters with great sensitivity (Batley and Florence 1976; Raspor et al. 1980; Ostapczuk et al. 1986). There are ASV procedures to determine other metals such as copper, zinc and manganese, but these have drawbacks related to poor solubility of these elements in the mercury and their use is restricted.

A major advantage of ASV is the relative simplicity of the technique: the sample can be analysed without any reagent addition, although it may be helpful to add a pH buffer. This method is used to determine the "labile metal" concentration: "labile" is that fraction which is plated during the deposition step. The labile metal which is detected when the unadulterated sample is analysed includes the "free" or inorganic fraction of the metal (also called [M']); however, it may also include any organic metal complexes which happen to dissociate at the applied potential in the diffusion layer at the electrode surface, which is, after all, depleted of free metal. The dissociation of organic metal complexes constitutes a problem in the ASV method to determine organic complexation: its extent has been investigated (Shuman and Michael 1978) and is somewhat contentious, as it is difficult to prove or disprove in the absence of a good alternative technique. The idea is that the complex dissociation is minimised by using a fast rotating disk electrode and a plating potential just negative of the reduction potential. To be sure, the labile fraction should be assumed to include labile organic metal complexes in addition to the free metal. The non-labile fraction is formed by the more inert organic metal complexes. Therefore, using ASV the concentrations of lead and cadmium can be subdivided into labile and non-labile species. This may be sufficient for the purpose of your speciation study.

8.4 Principle of Adsorptive Cathodic Stripping Voltammetry

In cathodic stripping voltammetry (CSV) the element is reduced, instead of oxidized as in ASV, during the potential scan. The preconcentration step typically involves adsorption of the element after complexation with a specific complexing ligand causing the formation of a complex with adsorptive properties. Several ligands are available for the preconcentration step including 8-hydroxyquinoline (van den Berg 1986), catechol (van den Berg 1984a), pyrrolidine dithiocarbamate (van den Berg 1984b) and others, and some twenty elements can be determined by this technique (van den Berg 1991).

Prior to the metal determination a specific ligand is added (such as salicylaldoxime for copper or iron, or nitrosonaphthol for iron (Yokoi and van den Berg 1992; Campos and van den Berg 1994; Rue and Bruland 1995)) which forms a complex with reactive metal ions. The complex is allowed to adsorb on the electrode (usually a mercury drop electrode), and is quantified by reduction of the metal in the adsorbed complex during the potential scan. CSV scans for copper in the presence of salicylaldoxime are shown in Fig. 8.1. It can be seen that an almost symmetrical peak is obtained because all adsorbed copper is reduced during the scan, and the current therefore drops back to the background level once the reduction potential has been passed. Only about 0.1% of the dissolved copper participates in the measurement so the scan can be repeated without a change in the peak height. The sensitivity is calibrated by a standard copper

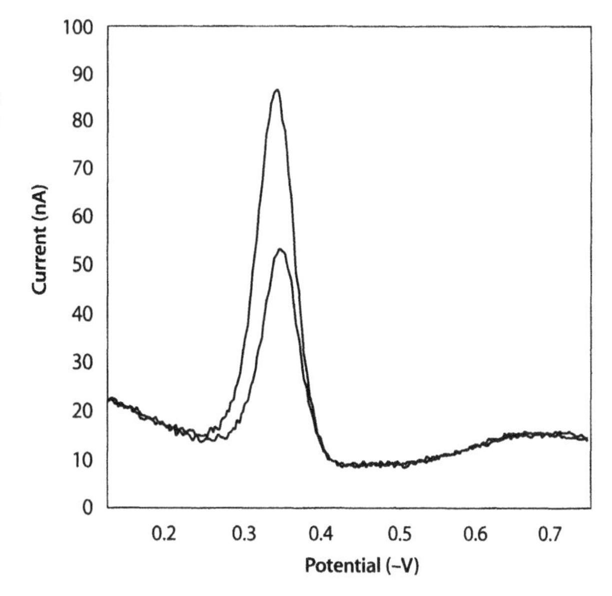

Fig. 8.1. Scans for copper in UV-digested sea water. Conditions: 25 μM SA, HEPPS pH buffer, 60 s plating at –1 V, scan initiated from –0.15 V. The second scan is after a standard addition of 3 nM Cu to the sea water

addition (in this case 3 nM was added) to the sample, followed by a repeat of the measurement.

The sensitivity of CSV is very high, allowing the determination of some twenty elements directly in sea water. For some metals (iron and cobalt for instance (Yokoi and van den Berg 1992; Vega and van den Berg 1997)) the sensitivity is enhanced using catalysis: an oxidant is added to the seawater which reoxidizes the metal which is being reduced at the electrode, thus allowing a catalytic cycle to develop. Catalytic CSV has detection limits at low pM to sub-pM levels for some metals.

8.5 Principle of Ligand Competition

Metals complexed by natural complexing ligands can be detected by adding a complexing ligand (AL) which competes for the metal ions. The ligand addition causes a shift in the equilibrium (Table 8.2) leading to the increased formation of a complex of the metal with the added ligand. The stability of the natural complexes is then evaluated from the comparative complexing ability of the two ligands (the natural one and

Table 8.2. Theory of metal speciation using competitive ligand equilibration with detection by CSV: labile metal concentrations

---

Addition of AL causes competition with the natural ligand L with a subsequent re-distribution of metal species:

$$MAL \xleftarrow{\alpha_{MAL}} M^{n+} \xrightarrow{\alpha_{ML}} ML$$

where $\alpha_{MAL} = K'_{MAL}[AL']$ and $\alpha_{ML} = K'_{M^{n+}L}[L']$.

Definition of the CSV-labile metal concentration:

$$[\text{labile metal}] = [MAL] + [M']$$

where $[M'] = \alpha_M[M^{n+}]$ and $[MAL] = \alpha_{MAL}[M^{n+}]$

The labile metal concentration is directly related to the CSV reduction current i_p:

$$[\text{labile metal}] = i_p S$$

where the sensitivity S is calibrated by a standard metal addition.

Relationship between the labile and total dissolved metal concentrations:

$$[\text{labile metal}] = [M_t](\alpha_M + \alpha_{MAL}) / (\alpha_M + \alpha_{MAL} + \alpha_{ML})$$

Calculation of α_{ML} from the ratio of the measured dissolved and labile metal concentrations (valid if $[L_t] > [M_t]$):

$$\alpha_{ML} = ([M_t] / [\text{labile metal}] - 1)(\alpha_{MAL} + \alpha_M)$$

Calculation of the metal speciation in the original seawater prior to the addition of a competing ligand:

$$[M'] = [M_t] / (\alpha_M + \alpha_{ML}) \qquad [ML] = [M_t] - [M'] \qquad \text{and} \qquad [M^{n+}] = [M'] / \alpha_M$$

---

the known, added, ligand). The principle of the method is that the added ligand is equilibrated with the sample containing the natural ligands, and the concentration of metal bound with the added ligand is detected. This method has been called "competitive ligand equilibration" (Rue and Bruland 1995).

The analytical problem is how to specifically detect the metal complexed with the added ligand. One possibility is to extract the complex with the added ligand (Moffett and Zika 1987), but it is likely that some of the natural complex is extracted too. CSV utilizes the principle of complex adsorption in the preconcentration step, and the complex with the added ligand is detected specifically. For this reason CSV is an excellent method to investigate metal speciation with ligand competition.

8.6 Determination of Labile Metal Concentrations Using CSV

The speciation in the sea water is altered by the added competing ligand (AL) (see Table 8.2) as the formation of the metal complex MAL causes shifts in the original speciation. The magnitude of the alteration depends on the stability of the MAL complexes, and can be regulated by varying the concentration of AL or by selecting ligands forming more or less stable complexes. Thus it is possible to investigate weak or strong natural complexes selectively by varying the "detection window." The labile metal concentration in CSV therefore equals all metal bound by the added ligand, and the fraction of the total metal that is labile depends on the relative stabilities of the complexes, and therefore on the magnitudes of the α-coefficients (α_{MAL} and α_{ML}). When the complex of the metal with the added ligand is more stable than that with the natural ligand ($\alpha_{MA} > \alpha_{ML}$), most of the metal will be bound by AL, and therefore most of the metal will be "labile;" half of the metal is labile when $\alpha_{MAL} = \alpha_{ML}$. The labile metal concentration can be made to approach the total metal concentration by using a complexing ligand which forms very stable complexes and thus outcompetes the natural species.

The theory of CSV of metal species is summarised in Table 8.2. It can be seen that the labile metal concentration is exactly defined and that a clear relationship exists between the labile metal concentration and the natural complexing ligand(s) in the seawater. The magnitude of the shifts in the equilibria by the addition of the competing ligand AL are exactly known and are made use of to calculate the complex stability of the natural species.

Measurements of the labile metal concentration are carried out in equilibrium conditions by allowing the added complexing ligand to equilibrate with the metal species (this can take several hours). The concentration of MAL is normally a significant fraction of the total dissolved metal concentration (typically 50% is complexed by MAL) and only a very small amount is adsorbed on the electrode, so no changes in the speciation occur during the actual measurement of MAL: the measurement is therefore independent of kinetic effects of metal species up to time-scales of hours.

8.7 Calculation of α_{ML}

The stability of the natural metal complexes is evaluated to a first approximation from the ratio of the labile and total dissolved metal concentrations (Table 8.2):

$$[\text{labile metal}] / [M_t] = (\alpha_M + \alpha_{MAL}) / (\alpha_{MAL} + \alpha_{ML} + \alpha_M)$$

This ratio is directly related to the ratio of the α-coefficients, where only the one for ML is unknown. A tentative value for α_{ML} can therefore be calculated from the ratio of the labile over the total metal concentration. This value is only a first approximation as its validity depends on whether the concentration of the natural ligand, $[L_t]$, is greater than that of the metal. The ligand concentration itself, and a value for the conditional stability constant, K'_{ML}, is determined by titration of the sample with metal ions.

8.8 Calibration of α_{MAL}

The complex stability of MAL can be calculated if the value for the conditional stability constants, K'_{MAL}, is known. If this is not known, the conditional stability constant has to be calibrated against a known ligand. This calibration is done by adding EDTA and measuring the decrease in the peak height for the metal, which itself is a measure of the concentration of MAL. EDTA is normally selected for this calibration because its complex stability is known for a large number of metals and the extent of side-reactions with the major ions in seawater can be fairly accurately calculated. The theory for the calibration is presented in Table 8.3.

The effect of adding EDTA to sea water also containing the competing ligand nitrosonaphthol (NN) on the peak height for iron is shown in Fig. 8.2: it can be seen that the peak height (representing the concentration of FeNN) diminishes when EDTA is added because the EDTA is competing with the ligand NN for a limited amount of iron. The decrease in the peak height causes the ratio of i_p / i_0 to decrease (see Table 8.3), and this can be used to calculate a value for α_{FeNN}, and then for β'_{FeNN3} as the concentration of NN is known (in this case it was assumed that complexes of the type $FeNN_3$ are formed). The value for α_{MAL} (α_{FeNN}) is then used to determine the complex stability of the unknown ligands.

8.9 Determination of Ligand Concentrations and Conditional Stability Constants Using CSV with Ligand Competition

Addition of the competing ligand AL to a sample causes some of the metal initially complexed by L to become labile by complexation with AL until the tendency to form complexes between AL and M is exactly balanced by that between the natural ligand (L) and M. At that point the ratio of the labile over the dissolved metal concentration is given by (Table 8.2):

$$[\text{labile metal}] / [M_t] = (\alpha_M + \alpha_{MAL}) / (\alpha_{MAL} + \alpha_{ML} + \alpha_M)$$

The fraction of metal which is made labile (i.e. that which is bound by AL) by the addition of the competing ligand, is determined by the ratio of the complex stability of MAL, which is defined by α_{MAL} (Table 8.2), over the complex stability of all the complexes in the water, which is defined by the sum of all the α-coefficients. The theory is shown in Table 8.2.

Table 8.3. Theory of complexing ligand titrations using ligand competition with CSV

Calibration of α_{MAL} by ligand competition against a known complex like EDTA:

EDTA is added to seawater containing the metal M and the competing ligand AL. The decrease in the peak height is determined. The ratio (X) of the peak current for the metal in the presence (i_p) over that in the absence (i_0) of EDTA is given by (van den Berg, 1985):

$$X = i_p / i_0 = [\mathrm{MAL}] / ([\mathrm{MAL}] + [\mathrm{MEDTA}])$$

This is equivalent to:

$$X = \alpha_{MAL} / (\alpha_{MAL} + \alpha_{MEDTA})$$

where the only unknown is α_{MAL}

Values for α_{MAL} are calculated from:

$$\alpha_{MAL} = [(\alpha_M + \alpha_{MEDTA})X - \alpha_M] / (1 - X)$$

Data treatment of labile metal concentrations resulting from a ligand titration with metal ions. The ratio of [labile metal] / [ML] is plotted as a function of the labile metal concentration. According to the van den Berg/Ruzic equation (Ruzic 1982; van den Berg 1982) the data for a single ligand will lie on a straight line:

$$[\text{labile metal}] / [\mathrm{ML}] = [\text{labile metal}] / [\mathrm{L_t}] + (\alpha_M + \alpha_{MAL}) / ([\mathrm{L_t}]K'_{ML})$$

where the concentrations of ML are calculated from $[\mathrm{ML}] = [\mathrm{M_t}]$ –[labile metal].

Values for C_L and K'_{ML} can be calculated from respectively the slope ($= [\mathrm{L_t}]^{-1}$) and the Y-axis intercept ($= (\alpha_M + \alpha_{MAL}) / ([\mathrm{L_t}]K'_{ML})$) of a linear least squares regression of [labile metal] as a function of [labile metal / [ML] (Ruzic 1982; van den Berg 1982), or from a non-linear data treatment (e.g. van den Berg 1984a).

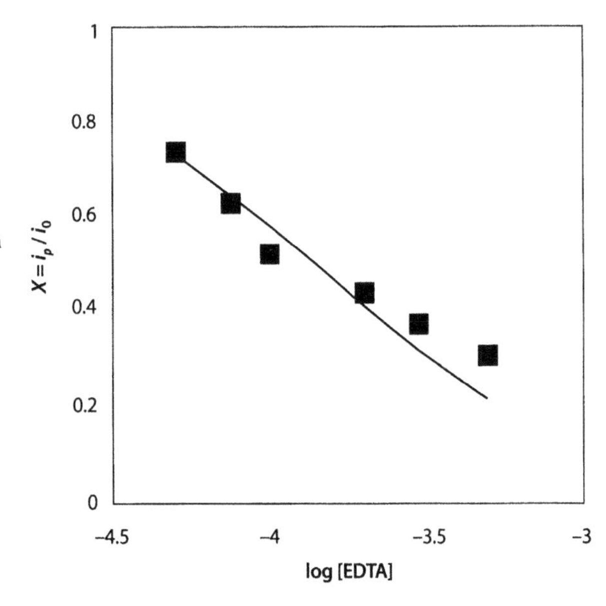

Fig. 8.2. Decrease in the peak height for iron when EDTA is added to seawater containing 5 μM of the competing ligand nitrosonaphthol. The ratio, $X = i_p / i_0$, has been plotted as a function of the EDTA concentration. The curve represents the best fit to the data using $\log \beta_{FeNN_3} = 28.4$ (data from van den Berg 1995)

The part of the natural ligands L which was initially complexed by M is thus liberated by the addition of the competing ligand. Titration of the ligands L with metal ions while monitoring the concentration of MAL (= the labile metal concentration) with CSV causes the response to increase in a non-linear fashion until the ligand is saturated when a linear response is obtained: first the ligand is titrated which was freed upon addition of AL along with any excess ligand. The total ligand concentration is determined. The curved response is confirmation that the complexes are chemically reversible: metal is released when the competing ligands are added, and the complex is formed again when metal is added.

The ligand titration is carried out with equilibration after each addition. For this reason the sample is typically subdivided into ten 10 ml aliquots. The procedure is as follows: to 100 ml seawater pH buffer is added (usually pH around 8), and a low concentration of the competing ligand is added sufficient to make less than half the metal concentration labile. Metal additions are made to ten Teflon® or polystyrene cups or vials to cover a concentration range between 0 and at least twice the expected ligand concentration, and 10 ml of the seawater/reagent mixture is pipetted into each cup. The cups are covered and allowed to equilibrate for at least 3 h, usually overnight. Starting from the lowest concentration, the aliquots are transferred to a voltammetric cell, oxygen is removed by 5 min purging with nitrogen gas, and the CSV response responding to the labile metal concentration is determined. The sensitivity is calibrated using the final (linear) part of the titration (where the natural ligands are saturated), and can be corroborated by further metal additions to the last aliquot.

8.10 How to Evaluate the Ligand Titrations: van den Berg/Ruzic Plots

An example of a complexing ligand titration is shown for a titration with iron of a seawater sample originating from a depth of 50 m in the north-western Mediterranean (van den Berg 1995). The labile iron concentration is monitored by CSV in the presence of nitrosonaphthol (NN) (Fig. 8.3). The CSV scans for this titration are shown in Fig. 8.3a. The peak height for iron increases slowly with the iron concentration until the natural ligand(s) have been saturated when the peak height increases more rapidly; at iron concentrations approximately twice the ligand concentration the increase of the peak height is approximately linear with the iron concentration. This part of the titration can be used to obtain the sensitivity.

There is a direct relationship between the sensitivity, the peak height and the labile metal concentration as the reduction current (i_p) is due to the reduction of adsorbed MAL (e.g. FeNN in the titration with iron). The sensitivity is calibrated by a standard metal addition which provides the link between the peak height and the concentration. Because the standard addition distributes itself over complexes with AL as well as slightly increasing the concentration of inorganic metal, the sensitivity (S) is calibrated with respect to the sum of these two species rather than just to the concentration of MAL:

$$S = i_p / ([\mathrm{MAL}] + [\mathrm{M'}])$$

For this reason the labile metal concentration is defined as [labile metal] = ([MAL] + [M']), which is calculated directly from [labile metal] = i_pS (Table 8.3).

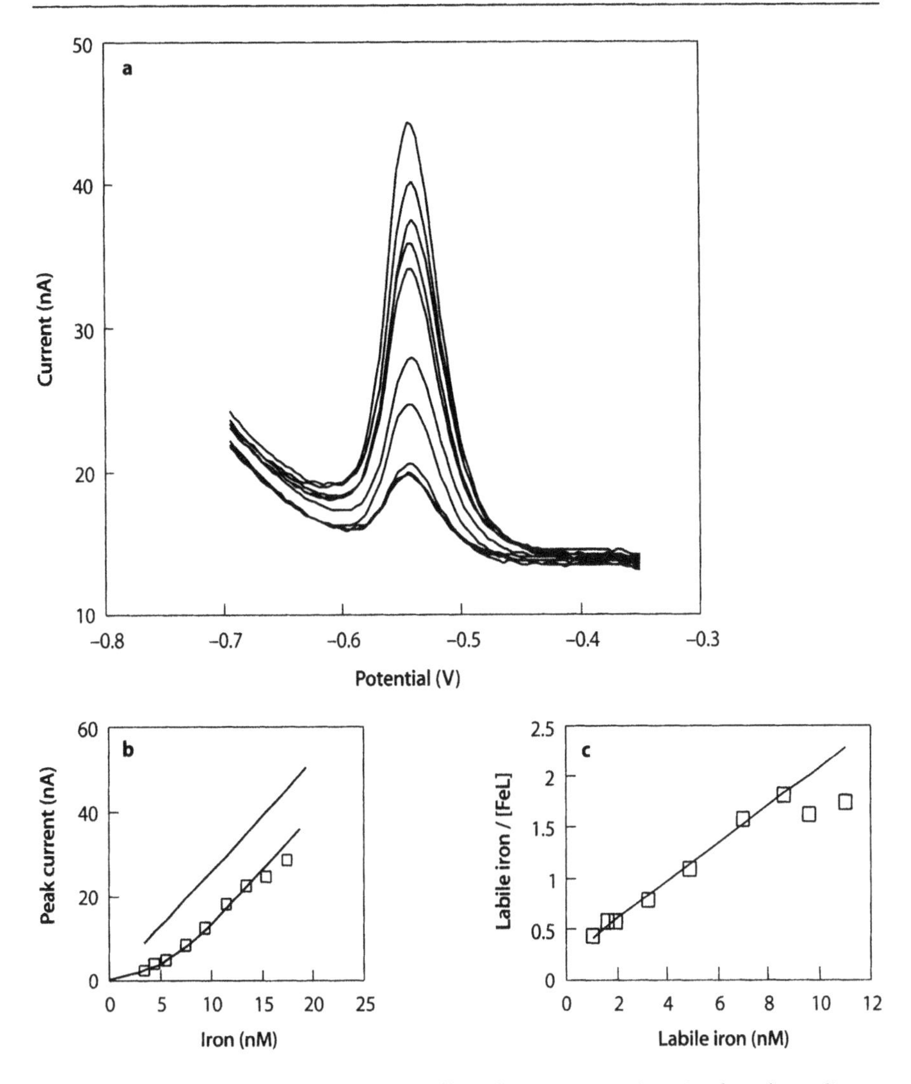

Fig. 8.3. Complexing ligand titration with iron of ligands in sea water originating from the Mediterranean (data from van den Berg 1995)

A first estimate of the ligand concentration can be obtained graphically from extrapolation of the linear part of the titration (Fig. 8.3) to the x-axis. Using the equations for the conditional stability constant, and mass balances for the metal and ligand concentrations, the data has been represented by a single equation (Ruzic 1982; van den Berg 1982), later called the van den Berg/Ruzic equation (Kramer 1986). The unknowns ($[L_t]$ and K'_{ML}) in this equation can be calculated using a linear least-squares regression of the ratio of [labile metal]/[ML] as a function of the [labile metal] (see Table 8.3).

A plot of the data using this equation is straight if a single ligand dominates the metal speciation. For instance, linearisation of the iron titration in Fig. 8.3 using the equation shown in Table 8.3 was used to calculate the ligand concentration ($[L_t] = 9.5$ nM) and the conditional stability constant ($\log K'_{FeL} = 21.5$) (Fig. 8.3c). The plot did not show evidence for the presence of more than one ligand which would manifest itself in two linear segments joined by a curve. In that case the data can be fitted using two linear segments, or by using non-linear fitting of the data.

In the example of the titration with iron shown in Fig. 8.3, the dissolved iron concentration was ~3 nM, well below the ligand concentration. The stability of the complex was very great with a value for $\log \alpha_{FeL} = 13.5$, much greater than that for inorganic complexation of iron: $\log \alpha_{Fe} = 11.3$ at pH 8). Most iron therefore occurred fully complexed by organic matter in this sample, in spite of a very great stability of the inorganic (hydroxide) species.

8.10.1 Example of a Ligand Titration with Copper

The organic complexation of copper in natural waters is probably most studied of all metals because copper is known to be more readily complexed than other divalent metals by organic ligands. Copper is known to occur strongly complexed by organic matter throughout the oceanic water column (Buckley and van den Berg 1986; Coale and Bruland 1988). Copper complexing ligands are known to be released by algae and bacteria in cultures (Sueur et al. 1982; Harwood-Sears and Gordon 1990).

A complexing ligand titration with copper of a culture of *Emiliania huxleyi* with detection of labile copper by CSV and ligand competition clearly shows curvature. The scans for this titration are shown in Fig. 8.4a, whereas a plot of the peak height as a function of the metal concentration reveals the curvature. The ligand concentration calculated from the data (Fig. 8.3c) shows a ligand concentration of 54 nM, much greater than the copper concentration of 8 nM originally present in the water (data from Leal et al. 1999). The complex stability is quite high: in this case the value for $\log K'_{CuL}$ was 12.0 but complexes of greater stability have been found in oceanic waters (Coale and Bruland 1990) and in estuarine waters (van den Berg et al. 1990).

8.11 Effects of the Detection Window

Separate investigations have shown that copper, zinc and nickel occur to various extents complexed in sea water. Different ligand concentrations and different complex stabilities are detected for different metals, and different ligand concentrations using different methods for the same metal. This has led to the suggestion that the detected ligand concentration depends on the detection window of the technique (van den Berg et al. 1990; van den Berg and Donat 1992). It is likely that natural waters contain several organic complexing ligands and what is detected depends somewhat on the technique. Mostly the stronger complexes should be detected first unless these have been saturated with metal due to sample contamination. Very strong complexes can be detected by using ligand competition using either a ligand which forms very stable complexes and can therefore compete with strong, natural, complexing ligands, or by us-

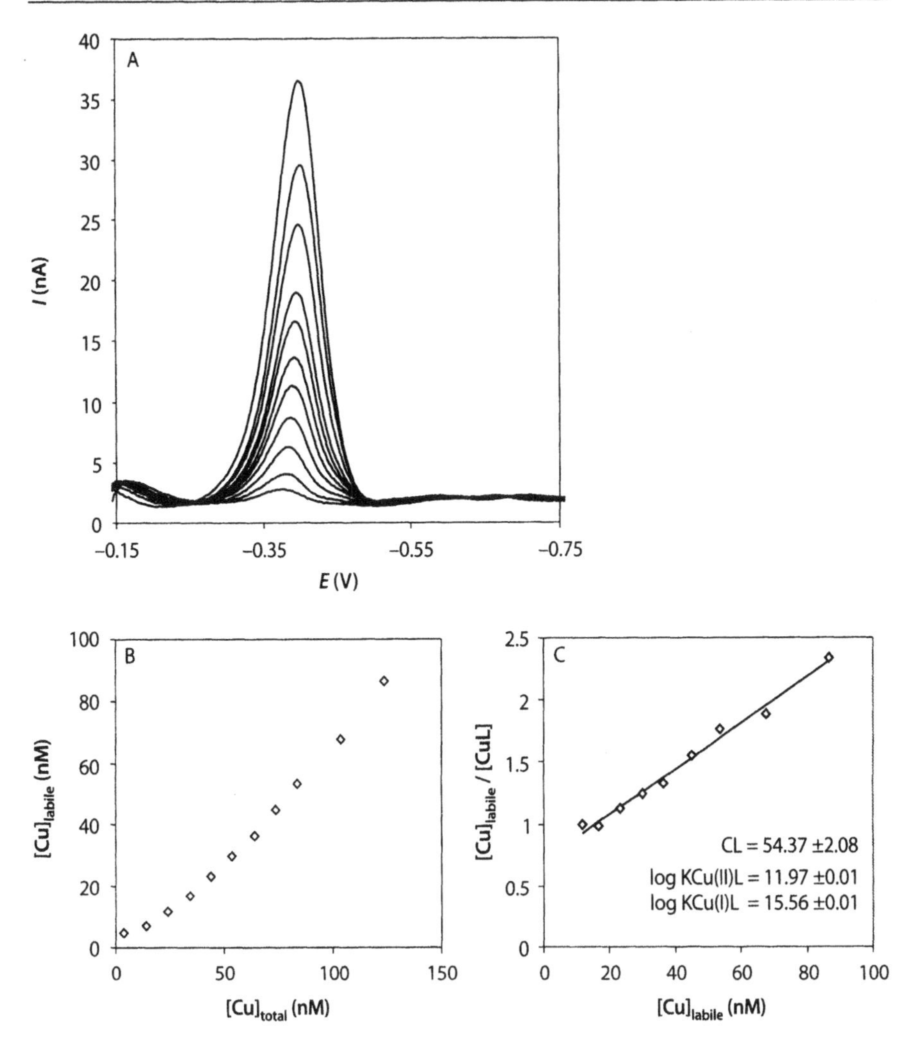

Fig. 8.4. Complexing ligand titration with copper of sea water containing ligands from a culture of *Emiliania huxleyi* (data from Leal et al. 1998)

ing a higher concentration of the added ligand; the detection window is moved in either case.

In CSV with ligand competition the detection window is determined by the complex stability of the complexes with the added ligand: complexes either much stronger or much weaker are not detected. The detection window is centred on α_{MAL}, and metal complexes with values for α_{ML} within a decade of α_{MAL} are detected: complex stabilities outside this range are not detected because the metal complexes either do not dissociate or all the metal is labile.

The concept of the detection window has to be kept in mind when discussing or comparing metal complexing ligands in natural waters.

Acknowledgements

The speciation research of the author is financially supported by a grant from the NERC ("PRIME" GST/02/1 058), and contracts with EU ("MERLIM" MAS3-CT95-0005, and "MOLAR" ENV4-CT95-0007).

References

Batley GE, Florence TM (1976) Determination of the chemical forms of dissolved cadmium, lead and copper in seawater. Mar Chem 4:347–363

Berger P, Ewald M, Liu D, Weber JH (1984) Application of the fluorescence quenching titration method to the complexation of copper(II) in the Gironde estuary (France). Mar Chem 14:289–295

Buckley PJM, van den Berg CMG (1986) Copper complexation profiles in the Atlantic Ocean. Mar Chem 19:281–296

Campbell PGC, Bisson M, Gagne R, Tessier A (1977) Critical evaluation of the copper (II) solubilization method for the determination of the complexation capacity of natural waters. Anal Chem 49:2358–2363

Campos MLAM, van den Berg CMG (1994) Determination of copper complexation in sea water by cathodic stripping voltammetry and ligand competition with salicylaldoxime. Anal Chim Acta 284:481–496

Chau YK, Gachter R, Lum-Shue-Chan K (1974) Determination of the apparent complexing capacity of lake waters. J Fish Res B Canada 31:1515–1519

Coale KH, Bruland KW (1988) Copper complexation in the Northeast Pacific. Limnol Oceanogr 33:1084–1101

Coale KH, Bruland KW (1990) Spatial and temporal variability in copper complexation in the North Pacific. Deep-Sea Res 47:317–336

Harwood-Sears V, Gordon AS (1990) Copper-induced production of copper-binding supernatant proteins by the marine bacterium *Vibrio alginolyticus*. Appl Environ Microbiol 56:1327–1332

Kramer CJM (1986) Apparent copper complexation capacity and conditional stability constants in North Atlantic waters. Mar Chem 18:335–349

Leal MFC, Vasconcelos MTSD, van den Berg CMG (1999) Copper induced release of complexing ligands similar to thiols by Emiliania huxleyi in seawater cultures. Limnol Oceanogr, in proof

Mills GL, Quinn JG (1981) Isolation of dissolved organic matter and copper-organic complexes from estuarine waters using reverse-phase liquid chromatography. Mar Chem 10:93–102

Moffett JW, Zika RG (1987) Solvent extraction of copper acetylacetonate in studies of copper(II) speciation in seawater. Mar Chem 21:301–313

Ostapczuk P, Valenta P, Nürnberg HW (1986) Square wave voltammetry – a rapid and reliable determination method of Zn, Cd, Pb, Cu, Ni and Co in biological and environmental samples. J Electroanal Chem 214:51–64

Raspor B, Nurnberg HW, Valenta P, Branica M (1980) The chelation of lead by organic ligands in sea water. In: Branica M, Konrad Z (eds) Lead in the marine environment. Pergamon Press, Oxford, pp 181–195

Rue EL, Bruland KW (1995) Complexation of iron(III) by natural organic ligands in the Central North Pacific as determined by a new competitive ligand equilibration/adsorptive cathodic stripping voltammetric method. Mar Chem 50:117–138

Ruzic I (1982) Theoretical aspects of the direct titration of natural waters and its information yield for trace metal speciation. Anal Chim Acta 140:99–113

Shuman MS, Michael LC (1978) application of the rotating disk electrode to measurement of copper complex dissociation rate constants in marine coastal samples. Env Sci Technol 12:1069–1072

Sueur S, van den Berg CMG, Riley JP (1982) Measurement of the metal complexing ability of exudates of marine macroalgae. Limnol Oceanogr 27:536–543

van den Berg CMG (1982) Determination of copper complexation with natural organic ligands in seawater by equilibration with MnO_2. I. Theory. Mar Chem 11:307–322

van den Berg CMG (1984a) Determination of the complexing capacity and conditional stability constants of complexes of copper(II) with natural organic ligands in seawater by cathodic stripping voltammetry of copper-catechol complex ions. Mar Chem 15:1–18

van den Berg CMG (1984b) Direct determination of sub-nanomolar levels of zinc in seawater by cathodic stripping voltammetry. Talanta 31:1069–1073

van den Berg CMG (1985) Determination of the zinc complexing capacity in seawater by cathodic stripping voltammetry of zinc-APDC complex ions. Mar Chem 16:121–130

van den Berg CMG (1986) Determination of copper, cadmium and lead in seawater by cathodic stripping voltammetry of complexes with 8-hydroxyquinoline. J Electroanal Chem 215:111–121
van den Berg CMG (1991) Potentials and potentialities of cathodic stripping voltammetry of trace elements in natural waters. Anal Chim Acta 250:165–276
van den Berg CMG (1995) Evidence for organic complexation of iron in seawater. Mar Chem 50:139–157
van den Berg CMG, Donat JR (1992) Determination and data evaluation of copper complexation by organic ligands in sea water using cathodic stripping voltammetry at varying detection windows. Anal Chim Acta 257:281–291
van den Berg CMG, Kramer JR (1979) Determination of complexing capacities and conditional stability constants for copper in natural waters using MnO_2. Anal Chim Acta 106:113–120
van den Berg CMG, Nimmo M, Daly P, Turner DR (1990) Effects of the detection window on the determination of organic copper speciation in estuarine waters. Anal Chim Acta 232:149–159
Vega M, van den Berg CMG (1997) Determination of cobalt in sea water by catalytic cathodic stripping voltammetry. Anal Chem 69:874–881
Yokoi K, van den Berg CMG (1992) The determination of iron in seawater using catalytic cathodic stripping voltammetry. Electroanalysis 4:65–69

Chapter 9

Organic Complexation of Metals in Sea Water

C.M.G. van den Berg

9.1 Introduction

Metals dissolve in water by ionisation, and their solubility is usually restricted by their solubility product with one of the major anions in sea water, such as with hydroxide, carbonate or phosphate. Their solubility is increased by the formation of soluble complexes with the major anions. The concentrations of the vast majority of the metals in the oceans are usually much less than their solubility because they are scavenged from solution by interaction with settling particles, including microorganisms, faecal pellets and inorganic hydroxides or oxides of poorly soluble metals. This interaction, and therefore the residence time of the elements, could be modelled using chemical equilibria if we knew the chemical speciation of the metals, and the interaction of the various species with the particles including their availability for uptake by the microorganisms. For some metals it is now known that their chemical speciation is dominated by organic complexation. This aspect will be discussed in detail in this chapter.

It is relatively easy to calculate the inorganic speciation of metals in sea water, as most reactions have been investigated and stability constants exist. However, recently it has been discovered that the metal speciation is more complicated and that several metals occur in species which react differently from the known inorganic species. By destruction of all organic matter in sea water, it has been demonstrated that these species are of organic nature, and it is now known that copper, zinc and iron exist predominantly bound by organic matter, whereas nickel and cobalt may be only partially associated with organic matter. Even metals which are expected to form hydroxide species (titanium, and aluminium) are known to occur at least in part as non-reactive species although for these elements the nature of these species has not yet been elucidated (van den Berg et al. 1994).

Using electroanalytical techniques it has been demonstrated that the organic species of copper, zinc, iron, nickel and cobalt are chemically reversible, meaning that they are complexes between the metal ions and organic ligands which can dissociate when the free metal ion concentration is lowered, and the concentration of organic complexes is increased when either the ligand or the metal concentration is raised.

Organic complexing ligands occur in all natural waters including sea water. The concentrations of these ligands are much less than those of the major ions in sea water but usually greater than those of the trace metals. Their concentrations have been determined using titrations with metal ions with detection of the reactive metal species using various techniques, mostly using voltammetry.

9.2 Distributions of Organic Metal Complexing Ligands

The distributions of organic complexing ligands in the oceanic water column are shown in Fig. 9.1a for copper, Fig. 9.1b for zinc, and Fig. 9.2 for iron.

Copper. The chemical speciation of copper was determined in the water column of the Pacific, and the data (Coale and Bruland 1990) were fitted to a model of two complexing ligands: L1 with log K'_{CuL} values around 13.0, and L2 with log K'_{CuL} values around 9.8 (K'_{CuL} is defined as $K'_{CuL} = [CuL] / [Cu^{2+}][L']$, and $\alpha_{Cu} = 24$ is used for inorganic complexation of copper). The data are mostly for the upper water column, and are shown in Fig. 9.1a. Deep water data are also presented in Coale and Bruland's paper for one station, but unfortunately these are rather "noisy." The concentration of the stronger, L1, ligand is less than that of dissolved copper at depths greater than ~200 m. In the upper water column L1 binds most copper because its complexes with copper are much more stable than those with L2. The dissolved copper concentration is just below 1 nM showing comparatively little variation down the water column, but the free, cupric, ion concentration of copper is lowered to ~10^{-14} M in the upper water column, whereas in the deeper waters it is higher at ~10^{-12} M. The concentration of the free copper ion, and therefore its availability to microorganisms, shows great variability down the water column due to the organic complexation of copper. This illustrates the importance of determining the chemical speciation in addition to the concentration of metals in water.

The concentration of L2 shows a maximum in the upper water column at 100 m depth (Fig. 9.1a), which would be somewhat below the chlorophyll maximum. A maximum ligand concentration coincident with the chlorophyll maximum would indicate that the ligands are released by the algae; a maximum below that of the chlorophyll would suggest that the ligands could be produced by bacteria or due the breakdown of the algae.

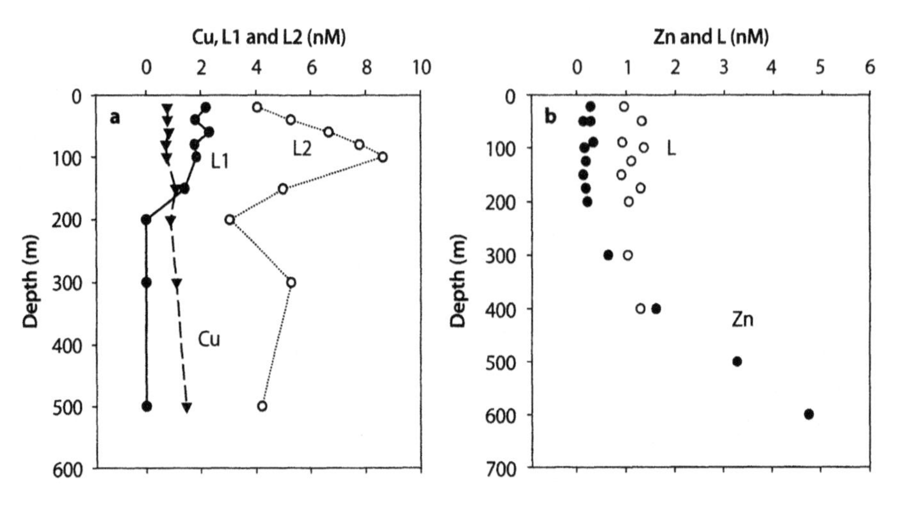

Fig. 9.1. Copper and zinc complexing ligands in the Pacific (copper data from Coale and Bruland 1990, zinc data from Bruland 1989)

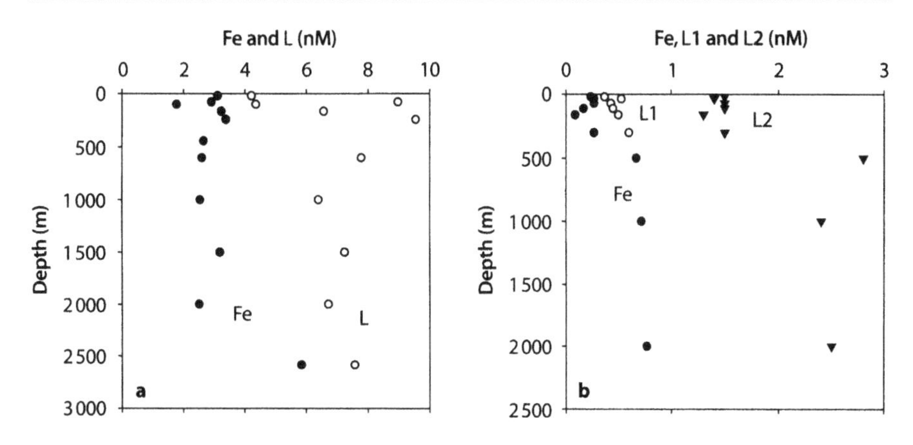

Fig. 9.2. Iron complexing ligands; **a** in the Mediterranean; **b** in the Pacific (data from Rue and Bruland 1995; van den Berg 1995)

Zinc. The speciation of zinc in samples from the water column of the central North Pacific is shown in Fig. 9.1b and Table 9.1. It can be seen that the dissolved zinc concentration shows a nutrient-like profile: the zinc concentration is depleted to 0.2–0.3 nM in the upper water column, whereas it is increasingly regenerated down the water column where it reaches 4.8 nM at 600 m depth. Only one group of ligands which bind zinc was detected, occurring at levels of around 1.2 nM. The ligand concentrations (Fig. 9.1b) show some variability but not a systematic trend. The zinc concentration is therefore less than the ligand concentration in the upper water column, whereas it is greater than the ligand concentration in waters deeper than ~400 m. The organic ligands could not be detected in the deep waters because of the high zinc concentrations which saturated the ligands. It appears therefore that zinc is predominantly complexed by organic matter in the upper water column, and it is mostly in the free, inorganic, form in the waters below this. This is perhaps a little unexpected because the free zinc should be relatively easily scavenged by settling particles. This scavenging should leave its imprint on the zinc distribution which should have lowered the regeneration profile apparent now. It is therefore possible that some ligands were present which have not yet been detected.

The organic complexation of zinc causes the concentration of inorganic zinc ([Zn']) to be lowered to 2–8 pM, and the concentration of Zn^{2+} to approximately half that (1–4 pM). Such low concentrations of zinc have been shown to limit growth of a neritic algal species in culture (where the metal speciation was EDTA-controlled), whereas an open oceanic species was not affected at Zn^{2+} concentrations above 10^{-13} M (Brand et al. 1983). It appears therefore that the oceanic algal species was adapted to the low free zinc concentrations occurring in the presence of organic complexing matter.

9.2.1 Competition between Copper and Zinc

It is likely that metals compete with each other for the complexing ligands unless the ligands form complexes specifically with certain metal ions. No experimental evidence

Table 9.1. Zinc speciation in the central North Pacific, at Vertex-IV station 28° N/158° W (Bruland 1989) (The concentrations of inorganic zinc (Zn') were measured in the deep waters whereas they were calculated from the speciation data in the other samples)

Depth	Zn_t (nM)	L (nM)	$\log K'_{Zn^{2+}L'}$	Zn' (nM)	
22	0.3	0.98	11.0	0.008	calc.
50	0.15	1.34	11.1	0.002	calc.
50	0.3	1.33	11.2	0.004	calc.
90	0.35	0.94	10.9	0.014	calc.
100	0.17	1.37	11.0	0.003	calc.
125	0.2	1.12	10.8	0.007	calc.
150	0.15	0.92	10.9	0.005	calc.
175	0.2	1.31	10.9	0.004	calc.
200	0.23	1.07	10.8	0.008	calc.
300	0.65	1.05	10.7	0.033	meas.
400	1.62	1.3	10.7	0.44	meas.
500	3.3	–		1.9	meas.
600	4.77	–		3.5	meas.

has been obtained to demonstrate that competition occurs, but this could be due to experimental difficulties. There are some examples where ligands were titrated with one metal in the presence of an increased concentration of another metal in an attempt to demonstrate such competition. For instance, titrations with zinc of sea water in which the background copper concentration had been doubled to 1.2 nM were not affected and the same zinc binding ligand concentration was found as before (Bruland 1989). Apparently the increased copper concentration did not decrease the concentration of ligands available for complexation by zinc. The stability of the copper complexes is much greater than those with zinc, so the apparent zinc-binding ligand concentration would have been lowered if the same ligands bind copper too. Such data suggests that these ligands are specific for zinc, and that copper is bound by different ligands.

9.2.2 Iron Complexing Ligands

Depletion of dissolved iron in the upper water column of the Pacific and Atlantic oceans, and coincidence of a maximum in the iron concentration and a minimum in the oxygen concentration, indicate that the iron is taken up by phytoplankton and released again at greater depth upon death and sinking of the organisms (Martin and Gordon 1988; Martin et al. 1993). Partial depletion of nitrate (as opposed to complete depletion which would indicate nitrate limitation) in surface sea water from the eastern equatorial Pacific, the Southern Ocean and the offshore Gulf of Alaska, indicate that primary production here is not nitrate limited (Martin et al. 1989). These three areas have in common very low levels of dissolved iron and very low atmospheric inputs of iron due to their remoteness from land (Martin et al. 1991). These observations have resulted in the argument that iron availability rather than the concentration of any of the major nutrients limits primary production in large parts of the world's oceans (Martin et al. 1991). This is very unexpected because iron is one of the more abundant elements on earth.

The iron limitation argument, and the fertilization experiments, have thus far been based on the premise that the solubility of iron in sea water is extremely small (e.g. Martin and Gordon 1988) and restricted to low nanomolar to subnanomolar levels, the dissolved iron occurring predominantly as hydroxy complexes (Byrne and Kester 1976; Turner et al. 1981). Laboratory experiments testing the availability of iron to phytoplanktonic organisms are usually based on excess iron levels, where the available iron is limited by chelation with a synthetic ligand (EDTA) (Brand 1991), and it is assumed that only the inorganic iron is available to the organisms, the uptake rate being controlled by complex formation at the transport site into the cells (Hudson and Morel 1990). Although these experiments clearly indicate that iron complexed by EDTA is not available to microorganisms, there is no evidence that iron complexed by other organic ligands is also unavailable. Indeed, siderophores are known to increase the availability of iron to certain marine bacteria (Reid and Butler 1991; Reid et al. 1993). It is therefore not impossible that organic complexing ligands are released by marine microorganisms for strategic reasons to convert the iron to a species which is specifically available to them and possibly not to others.

For this reason it is interesting that measurements of the chemical speciation of iron have shown that this element too is strongly complexed by organic matter in sea water (Gledhill and van den Berg 1994; Rue and Bruland 1995; van den Berg 1995). Examples for the distribution of organic ligands for iron are shown for the Mediterranean (Fig. 9.2a) and the Pacific (Fig. 9.2b). There are large differences between these areas: the Mediterranean receives large amounts of iron as atmospheric inputs which can vary significantly due to episodic variability. The iron concentration in the water column of the Mediterranean can therefore be relatively high (2–3 nM) compared to the subnanomolar levels more typical for open oceanic conditions (Johnson et al. 1997). The ligand concentration was found to be greater than the dissolved iron levels in the Mediterranean, and was found to show increases in the upper water column associated with, or just below, the chlorophyll maximum. The location of the maximum in the ligand concentration suggests that it could be the result of breakdown of algal matter, or production by bacteria, or production by algae, or all of these.

In the Pacific the iron concentrations are much lower, and the ligand concentrations are lower too (Fig. 9.2b): the data were fitted to a model of two complexing ligands, a strong one with concentrations of 0.5 to 0.6 nM detectable in the upper water column only due to the higher iron concentrations in the deeper waters, and a weaker one with concentrations of 1.5 nM in the upper water column to 2.4 nM in the deeper waters (Rue and Bruland 1995). It is possible that the strong ligand was present throughout the oceanic water column as it could not be detected due to its saturation by the higher iron concentrations. It is therefore likely that the 2.4 nM of the ligands in the deeper waters included this strong ligand, and that the weaker ligand occurred at levels between 1.5 and 1.8 nM throughout the water column.

The stability of the iron complexes was very high, with values for the stability constant ($K'_{Fe^{3+}L'}$) of the strong complexes in the Pacific of around 10^{23}, whereas for the ligands in the Mediterranean the constant was just below 10^{22}, or about an order of magnitude lower. The organic complexation causes the concentration of inorganic iron to be lowered by a factor of ~500 in the upper water column of the Pacific and throughout the water column of the Mediterranean, to about ~0.5 pM in the Pacific, and to 5 pM in the Mediterranean. An oceanic cyanobacterium in an EDTA-controlled culture has been shown to be limited by inorganic iron levels around 1 pM (Brand et al. 1983), suggesting that the organic complexation in combination with the low iron concentrations (0.1 nM) in the upper water column may well be the cause of growth limitation of the algae by iron.

9.3 Sources of Ligands

The distributions of the ligands in the oceanic water column (Fig. 9.1 and 9.2) tend to show a maximum in the upper water column, suggesting that microorganisms are the origin. Marine as well as fresh-water algae in cultures are known since a long time to produce ligands for copper and other metals (van den Berg et al. 1979; Morelli et al. 1989). Fungi and bacteria are also known to produce ligands for copper and iron (Neilands 1967; Sunda and Gessner 1989). The function of the ligands for iron produced by bacteria, fungi and some plants is reasonably clear: they are siderophores and improve the availability of iron to the organisms. This function has not (yet) been demonstrated for the marine system. It is likely that the ligands change the availability of metals to microorganisms, usually making them less available but sometimes increasing their availability, for instance if the complexes are lipid-soluble (Florence and Stauber 1986; Florence et al. 1992).

9.4 Composition of Ligands

Little is known about the composition of the ligands in the oceanic water column related to difficulties in their recovery. The free ligands are known to adsorb poorly on hydrophobic surfaces such as those in C-18 cartridges whilst they adsorb reasonably well once saturated with copper (Buckley and van den Berg 1986; Donat et al. 1986): this suggests that the ligands are quite hydrophilic. Several attempts have been made to get closer to some identification (Zhou and Wangersky 1989; Donat et al. 1997) but with no success besides general classifications. On the other hand several ligands have been proposed as potential models for the ligands occurring in the natural system. Phytochelatins are glutathione-rich peptides which are known to be produced by plants and marine algae in response to high metal concentrations (Ahner et al. 1995; Scarano and Morelli 1996). Algae are rich in thiol compounds like glutathione and cysteine, which are known to form very stable complexes with copper(I) (Leal and van den Berg 1998). Glutathione is known to occur throughout the oceanic water column at levels as high as 10 nM (Le Gall and van den Berg 1998) and, along with other thiol compounds, is therefore a strong candidate for the identity of copper binding ligands.

9.5 Calculation of Metal Complexation by the Ligands

The concentrations of the organic complexing ligands are much less than those of the major ions but usually greater than that of the trace metals. The ligands participate in side-reactions with the major cations and with hydrogen ions. These side-reactions lower the effective stability of the complexes with trace metals. Use is made of conditional stability constants valid for sea water of a given salinity and pH as the extent of the side-reactions of the natural ligands is normally not known but the stability of their metal complexes can be determined in the sea water. The concept of the conditional stability constant facilitates the calculation of the metal speciation by the complexing ligands in sea water which has otherwise a generally constant composition with respect to the major ions and pH. The conditional stability constant can then be used to calculate the complexation of trace metals by the ligand whilst automatically taking the side-reactions into account.

The conditional stability constant is defined by

$$K'_{ML} = [ML] / ([M^{n+}][L']) \qquad (9.1)$$

where [L'] is the concentration of all L (including protonated species and those bound by Ca^{2+}, Mg^{2+} and the other major cations) not complexed by the metal M.

Before the metal speciation is calculated it is useful to introduce the concept of α-coefficients which reduce the degree of metal complexation to a fraction

$$\alpha_M = [M_t] / [M^{n+}] \qquad (9.2)$$

meaning that the α-coefficient of metal M is the ratio of the total metal concentration over the free metal ion concentration. In addition to the overall α-coefficient, there is an α-coefficient for each individual metal species to indicate the ratio of its concentration over that of the free metal ion:

$$\alpha_{ML} = [ML] / [M^{n+}] \qquad (9.3)$$

The overall α-coefficient (α_M) is always >1 as it includes the concentration of the free metal ion; $\alpha_M = 1$ when there is no complexation at all. However, the individual α-coefficients (α_{ML}) can be anywhere from <1 to >>1: when these are smaller than 1 there is very little complexation, whilst when these are >>1 the complexation is very strong.

α-coefficients are additive which simplifies the calculation of metal speciation in the presence of several complexing ligands if these are all present in excess. It will be shown here how one calculates the complexation of Cu^{2+} by inorganic ligands and EDTA in sea water. There are two possibilities with regard to the concentration of the organic ligand:

- the ligand concentration is much greater than that of the metal (copper in this case) so the ligand concentration is not significantly decreased by its complexation with the metal, and
- the ligand concentration is similar to that of the metal.

A mass balance is set up which is the same for both cases:

$$[M_t] = [M^{n+}] + [\text{inorganic complexes}] + [ML] \qquad (9.4)$$

The inorganic species are grouped together in [M']:

$$[M'] = [M^{n+}] + [\text{inorganic complexes}] \tag{9.5}$$

The free metal ion concentration is related to the combined inorganic metal concentration via $\alpha_{M'}$:

$$[M'] = \alpha_{M'}[M^{n+}] \tag{9.6}$$

The fraction of metal occurring as ML is obtained by substitution for ML using the conditional stability constant:

$$[ML] = [M^{n+}]K'_{ML}[L'] \tag{9.7}$$

The fraction of M occurring as ML is calculated using its individual α-coefficient which is obtained from

$$\alpha_{ML} = K'_{ML}[L'] \tag{9.8}$$

An α-coefficient is therefore simply calculated from the product of the ligand concentration and the conditional stability constant. Then:

$$[ML] = \alpha_{ML}[M^{n+}] \tag{9.9}$$

Substitution into the mass balance for the metal:

$$[M_t] = [M^{n+}](\alpha_{M'} + \alpha_{ML}) \tag{9.10}$$

The concentration of the free metal ion concentration can thus be calculated by dividing the total metal concentrations by the sum of the α-coefficients.

The total metal concentration and the ligand concentrations can be determined. The problem is now reduced to finding the concentration of L' (which is L not complexed by M). There are two possibilities: high concentrations of L (L >> M), and the other situation of lower concentrations of L.

9.5.1 At High Natural Ligand Concentrations

From the mass balance of L:

$$[L_t] = [ML] + [L'] \tag{9.11}$$

it can be seen that $[L'] \approx [L_t]$ at high concentrations of L. In this situation the total concentration, ($[L_t]$) can be used instead of [L']. Then

$$\alpha_{ML} \approx K'_{ML}[L_t] \tag{9.12}$$

and can be readily calculated. The concentration of M^{n+} in the sea water is then calculated from

$$[M^{n+}] = [M_t] / (\alpha_{M'} + \alpha_{ML}) \tag{9.13}$$

An example of how to calculate the concentration of Cu^{2+} in sea water containing a high concentration of an organic complexing ligand (similar to EDTA) is shown in Table 9.2. It can be seen that 97% of the copper is complexed by a ligand similar to EDTA in sea water.

9.5.2 At Low Ligand Concentrations

The calculation of the metal speciation in the presence of a low concentration of a strong complexing ligand is a little more complicated, as a correction has to be made for the fraction of ligand bound by the metal when α_{ML} is calculated. Equation 9.8:

$$\alpha_{ML} = K'_{ML}[L']$$

At high ligand concentrations this equates to $\alpha_{ML} = K'_{ML}[L_t]$, but at low ligand concentrations the concentration of L' is calculated from

$$[L_t] = [L'] + [ML] \longrightarrow [L'] = [L_t] - [ML] \tag{9.14}$$

Table 9.2. Calculation of the speciation of copper in the presence of a high concentration of ligand

The speciation of copper is calculated for sea water containing 100 nM of ligands similar to EDTA (This concentration is a little high for oceanic conditions). The complex stability is given by $\log K'_{CuEDTA} = 10.08$ (which is a little low for the organic ligands known to occur in sea water).Then

$$\alpha_{CuEDTA} = 1202.$$

The free copper ion concentration is calculated from:

$$[Cu^{2+}] = [Cu_t] / (\alpha_{Cu'} + \alpha_{CuEDTA})$$

Using a copper concentration of $[Cu_t] = 2$ nM and an inorganic α-coefficient for copper complexation of $\alpha_{Cu'} = 37$ it is found that

$$[Cu^{2+}]=2 \times 10^{-9} / (37 + 1203) = 1.6 \times 10^{-12} \text{ M.}$$

The total *inorganic* copper concentration, [Cu'], is then calculated from

$$[Cu'] = [Cu^{2+}]\alpha_{Cu'} = 1.6 \times 10^{-12} \times 37 = 0.059 \text{ nM, the remainder, 1.94 nM (97\%)}$$

being complexed by EDTA.

This correction can be made by using iterative calculations, i.e. by first calculating the complexation of a metal by the ligand using the total ligand concentration for [L'] and thereafter repeating this after substitution of $[L_t] - [ML]$ for [L']. Alternatively the correction can be made mathematically as shown in the section here.

The conditional stability constant is used to obtain a relationship with the free metal ion concentration:

$$K'_{ML} = [ML] / ([M^{n+}][L']) \longrightarrow [ML] = K'_{ML}[M^{n+}][L'] \quad (9.15)$$

Substitution into the mass balance equations, and solving for [L'] gives

$$[L'] = [L_t] / (1 + K'_{ML}[M^{n+}]) \quad (9.16)$$

Here the free ligand concentration [L'] is calculated. This is substituted into Eq. 9.8 to obtain the value for α_{ML}:

$$\alpha_{ML} = K'_{ML}[L'] = K'_{ML}[L_t] / (1 + K'_{ML}[M^{n+}]) \quad (9.17)$$

Substitution into the mass balance of the metal (Eq. 9.10) and rearranging gives a quadratic equation for $[M^{n+}]$:

$$[M^{n+}]^2\alpha_{M'}K'_{ML} + [M^{n+}](\alpha_{M'} + K'_{ML}[L_t] - K'_{ML}[M_t]) - [M_t] = 0 \quad (9.18)$$

The free metal ion concentration is then calculated from

$$[M^{n+}] = (-b + (b^2 - 4ac)^{½}) / 2a \quad (9.19)$$

where

- $a = \alpha_{M'}K'_{ML}$
- $b = \alpha_M + K'_{ML}[L_t] - K'_{ML}[M_t]$
- $c = -[M_t]$

Table 9.3. Calculation of the speciation of copper at a low natural ligand concentration

The chemical speciation is calculated of 2 nM copper in pH 8 sea water in the presence of 5 nM of complexing ligands similar to EDTA.

Use $\alpha_{Cu} = 37$ and $\log K'_{CuEDTA} = 10.08$.

The concentration of Cu^{2+} is calculated directly from the quadratic equation:

$$[Cu^{2+}] = (-b + (b^2 - 4ac)^{½}) / 2a$$

giving a free cupric ion concentration of 0.024 nM. This compares with an overall inorganic copper concentration ($[Cu'] = \alpha_{Cu}[Cu^{2+}]$) of 0.88 nM and with 1.11 nM CuEDTA. 56% of copper is therefore complexed by EDTA, the remainder being mainly complexed by inorganic ligands, and the remainder of the EDTA (78%) being complexed by calcium and magnesium ions.

Table 9.4. Conditional stability constants reported for organic complexes occurring in sea water (some of these conditional stability constants were originally reported as $K''_{M'L'}$ values and have been converted here to $K'_{ML'}$ values)

Metal	Ligand concentrations (nM)	Log K'_{ML}	Reference
Copper		13.0	Coale and Bruland 1990
L1	2	10.0	
L2	5 - 10		
Iron	4 - 12	21.8	van den Berg 1995
	0.3	23.7	Rue and Bruland 1997
Lead	0.2- 0.5	10.6	Capodaglio et al. 1990
Zinc	1	11.0	Bruland 1990

Use of this theory to calculate the complexation of copper by an organic ligand occurring at low concentration in sea water is demonstrated in Table 9.3. It can be seen that about half (56%) of the copper is complexed by this low concentration of a ligand like EDTA, the remainder occurring as inorganic species.

On the whole the natural ligands form stable complexes, much more stable than those with EDTA, with trace metals in seawater. The conditional stability constants (as K'_{ML}) observed for organic complexes are shown for several metals in Table 9.4. Because of the high stability of the complexes several metals (copper, zinc and iron) occur predominantly complexed by the organic matter, depending on the distribution of the organic ligands in the oceanic water column.

Acknowledgements

The research is supported by a grant from the NERC ("PRIME", GST/02/1 058, for which this is publication no. R32) and a contract with the EU ("MERLIM", MAS3-CT95-0005).

References

Ahner BA, Kong S, Morel FMM (1995) Phytochelatin production in marine algae. 1. An interspecies comparison. Limnol Oceanogr 40:649–657

Brand LE (1991) Minimum iron requirements of marine phytoplankton and the implications for the biogeochemical control of new production. Limnol Oceanogr 36:1756–1771

Brand LE, Sunda WG, Guillard RRL (1983) Limitation of marine phytoplankton reproductive rates by zinc, manganese and iron. Limnol Oceanogr 28:1182–1198

Bruland KW (1989) Complexation of zinc by natural organic ligands in the central North Pacific. Limnol Oceanogr 34:269–285

Buckley PJM, van den Berg CMG (1986) Copper complexation profiles in the Atlantic Ocean. Mar Chem 19:281–296

Byrne RH, Kester DR (1976) Solubility of hydrous ferric oxide and iron speciation in seawater. Mar Chem 4:255–274

Capodaglio G, Coale KH, Bruland KW (1990) Lead speciation in surface waters of the Eastern North Pacific. Mar Chem 29: 221–233

Coale KH, Bruland KW (1990) Spatial and temporal variability in copper complexation in the North Pacific. Deep-Sea Res 47:317–336

Donat JR, Statham PJ, Bruland KW (1986) An evaluation of C-18 solid phase extraction technique for isolating metal-organic complexes from central North Pacific Ocean waters. Mar Chem 18:85–99

Donat JR, Kango RA, Gordon AS (1997) Evaluation of immobilized metal affinity chromatography (IMAC) for isolation and recovery of strong copper-complexing ligands from marine waters. Mar Chem 57:1–10
Florence TM, Stauber JL (1986) Toxicity of copper-complexes to the marine diatom *Nitzschia closterium*. Aquatic Toxicology 8:11–26
Florence TM, Powell HKJ, Stauber JL, Town RM (1992) Toxicity of lipid-soluble copper(II) complexes to the marine diatom *Nitzschia closterium* – amelioration by humic substances. Wat Res 26:1187–1193
Gledhill M, van den Berg CMG (1994) Determination of complexation of iron(III) with natural organic complexing ligands in sea water using cathodic stripping voltammetry. Mar Chem 47:41–54
Hudson RJM, Morel FMM (1990) Iron transport in marine phytoplankton: Kinetics of cellular and medium coordination reactions. Limnol Oceanogr 35:1002–1020
Johnson KS, Gordon RM, Coale KH (1997) What controls dissolved iron concentrations in the world ocean? Mar Chem 57:137–161
Leal MFC, van den Berg CMG (1998) Evidence for strong copper(I) complexation by organic complexing ligands in seawater. Aquat Geoch 4:49–75
Le Gall A-C, van den Berg CMG (1998) Folic acid and glutathione in the water column of the North East Atlantic. Deep-Sea Res I 45:1903–1918
Martin JH, Gordon RM (1988) Northeast Pacific iron distributions in relation to phytoplankton productivity. Deep-Sea Res 35:177–196
Martin JH, Gordon RM, Fitzwater SE, Broenkow WW (1989) VERTEX: Phytoplankton/iron studies in the Gulf of Alaska. Deep-Sea Res 35:649–680
Martin JH, Gordon RM, Fitzwater SE (1991) The case for iron. Limnol Oceanogr 36:1793–1802
Martin JH, Fitzwater SE, Gordon RM, Hunter CN, Tanner SJ (1993) Iron, primary production and carbon-nitrogen flux studies during the JGOFS North Atlantic Bloom Experiment. Deep-Sea Res II 40:115–134
Morelli E, Scarano G, Ganni M, Nannicini L, Seritti A (1989) Copper binding ability of the extracellular organic matter released by *Skeletonema costatum*. Chemical Speciation and Bioavailability 1:71–76
Neilands JB (1967) Hydroxamic acids in nature. Science 156:1443–1447
Reid RT, Butler A (1991) Investigation of the mechanism of iron acquisition by the marine bacterium *Alteromonas luteoviolaceus*: Characterization of siderophore production. Limnol Oceanogr 36:1783–1792
Reid RT, Live DG, Faulkner DJ, Butler A (1993) A siderophore from a marine bacterium with an exceptional ferric ion affinity constant. Nature 366:455–458
Rue EL, Bruland KW (1995) Complexation of iron(III) by natural organic ligands in the Central North Pacific as determined by a new competitive ligand equilibration/adsorptive cathodic stripping voltammetric method. Mar Chem 50:117–138
Rue EL, Bruland KW (1997) The role of organic complexation on ambient iron chemistry in the equatorial Pacific Ocean and the response of a mesoscale iron addition experiment. Limnol Oceanogr 42: 901–910
Scarano G, Morelli E (1996) Determination of phytochelatins by cathodic stripping voltammetry in the presence of copper(II). Anal Chim Acta 319:13–19
Sunda WG, Gessner RV (1989) The production of extracellular copper-complexing ligands by marine and estuarine fungi. Chemical Speciation and Bioavailability 1:65–70
Turner DR, Whitfield M, Dickson AG (1981) The equilibrium speciation of dissolved components in freshwater and seawater at 25 °C at 1 atm pressure. Geochim Cosmochim Acta 45:855–882
van den Berg CMG (1995) Evidence for organic complexation of iron in seawater. Mar Chem 50:139–157
van den Berg CMG, Wong PTS, Chau YK (1979) Measurement of complexing materials excreted from algae and their ability to ameliorate copper toxicity. J Fish Res B Canada 36:901–905
van den Berg CMG, Boussemart M, Yokoi K, Prartono T, Campos MLAM (1994) Speciation of aluminium, chromium and titanium in the NW Mediterranean. Mar Chem 45:267–282
Zhou M, Wangersky PJ (1989) Study of copper-complexing organic ligands: Isolation by a Sep-pak C18 column extraction technique and characterization by chromarod thin-layer chromatography. Mar Chem 26:21–40

Occurence, Pathways and Bioaccumulation of Organometallic Compounds in Marine Environments

R. Frache · P. Rivaro

10.1 Introduction

The last fifteen years have seen a growing concern for several classes of hydrophobic contaminants such as polychlorobiphenyls, polycyclic aromatic hydrocarbons and more lately organometallic compounds. Organometallic compounds occur in the environment as a result of direct anthropogenic inputs or because they are naturally formed there. They are used in a wide variety of industrial processes sometimes at the percent level. Some organometallic forms of Hg, As and now Pb have been observed to have efficient biocidal properties. This fact has largely been applied to the synthesis of a large variety of pesticides, leading then to direct introduction in the environment. In this paper the marine environmental impact of some organotin compounds will be examined.

10.2 Organotin Compounds: Environmental Chemical Aspects

Organotin compounds consist of a single central Sn atom covalently bound to one to four organic groups [$R_xSn_{(4-n)}$]. In Fig. 10.1 general forms of organotin compounds are shown. The introduction of organic groups increases the toxicity of the molecule reaching a maximum for trisubstituted compounds. The sensitivity shown by different types of organisms also depends on the nature of the organic substituent. For example, marine organisms are very sensitive to tributyltin (TBT) and triphenyltin (TPhT) substitutes. Organotin compounds are generally of anthropogenic origin apart from methyltins, which may be produced by environmental methylation as well. After recognising the biocidal properties of trialkylated organotins in the 1950s, the variety of applications increased considerably. Total worldwide use of organotin compounds increased from about 5 000 tons annually at the beginning of the sixties to over 60 000 tons annually in the late eighties. To date, organotins – butyltin and phenyltin – are the most widely used organometallic compounds (Morabito et al. 1995; Ritsema 1997). Butyltin compounds are mainly used as stabilizers for polyvinylchloride and as biocides. About 23% of the total worldwide organotin production is used as agrochemicals and as biocides in a broad spectrum of applications. However their environmental impact is far more than for organotins used as stabilizers due to direct introduction into the environment. The organotins used as antifouling agents (both tributyltin and triphenyltin) or agrochemicals (triphenyltin) will give direct input into the (marine) environment. Tributyltin (TBT) was in the 1970s and 1980s mainly used as a biocide in antifouling paints for ships, pleasure boats and docks. Since the eighties there

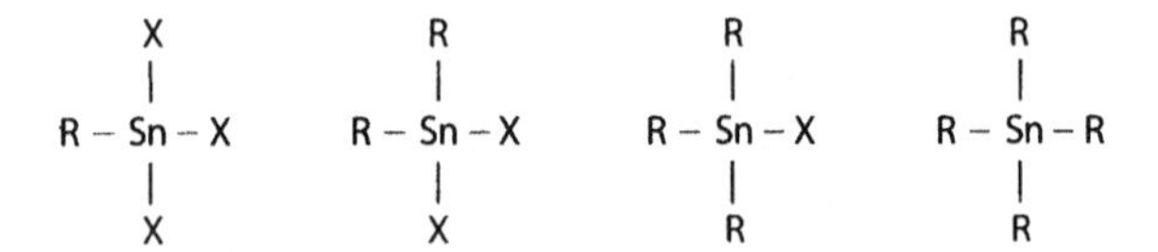

Fig. 10.1. General forms of organotin compounds

has been an increased environmental interest for the consequences of organotin compounds, especially for the aquatic environment. This increased interest was induced by the observed deleterious effects causing shell malformations and reduced growth of the Pacific oyster, *Crassostrea gigas* (Morabito et al. 1995). The observed malformations and reduced growth were ascribed to the water pollution by tributyltin compounds used as antifouling agents on the hulls of ships. The serious problems encountered in the commercial oyster cultures were soon followed by similar reports in United Kingdom (Waldock and Miller 1983). Similar results were reported for the Dutch coastal zone (Ritsema and Laane 1991). Moreover, the imposition of male sex organs on the female mud snail *Nucella lapillus* (a phenomenon called imposex), which led to a decline of this gasteropod population with serious ecological consequences, was related to the presence of TBT in sea water. The use of TBT-containing antifouling paints is now controlled or banned in many countries, in many cases resulting in a decrease of TBT contamination in marine and harbour waters.

Therefore, several studies were carried out to test on different species the toxicity of TBT. In Table 10.1 the acute toxicity values of tributyltin oxide (TBTO) for a few aquatic organisms are reported (UNEP 1988).

As it can be seen, larvae and juveniles are most sensitive to the TBTO toxic action than adults, but sensitivity of larvae may not be related to sensitivity in adults. For example, the adult Pacific oyster *Crassostrea gigas* is the most resistant to TBTO among the tested mollusks, showing a 48-h LC_{50} at 1 800 $\mu g\,l^{-1}$, whereas its larva stadium is the most sensitive with a 1.6 $\mu g\,l^{-1}$ 48-h LC_{50}.

Knowledge of the fate of organotins in the aquatic environments is essential for the prediction of environmental concentrations and understanding of possible ecotoxicological effects. The fate of organotins is closely linked to the partitioning in aqueous media. Several studies are carried out in this field and it is now possible to show a realistic hypothesis. These compounds either adsorb from water onto particulate matter, and are thus likely to be removed from the water into sediments, or stay dissolved in the water where they are susceptible to degradation processes or to be accumulated by aquatic organisms (Fig. 10.2).

Under normal conditions TBT can occur in various chemical speciation forms which are all in equilibrium with each other. The predominant TBT species at $pH < pK_a$ (6.5) is the cation, whereas at pH > 6.5 TBT predominate as a neutral complex. At seawater pH (8.1 ±0.2) the chloride, hydroxide and carbonate forms can occur. For the aquatic toxicity the nature of the TBT counter-ion does not play an important role. However when TBT occurs as a neutral chloride, hydroxide or carbonate complex, this should facilitate uptake into organisms (Ritsema 1997). Organotin compounds, particularly TBT to its lipophilicity, enter organisms via food or water, through lipid membranes.

Table 10.1. Acute toxicity values of tributyltin oxide (TBTO) for non target organism (from UNEP 1988, modified)

Test organism	TBTO concentration in water (μg l^{-1})	Effect
Algae		
Skeletonema costatum	0.1	No growth
Mollusks (bivalves)		
Crassostrea gigas (larvae)	1.6	48-h LC50
Crassostrea gigas (adult)	1 800	48-h LC50
Mytilus edulis(larvae)	2.3	48-h LC50
Mytilus edulis (adult)	300	48-h LC50
Crustaceans		
Acanthomysis sculpta (juveniles)	0.42	96-h LC50
Arcantia tonsa (adult)	1.1	48-h LC50
Crangon crangon (adult)	41	96-h LC50
Fish		
Solea solea (larvae)	8.5	48-h LC50
Solea solea (adult)	88	48-h LC50

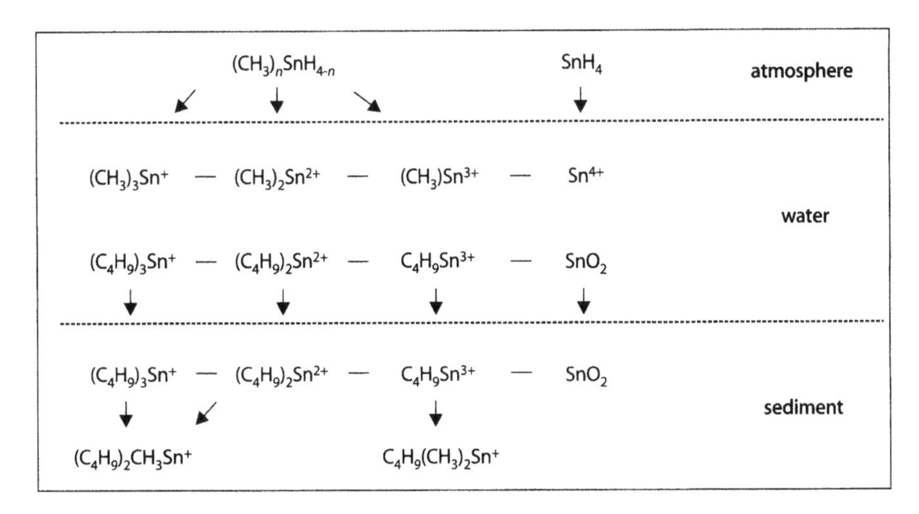

Fig. 10.2. Biogeochemical cycle for Sn (adapted from Ritsema 1997)

Marine animals are able to store or eliminate it; the kinetics of both processes are affected by metabolism or lack of it by an animal. TBT is metabolized in two phases, the first of which involves the cytocrome P-450 dependent mono oxigenase system, and transforming it in DBT, and later on in MBT, which are less toxic than TBT (Lee 1991). Among various invertebrate groups, the mollusks appear to be the most affected by TBT. In fact mollusks are characterized by very low cytochrome P-450 content and mono oxigenase activity; this low detoxifying activity may explain both the high TBT bioaccumulation factors (cf. Table 10.2) (Quevauvillier 1991) and the susceptibily of this group.

Table 10.2. Bioconcentration values found for aquatic organisms exposed to sublethal dose of TBT (from Quevauvillier 1991, modified)

Organisms	Exposition	Bioaccumulation factor
Micro-organisms		
Estuarine bacteria	Water	356 – 1 039
Pseudomonas 244	Water	438 – 487
Phytoplankton		
Isochrysis galbana	Water	5 500
Ankistrodesmus falcata	Water 20 μg l^{-1}	30 000
	Water 40 μg l^{-1}	860
Mollusks		
Crassostrea gigas	Water 0.15 μg l^{-1}	6 000
	Water 1.25 μg l^{-1}	2 000
Mytilus edulis	Water 5 μg l^{-1}	5 000
	Phytoplankton	<2
Fish		
Cyprinodon variegans	Water 2 μg l^{-1}	118 – 4 580
Salmo salar	Water 0.1 μg l^{-1}	3.9 (liver)
	Water 1 μg l^{-1}	1.6 (liver)

10.3 High Performance Liquid Chromatography – Hydride Generation – Inductively-Coupled Plasma Atomic Emission Spectrometry (HPLC-HG-ICP/AES) Hyphenated System to Organotin Analysis

The organometallic species determination has been solved most generally by using a combination of different analytical processes and analytical techniques. Most successful combinations result from the interface between chromatographic techniques, providing the species differentiation ability, to atomic spectrometry detectors, accounting for selectivity (Frache 1997). Their hyphenation allows mastery of the technical skills necessary to perform organometallic determination. In most cases, analytical speciation schemes rely on the combination of three basic steps and their interfacing design at the instrumental level:

- analyte pre concentration
- chromatographic separation
- selective detection

All techniques first require a preconcentration step. This procedure can be performed off-line from the instrumentation or on-line, as a part of the instrumentation set-up. The first range of techniques using off-line preconcentration most often use standard chromatographic procedures. Liquid chromatography separation techniques are very powerful, as they generally allow the derivatization stage to be by-passed and provide a large panel of possible chromatographic procedures, enabling a larger range of organometallic compounds to be determined. An on-line technique using direct interfacing between hydride generation methods and combining simultaneously the preconcentration step by cryofocusing and later chromatography by gentle warming of the trap can be very easily interfaced with atomic absorption or emission spectrometry. Techniques derived from these procedures achieve the highest sensitivity and ease of operation in comparison to other methods using off-line preconcentration methods with considerably fewer analytical steps.

In our studies we employed an HPLC-HG-ICP-AES hyphenated method in order to obtain the determination of mono-, di- and tributyltin species in biotic and abiotic matrices (Rivaro 1998; Rivaro et al. 1995). Separations were performed on a Partisil SCX 10 analytical column (10 µm particle size, 25 cm × 4.6 mm i.d.) (Whatman, Englewood Cliffs, NJ, USA). The flow rate was 1 ml min^{-1} and no gradient elution devices were used.The mobile phase used was 0.1 M ammonium acetate in 80% methanol/20% water containing 0.1% m/v tropolone. The eluent coming out from the HPLC column was mixed in a PTFE (polytetrafluoroethylene) "T" piece with hydrochloric acid supplied by a peristaltic pump, with a 0.7 ml min^{-1} flow. The organotin compounds were converted into their hydrides by addition of $NaBH_4$, supplied by a peristaltic pump with a 0.7 ml min^{-1} flow in another PTFE "T" piece. Evolved hydrides were drained to the gas-liquid separator, where an argon flow carried the tin vapours in to the ICP-AES torch, while the organic eluent was washed out. The HPLC-ICP determination was carried out detecting the tin signal every 300 ms with a scanning step of 0 nm, i.e. the ICP detector was fixed at the analytical tin wavelength Sn(II) 189.989 nm. A schematic diagram of the HPLC-HG-ICP/AES system is reported in Fig. 10.3.

Figure 10.4 shows the result of a chromatographic separation of a mixture of TBT, DBT, MBT, and standard solutions.

This analytical method was improved developing a new method for the determination of TBT and DBT by means of LC/MS with particle beam interface (Magi and Ianni 1998).

10.4
An Example of a "Field Study": The Butyltin Distribution in the Genoa Oil Port

As an example of the distribution of organotin compounds in the marine environment, a study carried out in the Genoa oil port is shown. The aim of the present study has been to follow the seasonal variations of butyltin compounds in water, suspended particulate matter and mussels, under natural conditions in the Genoa oil port. Mussels are often used as a biomonitor in monitoring programs assessing the quality of marine environment, because they can provide a time-integrated estimation of the contaminants in water. A growing number of studies valued the levels of alkyltins in mussel tissues from a variety of locations worldwide. Nevertheless, the input of organotins from large vessels in coastal waters has been seldom assessed, because these vessels spend the majority of the time offshore, where diluition and dispersion processes can reduce the release of TBT in the water.

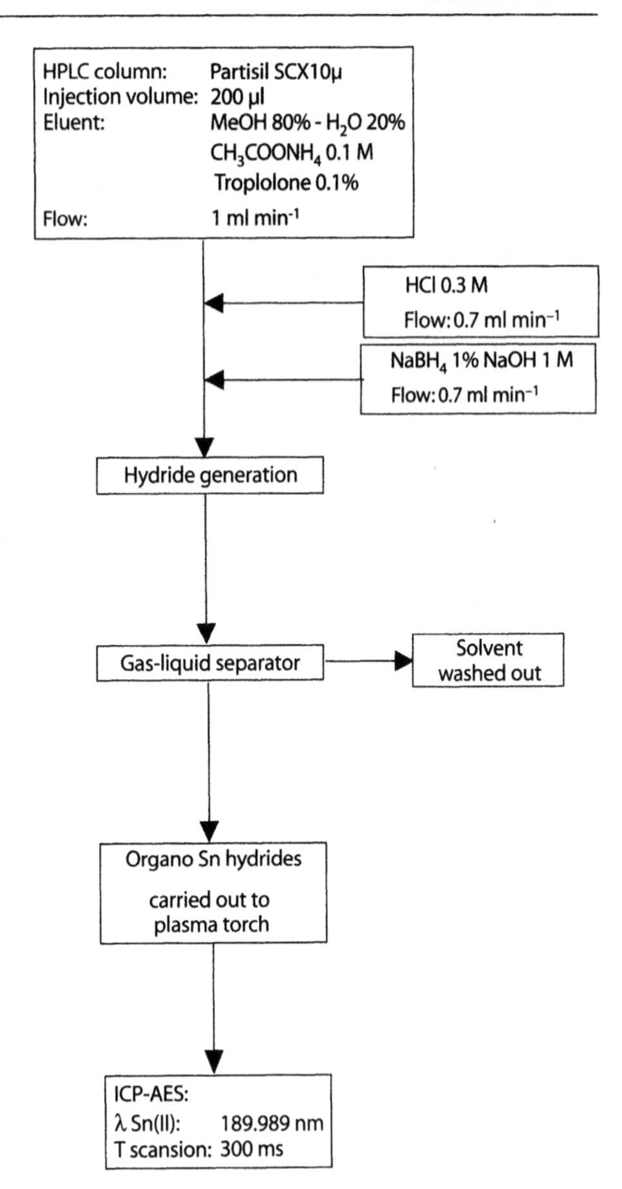

Fig. 10.3. Schematic diagram of HPLC-HG-ICP/AES system used for organotins determination

The Genoa oil port is on the west side of the town; there are 4 wharves, 250–500 m long, and it is 11–14 m deep. The terminal handles 500–600 oil tankers each year, which visit the port for one to two days. Sea water and mussels of a natural population, 5–6 cm in length, were sampled at two month intervals between July 1997 and May 1998. At each sampling, 10 l of sea water was collected by a LDPE container, and kept at 4 °C until organotin extraction. At the same time, 25 mussels were collected: the whole soft tissues of 8 individuals and the digestive glands and the gills of the remainder were dissected, homogenized and frozen at −20 °C until analysis. Organotins extractions from seawater and mussel tissues and determination procedures are schematized in

Fig. 10.4. Chromatographic separation of monobutyltin chloride (MBT), dibutyltin chloride (DBT) and tributyltin chloride (TBT) standard solutions

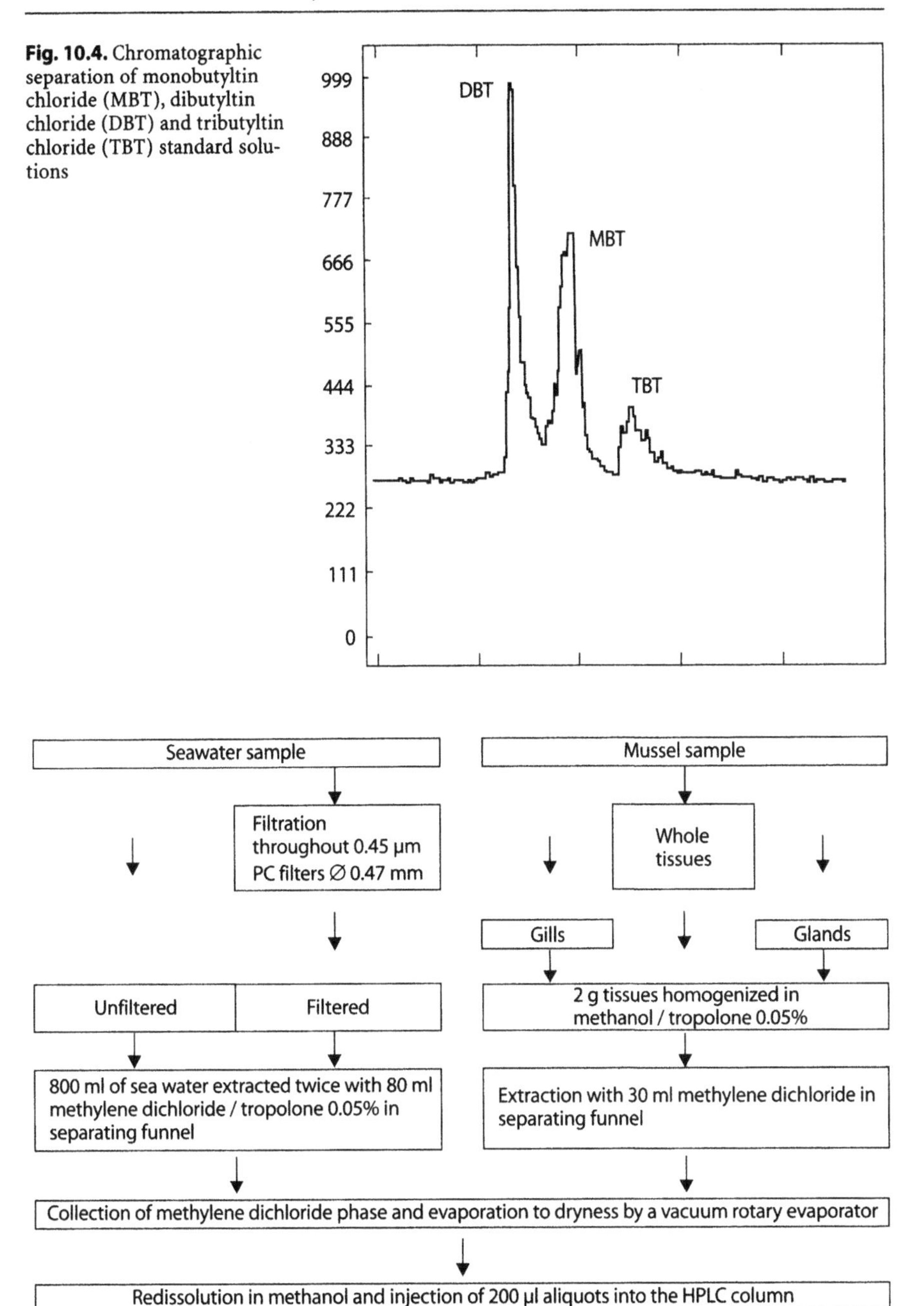

Fig. 10.5. Analytical schema for the extraction and preconcentration steps of butyltins from sea water and mussel tissues

Fig. 10.5. The chromatographic elution and following Hg-ICP/AES determination were then carried out.

The general results of this investigation are the following. In the sea water in the examined months, butyltin compounds were detected in all the samples; no phenyltin species were revealed. As shown in Fig. 10.6, seasonal variation in concentrations was found: TBT showed highest concentrations in summer while DBT and MBT had their maxima in February.

The species distribution pointed out that TBT represented 59–70% of the total organic Sn in autumn and winter, and ranged from 33 to 43% in spring. Looking at the partitioning between dissolved and particulate phases, reported in Fig. 10.7, it can be seen that butyltin compounds were predominantly in the dissolved matter in all the months excepting TBT in July, and the total organic Sn was found inversely correlated with particulate matter in the samples.

In Fig. 10.8 the butyltin trends found in the mussel tissues are reported. Among the tissues, gills showed the highest organotin concentrations. Total tissues and gills presented the same TBT trends with a maximum in September and February and a minimum in December, while digestive glands had a different trend in summer and au-

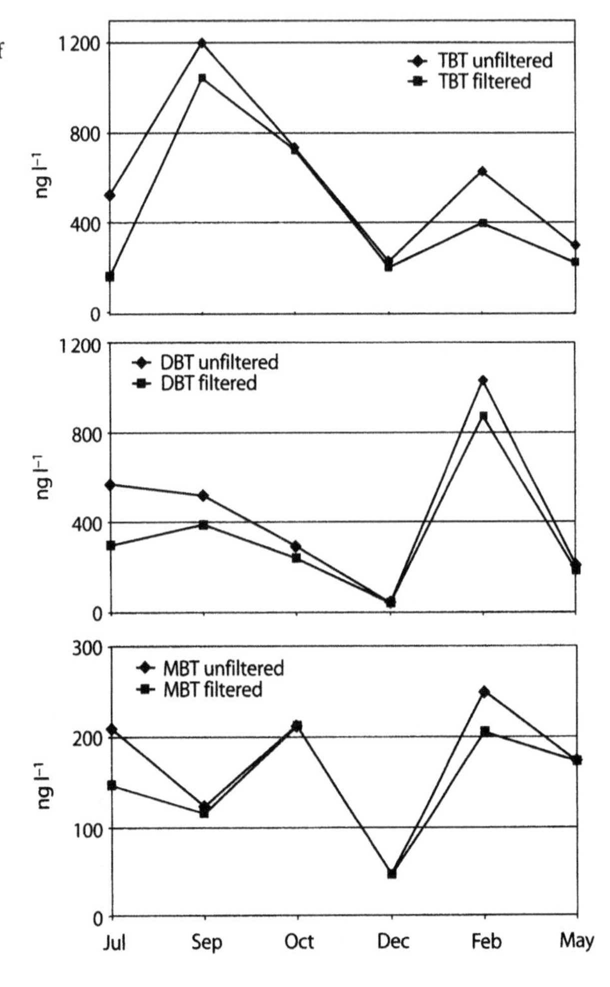

Fig. 10.6. Butyltin compounds trends found in the sea water of Genoa oil port. Data represent the mean ±standard deviation of 3 replicates

tumn. Even if seasonal variation in concentrations was detected, the speciation was not greatly different from summer to winter, TBT being prevalent in all seasons, both in gills and digestive glands.

The obtained data allowed us to draw some preliminary observations on relationships between butyltin trends in sea water and in mussel tissues. TBT and DBT trends showed the best correlation between seawater and gill levels. Therefore, this organ is the most suitable to be considered as bioindicator of organotin seasonal variations. Moreover the gills showed the highest TBT concentration factors (dry tissues/water), which was estimated at 9 500. In winter a TBT minimum was found both in sea water and in tissues. Looking at speciation, it is interesting to point out that in seawater

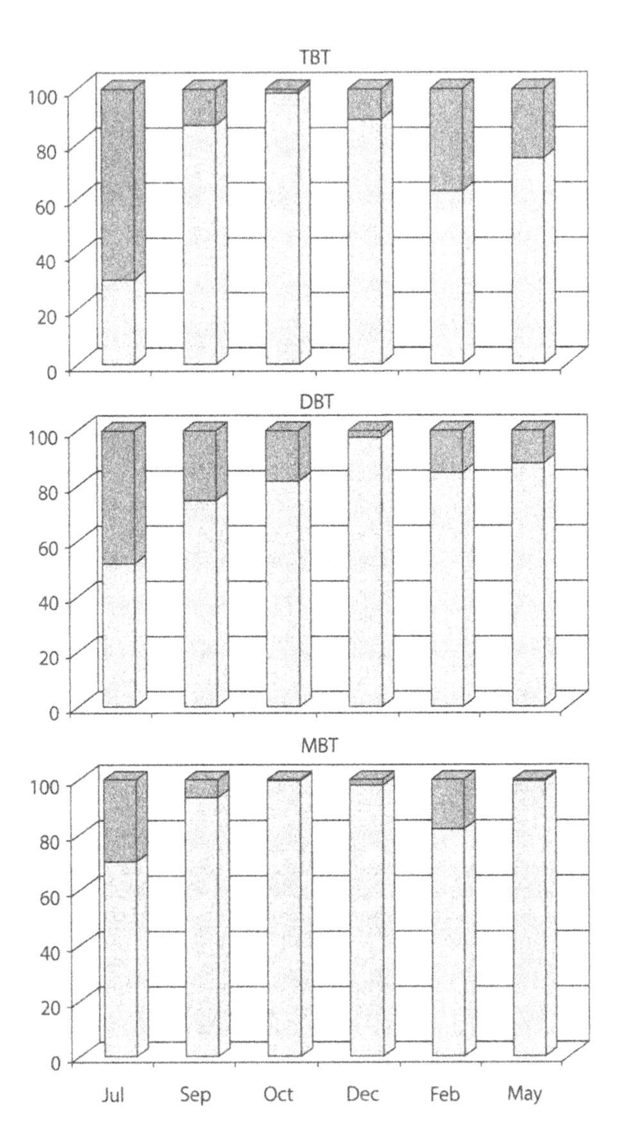

Fig. 10.7. Butyltin compounds partitioning (%) between dissolved (*grey*) and particulate (*dark grey*) phases

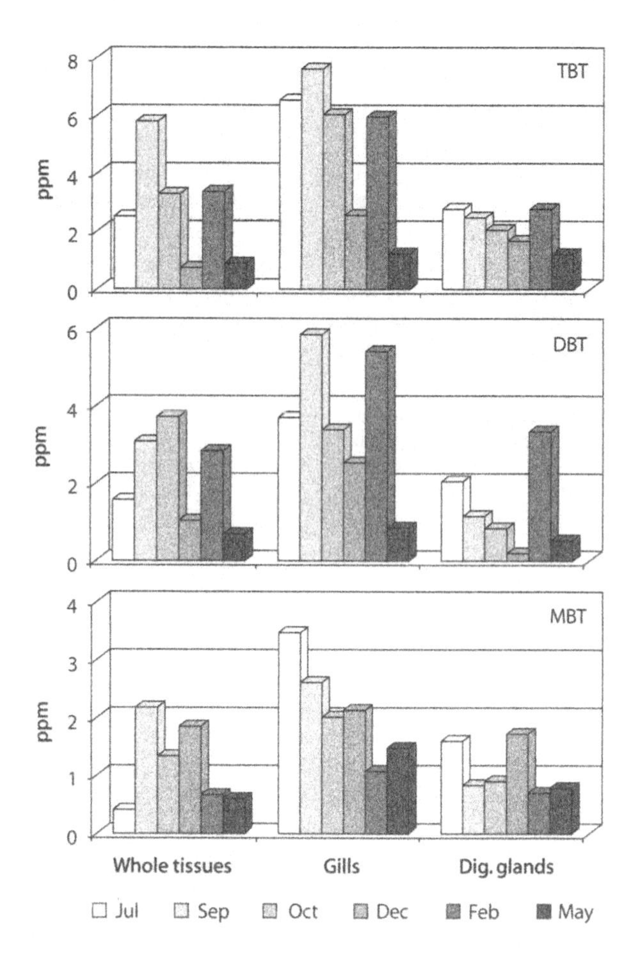

Fig. 10.8. Butyltin compound trends in mussel tissues ($\mu g\ g^{-1}$ dry wt). Data represent the mean ±standard deviation of 3 replicates

samples, TBT lowest concentration values (225 ng l^{-1}) corresponded to its highest percentage (71%), while in gills (2,5 $\mu g\ g^{-1}$) corresponded to its lowest (34%). The presence of degradation products in tissues is due not only to metabolic processes but to direct taking up from sea water, where they were detected, also.

References

Frache R (1997) Hyphenated instrumental methods for the detection of heavy metals in marine environment. In: Gianguzza A, Pelizzetti E, Sammartano S (eds) Marine chemistry – an environmental analytical chemistry approach. Kluwer Academic Publishers, Dordrecht, Boston, London, p 149–159

Lee RF (1991) Methabolism of tributyltin by marine animals and possible linkage to effects. Mar Environ Res 32:29–32

Magi E, Ianni C (1998) Determination of tributyltin in marine environment by means of liquid chromatography mass spectrometry with a particle beam interface. Analyt Chim Acta 359:237–244

Morabito R, Chiavarini S, Cremisini C (1995) Speciation of organotin compounds in environmental samples by Gc-MS. In: Quevauviller P, Maier EA, Griepink B (eds) Quality assurance for environmental analysis. Elsevier, Amsterdam, p 435

Quevauviller Ph (1991) Speciation de l'étain dans le milieux estuariens et cotiers. Ph.D. Thèse, Université de Bordeaux, France

Ritsema R (1997) Environmental applications of hyphenated techniques for the speciation of tin, arsenic and mercury. Ph.D. Thèse, Université de Pau et des Pays de l'Adour, France
Ritsema R, Laane RWPM (1991) Butyltins in marine waters of the Netherlands in 1988 and 1989; concentrations and effects. Mar Environ Res 32:243–260
Rivaro P (1998) Proceedings of International Confer. on Environm. and Biolog. Aspects of main Group Organometals. Odense, Denmark, 28 June–1 July 1998
Rivaro P, Zaratin L, Frache R, Mazzucotelli A (1995) Determination of organotin compounds in marine mussel samples by using high performance liquid chromatography – hydride generation inductively-coupled plasma atomic emission. Analyst 120:1937–1939
UNEP, United Nations Environment Programme (1988) Assessment of organotin compounds as marine pollutants and proposed measures for the Mediterranean
Waldock MJ, Miller D (1983) The determination of total and tributyltin in sea water and oysters in areas of high pleasure craft activities. ICES paper 1983/E:12

Chapter 11

Hydrolysis Processes of Organotin(IV) Compounds in Sea Water

C. De Stefano · C. Foti · A. Gianguzza · S. Sammartano

11.1 Introduction

Organotin(IV) compounds include a variety of organometallic moieties characterized by a central tin atom covalently bonded to various organic groups (methyl, ethyl, propyl, butyl, octyl, phenyl, etc.) through one or more carbon atoms. Their general formula can be indicated as $R_nSnX_{(4-n)}$ (R = organic group; X = halide, nitrate, acetate, hydroxide, etc.; $n = 1$ to 4). Because the tin-carbon bond is reported to be stable up to 200 °C (Zuckerman et al. 1978), these compounds must be considered thermally stable under environmental conditions. Some factors, such as ultraviolet (UV) and gamma (γ) irradiation and biological and chemical cleavage, influence the degradation of organotin compounds by a progressive removing of organic groups, according to the following scheme:

$$R_4Sn \longrightarrow R_3SnX \longrightarrow R_2SnX_2 \longrightarrow RSnX_3 \longrightarrow SnX_4$$

On the other hand, the inverse process can occur through the bioalkylation of inorganic tin, often leading to the formation of unsymmetrical organotin compounds. Biomethylation of tin compounds occurs both in aerobic and anaerobic conditions, by means of a variety of bacterial substrates (Thayer 1993). In particular, sulfate-reducing bacteria form mono and dimethyltin from inorganic tin(IV) in the absence of sediment, whilst the abiotic methylation of tin(IV) is favoured by low pH values and low ionic strengths of the aquatic environment. Di- and monobutyltin species have been detected as breakdown degradation products of bis(tributyltin)oxide. Moreover, organotin compounds are widely distributed in the environment owing to their industrial applications, i.e. as fungicides and acaricides in agriculture, as wood and stone preservatives, as stabilizers and catalysts in PVC and in foam production, etc. (Blunden and Chapman 1986). One of the most important applications is the use of tributyltin derivatives (mainly polytributyltin methacrilate, coupled with Cu_2O) in the "antifouling" paints for ships. For this reason, tributyltin compounds are often found in seawater, in sediments and in biota, particularly in harbour zones where naval traffic is more intensive (Champ and Seligman 1996; see also Frache and Rivaro, Chapter 10 of this volume). Due to their well known toxicity, which depends on the number ($R_3Sn^+ > R_2Sn^{2+} > RSn^{3+} > Sn^{4+}$, toxicity scale), the kind of organic groups bonded to tin(IV) atom, and on their structure, organotin(IV) derivatives have attracted the attention not only of environmental protection agencies, but also of a number of research groups all over the world. The toxicity of the organotin(IV) halides, oxides, carboxylates, and of many newly synthesized organotin(IV) complexes, has been extensively studied by different authors (Arakawa and Wada 1993; Barnes and Magos 1968; Mennie and Craig 1993; Thayer 1978, 1984; Gianguzza et al. 1994).

11.2 Aqueous Solution Chemistry of Organotin(IV) Compounds

In contrast with the significant number of studies on toxicity of organotin compounds, there are relatively few data in literature about the chemical behaviour of these compounds in aqueous solutions.

Organotin cations are considered as acids, in the Lewis scale, of different hardness, depending on the groups bonded to the tin(IV) (Tobias 1966). Consequently they show a strong tendency to hydrolysis in aqueous solution, as demonstrated by Tobias et al. (1966). The majority of investigations into the aqueous solution chemistry of organotin compounds are reported in the literature in the sixties, when the use of these compounds was widespread. Unfortunately, the thermodynamic parameters reported in all these studies refer to a non-interacting ionic medium and a single value of ionic strength. From these data, therefore, it is impossible to know the behaviour of organotin compounds in a multicomponent solution, which is necessary to describe the chemical speciation of these compounds in natural fluids. Moreover, a number of these investigations refer to the interactions of organotin moieties with organic ligands of biological interest (with the aim of contributing to knowledge of their biological activity), often neglecting the hydrolysis processes of organotin cations. However, the majority of investigations carried out into the aqueous chemistry of some alkyltin compounds demonstrated that hydrolysis processes are very important in the chemical speciation studies of this class of compounds. *After the hydrolytic equilibria have been examined in the presence of non-complexing agents, it is possible to study the interaction of the organometallic cations with other ligands in order to define their chemical speciation in natural waters.*

Since most of the organometallic cations have more than one water molecule in the first coordination sphere, there are several stepwise proton transfer equilibria. Condensation reactions of the monomeric conjugate bases often lead to the formation of polynuclear hydroxo-complexes in solution giving very complicated systems.

Briefly, the problem is first of determining which aqueous species $[(R_nSn)_q(OH)_p]^{(qz-p)}$ are formed and then the equilibrium constants for formation reactions of type 11.1 using measurements of the equilibrium ion concentration as a function of solution composition.

$$qR_nSn(OH_2)_x^{z+} + pH_2O \rightleftharpoons pH_3O^+ + (R_nSn)_q(OH)_p(OH_2)_y{}^{(qz-p)+} \qquad (11.1)$$

The most precise method for studying these equilibria is potentiometry, and the procedures are essentially the same as those used in the study of simple aquo-metal ions.

A detailed analysis of data available in literature on the aqueous chemistry of mono-, di- and triorganotin compounds, with particular reference to hydrolysis products formation, is reported below.

11.2.1 Trialkyltin Compounds

Among trialkyltin compounds, tributyl derivatives are the most widely distributed in the aquatic environment, owing to their industrial applications (see Table 11.1). A number of articles on their toxicological effects exist in literature. Useful updated information on this subject can be found in the very recent book by Champ and Seligman (1996).

The acidity of some low water soluble trialkyltin(IV) ions was first studied by Janssen and Luijten (1963), in mixed ethanol-water solvent. Differences in the hydrolysis constants of various trialkyltin(IV) cations were discussed by the authors in terms of steric and inductive effects. Surprisingly, the values obtained for tributyltin cation and other systems in the mixed solvent are very close to those found by Tobias et al. (1966) for $(CH_3)_3Sn^+$ and $(C_2H_5)_3Sn^+$ in 3 mol dm^{-3} aqueous $NaClO_4$ solution. Apart from the difficulty of interpretation of acidity measurements in mixed water organic solvents, it must be said that the good agreement with hydrolysis constant values of analogous water soluble systems, such as $(CH_3)_3Sn^+$, allows us to extrapolate the same considerations for low water soluble systems too. Acid-base properties of $(CH_3)_3Sn^+$ and $(C_2H_5)_3Sn^+$ cations in aqueous solutions were also studied by Asso and Carpeni (1968), by means of potentiometry, at 25 °C and different ionic strengths. Hynes and O'Dowd (1987) reported the hydrolysis constant value of $(CH_3)_3Sn(OH)$ in a study of the interactions of trimethyltin cation with carboxylic acids and aminoacids. Further investigations into the hydrolysis process of $(CH_3)_3Sn(H_2O)_2^+$ were conducted by Hynes et al. (1991) by means of ^{119}Sn-NMR spectroscopy. Investigations into the hydrolysis processes of trimethyltin(IV) cation, with the aim of defining the chemi cal speciation of that cation in natural fluids, have been very recently reported by Takahashi et al. (1997) ($NaClO_4$ ionic medium, $I = 0.1$ mol dm^{-3}, $T = 25$ °C) and by Cannizzaro et al. (1998) ($NaNO_3$, NaCl and Na_2SO_4 aqueous media, $0.1 \leq I \leq 1.0$ mol dm^{-3} and $5 \leq T$ (°C) ≤ 45). Hydrolysis constants are expressed as $\log \beta_{pq}$ according to the reaction:

$$pM^{z+} + qH_2O = M(OH)^{(z-q)} + qH^+ \tag{11.2}$$

Table 11.1. Literature data for the hydrolysis constants of the $(CH_3)_3Sn^+$ at $T = 25$ °C

Ionic medium	I (mol dm^{-3})	$-\log \beta_{11}$	$-\log \beta_{12}$	Reference
$NaClO_4$	3	6.59	-	Tobias et al. (1966)
KCl	2	6.40	-	Asso and Carpeni (1968)
$NaClO_4$	0.3	6.26	-	Hynes and O'Dowd (1987)
KNO_3	0.5	6.35	-	Hynes et al. (1991)
KCl	0.5	6.38	-	Hynes et al. (1991)
$NaNO_3$	0.5	6.21	18.95	Cannizzaro et al. (1998)
NaCl	0.5	6.25	18.99	Cannizzaro et al. (1998)

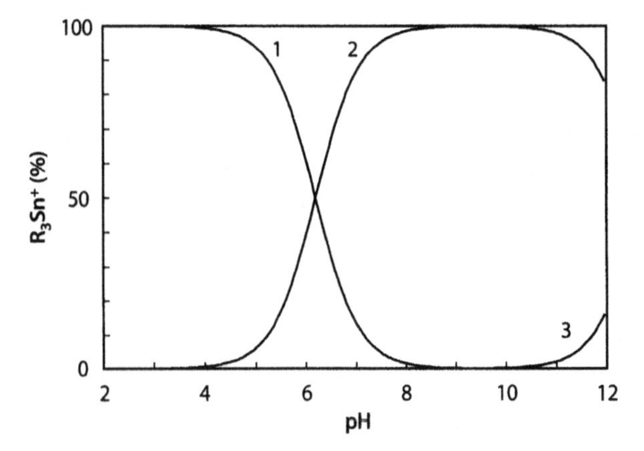

Fig. 11.1. Distribution diagram of the R_3Sn^+ species vs. pH in $NaNO_3$ (0.5 mol dm^{-3}) at T = 25 °C; [R = CH_3] (Cannizzaro et al. 1998). Species: (*1*) $[(CH_3)_3Sn]^+$, (*2*) $[(CH_3)_3Sn(OH)]^0$, (*3*) $[(CH_3)_3Sn(OH)_2]^-$

The most significant published data on the hydrolysis constants determined in different aqueous media are set out in Table 11.1. All the published data concerning the hydrolysis process of trialkyltin(IV) cations confirm that trimethyltin(IV) cation is present, at pH < 5, as aquo-cation $(CH_3)_3Sn(H_2O)_2^+$ with a bipyramidal trigonal structure with respect to the central atom of tin and with water molecules in axial position, according to Wada and Okawara (1965) and Barbieri and Silvestri (1991).

At pH > 5 the hydrolysis process occurs and, in the pH range 7–10 ($NaNO_3$ medium), only the species $(CH_3)_3Sn(OH)$ is present, according to the reaction $(CH_3)_3Sn(H_2O)_2^+ \rightleftharpoons (CH_3)_3Sn(H_2O)(OH) + H^+$ (see distribution diagram in Fig. 11.1). A second 1 : 2 hydrolytic species, $[(CH_3)_3Sn(OH)_2]^-$, is formed at pH > 10, reaching about 15% formation at pH = 11.5.

11.2.2 Dialkyltin Compounds

On the basis of results first obtained by Rochow and Seyferth (1953) on the hydrolysis of dialkyltin, Tobias and co-workers investigated the formation and structure of the hydrolysis products of dimethyltin cation in aqueous nitrate solution. The authors proposed the formation of the following hydrolytic species: $[(CH_3)_2Sn(OH)]^+$, $[(CH_3)_2Sn(OH)_2]^0$, $[((CH_3)_2Sn)_2(OH)_2]^{2+}$ (Tobias et al. 1962) and $[((CH_3)_2Sn)_3(OH)_4]^{2+}$ and $[((CH_3)_2Sn)_4(OH)_6]^{2+}$ (Tobias and Yasuda 1964). Owing to the very low percentage formation, the last two species have to be considered as uncertain, even if their formation has also been reported more recently by other authors (Arena et al. 1989). A confirmation of the hypothesised structure of some dimethyltin hydroxo compounds and aquo ions comes from Raman and NMR spectroscopic studies by Tobias and Freidline (1965) and McGrady and Tobias (1964). The latter authors extended their studies to diphenyltin compounds (McGrady and Tobias 1965). Spectroscopic investigations about steric effects on the dissociation of $(C_2H_5)_2Sn^{2+}$ and $(C_3H_7)_2Sn^{2+}$ aquo cations were carried out by Tobias et al. (1966). The results of all these investigations confirm that hydrolysis processes of dimethyltin cation occur at pH > 4, while the predominant species before that pH is the aquo cation $[(CH_3)_2Sn(H_2O)_4]^{2+}$ which shows an octahedral structure with methyl groups in *trans* position. After the sixties, interest in the aqueous chemistry of organotin compounds considerably decreased. Only in the nineties have the hydrolytic products and complex species formation of dialkyltin(IV) cations, as well as structural investigations, been reconsidered by some authors (Cunningham et al.1990; Cucinotta et al. 1992; Burger et al. 1993; Natsume et al. 1994). By using the most recent developments in data fitting analysis, definitive results on the hydrolysis products of dimethyltin cation in a wide range of ionic strengths and in different ionic media have been reported by De Stefano et al. (1996). Stability constants of hydrolysis products by different authors are reported in Table 11.2.

On the basis of results obtained by De Stefano et al. (1996), a distribution diagram of the hydrolytic species in a non-interacting medium ($NaNO_3$) has been drawn and is shown in Fig. 11.2.

Table 11.2. Literature data for the hydrolysis constants of $(CH_3)_2Sn^{2+}$ at $T = 25$ °C and $I = 0.1$ and 3 mol dm^{-3}

Medium	$-\log \beta_{11}$ [a]	$-\log \beta_{12}$ [a]	$-\log \beta_{13}$ [a]	$-\log \beta_{22}$ [a]	$-\log \beta_{23}$ [a]	Reference
KNO_3 [b]	3.12	8.43	19.45	5.05	9.75	Arena et al. (1989)
NaCl [b]	3.25	8.54	–	5.05	9.81	Tobias and Yasuda (1964)
$NaNO_3$ [b]	3.18	8.42	–	4.69	9.64	Natsume et al. (1994)
$NaNO_3$ [b]	3.06	8.36	19.36	5.16	9.44	De Stefano et al. (1996)
NaCl [b]	3.12	8.45	19.48	5.20	9.70	De Stefano et al. (1996)
$NaClO_4$ [c]	3.54	8.98	–	4.60	9.76	Tobias and Yasuda (1964)
$NaNO_3$ [c]	3.52	9.07	20.10	5.10	10.20	De Stefano et al. (1996)
$NaClO_4$ [c]	3.30	9.08	20.30	5.10	9.70	De Stefano et al. (1996)

[a] Log β_{pq} refers to the reaction $pM^{2+} = M_p(OH)_q^{(p-q)} + qH^+$.
[b] $I = 0.1$ mol dm^{-3}.
[c] $I = 3.0$ mol dm^{-3}.

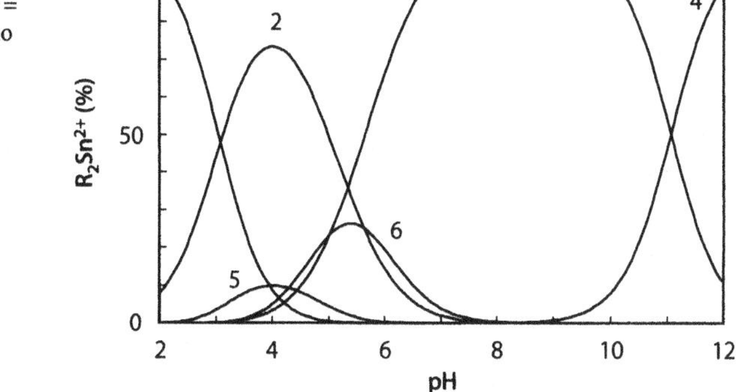

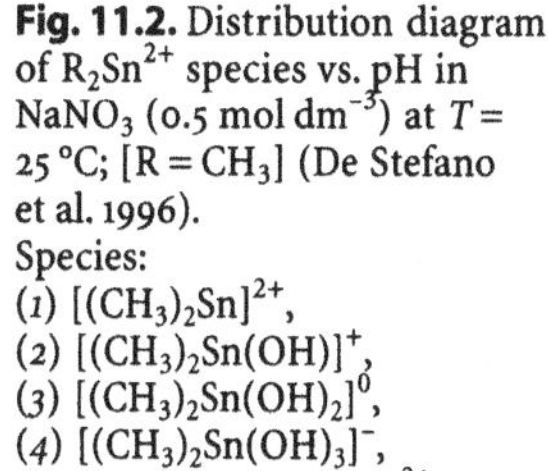

Fig. 11.2. Distribution diagram of R_2Sn^{2+} species vs. pH in $NaNO_3$ (0.5 mol dm^{-3}) at $T = 25$ °C; [$R = CH_3$] (De Stefano et al. 1996).
Species:
(*1*) $[(CH_3)_2Sn]^{2+}$,
(*2*) $[(CH_3)_2Sn(OH)]^+$,
(*3*) $[(CH_3)_2Sn(OH)_2]^0$,
(*4*) $[(CH_3)_2Sn(OH)_3]^-$,
(*5*) $[((CH_3)_2Sn)_2(OH)_2]^{2+}$,
(*6*) $[((CH_3)_2Sn)_2(OH)_3]^+$

As can be seen, in the pH range of natural fluids (6.5–9.5), the species $(CH_3)_2Sn(OH)_2$ predominates (curve 3), while the other hydrolytic species [i.e. $(CH_3)_2Sn(OH)$ and $((CH_3)_2Sn(OH)_3]$ are negligible in the same pH range.

11.2.3 Monoalkyltin Compounds

Mono-organotin(IV) compounds have not found so many commercial applications as diorgano and triorgano derivatives, and are considered the least toxic among organotin(IV) derivatives ($R_3Sn^+ > R_2Sn^{2+} > RSn^{3+} > Sn^{4+}$, *toxicity scale*). However, they are often used as hydrophobic agents for building materials and cellulosic matter (Blunden et al. 1985) and can be present in the aquatic environment as the first step of alkylation of inorganic tin (Mennie and Craig 1993).

The majority of studies, carried out many years ago, concern complex formation with N-donor molecules (Gielen and Sprecher 1966) and with chloro (Kriegsmann and Pauly 1964; Farrer et al. 1965; Van den Berghe and Van der Kelen 1965) and fluoro ions (Cassol et al. 1967). Raman measurements by Kriegsmann and Pauly (1964) showed the formation of $[CH_3SnCl_2(OH)_2]^-$ in the presence of high concentrations of chloride ions. Van den Berghe and Van der Kelen (1965), by means of NMR measurements made in very concentrated solutions of CH_3Sn^{3+} (from 210 up to 830 mmol dm^{-3}), report the formation of mixed chloro-hydroxo complexes $[CH_3SnCl_x(OH)_y]$ with x de-

Table 11.3. Log β of hydrolytic species for the systems CH_3Sn^{3+} at $I = 0$ mol dm^{-3} and $T = 25$ °C

Equilibrium	Species	$-\log\beta$
$CH_3Sn^+ + 2\,H_2O = CH_3Sn(OH)_2^+ + 2\,H^+$	$[CH_3Sn(OH)_2]^+$	3.36
$CH_3Sn^+ + 3\,H_2O = CH_3Sn(OH)_3^0 + 3\,H^+$	$[CH_3Sn(OH)_3]^0$	8.99
$CH_3Sn^+ + 4\,H_2O = CH_3Sn(OH)_4^- + 4\,H^+$	$[CH_3Sn(OH)_4]^-$	20.27
$2\,CH_3Sn^+ + 5\,H_2O = (CH_3Sn)_2(OH)_5^+ + 5\,H^+$	$[(CH_3Sn)_2(OH)_5]^+$	7.61

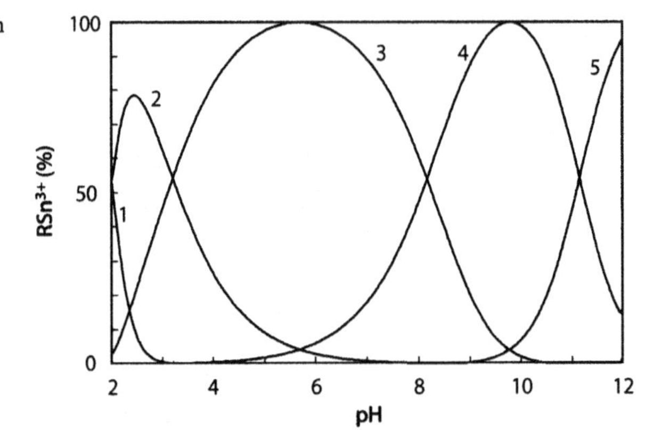

Fig. 11.3. Distribution diagram of RSn^{3+} species in $NaNO_3$ medium (0.5 mol dm^{-3}) at $T = 25$ °C; [R = CH_3]. Species: (1) $[CH_3Sn]^{3+}$, (2) $[CH_3Sn(OH)_2]^+$, (3) $[(CH_3Sn)_2(OH)_5]^+$, (4) $[CH_3Sn(OH)_3]^0$, (5) $[CH_3Sn(OH)_4]^-$

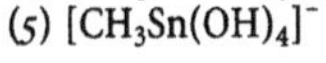

creasing in the dilute solution. Analogous behaviour was found by Tobias and co-workers (1962, 1964, 1965, 1966) and more recently by De Stefano et al. (1996, 1999) in studying the interactions of $(CH_3)_2Sn^{2+}$ and $(CH_3)Sn^{3+}$ in chloride solutions. Investigations into the hydrolysis products of ethyl-, butyl- and octyltin trichlorides carried out by Luijten (1966) report their properties and solid-state preparation. Some potentiometric studies of hydrolysis and sulfide complex formation of ethyl-trichlorotin(IV) in mixed water/methanol solution have been carried out by Devaud (1969, 1970, 1971, 1972) and by Langlois and Devaud (1974), in an organometallic compound concentration range of 6–60 mmol dm^{-3}. The results obtained by these authors are not altogether consistent with findings for monomethyltin trichloride (Van den Berghe and Van der Kelen 1965), probably owing to the different solvent and concentration ranges used. The results of ^1H- and ^{119}Sn-NMR and ^{119}Sn-Mössbauer spectroscopic studies on the hydrolysis of methyl- and butyltin trichloride (0.5 mol dm^{-3}) have been reported by Blunden et al. (1982) and Blunden and Hill (1990). From analysis of published data it can be affirmed that there are, in general, great discrepancies among the results of investigations into the hydrolysis of mono-organotin(IV) derivatives, probably owing to the difficulty of investigating a very reactive species such as RSn^{3+}, which presents a hardness in a Lewis acid scale, which is higher than that of mono- and di-charged cations. Moreover, the majority of studies have been carried out by using a very high concentration of monoalkyltrichloride, and, therefore, the formation of mixed hydroxo-chloride complexes cannot be avoided. Consequently, the picture of the chemical speciation of this system is not clear. It has been suggested that methyltin trichloride exists in aqueous solution only as a hydroxide species according to the following pH dependent equilibrium

$$\underset{\text{Species 1}}{CH_3Sn(OH)Cl_2 \cdot 2\,(H_2O)} \rightleftharpoons \underset{\text{Species 2}}{CH_3Sn(OH)_2Cl \cdot H_2O} \rightleftharpoons \underset{\text{Species 3}}{CH_3Sn(OH)_3} \rightleftharpoons \underset{\text{Species 4}}{CH_3Sn(OH)_4}$$

$$\text{pH increasing} \longrightarrow$$

Blunden et al. (1982) report the formation of the Species 1 at pH 1.4. It might be expected that, at environmental pH, Species 3 will predominate. Potentiometric investigations by De Stefano et al. (1999a) on the CH_3Sn^{3+} system ($NaNO_3$, NaCl and Na_2SO_4 ionic media) show the formation of the following simple hydrolytic species: $[CH_3Sn(OH)_2]^+$, $[CH_3Sn(OH)_3]^0$, $[(CH_3Sn)_2(OH)_5]^+$ and $[CH_3Sn(OH)_4]^-$ (log β in Table 11.3), whose formation percentages, in the pH range 2–12, are reported in Fig. 11.3.

As can be seen, the species $[((CH_3)Sn)_2(OH)_5]^+$ and $[CH_3Sn(OH)_3]^0$ predominate in the pH range of natural fluids (6.5–9.5), both reaching about 50% formation at seawater pH value.

11.3 Salt Effect on the Hydrolysis Process

In order to establish the effects of seawater salt on the hydrolysis process of mono- di- and trimethyltin(IV) cations, we carried out investigations into these systems both in chloride ($0 \le I \le 3$ mol dm^{-3}) and sulfate media ($0 \le I \le 1$ mol dm^{-3}), and in synthetic sea water (5–45‰ salinity range, see Section 11.6). The formation constants of chloride and sulfate species for all the organotin systems investigated are set out in Table 11.4 (Cannizzaro et al. 1998; De Stefano et al. 1996, 1999a, 1999b; Foti et al. 1999).The formation of mixed $(CH_3)_xSn_{(4-x)}(OH)_yCl_z$ and/or $(CH_3)_xSn_{(4-x)}(OH)_y(SO_4)_z$ species lowers, to different extents, the formation percentages of the main hydrolytic species (i.e. $[(CH_3)_3Sn(OH)]^0$ for trimethyltin, $[(CH_3)_2Sn(OH)_2]^0$ for dimethyltin and $[(CH_3Sn)_2(OH)_5]^+$ and $[CH_3Sn(OH)_3]^0$ for monomethyltin cations) formed in the pH range of sea water.

11.4 Dependence on Ionic Strength of Hydrolysis Constants

The dependence on ionic strength of hydrolysis constants in nitrate and in perchlorate media is as expected for reactions where reactants do not undergo weak interactions with the anion and the cation of the background salt (Cannizzaro et al. 1998; De Stefano et al. 1996, 1999a). Chloride (and sulfate) salts have a noticeable influence on hydrolysis constants as shown in Fig. 11.4 (hydrolysis of trimethyltin(IV) in nitrate and chloride salts), due to the complexation of these anions. The dependence of log β_{pq} (Reaction 11.2) in different salt media has been recently studied in these laboratories (Cannizzaro et al. 1998; De Stefano et al. 1996, 1999a). Hydrolysis constants obtained at different ionic strengths in different salt solutions are useful for determining I-dependence parameters, such as Pitzer interaction parameters (Pitzer 1991), which have been widely used by Millero (1997) for the speciation of metal ions and inorganic ligands in seawater. Pitzer parameters have been recently used also for the speciation of $(CH_3)_2Sn^{2+}$ and $(CH_3)_3Sn^+$ cations in presence of major components of natural waters (Foti et al. 1999 and De Stefano et al. 1999b, respectively).

Table 11.4. Formation constants of R_xSn-Cl, and R_xSn-SO_4 complex species [R = CH_3] (I = 0 mol dm^{-3}, T = 25 °C)

Chloride species	log β	Sulphate species	log β
R_3Sn^+ species			
$[R_3SnCl]^0$	-0.60[a]	$[R_3SnSO_4]^-$	0.35[c]
R_2Sn^{2+} species			
$[R_2SnCl]^+$	0.92[b]	$[R_2SnSO_4]^0$	2.53[b]
$[R_2SnCl_2]^0$	1.07[b]	$[R_2Sn(SO_4)_2]^{2-}$	2.98[b]
$[R_2SnCl(OH)]^0$	-2.60[b]	$[R_2Sn(SO_4)(OH)]^-$	-1.22[b]
$[R_2SnCl(OH)_2]^-$	-8.85[b]	$[R_2Sn(SO_4)(OH)_2]^{2-}$	-8.27[b]
RSn^{3+} species			
$[RSnCl(OH)]^+$	-1.40[c]	$[RSn(SO_4)(OH)]^0$	1.29[c]
		$[RSn(SO_4)(OH)_2]^-$	-0.92[c]
		$[(RSn)_2(SO_4)(OH)_5]^-$	-5.41[c]

[a] Cannizzaro et al. 1998.
[b] De Stefano et al. 1996.
[c] De Stefano et al. 1999b.

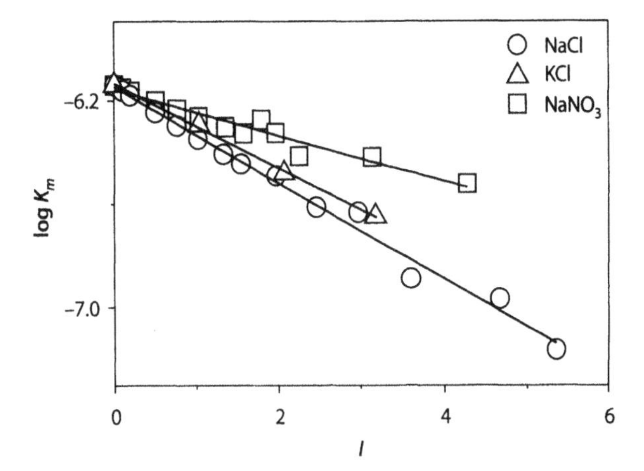

Fig. 11.4. Hydrolysis constants of $[(CH_3)_3Sn]^+$ vs. I, in NaCl, KCl and $NaNO_3$, at $T = 25\,°C$

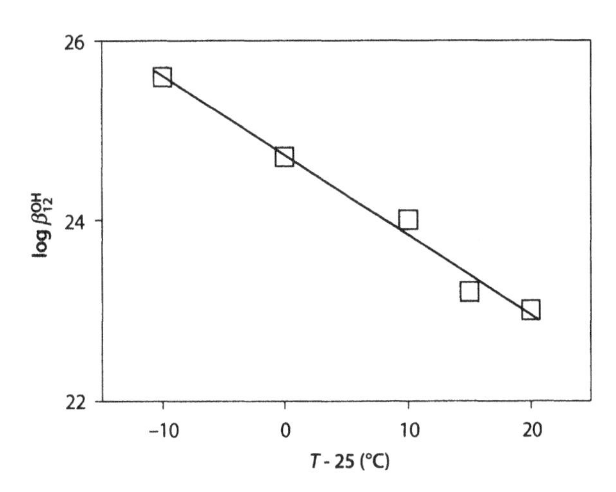

Fig. 11.5. Log β_{12}^{OH} values vs. T–25 (°C)

11.5 Dependence on Temperature of Hydrolysis Constants

Hydrolysis constants strongly depend on temperature as shown in Fig. 11.5, where log β_{12}^{OH} of CH_3Sn^{3+} is reported vs. (T (°C) – 25).

ΔH^0 values have been determined for the hydrolysis of mono-, di- and trimethyltin(IV), and are reported in Table 11.5. ΔH^0 values for trimethyltin(IV) hydrolysis have been studied both calorimetrically and by temperature dependence of hydrolysis constants in a wide ionic strength range, in NaCl and $NaNO_3$ (De Stefano et al. 1999b). The hydrolysis of dimethyltin(IV) has been studied calorimetrically at one ionic strength value (0.1 mol dm^{-3}). Hydrolysis constants of monomethyltin(IV) have been determined at different temperatures and the ΔH^0 values obtained are affected by large errors. ΔH^0 values depend on ionic strength, as shown in Fig. 11.6, where this parameter is reported vs. $I^{1/2}$ for two background salts. Thermodynamic parameters have also been reported for dimethyltin(IV) at one ionic strength. No other data are reported in literature.

Table 11.5. ΔH^0 values for the formation of alkyltin(IV) hydrolytic species, at 25 °C and $I = 0$ mol dm^{-3}

Compound	ΔH^0 values	Compound	ΔH^0 values	Compound	ΔH^0 values
$(CH_3)_3Sn^+$	$\Delta H^{0\,a}_{11} = -30.0$	$(CH_3)_2Sn^{2+\,b}$	$\Delta H^0_{11} = -27.5$	CH_3Sn^{3+}	$\Delta H^0_{12} = -150$
	$\Delta H^0_{12} = -57.0$		$\Delta H^0_{12} = -53.0$		$\Delta H^0_{13} = -230$
			$\Delta H^0_{13} = -71.8$		$\Delta H^0_{14} = -260$
			$\Delta H^0_{22} = -50.6$		$\Delta H^0_{25} = -380$
			$\Delta H^0_{23} = -87.0$		

[a] ΔH^0(kJ mol^{-1}), relative to the reaction ΔH_{pq}: $pM^{z+} + q\,H_2O = M_p(OH)_q^{(p-q)} + q\,H^+$.
[b] $I = 0.1$ mol KNO_3 dm^{-3}.

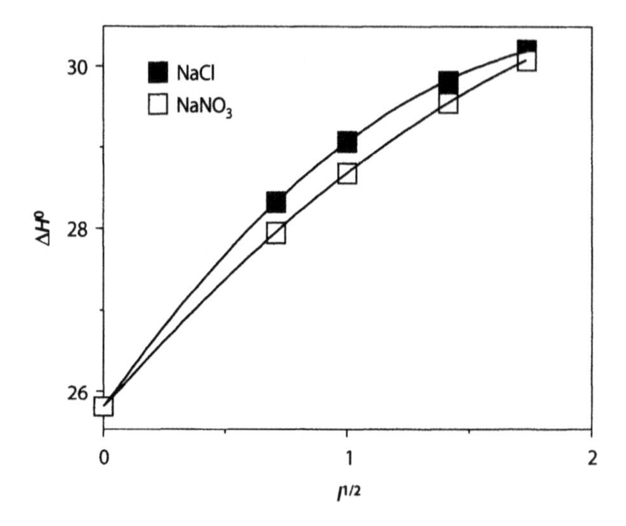

Fig. 11.6. ΔH^0 values vs. $I^{1/2}$ for the hydrolysis reaction of $[(CH_3)_3Sn]^+$ in NaCl and $NaNO_3$, at $T = 25$ °C

11.6 Hydrolysis of Organotin(IV) Compounds in Sea Water

When dealing with a multicomponent electrolyte solution such as sea water, the chemical speciation studies of an acid/base system are complicated owing to the network of interactions between all the solution components. Among these, the macro-components are present in constant concentration ratios and, for this reason, it is possible to build up a chemical base model for sea water by using a synthetic sea water [(SSWE), containing Na^+, K^+, Ca^{2+}, Mg^{2+}, Cl^- and SO_4^{2-}] as an ionic medium representative of the macro-components of the natural sea water whose composition[1] at different sa-

[1] Many other recipes for artificial seawater have been proposed (see refs. in De Stefano et al. 1994) with very similar composition. Studies on the complexing ability of the inorganic content of sea water (as single salt BA) towards various acid-base systems, showed that the single salt approximation can be successfully applied (De Stefano et al. 1998b).

linities is reported in Table 11.6 (De Stefano et al. 1994 and refs. therein). Therefore, besides the hydrolysis processes of organotin cations, interactions with seawater anions (chloride and sulfate) must also be considered, *after all the internal interactions of background salt components have been taken into account* (De Robertis et al. 1994). In order to simplify equilibrium calculations, and to give a cumulative picture of the complexing capacity of seawater toward this class of compounds, the composition of SSWE can be expressed in terms of a single salt, BA, whose ions have a charge $z = \pm 1.117$ (De Stefano et al. 1998a). Concentrations of BA (mean ionic concentration) at different salinities are reported in Table 11.6, too. The anion (A) of this salt forms the species BA^0, $HA^{(z-1)}$.

The chemical speciation model for artificial seawater (expressed both as six components and as a single salt, BA) and the equilibrium constants and thermodynamic parameters for the system BA are reported in Table 11.7.

By using the single salt approximation (BA), a chemical model where, as well as hydrolysis processes, the binary and ternary interactions of each $[R_xSn_{(4-x)}]^{z+}$ cation with A^{z-} and B^{z+} and the internal association of BA have also been taken into account, has been built up according to the scheme in Table 11.8.

Table 11.6. Composition of SSWE at 35‰ and $T = 25$ °C (concentrations in mol dm^{-3}, at 25 °C)

Component	S (‰)				
	5	15	25	35	45
NaCl	0.0590	0.1782	0.2992	0.4221	0.5467
Na_2SO_4	0.0040	0.0122	0.0204	0.0288	0.0373
KCl	0.0015	0.0046	0.0078	0.011	0.0142
$CaCl_2$	0.0015	0.0047	0.0078	0.0111	0.0143
$MgCl_2$	0.0077	0.0231	0.0389	0.0548	0.0710
BA	0.0803	0.2428	0.4078	0.5751	0.7449
I	0.1000	0.3030	0.5090	0.717	0.929

Table 11.7. Speciation model for SSWE

Equilibrium	$\log K$[a]					
	H^+	Na^+	K^+	Mg^{2+}	Ca^{2+}	$B^{1.117+}$
$M + H_2O = M(OH) + H^+$	–	–14.0	–14.3	–11.7	–12.9	–12.9
$M + Cl = MCl$	–	–0.6	–0.5	0.1	0.0	(–0.39)[b]
$M + SO_4 = MSO_4$	1.70	0.5	0.6	1.6	1.5	(–0.04)[c]

[a] Formation constants at $I_{eff} = 0.56$ mol dm^{-3} (35‰ salinity) and $T = 25$ °C.

[b] $B^{1.117+} + A^{1.117-} = AB^0$.

[c] $H^+ + A^{1.117-} = HA^{0.117-}$.

On the basis of the above model, distribution diagrams have been drawn to evaluate the formation percentages of the various hydrolytic species in synthetic sea water containing anion macro-components by also considering the association species with BA for the different organotin systems (Fig. 11.7–11.9).

In Fig. 11.7 the distribution diagram of the system is reported $(CH_3)_3Sn^+$ in SSWE single salt (BA). As one can see, the formation of the species $[(CH_3)_3SnA]^{-0.117}$ (curve 3) is appreciable only in the acidic pH range and determines a lowering of the free cation $(CH_3)_3Sn^+$ percentage. Association with the seawater anion (A) does not significantly influence the formation of the hydrolytic species. Analogously, the formation of ternary species $[(CH_3)_3SnB(OH)_2]^{+0.117}$ (curve 4), achieving a very high percentage

Table 11.8. Chemical speciation model of tri-, di- and monomethyltin in SSWE as single salt (BA)[a]

System	Hydrolytic species[b]	Association with BA[c]	log β
$(CH_3)_3Sn-BA$	$[(CH_3)_3Sn(OH)]^0$	$[(CH_3)_3Sn(A)]^{-0.117}$	0.08
	$[(CH_3)_3Sn(OH)_2]^-$	$[(CH_3)_3SnB(OH)_2]^{+0.117}$	-16.67
$(CH_3)_2Sn-BA$	$[(CH_3)_2Sn(OH)]^+$	$[(CH_3)_2SnA]^{+0.883}$	0.77
	$[(CH_3)_2Sn(OH)_2]^0$	$[(CH_3)_2SnA(OH)]^{-0.117}$	-3.09
	$[(CH_3)_2Sn(OH)_3]^-$	$[(CH_3)_2SnB(OH)_3]^{+0.117}$	-18.79
	$[((CH_3)_2Sn)_2(OH)_3]^+$		
CH_3Sn-BA	$[CH_3Sn(OH)_2]^+$	$[CH_3SnA(OH)]^{+0.117}$	-1.00
	$[CH_3Sn(OH)_3]^0$	$[CH_3SnA(OH)_2]^{-0.117}$	-1.95
	$[(CH_3Sn)_2(OH)_5]^+$	$[(CH_3Sn)_2A(OH)_5]^{-0.117}$	-6.28
	$[CH_3Sn(OH)_4]^-$	$[(CH_3Sn)(B)(OH)_4]^{+0.117}$	-17.62

[a] Log β values of self association species of BA in Table 11.7.
[b] Log β values in Tables 11.1–11.3.
[c] Log β values at $I=0$ and $T=25$ °C.

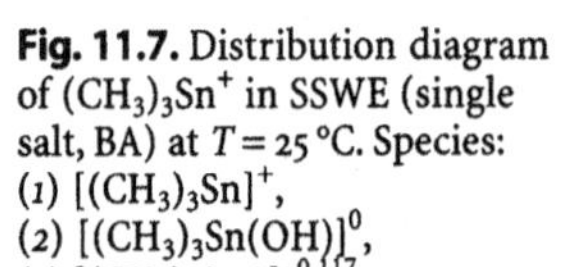

Fig. 11.7. Distribution diagram of $(CH_3)_3Sn^+$ in SSWE (single salt, BA) at $T=25$ °C. Species: (1) $[(CH_3)_3Sn]^+$, (2) $[(CH_3)_3Sn(OH)]^0$, (3) $[(CH_3)_3SnA]^{-0.117}$, (4) $[(CH_3)_3SnB(OH)_2]^{0.117+}$

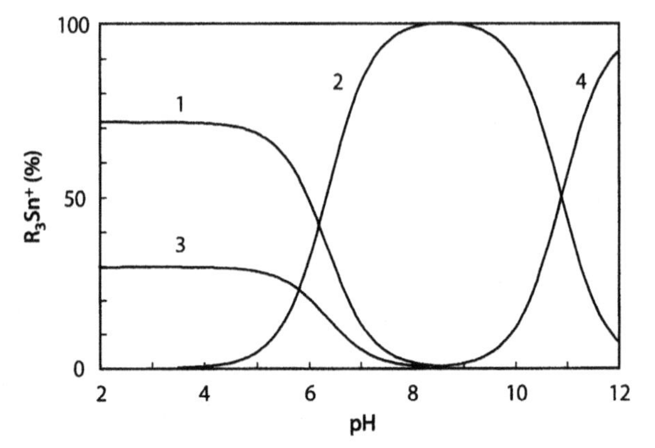

formation at pH > 10 (completely replacing the simple hydrolytic species 1 : 2), does not influence the formation of the main species at the pH values of sea water.

The distribution diagram of dimethyl cation in SSWE (as single salt, BA) is set out in Fig. 11.8. The interactions with anion components of artificial seawater lead to the formation of the species $[(CH_3)_2SnA]^{+0.883}$ and $[(CH_3)_2SnA(OH)]^{-0.117}$ (curves *5* and *6* respectively, Fig. 11.8) which do not influence the main hydrolytic species $[(CH_3)_2Sn(OH)_2]^0$ in the pH range of natural waters (see for comparison curve *2* in Fig. 11.2 and curve *3* in Fig. 11.8). Also in this case a ternary species $[(CH_3)_2SnB(OH)_3]^{+0.117}$, deriving from the interaction of the $[(CH_3)_2Sn(OH)_3]^-$ species with cation component of sea water, is formed and decreases the percentage formation of the simple hydrolytic one, at pH > 10 (curves *4* and *7* in Fig. 11.8).

On the basis of these results, De Stefano et al. (1997) also considered the complex formation of dimethyltin with mono- and polycarboxylate ligands. Complexation equilibria and structures of dimethyltin(IV) complexes with N- and O- donor ligands have also been discussed by Aizawa et al. (1996).

The distribution diagram of the CH_3Sn^{3+} species in SSWE (Fig. 11.9) is quite different from the one showing the species in a non-interacting medium (Fig. 11.3). The for-

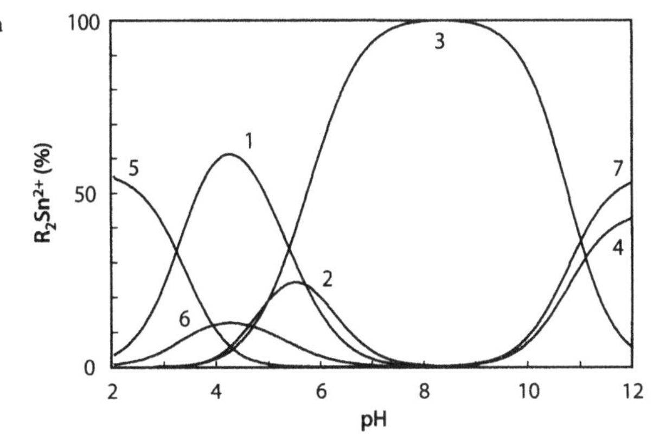

Fig. 11.8. Distribution diagram of R_2Sn^{2+} in SSWE (single salt BA) at $T = 25$ °C; [R = CH_3]. Species:
(*1*) $[(CH_3)_2Sn(OH)]^+$,
(*2*) $[((CH_3)_2Sn)_2(OH)_3]^+$,
(*3*) $[(CH_3)_2Sn(OH)_2]^0$,
(*4*) $[(CH_3)_2Sn(OH)_3]^-$,
(*5*) $[(CH_3)_2SnA]^{+0.883}$,
(*6*) $[(CH_3)_2SnA(OH)]^{-0.117}$,
(*7*) $[(CH_3)_2SnB(OH)_3]^{+0.117}$

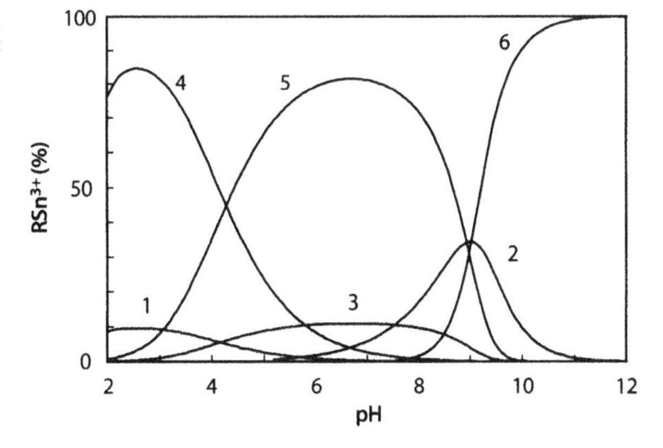

Fig. 11.9. Distribution diagram of CH_3Sn^{3+} in SSWE (single salt BA) at $T = 25$ °C. Species:
(*1*) $[CH_3Sn(OH)_2]^+$,
(*2*) $[CH_3Sn(OH)_3]^0$,
(*3*). $[(CH_3Sn)_2(OH)_5]^+$,
(*4*) $[CH_3SnA(OH)_2]^{-0.117}$,
(*5*) $[(CH_3Sn)_2A(OH)_5]^{-0.117}$,
(*6*) $[CH_3SnB(OH)_4]^{+0.117}$

mation of the anion species $[(CH_3Sn)_2A(OH)_5]^{-0.117}$ (curve 5) considerably lowers the formation percentage of the corresponding simple hydrolytic species, $[(CH_3Sn)_2(OH)_5]^+$, and becomes predominant by achieving a formation percentage of over 70%, in the pH range 6–7. Association with the anion components of sea water does not influence the formation of the other important species $[CH_3Sn(OH)_3]^0$ formed at higher pH values. On the contrary, the interaction with the cation component (B) of sea water leads to the formation of the very strong species $[CH_3SnB(OH)_4]^{+0.117}$ which becomes the only present at pH > 10. In conclusion, at seawater pH value, hydrolytic species of monomethyltin(IV) are $[(CH_3Sn)_2(OH)_5]^+$, $[CH_3Sn(OH)_3]^0$ and $[(CH_3Sn)_2A(OH)_5]^{-0.117}$ with formation percentages achieving 10%, 20% and 75%, respectively.

11.7 Conclusions

The most important feature of the solution chemistry of organotin(IV) compounds lies in their very strong hydrolysis. No study of the speciation of these cations can be made without deep knowledge of their hydrolysis thermodynamic parameters. In the absence of interacting anions, in the pH range of interest for natural fluids, $(CH_3)_xSn^{(4-x)}$ is fully hydrolyzed. The interaction of chloride and sulfate complexes is significant, but no very strong species are formed. However, owing to the high concentration of these anions in sea water, the speciation of $(CH_3)_xSn^{(4-x)}$ cations is altered, as shown in Fig. 11.7–11.9. Note that at pH > 10 the cation of sea salt also interacts with the anionic species of dimethyltin(IV). Other major anionic components of sea water were not considered in this report, i.e. F^- and CO_3^{2-} (or HCO_3^-). As regards carbonate, recent measurements from these laboratories showed that it forms strong complexes which may have great importance in the speciation of alkyltin(IV) cations.

The final remarks must deal with the scarceness of data we can found in literature. In particular we need:

a. hydrolysis constants of other alkyltin(IV) cations (in particular unsimmetrical);
b. dependence on medium in large ionic strength range;
c. more calorimetric measurements to obtain reliable thermodynamic parameters;
d. formation constants for inorganic complexes, in particular with fluoride and carbonate.

References

Aizawa S, Natsume T, Hatano K, Funahaschi S (1996) Complexation equilibria and structure of dimethyltin(IV) complexes with N-methyliminodiacetate, pyridine-2,6-dicarboxylate, ethylenediamine-N-N'-diacetate and ethylenediamine-N,N,N',N'-tetracetate. Inorg Chim Acta 248:215–224

Arakawa Y, Wada O (1993) Biological properties of alkyltin compounds. In: Siegel H, Siegel A (eds) Metal ions in biological systems, vol XXIX. Marcel Dekker, Inc., New York, Basel, p 101–136

Arena G, Gianguzza A, Pellerito L, Purrello R, Rizzarelli E (1989) Thermodynamics of hydroxo complex formation of dialkyltin(IV) ions in aqueous solution. J Chem Soc Dalton Trans 773–777

Asso M, Carpeni G (1968) Recherches sur le point isohydrique et les équilibres acido-basique de condensation ou association en chimie. XXVII. Les hydroxocomplexes des sels organostanniques $(CH_3)_3SnCl$ et $(C_2H_5)_2SnCl$, en solutions aqueuses, à 25 °C et à differentes forces ioniques. Can J of Chem 46:1795–1802

Barbieri R, Silvestri A (1991) The hydrolysis of $Me_2Sn(IV)$ and $Me_3Sn(IV)$ moieties monitored through ^{119}Sn Mössbauer spectroscopy. Inorganica Chimica Acta 188:95–98

Barnes J, Magos L (1968) The toxicology of organometallic compounds. Organometal Chem Rev 3:137–150

Blunden SJ, Chapman A (1986) Organotin compounds in the environment. In: Craig PJ (ed) Organometallic compounds in the environment. Longman, Harlow Essex (England)

Blunden SJ, Hill R (1990) An investigation of the base hydrolysis of methyl- and butyl-tin trichloride in aqueous solution by ^{1}H and ^{119}Sn NMR spectroscopy. Inorganica Chim Acta 177: 219–223

Blunden SJ, Smith PJ, Gillies DG (1982) An investigation of the hydrolysis products of monoalkyltin trichlorides by ^{119}Sn Mossbauer and ^{1}H and ^{119}Sn NMR spectroscopy. Inorganica Chim Acta, 60:105–109

Blunden SJ, Cusack PA, Hill R (1985) The industrial use of tin chemicals. Royal Society Chemistry, London

Burger K, Nagy L, Buzas N, Vertes A, Menher H (1993) Co-ordination number, symmetry of the co-ordination sphere of tin(IV) and oligomerisation in carbohydrate complexez of dibutyltin(IV). J Chem Soc Dalton Trans 2499–2504

Cannizzaro V, Foti C, Gianguzza A, Marrone F (1998) Hydrolysis of trimethyltin(IV) cation in $NaNO_3$ and NaCl aqueous media at different temperatures and ionic strengths. Ann Chim (Rome) 88:45–54

Cassol A, Magon L, Barbieri R (1967) Complexes of organometallic compounds. Stability constants of organotin(IV)-fluoride and -chloride complexes in aqueous solution. Inorg Nucl Chem Letters, 3:25–29

Champ MA, Seligman PF (1996) Organotin. Environmental fate and effects. Chapman & Hall, London

Cucinotta V, Gianguzza A, Maccarrone G, Pellerito L, Purrello R, Rizzarelli E (1992) Thermodynamic and multinuclear magnetic resonance study of dimethyltin(IV) complexes with tridentate ligands in aqueous solution. J Chem Soc Dalton Trans 2299–2303

Cunningham D, McManus J, Hynes MJ (1990). Nuclear magnetic resonance studies and structural investigations of the chemistry of organotin compounds. J Organomet Chem 393:69–82

De Robertis A, De Stefano C, Gianguzza A, Sammartano S (1994) Equilibrium studies in natural fluids: A chemical speciation model for the major constituents of sea water. Chem Speciation and Bioavailability 6:65–84

De Stefano C, Foti C, Gianguzza A, Rigano C, Sammartano S (1994) Equilibrium studies in natural fluids. Use of synthetic seawater and other media as background salts. Ann Chim (Rome) 84:159–175

De Stefano C, Foti C, Gianguzza A, Martino M, Pellerito L, Sammartano S (1996) Hydrolysis of $(CH_3)_2Sn^{2+}$ in different ionic media: salt effects and complex formation. J Chem Eng Data 41:511–515

De Stefano C, Gianguzza A, Marrone F, Piazzese D (1997) Interaction of alkyltin(IV) compounds with ligands of interest in the speciation of natural fluids: Complexes of $(CH_3)_2Sn^{2+}$ with carboxylates. Appl Organomet Chem 11:683–691

De Stefano C, Foti C, Gianguzza A, Sammartano S (1998a) The single salt approximation for the major components of seawater: Association and acid-base properties. Chem Speciation and Bioavailability 10(1):27–29

De Stefano C, Foti C, Gianguzza A, Piazzese D, Sammartano S (1998b) Equilibrium studies in natural fluids: Interactions of PO_4^{3-}, -$P_2O_7^{4-}$ and -$P_3O_{10}^{5-}$ with the major constituents of sea water. Chem Speciation and Bioavailability 10(1):19–26

De Stefano C, Foti C, Gianguzza A, Marrone F, Sammartano S (1999a) Hydrolysis of monomethyltin(IV) in aqueous NaCl and $NaNO_3$ solutions at different ionic strengths and temperatures. Appl Organomet Chem 13:1–7

De Stefano C, Foti C, Gianguzza A, Millero FJ, Sammartano S (1999b) Hydrolysis of $(CH_3)_3Sn^+$ in different salt media. J Solution Chem 28:959–972

Devaud M (1969) Etude en solution hydroalcoolique de l'ethyltrichloroetain. J Chimie Phys 66:302–312

Devaud M (1970) Etude potentiometrique en milieu hydroalcoolique du trichlorure e du triperchlorate d'ethyletain. J Chimie Phys 67:270–278

Devaud M (1971) Etude de complexes ethyltrichlorostannane-acetate. J Chimie Phys 68:1043–1054

Devaud M (1972) Comportement de l'ethyltrichlorostannane en milieu hydroalcoolique basique. J Chimie Phys 69:460–469

Farrer HN, Mc Grady MM, Tobias RS (1965) Electromotive force, Raman and Nuclear Magnetic Resonance studies on the interaction of chloride and bromide ions with the dimethyltin(IV) ion. Inner- and outher-sphere complexes. J Am Chem Soc 20:5019–5026

Foti C, Gianguzza A, Millero FJ, Sammartano S (1999) The speciation of $(CH_3)_2Sn^{2+}$ in electrolyte solution containing the major components of natural waters. Aquatic Geochemistry, in press

Frache R, Rivaro P (2000) Occurrence, pathways and bioaccumulation of organometallic compounds in the marine environment. This Volume

Gianguzza A, Maggio F, Mansueto C, Pellerito A, Pellerito L, Vitturi R (1994) Organometallic complexes with biological molecules. In vivo cytotoxicity of diorganotin(IV) chloro and triorganotin(IV) chloro derivatives of Penicillin G on chromosomes of *Aphanius fasciatus* (Pisces, Cyprinodontiformes). Appl Organomet Chem 8:509–515

Gielen M, Sprecher N (1966) Coordination au niveau des atomes de metal du groupe IVB. Intervention des orbitales d dans la reactivité des composes organometalliques. Organomet Chem Rev 1:455–489
Hynes MJ, O'Dowd M (1987) Interactions of trimethyltin(IV) cation with carboxylic acids, amino acids, and related ligands. J Chem Soc Dalton Trans 563–566
Hynes MJ, Keely JM, McManus J (1991) Investigation of $[Sn(CH_3)_3(H_2O)_2]^+$ in aqueous solution by Tin-119 Nuclear Magnetic Resonance Spectroscopy. J Chem Soc Dalton Tran 3427–3429
Janssen MJ, Luijten JG (1963) Investigations on organotins compounds. XVIII The basicity of triorganotin hydroxides. Rec Trav Chim 85:1008–1014
Langlois MC, Devaud M (1974) Etudes des complexes ethyltrichlorostannane-ion sulfure. Bull Soc Chim France 5/6:789–792
Luijten JGA (1966) Investigations on organotin compounds. XXII. Hydrolysis products of alkyltin trichlorides. Recueil 85:873
Kriegsmann H, Pauly S (1964) Die IR- und Ramanspektren von CH_3SnCl_3 und CH_3SnBr_3 und Komplexbildung dieser Verbindungen in verschiedenen Lösungsmitteln. Z Anorg Allg Chem 330: 275
McGrady MM, Tobias RS (1964) Raman, Infrared and Nuclear Magnetic Resonance Spectroscopic studies on aqueous solutions of dimethyltin(IV) compounds: Structure and bonding of the aquo dimethyltin ion. Inorg Chem (vol III) 8:1157–1163
McGrady MM, Tobias RS (1965) Synthesis, structure and bonding of complexes of dimethyltin ions with bidentate ligands. J Am Chem Soc 5:1909–1916
Mennie D, Craig PJ (1993) Analysis of organometallic compounds in the environment. In: Siegel H, Siegel A (eds) Metal ions in biological systems, vol XXIX. Marcel Dekker Inc., pp 37–78
Millero FJ (1997) The effect of ionic interactions on thermodynamic and kinetic processes in natural waters. In: Gianguzza A, Pelizzetti E, Sammartano S (eds) Marine chemistry – an environmental analytical chemistry approach. Kluwer Academic Publisher, Amsterdam, pp 11-31
Natsume T, Aizawa S, Hatano K, Funahaschi S (1994) Hydrolysis, polymerisation and structure of dimethyltin(IV) in aqueous solution. Molecular structure of the polymer $[(SnMe_2)_2(OH)_3]ClO_4$. J Chem Soc Dalton Trans 2749–2753
Pitzer KS (1991) Activity coefficients in electrolyte solutions, 2nd edn. CRC Press, Boca Raton
Rochow EG, Seyferth D (1953) The electrolytic dissociation of dimethyltin dichloride. J Am Chem Soc 75:2877–2878
Takahashi T, Natsume N, Koshino S, Funahashi Y, Takagi D (1997) Speciation of trimethyltin(IV): hydrolysis, complexation equilibria, and structures of trimethyltin(IV) ion in aqueous solution. Can J Chem 75:1084–1092
Thayer JS (1978) Organotin. In: Brinckman FE, Bellama JM (eds) Organometals and organometalloids: Occurrence and fate in the environment. ACS Symposium Series 82:88–204
Thayer JS (1984) Organometallic compounds and living organisms. Academic Press, New York
Thayer JS (1993) Global bioalkylation of the heavy elements. In: Siegel H, Siegel A (eds) Metal ions in biological systems, vol XXIX. Marcel Dekker Inc., pp 1–30
Tobias RS (1966) Bonded organometallic cations in aqueous solutions and crystals. Organometal Chem Rev 1:93–129
Tobias RS, Freidline CE (1965) Raman spectroscopic and e.m.f. studies on aqueous solutions of the trans-tetrahydroxidodimethylstannate(IV) ion. The four acid dissociation constants of dimethyltin(IV) aquo ion. Inorg Chem (vol IV) 2:215–220
Tobias RS, Yasuda M (1964) Studies on the soluble intermediates in the hydrolysis of dimethyltin dichloride. Can J Chem 42: 781–791
Tobias RS, Ogrins I, Nevett BA (1962) Studies on the mono- and polynuclear hydroxo complexes of dimethyltin(IV) ion in aqueous solution. Inorg Chem (vol I) 3:638–646
Tobias RS, Farrer H, Hughes M, Nevett BA (1966) Hydrolysis of the aquo ions R_3Sn^+ and R_2Sn^{2+}: steric effects on the dissociation of aquo acids. Inorganic Chemistry 5:2052–2055
Van den Berghe EV, van der Kelen GP (1965) On the NMR spectra of methyltintrichloride in solution. Bull Soc Chim Belges 74:479–480
Wada M, Okawara R (1965) Isolation of a compound containing the trimethyltin cation dihydrated. J Organometal Chem 4:487–488
Zuckerman JJ, Reisdorf RP, Ellis HV, Wilkinson RR (1978) Organotin compounds. In: Brinckman FE, Bellama JM (eds) Organometals and organometalloids: Occurrence and fate in the environment. ACS Symposium Series 82:388–424

^{119}Sn Mössbauer Spectroscopy Studies on the Interaction of Organotin(IV) Salts and Complexes with Biological Systems and Molecules

R. Barbieri · L. Pellerito · G. Ruisi · A. Silvestri · A. Barbieri-Paulsen · G. Barone
S. Posante · M. Rossi

12.1 Introduction

Organotin(IV) compounds are widely spread out in the environment (Blunden and Chapman 1986), owing to their industrial uses (Blunden et al. 1985; Evans 1998). Interacting with living organisms, effects are provoked on biological systems and functions (Thayer 1984; Arakawa 1998; Smith 1998).

^{119}Sn Mössbauer spectroscopy (whose principles are sketched in a recent review by Barbieri et al. 1998) has been employed in a number of studies on bonding and structure of organotins in biological systems, as well as bound to biological molecules. The speciation of triorganotins spiked to estuarine sediments (see Section 12.2) was effected by comparison of the spectra of pure triorganotins, in the solid state, with spectra of spiked samples, obtaining information about the nature of the tin derivative in a given sediment. In fact, Mössbauer spectroscopy cannot be employed as a microanalytical technique for the determination of trace elements (Barbieri et al. 1998); organotins in relatively low amounts could be investigated employing $R_n{}^{119}SnX_{4-n}$ derivatives, synthesized from ^{119}Sn metal samples. In the case of organotin(IV) complexes with biological molecules, possibly with known stoichiometry, the structure and bonding of the tin environment are generally determined at a high degree of certainty by the point-charge model treatment of the parameter nuclear quadrupole splitting, ΔE (Barbieri et al. 1998, and refs. therein). The eventual degradation to inorganic derivatives, and in a special way the formation of Sn(II) compounds, is detected by the magnitude of the parameter isomer shift, δ (Barbieri et al. 1998). Structural information is also extracted from δ / Q_{Sn} correlations, Q_{Sn} being the partial charge on tin atoms (Barbieri and Silvestri 1984).

In the present review, besides the speciation studies above mentioned, ^{119}Sn Mössbauer determinations of structure and bonding are reported and discussed for organotins coordinated to biological system components, to hemoglobin and to nucleic acids. Particular emphasis is given to unpublished work on the interaction with DNA, and tin bonding and structures in the obtained DNA condensates, by mono-organotin derivatives as well as by $R_nSn(IV)$ complexes; systems $R_nSn(IV)$-aminoacid-DNA are reported and discussed in the context of protein-nucleic acids interactions. [*Annotation:* It seems worthwhile to note that the actual scientific activity at the Department of Inorganic Chemistry, University of Palermo, concerning research on organotin chemistry (extended to environmental interactions) as well as Mössbauer spectroscopy investigations, originates from studies started at the University of Padua, as far as Italian sources are concerned. In fact, pioneering work on organometals was effected by G. Semerano and L. Riccoboni around 1935–1941 at the Institute of Physical Chemistry, University of Padua, possibly suggested by their head A. Miolati (see e.g. the study on organotins, Riccoboni 1937); a number of papers in the field were successively published by L. Riccoboni, G. Tagliavini and their co-workers (Tagliavini et al. 1962, Barbieri et al. 1958). U. Croatto introduced Mössbauer spectroscopy at the Institute of General Chemistry, University of Padua, around 1960, as a consequence of a scientific tour in Germany; a homemade spectrometer was built, according to Croatto's design, which allowed the beginning of studies in the field. Lastly, the research actually effected in Palermo on the interaction of nucleic acids, and their constituents, with metal derivatives originates from the thesis of R. Barbieri, the supervisor being A. Turco (Barbieri 1956; Cessi and Turco 1956).]

12.2 Analysis and Speciation of Organotins in the Environment

Analytical methods and procedures for the determination of organotin(IV) derivatives in biological and environmental systems have been amply reported and discussed in a series of papers and review articles (Miller and Craig 1998, Maguire 1991, Maguire 1987).

Mössbauer spectroscopy, ^{119}Sn, has been employed in the "in vitro" speciation of tributyltin(IV) and triphenyltin(IV) derivatives spiked to estuarine sediments (Eng et al. 1986; Lucero et al. 1992; May et al. 1994; Whalen et al. 1993) mainly in view of the large use of triorganotins in antifouling coating of ship hulls (Kjaer 1992). The field has been reviewed, in relation also with environmental and biological aspects (May et al. 1993). The determined ^{119}Sn Mössbauer parameters δ (isomer shift) and ΔE (nuclear quadrupole splitting) are summarized in Table 12.1.

The procedure employed for the obtainment of absorber samples to be submitted to ^{119}Sn Mössbauer investigation was as follows (Lucero et al. 1992; Whalen et al. 1993; May et al. 1994):

i. Sediment samples (defined as anaerobic, anoxic) were collected from selected sites in bays and estuaries, and kept frozen until employed.
ii. Aerobic (oxic) samples were prepared by air-drying part of a given anoxic sample, and grinding.
iii. "Sediment spiking" was effected by adding solid R_3SnX (e.g. 3.3% w/w) to oxic and anoxic sediment samples (e.g. 5 g), covering the mixture with, e.g. 5–100 ml of synthetic sea water, and shaking in the dark at room temperature for about one week; the mixture was left in the dark for an additional two weeks. The pH of synthetic sea water, as well as other solution characteristics such as salinity, were adjusted before addition to the R_3SnX spiked sediments.

Comparison of the ^{119}Sn Mössbauer parameters of the "in vitro" systems described above, with data related to solid state triorganotins, Table 12.1, yielded the following conclusions:

i. $^{n}Bu_3SnF$ and $(^{n}Bu_3Sn)_2SO_4$: limited, or null, interaction with both oxic and anoxic sediments; the same holds for $^{n}Bu_3Sn(O_2CCH_3)$ in oxic sediments (May et al. 1994).
ii. $^{n}Bu_3SnCl$ and $^{n}Bu_3Sn(O_2CCH_3)$ interact with anoxic sediments in different ways in relation to the origin of the sediment (May et al. 1994).

Table 12.1. ^{119}Sn Mössbauer parameters of tributyltin(IV) and triphenyltin(IV) derivatives, interacted with estuarine sediments[a]

Compound	$\delta^{b,c}$ (mm s^{-1})	$\Delta E^{b,c}$ (mm s^{-1})	Interacted with sediments[d]			
			Anoxic[d]		Oxic[d]	
			$\delta^{b,d}$ (mm s^{-1})	$\Delta E^{b,d}$ (mm s^{-1})	$\delta^{b,d}$ (mm s^{-1})	$\Delta E^{b,d}$ (mm s^{-1})
$^nBu_3SnF$	1.34	3.63	1.47–1.55	3.34–3.63	1.43–1.57	3.26–3.55
$^nBu_3SnCl$	1.56	3.43	1.29–1.49	1.65–3.25	1.47–1.54	3.42–3.56
$^nBu_3Sn(O_2CCH_3)$	1.46	3.51	1.37–1.53	1.62–3.40	1.40–1.53	3.26–4.02
$(^nBu_3Sn)_2O$	1.11	1.55	1.29–1.50	1.58–2.90	1.35–1.50	2.20–3.54
$(^nBu_3Sn)_2S$	1.49	1.71	1.27–1.44	1.66–3.26	1.21–1.45	1.63–3.21
$(^nBu_3Sn)_2SO_4$	1.12	3.56	1.36–1.51	3.17–3.50	1.38–1.65	3.15–3.51
$(^nBu_3Sn)_2CO_3$	1.46	3.30	1.34–1.48	1.64–3.16	1.35[e]	3.28[e]
Ph_3SnF	1.36	3.62	1.22–1.36	2.99–3.62	1.21–1.30	3.09–3.63
Ph_3SnCl	1.35	2.52	1.14–1.28	2.43–2.79	1.13–1.25	2.50–2.77
$Ph_3Sn(O_2CCH_3)$	1.29	3.31	1.12–1.21	2.63–2.76	1.04–1.15	2.51–2.76
Ph_3SnOH	1.23	2.95	1.12–1.22	2.66–2.81	1.10–1.17	2.71–2.80

[a] Eng et al. 1986; May et al. 1994; Lucero et al. 1992; Whalen et al. 1993.
[b] δ = isomer shift with respect to $BaSnO_3$; ΔE = nuclear quadrupole splitting; T = 80 K.
[c] Data for solid state samples of the organotin(IV) salts employed in the interaction with sediments.
[d] See text. Ranges of δ and ΔE measured for compounds interacted with sediments from series of estuarine and marine sites. Neutral or basic pH was detected in suspensions of the anoxic sediments investigated (in artificial seawater, at room temperature, without added organotins) unless otherwise stated. See Refs. in note *a*.
[e] Data reported for one site. See May et al. 1994.

iii. $^nBu_3SnOH$ and $(^nBu_3Sn)_2O$ show analogous interactions with oxic sediments, while, when spiked to anoxic sediments, the oxide forms different compounds in samples taken from given sites (Eng et al. 1986; May et al. 1994).
iv. $(^nBu_3Sn)_2S$ and $(^nBu_3Sn)_2CO_3$ are converted to other derivatives when spiked to both type of sediments (May et al. 1994).
v. Ph_3SnOH and $Ph_3Sn(O_2CCH_3)$ are probably converted to Ph_3Sn^+, which possibly binds to both oxic and anoxic sediments; Ph_3SnCl and Ph_3SnF spike unaltered (Lucero et al. 1992).

12.3 Interaction of Organotin(IV) Compounds with Biological Systems

The structures of tin bonding environment, calculated by the point-charge model treatment of the ^{119}Sn parameter nuclear quadrupole splitting, have been reported for $R_2Sn(IV)$ and $R_3Sn(IV)$ (R = Alkyl, Aryl) derivatives interacted with human whole erythrocytes, erythrocyte membranes, ghost cells, cytoplasm, as well as with rat liver mitochondria and the mitochondrial membrane; data concerning systems organotin(IV)-cells, and cell walls, of the fungus *Ceratocystis ulmi*, are also given. Diorganotins exhibit possible (distorted) trans-R_2 octahedral configurations, as well as trigonal bipyramidal with equatorial C, C atoms; triorganotin derivatives would assume both trigonal bipyramidal (equatorial C atoms) and tetrahedral configurations. The nature of tin bonding atoms other than C has not been assigned. The field has been recently reviewed, bond angles and structures being extensively reported together with ^{119}Sn Mössbauer parameters (Barbieri et al. 1998; Musmeci et al. 1992).

12.4 Hemoglobin

Bonding of $Alk_3Sn(IV)$ and $Me_2Sn(IV)$ with cat and rat hemoglobin tetramers has been interpreted in terms of the primary coordination of tin by thiol functions of cysteine 13, and eventually by heterocyclic nitrogen of histidine 113; regular, as well as distorted, tetrahedral and trigonal bipyramidal structures of tin environments have been extracted from point-charge model treatment of ΔE parameters (Barbieri et al. 1998). It seems worthwhile to note the recent publication of an extended review on the classification and structure function analysis of metal binding sites in proteins (Holm et al. 1996).

12.5 Deoxyribonucleic Acid

The research field inherent to the interaction of metal ions with nucleic acids, as well as the consequent condensation of DNA, is amply investigated (Berthon 1995; Bregadze 1996, Bloomfield 1996). Interactions involving organometal derivatives have been also studied, such as organomercury-DNA (see e.g. Gruenwedel 1985; Gruenwedel and Cruikshank 1990), and the organotin-DNA systems treated below. The results of ^{119}Sn Mössbauer spectroscopy studies on structure and molecular dynamics of systems $Alk_nSn(IV)$-deoxyribonucleic acid (DNA; $n = 1$–3), are here summarized and discussed; for representative complexes and systems, values of hyperfine parameters are reported in Table 12.2, and molecular dynamics data and functions in Table 12.3. The discussion and structural assumptions in the following are based upon the research cited in Table 12.2 and 12.3, according to the Mössbauer spectroscopy principles and data treatment procedures sketched in a recent review (Barbieri et al. 1998). Further work in the field has been discussed previously (Piro et al. 1992; Barbieri et al. 1998).

The hyperfine parameters of the DNA condensates obtained by addition of $MeSnCl_3$ in ethanol solutions to aqueous buffered native DNA (calf thymus) appear to be constant at any ratio [Sn]-[DNA monomer] in the gel phases (Table 12.2, No. *1–4*). Data are invariant with temperature (in the ranges reported in Table 12.3, No. *6*), as well as with the H_2O content in the condensed phases, the values for lyophilized samples exactly corresponding to those for the gels. A possible structure is shown in Fig. 12.1a; no definite attributions from ΔE data are possible, as often detected for RSn(IV) derivatives. The slopes of functions, total area under the resonant peaks vs. T, No. *6* in Table 12.3, are also invariant for gels and lyophilized condensates, and correspond to the general occurrence of polymeric species. It is then concluded that a unique tin coordination environment occurs in $MeSnCl_3(EtOH)_n$ interacted with aqueous DNA, independent from the water content in the condensed phases (gels and lyophilized specimens) as well as from the temperature; the tin site in Fig. 12.1a, would account for the experimental parameters and conditions, assuming the occurrence of bond-

Table 12.2. ^{119}Sn Mössbauer parameters of DNA condensates (gels) formed by interaction with mono-, di- and triorganotin(IV) moieties

No.	Compound or system[a]	δ^b (mm s^{-1})	ΔE^b (mm s^{-1})	$\Gamma_1{}^b$ (mm s^{-1})	$\Gamma_2{}^b$ (mm s^{-1})	Reference
1	$MeSnCl_3(EtOH)_n$ + DNA, $r = 0.33$	0.56	2.19	0.91	0.90	Posante 1996
2	id., $r = 0.5$	0.56	2.25	0.83	0.80	"
3	id., $r = 1.0$	0.56	2.17	0.86	0.72	"
4	id., $r = 0.2–1.2$[c]	0.56 ±0.01	2.17 ±0.04	0.85 ±0.07	0.75 ±0.05	"
5	$Me_2SnCl_2(EtOH)_n$ + DNA, $r = 0.4$	1.26	4.39	0.90	0.79	Barbieri and Silvestri 1991
6	id., $r = 1.0$	1.33	4.44	0.91	0.91	"
7	$Et_2SnCl_2(EtOH)_n$ + DNA, $r = 0.6$	1.39	4.49	0.89	0.77	Barbieri et al. 1992
8	id., r = 1.0	1.43	4.40	0.74	0.99	"
9	$Me_3SnCl(EtOH)_n$ + DNA, $r = 2.4$	1.31	3.77	0.89	0.82	"
10	$Et_3SnCl(EtOH)_n$ + DNA, $r = 1.0$; 1.2	1.49	3.86	0.85	0.88	"
11	id., $r = 2.4$	1.47	3.83	0.86	0.82	"
12	$Me_3Sn(H_2O)_2^+$ + DNA; aq. sol., $r = 1.20–1.56$	1.32	3.80	0.89	0.78	"
13	id., pellet, $r = 2.4$	1.38	3.84	0.93	0.75	"

[a] No. *1–11*, *13*: DNA condensed phases obtained by addition to aqueous (buffered) DNA of: *1–11*, ethanol solutions of organotin chlorides; *13*, $Me_3Sn(IV)$ in H_2O at acid pH; Mössbauer spectra were taken on the gelled phases, separated by centrifugation and frozen to 77.3 K. r = [Sn] / [DNA monomer] (DNA monomer = mononucleotide unit, average MW being 315 in calf thymus (native) DNA). No. *12*, aqueous solution frozen at 77.3 K.

[b] δ = isomer shift with respect to RT $CaSnO_3$ or $BaSnO_3$; ΔE = nuclear quadrupole splitting; Γ = full width at half height of the resonant peaks. Spectra were taken at 77.3 K (liquid N_2 temperature) unless otherwise stated.

[c] Average data from 27 spectra in the r range indicated, with standard error.

ing to tin by two adjacent phosphodiester groups of the DNA chain, and by a phosphodiester adjacent to Sn in the toroidal (Bloomfield 1996) DNA condensate, thus originating the apparent polymeric nature of tin bonding environment (Posante 1996).

Systems $Alk_2SnCl_2(EtOH)_n$ + DNA (Alk = Me, Et, No. 5–8, Table 12.2), appear to consist of species Alk_2Sn(DNA monomer)$_2$, according to molecular dynamics studies on lyophilized condensates (Barbieri et al. 1992; Barbieri et al. 1995; Posante 1996). Tin environment in gelled and lyophilized systems would be of trans-R_2 octahedral type, such as in Fig. 12.1b. In freeze-dried specimens, two coordination sites of type 12.1b would occur for each system, according to the fitting of each experimental spectrum with two symmetric doublets: Me_2Sn(DNA monomer)$_2$, $\Delta E_{exp} = 4.44$ mm s^{-1}, $C\hat{Sn}C = 180°$, and $\Delta E_{exp} = 3.52$ mm s^{-1}, $C\hat{Sn}C = 143°$; Et_2Sn(DNA monomer)$_2$, $\Delta E_{exp} = 4.35$ mm s^{-1}, $C\hat{Sn}C = 180°$, and $\Delta E_{exp} = 3.56$ mm s^{-1}, $C\hat{Sn}C = 145°$; angles are estimated by point-charge

Table 12.3. Molecular dynamics data for organotin(IV) complexes with adenosine-5'-monophosphate and the model ligand phenylphosphate, and for organotin(IV)-DNA condensates

No.	Compound or system[a]	$10^2 d\ln(A_T/A_{77.3})/dT$[b] ($K^{-1}$)	T range (K)	Reference
1	$Me_2Sn(AMP)\cdot 2\,H_2O$	−0.95	78 – 295	Barbieri et al. 1987
2	$^nBu_2Sn(AMP)\cdot 2\,H_2O$	−1.28	78 – 228	"
3	$Me_2Sn[PO_3(OPh)]$	−1.303	78 – 215	"
4	$Et_2Sn[PO_3(OPh)]$	−1.025	78 – 165.5	"
5	$^nBu_2Sn[PO_3(OPh)]$	−1.947	78 – 195	"
6	$MeSnCl_n(DNA\ monomer)_{3-n}$[c], gel I	−0.866 (0.996)	77.3 – 181.7	Posante 1996
	idem, gel II	−0.911 (0.999)	77.3 – 140.0	"
	idem, lyophil. II	−0.944 (0.988)	77.3 – 298	"
7	$Me_2Sn(DNA\ monomer)_2$, lyophil.	−1.220 (0.996)	77.3 – 293	Barbieri et al. 1995
8	$Me_3Sn(DNA\ monomer)$, lyophil.	−1.555 (0.989)	77.3 – 220	"
9	$Et_2Sn(DNA\ monomer)_2$, gel	−1.724 (0.984)	77.3 – 161.4	Posante 1996
10	idem, lyophil.	−1.250 (0.996)	77.3 – 269.5	Barbieri et al. 1995
11	$Et_3Sn(DNA\ monomer)$, gel	−3.177 (0.985)	77.3 – 130.5	Posante 1996
12	idem, lyophil.	−2.045 (0.993)	77.3 – 221	Barbieri et al. 1995

[a] Absorber samples: No. *1–5*: crystalline solid complexes of adenosine-5'-monophosphate (AMP) and phenylphosphate. No. *6–12*: organotin(IV)-DNA condensates in gel as well as lyophilized (freeze-dried) phases (see note *a* to Table 12.2). DNA monomer = DNA (calf thymus) monomeric unit base-ribose-phosphate; $MW_{av} = 315$.
[b] Slopes of normalized total areas under the resonant peaks vs. T, correlation coefficients (whenever reported) in parentheses. No. *1–5*: total areas are normalized to the area at 78 K. Functions $10^2 d\ln(A_T)/dT$ (K^{-1}) are reported for the MeSn-DNA condensates, No. *6*.
[c] "gel I" and "gel II" refer to two independent sets of measurements effected on different gel preparations; "lyophil. II" consists of the freeze-dried "gel II" sample. In all samples, $r = [Sn]/[DNA\ monomer] = 0.4$ in the gel preparation. See note *a* to Table 12.2.

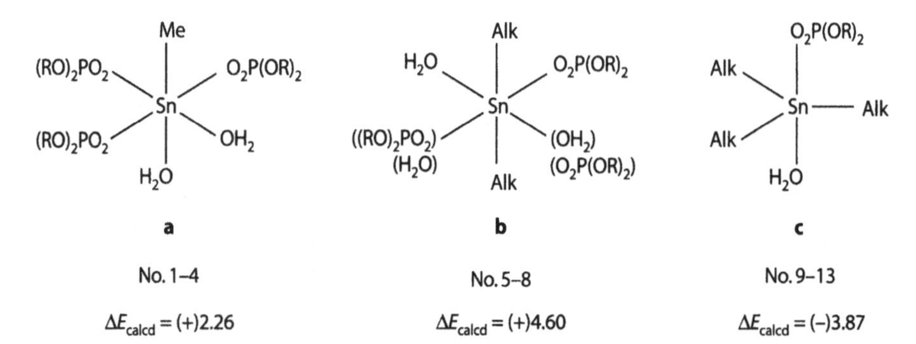

Fig. 12.1. Point-charge model simulation of the tin environment in the complexes and systems listed in Table 12.2. Regular structures assumed in the calculations. $(RO)_2PO_2^{(-)}$ stands for DNA phosphodiester

model treatment of ΔE parameters. The molecular dynamics parameters (e.g. the slopes of $\ln(A_T/A_{77.3})$ vs. T reported in Table 12.3, No. *7* and *10*) indicate the occurrence of polymeric species in freeze-dried specimens analogous to the $R_2Sn(IV)$ crystalline complexes of $Alk_2Sn(IV)$ with AMP and $[PO_3(OPh)]$, No. *1–5*, Table 12.3; consequently, the occurrence of structure in Fig. 12.1b with trans-phosphodiester groups could be assumed, so that polymericity would originate from interstrand bonding in the condensed, toroidal, DNA (vide supra). Slope of the function $\ln(A_T/A_{77.3})$ vs. T for gelled $Et_2Sn(DNA\ monomer)_2$, No. *9* in Table 12.3, is instead in the border zone monomer/polymer for $Alk_2Sn(IV)$ derivatives, so that presence of H_2O would influence bonding to DNA: a structure like in Fig. 12.1b with cis-phosphodiester groups could be assumed to occur, so that the $Et_2Sn(IV)$ moiety would be linked to two vicinal phosphodiester of a double helix of DNA. This behavior makes a consistent difference with that of systems MeSn(IV)-DNA (vide supra), where the persistence also in gelled phases of the assumed polymericity could be now attributed to the possible employment in bonding of a larger positive charge on MeSn(IV) with respect to $Et_2Sn(IV)$.

As far as systems $Alk_3Sn(IV)$-DNA are concerned (Alk = Me, Et), hyperfine parameters in gelled condensates and frozen aqueous solutions, No. *9–13* in Table 12.2, are invariant in freeze-dried specimens, which implies the persistence of a trigonal bipyramidal structure such as in Fig. 12.1c, as extracted from point-charge model treatment of the nuclear quadrupole splitting ΔE. From molecular dynamics studies, No. *8* and *12* in Table 12.3, it appears that slopes $d\ln(A_T/A_{77.3})\,/\,dT$ lie in the border zone between $Alk_3Sn(IV)$ monomers and polymers. As a consequence, structure in Fig. 12.1c would indicate the occurrence of $Alk_3Sn(IV)$ moieties appended to the DNA double helix in lyophilized specimens, showing the 1 : 1 composition Alk_3Sn(DNA monomer). In gelled systems $Et_3Sn(IV)$-DNA, such as No. *11* in Table 12.3, slope increases, being located in the monomers zone: then the presence of H_2O would provoke a further decrease in the interchain interactions by $Alk_3Sn(IV)$ bound to one phosphodiester group, the latter being perhaps due to interchain hydrogen bonds by the axial H_2O molecule (Fig. 12.1c) (Barbieri et al. 1992, 1995; Posante 1996).

12.6 Interaction of Organotin(IV) Complexes with Deoxyribonucleic Acid, and Ternary Systems $R_nSn(IV)$-Amino Acid-Nucleic Acid

The field of bonding of metal ion complexes to nucleic acids is very widely investigated, as inferred from recent reviews (Yamauchi et al. 1996; Dubler 1996; Clarke and Stubbs 1996; Kozelka 1996; Kimura and Shionoya 1996; Nordén et al. 1996; Draganescu and Tullius 1996; Sigman et al. 1996; Gravert and Griffin 1996; Burrows and Rokita 1996; Petering et al. 1996); studies on the argument are continuing (Cusumano et al. 1998; Navarro et al. 1998; Mandal et al. 1997; Wu et al. 1997; Tuite et al. 1997; Jacquet et al. 1997; Bauer and Wang 1997; Jin and Yang 1997; Magda et al. 1997; Lippert 1997; Yam et al. 1997; Kieft and Tinoco 1997; Cheatham and Kollman 1997). Cleavage of DNA by interaction with Fe(III)-bleomycin and ^{57}Fe Mössbauer irradiation (MIRAGE) has been investigated in relation to tumor therapy (Mac Donnell 1995).

In this context, a study on the interaction of the complex Et_2SnCl_2(o-phenanthroline) with mononucleotides and DNA appears to be the only one insofar effected in the field of R_nSn complexes-DNA systems (Li et al. 1996b, 1997). We report here studies on the ternary systems concerning the interaction with DNA by organotin(IV) moieties coordinated by 2-mercaptopyridine and 2-mercaptopyrimidine (Fig. 12.2), the complexes being characterized by chelation of tin by S, N donor ligand atoms (Schmiedgen et al. 1994, 1998; Schürmann 1994; Huber et al. 1997). These heterocycles belong to the same class as the nucleic acid bases; in fact, 2-mercaptopyrimidine derivatives have been detected in transfer RNA (Carbon et al. 1965; Lipsett 1965; Carbon et al. 1968; Baczynskyj et al. 1968). Studies in the field concern the synthesis, characterization and structure of a series of model systems (Wang et al. 1994; Singh et al. 1996; Boggon et al. 1996; Patel and Eckstein 1997), including metal binding to thionucleosides (Heitner et al. 1972; Kowalik-Jankowska et al. 1997).

Thiol sulfur-tin bonds in the organotin(IV)-2-mercapto-pyridine and -pyrimidine complexes are expected to persist in solution phases as well as in organotin(IV)-DNA condensates in consideration of the large values of stability constants detected, e.g. in $Me_3Sn(IV)$-SR derivatives (Hynes and O'Dowd 1987). In fact, molar conductance and vapour pressure osmometry data taken on a series of representative SPy, SPym complexes of organotins evidenced the occurrence of monomolecular, nearly undissociated, species in ethanol solution (Rossi 1994) in line with data for chloroform solutions (Schmiedgen et al. 1994, 1998; Schürmann 1994; Huber et al. 1997). The structure of the complexes in ethanol solution, as extracted from ^{119}Sn Mössbauer spectroscopy on frozen absorber samples, fully corresponds to the standard solid state structure, chelation to tin by S, N donor atoms being maintained (see e.g. Fig. 12.2) (Rossi 1994), in line with structures in $CDCl_3$ and $(CD_3)_2SO$ solutions, extracted from IR and ^1H, ^{13}C, ^{119}Sn NMR spectroscopic data (Schmiedgen et al. 1994, 1998; Schürmann 1994; Huber et al. 1997); the only structural variations in ethanol solutions would concern the eventual coordination to tin centres by C_2H_5OH molecules (Rossi 1994), in a special way for five-coordinated solid state complexes (Fig. 12.2), and partial Cl^- dissociation.

The DNA condensation by ethanolic organotin(IV)-2-mercaptopyridine and -pyrimidine complexes, Table 12.4, is then interpreted in terms of electrostatic interaction of the cationic complexes with the phosphodiester groups of calf thymus DNA,

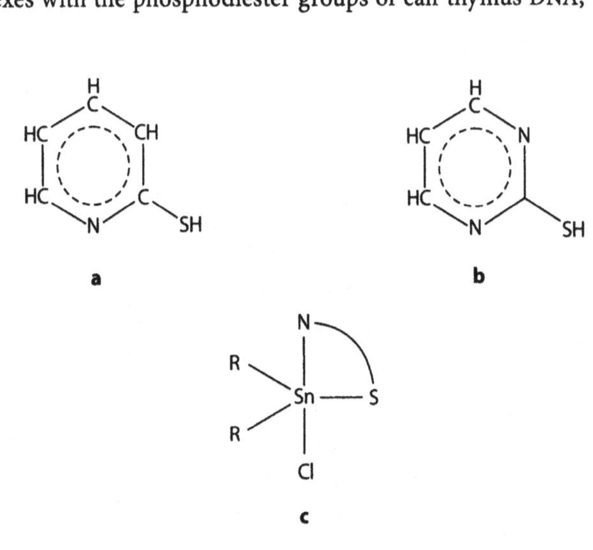

Fig. 12.2. a 2-mercaptopyridine (HSPy); **b** 2-mercapto-pyrimidine (HSPym); **c** Example of bonding of $^{(-)}$SPy, $^{(-)}$SPym to organotin(IV) acceptors: the idealized regular structure of R_2SnCl (SPy, SPym), according, e.g. to the crystal structure of $Ph_2SnCl(SPy)$ (Schmiedgen et al. 1994)

Table 12.4. ^{119}Sn Mössbauer parameters for DNA condensates (gels) obtained by interaction with DNA of complexes of $Me_2Sn(IV)$ and MeSn(IV) with 2-mercaptopyridine (HSPy)[a] and 2-mercaptopyrimidine (HSPym)[a]

No.	System[b]	δ^c (mm s^{-1})	ΔE^c (mm s^{-1})	$\Gamma_1{}^c$ (mm s^{-1})	$\Gamma_2{}^c$ (mm s^{-1})	References
1	$Me_2SnCl(SPy)$+DNA					Rossi 1994; Barone 1997; Barbieri et al. 1999
	doublet A[d]	1.31 – 1.26	3.96 – 4.55	0.70 – 0.86	0.70 – 0.86	
	doublet B[d]	1.07 – 1.19	2.89 – 3.28	0.84 – 1.06	0.84 – 1.06	
2	$Me_2SnCl(SPym)$+DNA					"
	doublet A[d]	1.24 – 1.38	4.41 – 4.47	0.75 – 0.81	0.75 – 0.81	
	doublet B[d]	1.18 – 1.27	3.13 – 3.18	0.92 – 1.12	0.92 – 1.12	
3	$MeSnCl_2(SPy)$+DNA	0.57	2.19	0.85	0.88	Rossi 1994; Barone 1997
4	$MeSnCl_2(SPym)$+DNA	0.55	2.21	0.87	0.87	"
5	$MeSnCl(SPy)_2$+DNA	0.53	2.01	1.07	1.02	"
6	$MeSnCl(SPym)_2$+DNA	0.55	2.07	1.18	0.93	Barone 1997 (unpublished data)

[a] See Fig. 12.2.
[b] DNA condensates obtained by addition of ethanol solutions of the $Me_nSn(IV)$-SPy, SPym complexes to aqueous buffered calf thymus DNA, the molar ratios r = [Sn] / [DNA monomer] = 1.0 being employed for all samples; % $C_2H_5OH \leq 50$ in the supernatants.
[c] See note *b* to Table 12.2. Average data, measured at 77.3 K, are reported, or data intervals. Data for lyophilized specimens essentially correspond to the values for gelled systems here reported.
[d] Spectra fitted with two Lorentzian doublets, indicating the occurrence of two coordination sites to Sn.

in line with systems organotin chlorides-DNA (Table 12.2), and in analogy to assumptions for Et_2SnCl_2(o-phenanthroline)-DNA (Li et al. 1996b). The structure of tin environments in the DNA condensates, as extracted from the point-charge model treatment of the experimental ΔE ^{119}Sn Mössbauer parameters (Table 12.4), are reported in Fig. 12.3, where a further dissociation of N$\longrightarrow$Sn bonds is also accounted for, in line with the orders of the magnitude of related stability constants ($Sn\text{-}S_{thiol}$ > Sn-phosphate > $Sn\text{-}N_{het}$; Hynes and O'Dowd 1987). Water coordination to tin is also taken into account (Rossi 1994; Barone 1997).

Additional information on the nature of the interaction of complexes organotin(IV)-2-mercaptopyridine and -pyrimidine with DNA is extracted from the dynamics of tin by variable temperature ^{119}Sn Mössbauer spectroscopy, effected on the representative DNA condensates listed in Table 12.5.

From the data in Table 12.5, the following is inferred:

i. The dynamics of tin in the assumed systems $Me_2Sn(SPym)$(DNA monomer) and $MeSn(SPy)_2$(DNA monomer) consistently differ from data for MeSn(IV) and $Me_2Sn(IV)$ complexes with S, N chelators (Barbieri A et al. 1995).
ii. Slopes of $\ln(A_T / A_{77.3})$ vs. T functions practically coincide in the gel and lyophilized absorbers $Me_2Sn(SPym)$(DNA monomer), being of the order of magnitude as $Alk_2Sn(\text{DNA monomer})_2$, gel and lyophilized (Table 12.3).

iii. $MeSn(SPy)_2$(DNA monomer) in the gel phase corresponds to the gel $Me_2Sn(SPym)$ (DNA monomer) as well as to Alk_2Sn(DNA monomer)$_2$ and $MeSnCl_n$(DNA monomer)$_{3-n}$ (Table 12.3).
iv. Lyophilized $MeSn(SPy)_2$(DNA monomer) shows a phase transition around $T = 210$ K; in the range 210–290 K, the $\ln A / T$ function shows a slope of the order of magnitude similar to that for the gel phase, as well as for $Me_2Sn(SPym)$(DNA monomer) and lyophilized Alk_2Sn(DNA monomer)$_2$ (Table 12.3).

It is concluded that the complex cation species $Me_2Sn(SPym)^+$ and $MeSn(SPy)_2^+$ bind DNA analogously to MeSn(IV) and Alk_2Sn(IV), Tables 12.2 and 12.3; the type of interaction does not depend upon the water content of the system for $Me_2Sn(SPym)$ (DNA monomer) and $MeSnCl_n$(DNA monomer)$_{3-n}$ (Table 12.3), Et_2Sn(DNA monomer)$_2$ (Table 12.3), Et_3Sn(DNA monomer) (Table 12.3), and $MeSn(SPy)_2$(DNA monomer) in the temperature range 77.3–207.5 K (Table 12.5). The variations detected in the gelled phases with respect to freeze-dried specimens could be ascribed to outer sphere DNA bonding (Nordmeier 1995) through Sn coordinated H_2O molecules, analogous to findings for $Mg(H_2O)_6^{2+}$ (Black et al. 1994).

The interaction of nucleic acids with proteins, and constituent molecules, is widely studied (Harrison and Sauer 1994; Phillips and Moras 1995; Sauer and Harrison 1996; Rhodes and Burley 1997; Richmond and Steitz 1998). The affinity between the protein-nucleic acid systems is demonstrated in peptide nucleic acid, PNA (Nielsen 1997; Nielsen and Haaima 1997), and its interaction with DNA (Nielsen 1997; Kosaganov et al. 1998). Metal ions and complexes interact with nucleic acids and proteins, and their constituents, in ternary systems (Sabat 1996; Long et al. 1996; Harada et al. 1996). Polypeptides bind also to toroidal DNA condensates (Reich et al. 1990). In this context, the interaction of organotins with nucleic acid in the presence of amino acids has been investigated by ^{119}Sn Mössbauer spectroscopy, and the results obtained are reported in Table 12.6 (Barbieri 1995).

The following considerations can be made:

i. The aqueous systems No. *1* and *2* consist of species $Me_2Sn[S(CH_2)_2COO]$ and $Me_2Sn[SCH_2CH(NH_2)COO]$, with trigonal bipyramidal tin environments (equatorial C_2SnS; axial 2 H_2O, H_2O and N, H_2O and OH^-, as function of pH; Silvestri et al. 1988; Barbieri and Musmeci 1988; Barbieri et al. 1990). Then, the complexes above do not interact with DNA in systems *1*, Table 12.6. In fact, the Mössbauer parameters of No. *1* strictly correspond to those of the complexes No. *2*, so that the latter, formally characterized by partial positive charge on tin (Silvestri et al. 1988; Barbieri and Musmeci 1988), do not induce DNA condensation nor bonding to DNA phosphodiester groups in solution phases (which would be expected to occur in 1/1 ratio by assuming the occurrence of R_2Sn^{+1}).
ii. The system No. *3*, Table 12.6, produces the condensed phase species Et_2Sn(DNA monomer)$_2$, with trans-Et_2 octahedral tin environment (No. *7* in Table 12.2, structure in Fig. 12.1b; in the supernatant, No. *4* in Table 12.6, both Et_2Sn(DNA monomer)$_2$ and $Et_2Sn[SCH_2CH(NH_2)COO]$ (No. *5* in Table 12.6) could be assumed to occur, according to the magnitude of the ΔE parameters. The complex $Et_2Sn[SCH_2CH(NH_2)COO]$, No. *5*, corresponds to the Me_2Sn(IV) complexes, No. *1*, the same hyperfine parameters being detected (Table 12.6).

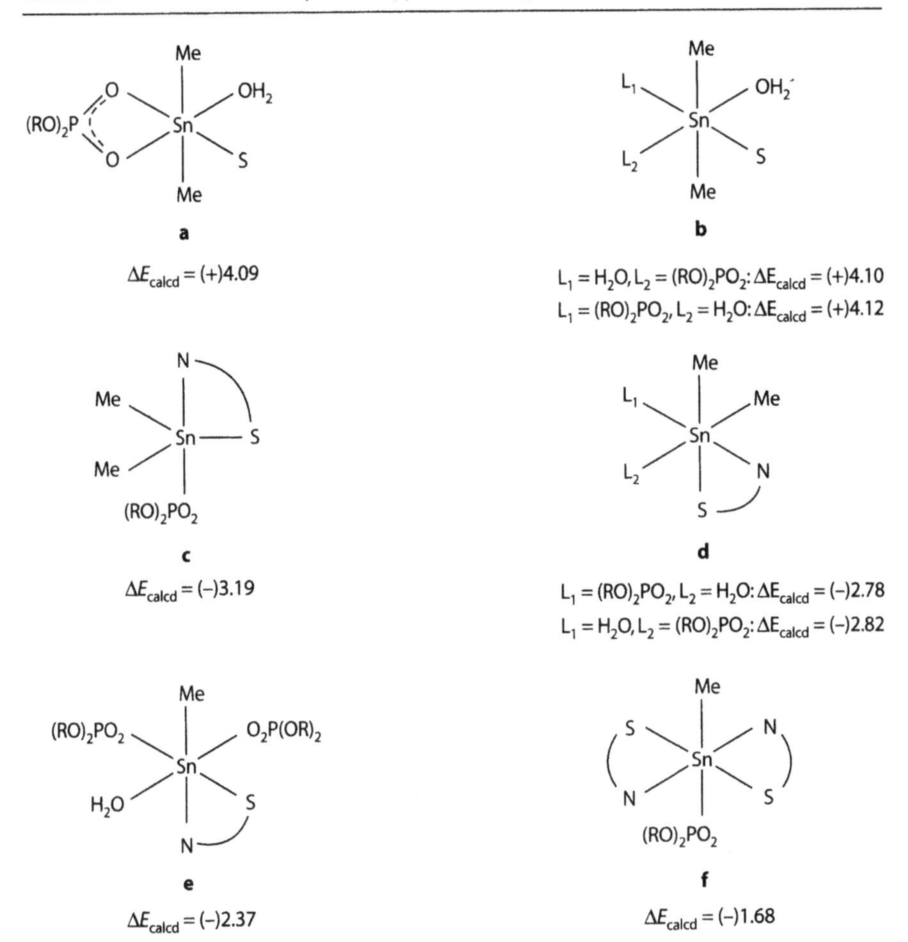

Fig. 12.3. Point-charge model simulation of the tin environment in the systems listed in Table 12.4. Regular structures are assumed in the calculations. $(RO)_2PO_2^{(-)}$ stands for DNA phosphodiester; S, N stand for thiol sulfur and heterocyclic nitrogen in the coordinated ligands 2-mercaptopyridine and 2-mercaptopyrimidine. **a,b** possible tin coordination sites indicated by doublets (A) in systems *1* and *2*, Table 12.4; **c,d** idem, doublets (B); **e** tin sites in systems *3* and *4*, Table 12.4; **f** tin sites in systems *5* and *6*, Table 12.4

From the experimentation summarized in Table 12.6, it would then appear that moieties $Alk_2Sn(IV)$ (which show a pronounced tendency to bind to DNA phosphodiester, Table 12.2) do not interact with DNA when coordinated by thiol sulfur in aqueous solution, even if a partial positive charge may be supposed to persist on tin (Silvestri et al. 1988; Barbieri and Musmeci 1988); in this context, it is worth mentioning that $Me_2Sn[SCH_2CH(NH_2)COO]$ reacts with a further cysteine thiol sulfur in rat hemoglobin (Barbieri and Musmeci 1988). On the other hand, Et_2SnCl_2 from ethanol solution (the latter mimicking biological non aqueous phases) condensates DNA even in the presence of previously added aqueous L-cysteine (No. 3 in Table 12.6). In order to inhibit DNA condensation, a large excess of cysteine must be present in the aqueous DNA phase ($Et_2Sn(IV)$/DNA monomer/L-cysteine = (0.5–1.0)/1/4) (Barbieri 1995).

Table 12.5. Molecular dynamics data for representative complexes of $Me_2Sn(IV)$ and $MeSn(IV)$ with 2-mercaptopyrimidine (HSPym)[a] and 2-mercaptopyridine (HSPy)[a], respectively, interacted with DNA, as determined by variable temperature ^{119}Sn Mössbauer spectroscopy

No.[b]	System[c]	$10^2 d\ln(A_T/A_{77.3})/dT$[d] ($K^{-1}$)	T range (K)	References
2	$Me_2Sn(SPym)$(DNA monomer)			Barbieri et al. 1999
	gel	−1.228 (0.989)	77.3–140.5	
	lyophil.	−1.309 (0.996)	77.3–240.75	
5	$MeSn(SPy)_2$(DNA monomer)			Rossi 1994; Barone 1997
	gel	−1.364 (0.983)	77.3–181.9	
	lyophil.	−0.639 (0.980)	77.3–207.5	
	lyophil.	−1.420 (0.982)	210.0–290.0	

[a] See Fig. 12.2.
[b] Correspond to No.s in Table 12.4.
[c] Composition is attributed on the basis of the value r = [Sn] / [DNA monomer] employed in the condensation process (see note b in Table 12.4) as well as of the composition of R_nSn-DNA condensates reported in Table 12.3. See also note a in Table 12.3.
[d] Slopes of normalized total areas under the resonant peaks vs. T; correlation coefficients in parentheses.

Table 12.6. ^{119}Sn Mössbauer parameters for 1/1 systems Alk_2SnCl_2/L-cysteine and Alk_2SnCl_2/3-mercaptopropionic acid, interacted with DNA (1/1), and Et_2SnCl_2 interacted with DNA-L-cysteine [1/(1/1)]

No.	System[a]	Absorber sample	δ^b ($mm\,s^{-1}$)	ΔE^b ($mm\,s^{-1}$)	Γ_1, Γ_2[b,c] ($mm\,s^{-1}$)	Reference
1	$[(Me_2SnCl_2)(HS\text{-}COOH)]$[d] + DNA	Solutions in H_2O pH = 4.80–5.94	1.24–1.34	2.98–3.23	0.81	Barbieri 1995
2	$(Me_2SnCl_2)(HS\text{-}COOH)$	Solutions in H_2O pH = 4.33–4.36	1.26–1.30	2.98–3.26	0.80	"
3	Et_2SnCl_2 + [(L-cys)(DNA)][e]	Pellet, gel	1.53	4.51	0.79	"
4	id.	Supernatant[f] pH = 3.23	1.47	3.67	0.98	"
5	Et_2SnCl_2 + (L-cys)[e]	Solutions in H_2O[f] pH = 3.42	1.45	3.26	0.75	"

[a] HS-COOH stands for both 3-mercaptopropionic acid, $HSCH_2CH_2COOH$, and L-cysteine, $HSCH_2CH(NH_2)COOH$. Molar ratios are generally 1 / 1, DNA standing for DNA monomer (MW = 315, calf thymus DNA). Aqueous solutions are employed, unless otherwise stated; DNA was buffered with Tris-EDTA.
[b] See note b to Table 12.2. Limiting data for series of spectra are reported, measured at 77.3 K.
[c] Average values.
[d] Includes a solution of $Me_2Sn[SCH_2CH(NH_2)COO]$ in H_2O.
[e] Ethanol solutions of Et_2SnCl_2 are employed.
[f] Ethanol ≤ 10% vol.

These results are interpreted as follows (Barbieri 1995):

i. Systems No. *1*, Table 12.6: the cysteine, as well as mercaptocarboxylic acid, moiety, interact with DNA, according to the general trends mentioned here in the preceding; ternary systems are formed, preferentially with respect to the interaction of Sn with DNA phosphodiester.
ii. In system No. *3*, cysteine binds to DNA in the general way for amino acids-nucleic acid systems (vide supra). Then, the subsequent addition of ethanolic Et_2SnCl_2 yields the common DNA condensate by $Et_2Sn(IV)$ phosphodiester binding.

References

Arakawa Y (1998) Recent studies on the mode of biological action of di- and tri-alkyltin compounds. In: Smith PJ (ed) Chemistry of tin. Blackie Acad and Prof, London, pp 388–428

Baczynskyj L, Biemann K, Hall RH (1968) Sulfur-containing nucleoside from yeast transfer ribonucleic acid: 2-thio-5 (or 6) -uridine acetic acid methyl ester. Science 159:1481–1483

Barbieri A (1995) Sintesi e caratterizzazione di complessi di Sn(IV), $R_nSn(IV)$ e Fe(II,III). Studio dell'interazione di $Alk_nSn(IV)$ con DNA nativo (Alk= Me, Et; $n = 2,3$). Thesis for Ph.D. degree, University of Palermo

Barbieri A, Giuliani AM, Ruisi G, Silvestri A, Barbieri R (1995) Tin(IV), monomethyltin(IV) and dimethyltin(IV) complexes with thiol sulphur and heterocyclic nitrogen donors: Molecular dynamics and structure by ^{119}Sn Mössbauer spectroscopy. Z Anorg Allg Chem 621:89-96

Barbieri R (1956) Separazione e dosaggio dei mononucleotidi degli acidi nucleici del fegato; ricerche sulla incorporazione di fosfato marcato con ^{32}P. Thesis for Master degree in Chemistry, University of Padua

Barbieri R, Musmeci MT (1988) A ^{119}Sn Mössbauer spectroscopic study on the interaction of dimethyltin(IV) derivatives with rat hemoglobin, and of related model systems in aqueous solution. J Inorg Biochem 32:89–108

Barbieri R, Silvestri A (1984) The correlation between tin-119 Mössbauer isomer shifts and atomic charges on tin in five-coordinated tin(IV) derivatives. J Chem Soc Dalton Trans 1019–1025

Barbieri R, Silvestri A (1991) The interaction of native DNA with dimethyltin(IV) species. J Inorg Biochem 41:31–35

Barbieri R, Belluco U, Tagliavini G (1958) Separazione cromatografica su carta di composti metallorganici di piombo e stagno. Annali di Chimica 48:940–949

Barbieri R, Alonzo G, Herber RH (1987) Configuration and lattice dynamics of complexes of dialkyltin(IV) with adenosine-5'-monophosphate and phenyl phosphates. J Chem Soc Dalton Trans 789–794

Barbieri R, Silvestri A, Filippeschi S, Magistrelli M, Huber F (1990) Studies on the antitumor activity of complexes of $R_2Sn(IV)$ with penicillamine enantiomers and with 3-thio-propanoic acid, and correlation with structural aspects. Inorg Chim Acta 177:141–144

Barbieri R, Silvestri A, Giuliani AM, Piro V, Di Simone F, Madonia G (1992) Organotin compounds and deoxyribonucleic acid. J Chem Soc Dalton Trans 585–590

Barbieri R, Ruisi G, Silvestri A, Giuliani AM, Barbieri A, Spina G, Pieralli F, Del Giallo F (1995) Dynamics of tin nuclei in alkyltin(IV)-deoxyribonucleic acid condensates by variable temperature tin-119 Mössbauer spectroscopy. J Chem Soc Dalton Trans 467–475

Barbieri R, Huber F, Pellerito L, Ruisi G, Silvestri A (1998) 119-m Sn Mössbauer studies on tin compounds. In: Smith PJ (ed) Chemistry of tin. Blackie Acad and Prof, London, pp 496–540

Barbieri R, Huber F, Silvestri A, Ruisi G, Rossi M, Barone G, Barbieri Paulsen A (1999) The interaction of S,N-coordinated dimethyltin(IV) derivatives with deoxyribonucleic acid: structure and dynamics by ^{119}Sn Mössbauer spectroscopy. Appl Organometal Chem 13:595-603

Barone G (1997) Interazione di acido deossiribonucleico (DNA), e molecole costituenti, con $(CH_3)_nSn(IV)$ ($n = 1$–3). Thesis for Ph.D. Degree in Chemistry, University of Palermo

Bauer C, Wang AHJ (1997) Bridged cobalt amine complexes induce DNA conformational changes effectively. J Inorg Biochem 68:129–135

Berthon G (1995) Handbook of metal-ligand interactions in biological fluids. M Dekker, New York

Black CB, Huang HW, Cowan JA (1994) Biological coordination chemistry of magnesium, sodium, and potassium ions. Protein and nucleotide binding sites. Coord Chem Revs 135/136:165–202

Bloomfield V A (1996) DNA condensation. Current Opinion in Struct Biol 6:334-341

Blunden SJ, Chapman A (1986) Organotin compounds in the environment. In: Craig PJ (ed) Organometallic compounds in the environment. Wiley, New York, pp 111–159
Blunden SJ, Cusack PA, Hill R (1985) The industrial uses of tin chemicals. The Royal Society of Chemistry, London
Boggon TJ, Hancox EL, Mc Auley-Hecht KE, Connolly BA, Hunter WN, Brown T, Walker RT, Leonard GA (1996) The crystal structure analysis of d(CGCGAASSCGCG)$_2$, a synthetic DNA dodecamer duplex containing four 4'-thio-2'-deoxythymidine nucleotides. Nucleic Acid Res 24:951–961
Bregadze VG (1996) Metal ion interactions with DNA: considerations on structure, stability, and effects from metal ion binding. In: Sigel A, Sigel H (eds) Metal ions in biological systems, vol XXXII. M Dekker, New York, pp 419–451
Burrows CJ, Rokita SE (1996) Nickel complexes as probes of guanine sites in nucleic acid folding. In: Sigel A, Sigel H (eds) Metal ions in biological systems, vol XXXIII. M Dekker, New York, pp 537–560
Carbon JA, Hung L, Jones DS (1965) A reversible oxidative inactivation of specific transfer RNA species. Proc NY Natl Acad Sci US 53:979–986
Carbon J, David H, Studier MH (1968) Thiobases in *Escherichia coli* transfer RNA: 2-thiocytosine and 5-methylaminomethyl-2-thiouracil. Science 161:1146–1147
Cessi C, Turco A (1956) Incorporazione del fosfato radioattivo negli acidi nucleici del fegato nell'intossicazione difterica. Giornale di Biochimica V:124–133
Cheatham TE, Kollman PA (1997) Insight into the stabilization of A-DNA by specific ion association: spontaneous B-DNA to A-DNA transitions observed in molecular dynamics simulations of d[ACCCGCGGGT]$_2$ in the presence of hexaamminecobalt(III). Structure 5:1297–1311
Clarke MJ, Stubbs M (1996) Interactions of metallopharmaceuticals with DNA. In: Sigel A, Sigel H (eds) Metal ions in biological systems, vol XXXII. M Dekker, New York, pp 727–780
Cusumano M, Di Pietro ML, Giannetto A, Nicolò F, Rotondo E (1998) Noncovalent interactions of platinum(II) square planar complexes containing ligands out-of-plane with DNA. Inorg Chem 37:563–568
Draganescu A, Tullius TD (1996) Targeting of nucleic acids by iron complexes. In: Sigel A, Sigel H (eds) Metal ions in biological systems, vol XXXIII. M Dekker, New York, pp 453–484
Dubler E (1996) Metal complexes of sulfur-containing purine derivatives. In: Sigel A, Sigel H (eds) Metal ions in biological systems, vol XXXII. M Dekker, New York, pp 301–338
Eng G, Bathersfield O, May L (1986) Mössbauer studies of the speciation of tributyltin compounds in seawater and sediment samples. Water Air Soil Pollut 27:191–197
Evans CJ (1998) Industrial uses of tin chemicals. In: Smith PJ (ed.) Chemistry of tin. Blackie Acad and Prof, London, pp 442–479
Gravert DJ, Griffin JH (1996) Specific DNA cleavage by manganese(III) complexes. In: Sigel A, Sigel H (eds) Metal ions in biological systems, vol XXXIII. M Dekker, New York, pp 515–536
Gruenwedel DW (1985) Circular dichroism of micrococcal nuclease-treated calf-thymus chromatin (soluble chromatin) in presence of CH_3HgOH. J Inorg Biochem 25:109–120
Gruenwedel DW, Cruikshank MK (1990) Mercury-induced DNA polymorphism: Probing the conformation of Hg(II)-DNA via staphylococcal nuclease digestion and circular dichroism measurements. Biochemistry 29:2110–2116
Harada W, Nojima T, Shibayama A, Ueda H, Shindo H, Chikira M (1996) How amino acids control the binding of Cu(II) ions to DNA. I. The role of the hydroxyl group of serine and threonine in fixing the orientation of the complexes. J Inorg Biochem 64:273–285
Harrison SC, Sauer RT (eds) (1994) Protein-nucleic acid interaction. Curr Opin Struct Biol 4:1–66
Heitner HI, Lippard SJ, Sunshine HR (1972) Metal binding by thionucleosides. J Amer Chem Soc 94:8936–8937
Holm RH, Kennepohl P, Solomon EI (1996) Structural and functional aspects of metal sites in biology. Chem Rev 96:2239–2314
Huber F, Schmiedgen R, Schürmann M, Barbieri R, Ruisi G, Silvestri A (1997) Mono-organotin(IV) and tin(IV) derivatives of 2-mercaptopyridine and 2-mercaptopyrimidine: X-ray structures of methyltris(2-pyridinethiolato)tin(IV) and phenyltris(2-pyridinethiolato)tin(IV)·1.5$CHCl_3$. Appl Organometal Chem 11:869–888
Hynes MJ, O'Dowd M (1987) Interaction of the trimethyltin(IV) cation with carboxylic acids, amino acids, and related ligands. J Chem soc Dalton Trans 563–566
Jacquet L, Davies RJH, Kirsch de Mesmaecker A, Kelly JM (1997) Photoaddition of $Ru(tap)_2(bpy)^{2+}$ to DNA: A new mode of covalent attachment of metal complexes to duplex DNA. J Am Chem Soc 119:11763–11768
Jin L, Yang P (1997) Synthesis and DNA binding studies of cobalt(III) mixed-polypyridyl complex. J Inorg Biochem 68:79–83
Kieft JS, Tinoco I Jr (1997) Solution structure of a metal-binding site in the major groove of RNA complexes with cobalt(III) hexammine. Structure 5:713–721

Kimura E, Shionoya M (1996) Zinc complexes as targeting agents for nucleic acids. In: Sigel A, Sigel H (eds) Metal ions in biological systems, vol XXXIII. M Dekker, New York, pp 29–52
Kjaer EB (1992) Bioactive materials for antifouling coatings. Progress Org Coatings, 20:339–352
Kosaganov YN, Stetsenko DA, Lubyako EN, Kvitko NP, Lazurkin YS (1998) Stability of DNA complexes with peptide nucleic acid. Molecular Biology 32:105–108
Kowalik-Jankowska T, Varnagy K, Swiatek-Kozlowska J, Jon A, Sovago I, Sochacka E, Malkiewicz A, Spychala J, Kozlowski H (1997) Role of sulfur site in metal binding to thiopurine and thiopyrimidine nucleosides. J Inorg Biochem 65:257–262
Kozelka J (1996) Molecular modeling of transition metal complexes with nucleic acids and their constituents. In: Sigel A, Sigel H (eds) Metal ions in biological systems, vol XXXIII. M Dekker, New York, pp 1–28
Li Q, Jin N, Yang P, Wan J (1997) Interaction of Et_2SnCl_2(phen) with nucleotides. Synth React Inorg Met-Org Chem 27:811–823
Li Q, Yang P, Hua E, Tian C (1996a) Diorganotin(IV) antitumor agents. Aqueous and solid-state coordination chemistry of nucleotides with R_2SnCl_2. J Coord Chem 40:227–236
Li Q, Yang P, Wang H, Guo M (1996b) Diorganotin(IV) antitumor agent. $(C_2H_5)_2SnCl_2$(phen)/ nucleotides aqueous and solid-state coordination chemistry and its DNA binding studies. J Inorg Biochem 64:181–195
Lippert B (1997) Effects of metal-ion binding on nucleobase pairing: Stabilization, prevention and mismatch formation. J Chem Soc Dalton Trans 3971–3976
Lipsett MN (1965) The behavior of 4-thiouridine in the *E. coli* s-RNA molecule. Biochem Biophys Res Commun 20:224–229
Long EC, Denney Eason P, Liang Q (1996) Synthetic metallopeptides as probes of protein-DNA interactions. In: Sigel A, Sigel H (eds) Metal ions in biological systems, vol XXXIII. M Dekker, New York, pp 427–452
Lucero RA, Otieno MA, May L, Eng G (1992) Speciation of some triphenyltin compounds in estuarine sediments using Mössbauer spectroscopy. Appl Organometal Chem 6:273–278
Mac Donnell FM (1995) Reexamining the Mössbauer effect as a means to cleave DNA. Biochemistry 34:12871–12876
Magda D, Wright M, Crofts S, Lin A, Sessler JL (1997) Metal complex conjugates of antisense DNA which display ribozyme-like activity. J Am Chem Soc 119:6947–6948
Maguire RJ (1987) Environmental aspects of tributyltin. Appl Organometal Chem 1:475–498
Maguire RJ (1991) Aquatic environmental aspects of non-pesticidal organotin compounds. Water Poll Res J Canada 26:243–360
Mandal SS, Varshney U, Bhattacharya S (1997) Role of the central metal ion and ligand charge in the DNA binding and modification by metallosalen complexes. Bioconjugate Chem 8:798–812
May L, Whalen D, Eng G (1993) Interaction of triorganotin compounds with Chesapeake Bay sediments and benthos. Appl Organometal Chem 7:437–441
May L, Berhane L, Berhane M, Counsil C, Keane M, Reed BB, Eng G (1994) The speciation of some tributyltin compounds using Mössbauer spectroscopy in different estuarine sediments. Water Air Soil Pollut 75:293–306
Miller DP, Craig PJ (1998) The analysis of organotin compounds from the natural environment. In: Smith PJ (ed.) Chemistry of tin. Blackie Acad and Prof, London, pp 540–565
Musmeci MT, Madonia G, Lo Giudice MT, Silvestri A, Ruisi G, Barbieri R (1992) Interactions of organotins with biological systems. Appl Organometal Chem 6:127–138
Navarro JAR, Salas JM, Romero MA, Vilaplana R, Gonzalez-Vilchez F, Faure R (1998) cis-[$PtCl_2$(4,7-H-5-methyl-7-oxo[1,2,4]triazolo[1,5-a]pyrimidine)$_2$]: A sterically restrictive new cisplatin analogue. Reaction kinetics with model nucleobases, DNA interaction studies, antitumor activity, and structure-activity relationships. J Med Chem 41:332–338
Nielsen PE (1997) Peptide nucleic acid (PNA). From DNA recognition to antisense and DNA structure. Biophysical Chemistry 68:103–108
Nielsen PE, Haaima G (1997) Peptide nucleic acid (PNA). A DNA mimic with a pseudopeptide backbone. Chemical Society Reviews 73–78
Nordén B, Lincoln P, Akerman B, Tuite E (1996) DNA interactions with substitution-inert transition metal ion complexes. In: Sigel A, Sigel H (eds) Metal ions in biological systems, vol XXXIII. M Dekker, New York, pp 177–252
Nordmeier E (1995) Advances in polyelectrolyte research: Counterion binding phenomena, dynamic processes, and the helix-coil transition of DNA. Macromol Chem Phys 196:1321–1374
Patel BK, Eckstein F (1997) 5'-deoxy-5'-thioribonucleoside-5'-triphosphate. Tetrahedron Lett 38:1021–1024
Petering DH, Mao Q, Li W, De Rose E, Antholine WE (1996) Metallobleomycin-DNA interactions: structures and reactions related to bleomycin-induced DNA damage. In: Sigel A, Sigel H (eds) Metal ions in biological systems, vol XXXIII. M Dekker, New York, pp 619–648

Phillips SEV, Moras D (eds) (1995) Protein-nucleic acid interactions. Current Opinion in Structural Biology 5:1–55
Piro V, Di Simone F, Madonia G, Silvestri A, Giuliani AM, Ruisi G, Barbieri R (1992) The interaction of organotins with native DNA. Appl Organometal Chem 6:537–542
Posante S (1996) Organostagno(IV) e acido deossiribonucleico: dinamica e struttura. Thesis for master degree in Chemistry, University of Palermo (Italy)
Reich Z, Ittah Y, Weinberger S, Minsky A (1990) Chiral and structural discrimination in binding of polipeptides with condensed nucleic acid structures. J Biol Chem 265:5590–5594
Rhodes D, Burley SK (eds) (1997) Protein-nucleic acid interactions. Current Opinion in Structural Biology 7:73–134
Riccoboni L (1937) Comportamento elettrolitico di alcuni composti metallorganici dello stagno. Atti Istituto Veneto Scienze, Lettere ed Arti, XCVI:183–192
Richmond TJ, Steitz TA (eds) (1998) Protein-nucleic acid interactions. Current Opinion in Structural Biology 8:11–63
Rossi M (1994) Interazione di DNA nativo con composti organometallici dello stagno. Thesis for master degree in chemistry, University of Palermo (Italy)
Sabat M (1996) Ternary metal ion-nucleic acid base-protein complexes. In: Sigel A, Sigel H (eds) Metal ions in biological systems, vol XXXII. M Dekker, New York, pp 521–555
Sauer RT, Harrison SC (eds) (1996) Protein-nucleic acid interactions. Current Opinion in Structural Biology 6:51–100
Schmiedgen R, Huber F, Preut H, Ruisi G, Barbieri R (1994) Synthesis and characterization of diorganotin(IV) derivatives of 2-mercaptopyridine and crystal structure of diphenyl pyridine-2-thiolatochlorotin(IV). Appl Organometal Chem 8:397–407
Schmiedgen R, Huber F, Silvestri A, Ruisi G, Rossi M, Barbieri R (1998) Diorganotin(IV)-2-mercaptopyrimidine complexes. Appl Organometal Chem. 12:861–871
Schürmann M (1994) Untersuchungen zur Darstellung und Struktur von Mono- und Diarylblei(IV)-Acetaten sowie von Monoarylblei(IV) und Monoorganozinn(IV)-Derivaten. Thesis for master degree in Chemistry, University of Dortmund (Germany)
Sigman DS, Landgraf R, Perrin DM, Pearson L (1996) Nucleic acid chemistry of the cuprous complexes of 1,10-phenantroline and derivatives. In: Sigel A, Sigel H (eds) Metal ions in biological systems, vol XXXIII. M Dekker, New York, pp 485–513
Silvestri A, Duca D, Huber F (1988) A study of dimethyltin(IV)-L-cysteinate in aqueous solution. Appl Organometal Chem 2:417–425
Singh K, Groth-Vasselli B, Farnsworth PN, Rai DK (1996) Effect of thiobase incorporation into duplex DNA during the polymerization reaction. Res Commun Mol Pathol and Pharmacol 94:129–140
Smith PJ (1998) Health and safety aspects of tin chemicals. In: Smith PJ (ed) Chemistry of tin. Blackie Acad and Prof, London, pp 429–441
Tagliavini G, Cattalini L, Belluco U (1962) Ricerche sui composti metallorganici dello stagno. Reazione tra stagno tetrametile, stagno tetraetile e nitrato mercuroso. La Ricerca Scientifica, 32(IIA):286–290
Thayer JS (1984) Organometallic compounds and living organisms. Acad Press, NewYork
Tuite E, Lincoln P, Nordén B (1997) Photophysical evidence that Δ- and Λ-$[Ru(phen)_2dppz)]^{2+}$ intercalate DNA from the minor groove. J Am Chem Soc 119:239–240
Wang H, Osborne SE, Zuiderweg ERP, Glick GD (1994) Three-dimensional structure of a disulfide-stabilized non-ground-state DNA hairpin. J Amer Chem Soc 116:5021–5022
Whalen D, Lucero R, May L, Eng G (1993) The effects of salinity and pH on the speciation of some triphenyltin compounds in estuarine sediments using Mössbauer spectroscopy. Appl Organometal Chem, 7:219–222
Wu JZ, Li L, Zeng TX, Ji LN, Zhou JY, Luo T, Li RH (1997) Synthesis, characterization and luminescent DNA-binding study of a series of ruthenium complexes containing 2-arylimidazo[f]1,10-phenanthroline. Polyhedron 16:103–107
Yam VWW, Lo KKW, Cheung KK, Kong RYC (1997) Deoxyribonucleic acid binding and photocleavage studies of rhenium(I) dipyridophenazine complexes. J Chem Soc Dalton Trans 2067–2072
Yamauchi O, Odani A, Masuda H, Sigel H (1996) Stacking interactions involving nucleotides and metal ion complexes. In: Sigel A, Sigel H (eds) Metal ions in biological systems, vol XXXII. M Dekker, New York, pp 207–270

Chapter 13

Mercury in Marine Environments

M.E. Farago

13.1 Introduction

Concentrations of elements vary depending on the compartment of the environment that is under consideration. Thus, although the inputs of elements to the oceans are very large in absolute terms, the dilution is extremely high, leading to very low concentrations. This situation is generalised in Fig. 13.1. The consequences of low concentrations in the open oceans are that the possibilities for deficiency are high whereas those for toxicity are low (Mcgrath 1997). In rivers, wetlands and closed seas, where the dilution is lower, there can be accumulation of potentially toxic elements. Because of the low concentrations in open oceans, analytical methods for the determination of an element and its chemical species must have very low detection limits, and it is only in recent years that such methods have emerged.

13.2 Mercury in the Environment

13.2.1 Chemical and Physical Properties

Elemental mercury Hg^0 is a heavy silvery liquid at ambient temperatures, with a very high vapour pressure. There are three oxidation states of mercury: Hg^0, Hg_2^{2+} (mercurous) and Hg^{2+} (Hg(II), mercuric). There are numerous inorganic and organic compounds, of which methyl-mercury species are important environmentally. The most important biochemical property of Hg^{2+} and alkylmercurials is their affinity for sulfhydryl groups (WHO/IPCS 1991).

Lindqvist et al. (1984) have suggested the following classification of environmentally important mercury species:

- *Volatile species*: Hg^0, $(CH_3)_2Hg$ (dimethyl mercury, DMHg)
- *Water soluble particle-borne reactive species*: Hg^{2+}; HgX_2, HgX_3^-, HgX_4^{2-} (where X = OH^-, Cl^- or Br^-); HgO on aerosol particles; Hg^{2+} complexes with organic acids
- *Non-reactive species*: CH_3Hg^+, CH_3HgCl, CH_3OH (monomethyl-mercury species, MMHg), and other organomercury compounds; $Hg(CN)_2$; HgS and Hg^{2+} bound to sulfur in fragments of humic matter

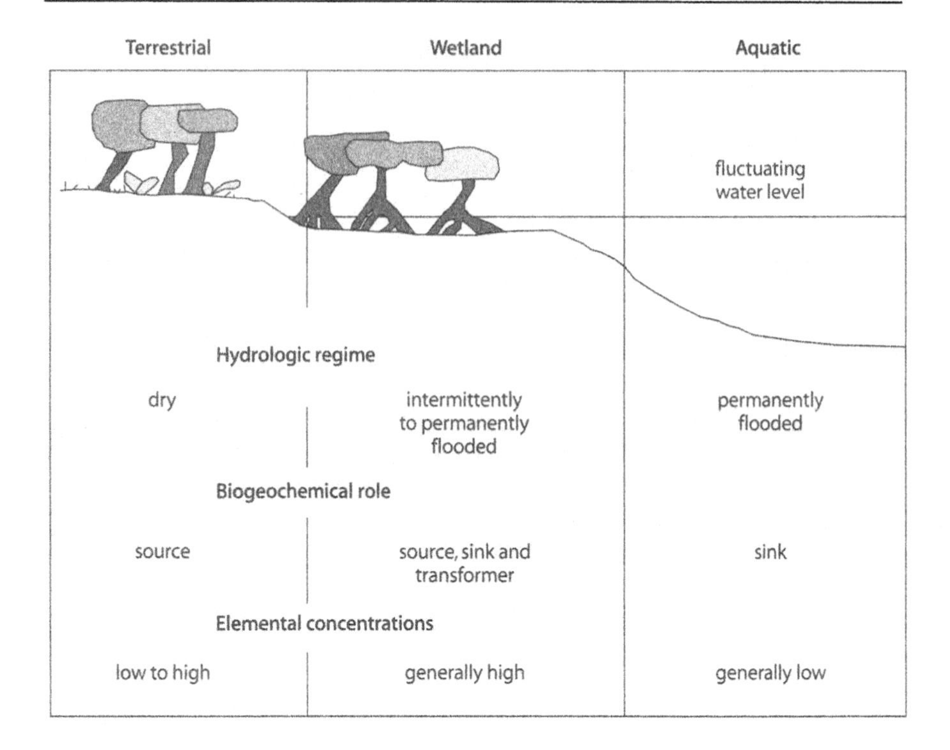

Fig. 13.1. Generalized diagram showing the hydrology and biogeochemical link between the terrestrial, wetland and aqueous systems (after Mitsch and Gosslink 1986 and McGrath 1997)

13.2.2 Mercury Emissions

The main natural sources of mercury in the environment are degassing of the earth's crust, emissions from volcanoes and vegetation, evaporation from hydrosphere and biomass burning. It has been pointed out by Rasmussen et al. (1998), that the estimates of the total natural emissions of mercury vary considerably. Emissions from natural sources are difficult to distinguish from secondary emissions and diffusive re-emissions from anthropogenic sources. Emissions from natural sources have been estimated to be between 2 700 and 6 000 tonnes yr^{-1} (Lindberg et al. 1987), 6 000 tonnes yr^{-1} (Nriagu and Pacyna 1988), and in the order of 4 000 tonnes yr^{-1} by Fitzgerald (1989).

Mercury is produced from its sulfide ore, cinnabar. Anthropogenic sources of mercury include inputs from chemical and electrical industries, fossil fuel burning and other combustion processes and mercury mining. The major uses of mercury are in the chlor-alkali industry, in batteries, in agricultural pesticides and fungicides, and in manometers, barometers and thermometers, although recently, solid state instruments are replacing this last use. Mercury also has an important use in dental amalgams. Because of the toxicity of mercury all its uses have come under increasing scrutiny (Herbert et al. 1980).

Global anthropogenic inputs into the environment have been estimated to be 2 000–30 000 tonnes yr^{-1} (Lindberg et al. 1987; Pacyna 1987) and by Nriagu and Pacyna

(1988) to be in the range 900–62 000 tonnes for the year 1983 with a mean value of 3 600 tonnes. Other estimates have been 2 000 tonnes yr^{-1} (Bernard 1997) or alternatively to be about 25% of the total input to the atmosphere (WHO/IPCS 1990). Annual emissions for Europe have been given as 726 tonnes for 1987 (Pacyna 1996) and 463 tonnes for 1990 (Berdowski et al. 1998). It has been pointed out (Jernelöv and Ramel 1994) that estimates of the geogenic and anthropogenic inputs of mercury were based on analyses of mercury in open-ocean water carried out in the 1970s, which were reported to be in the range 10–100 ng l^{-1}. These findings suggested that natural sources were the major contributors to mercury in sea water. However, recent estimations of total mercury concentrations in uncontaminated open oceans are in the range 0.1–1 ng l^{-1}. These revised values lead to the conclusion that the residence time for mercury in open-ocean surface waters is only decades, and therefore mercury fallout is important for mercury levels in water and biota. It was concluded that anthropogenic sources account for about 50% of total atmospheric mercury.

Emissions of mercury from coal combustion processes range between 20–50% elemental mercury, Hg^0, and 50–80% of Hg(II), which is predominantly $HgCl_2$, while those from waste incinerators contain 75–85% Hg(II) (Carpi 1997).

13.2.3
The Global Mercury Cycle

Hg(II) is particulate and water soluble and is removed from the atmosphere by dry and wet deposition close to its source. Mercury, as Hg^0, in the atmosphere, because of its volatility and low water solubility, follows a cycle characterized by long-range transport, and can thus be considered on the global scale. It can be deposited in remote locations, and the concentration of mercury has been found to be greater in the recently deposited layers of the Greenland ice cap than in older layers (Weiss et al. 1971), and has thus been increasing globally (Slemr and Langer 1992; Mason et al. 1994). Part of the cycle is illustrated in Fig. 13.2, where elemental mercury (Hg^0), emitted from both natural and anthropogenic sources, is removed from the atmosphere by wet deposition after oxidation to water soluble Hg(II), to land and water surfaces after which mercury can be reduced and returned to the atmosphere. Some Hg^0 may be eventually removed from the atmosphere by dry deposition (Carpi 1997).

Concentrations of mercury in various environmental compartments were estimated by Lindqvist et al. (1984). Representative values for total dissolved mercury, Hg_T are: open ocean, 0.5–3 ng l^{-1}; coastal sea water, 2–15 ng l^{-1}; rainwater, 2–25 ng l^{-1} (WHO/IPCS 1990; Lindqvist et al. 1984). Recently, Leermakers et al. (1997), reported that the concentrations of total gaseous mercury, (*TGM*), over the North Sea were between 0.7 and 2.6 ng m^{-3}, and that the range in rainwater was between 5 and 25 ng l^{-1}; these results are of the same order as those estimated by Lindqvist et al. (1984). More recent estimations of Hg_T in uncontaminated open oceans are in the range 0.1–1 ng l^{-1} (Jernelöv and Ramel 1994; Mason et al. 1995)

Mercury may be transformed by aquatic microorganisms into methyl mercury. This is an important part of the biogeochemical cycle in the consideration of human exposure, since methyl mercury has high lipophilicity and accumulates in the food chain (D'Itri 1991; Renzoni et al. 1998). Biological methylation by aquatic organisms was reported by Jensen and Jernelöv in 1969, when they demonstrated that lake sediments,

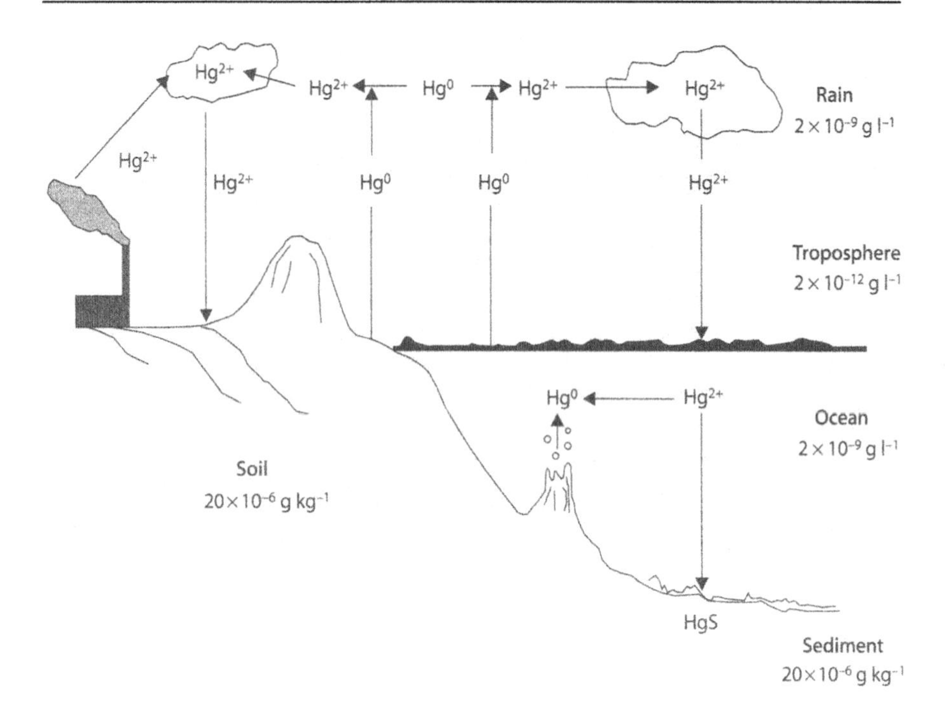

Fig. 13.2. Part of the global cycle of mercury (adapted from WHO/IPCS 1990, concentrations from Lindqvist et al. 1984)

when incubated with $HgCl_2$, produced methyl mercury. In is now known that such aquatic microbial reactions are responsible also for demethylation of mercury (Holm and Cox 1975) and for the methylation of a variety of other metals (Craig 1980). The methylation reaction occurs in a number of bacteria including methanogens (Wood et al. 1968; Schlesinger 1997) and sulfate reducing bacteria under anoxic conditions (Compeau and Bartha 1985). The reaction usually involves the methylated form of vitamin B_{12}, methylcobalamine (Wood et al. 1968), which transfers a carbanion, CH_3^-, to mercury(II), Hg^{2+} (Ridley et al. 1977). Microbial activity is also responsible for the oxidation of elemental mercury (Holm and Cox 1975) and for the reduction of inorganic mercury compounds (Summers and Silver 1978).

13.2.3.1
Factors Affecting Methylation

The methylation, both biotically and abiotically, can occur under a wide variety of conditions (Beijer and Jernelöv 1979; Gilmore and Henry 1991), but in the aquatic environment, biomethylation is the predominant path. Biomethylation has been reported to occur rapidly under anoxic conditions (Miskimmin et al. 1992; Regnell 1994). When Hg(II) and methylcobalamine react under mildly reducing conditions, the main product is dimethyl mercury, DMHg (Wood et al. 1968). When Hg(II) is present in excess, however, MMHg is formed (Craig 1986). The production of DMHg was demonstrated in the presence of dead fish tissue (Jensen and Jernelöv 1969). The methyl-mercury ion CH_3Hg^+ forms a very stable complex CH_3HgCl. Ahrland (1985) suggested that because of the very high concentrations of Cl^- in sea water, CH_3HgCl should be more abundant than CH_3HgCH_3 (DMHg).

The pH has a major effect: in alkaline anoxic sediments, DMHg is produced, whereas at lower pH values, MMHg is produced (Fagerstrom and Jenerlöv 1972). At higher pH ranges (>7) the dominant species of mercury in sediments are those bound to sulfides or oxides of iron and manganese, whereas in more acidic conditions mercury is associated with humic acids or sulfides. Each form of mercury has a different potential for methylation. Hg(II) is bound to sulfur in sulfate-rich anoxic marine sediments, and becomes less available for methylation (Capone and Kiene 1988), whereas Hg-humic complexes are more readily methylated than is HgS (Kannan and Falandysz 1998). Thus, in fresh-water sediments, high amounts of humic material favour methylation, and concentrations of MMHg are generally higher in fresh water than in marine sediments. There is an inverse correlation between methylation and sulfate levels which is strange, since the principal methylators are sulfate reducing bacteria (Compeau and Bartha 1985). This apparent anomaly was explained by Choi and Bartha (1994), who pointed out that the H_2S, formed in reducing sulfate rich environments, precipitates the Hg(II), thus rendering it unavailable for methylation.

Topping and Davies (1981) demonstrated the production of methyl mercury in coastal marine surface waters, thus sediment is not the exclusive natural source of methyl mercury entering the food chain.

13.3 Mercury in Coastal Marine Sediments

Kannan and Falandysz (1998) measured total mercury, (Hg_T), methyl mercury and Hg(II) in coastal marine sediments collected from the Baltic, South China and Bering Seas (Table 13.1). They reported that in most sediments only 10% of the total mercury (Hg_T) was accounted for by the sum of Hg(II) and methyl mercury. They concluded that most of the mercury was strongly bound to sulfide and thus unavailable for methylation. Since there are generally more humic materials in fresh-water sediments, and methylation has been positively correlated with humic content and negatively correlated with salinity (Choi and Bartha 1994; Compeau and Bartha 1983), a larger percentage of Hg_T was found to be in the methylated form in the fresh-water sediments (Table 13.1).

The percentage of the total mercury present in some marine sediments, in the methylated form, is shown in Table 13.2. In unpolluted locations the percentage is usually <1, and some authors have suggested a maximum value of 1%. The highest values were found in the Elbe Estuary and in the Mulde, a tributary of the Elbe (Wilken and Hintlemann 1991). The Elbe Estuary includes Hamburg Harbour, and is one of the most polluted locations in the world. The mercury pollution, resulting from chlor-alkaline plants, has led not only to high percentages of methyl mercury, but also to high total Hg in sediments and particulate matter (Wilken and Hintlemann 1991). These authors suggest that the very high percentage of methylated mercury might be accounted for by the presence, in the polluted Elbe environment, of specially adapted bacteria, which can methylate Hg^{2+} very efficiently. The results of Kannan and Falandysz (1998) indicate that the percentage of methyl mercury does not depend on climatic zone (Table 13.1).

Table 13.1. Total mercury ($ng\,g^{-1}$ dw) and % methyl mercury in coastal marine sediments (Kannan and Falandysz 1998

Location	Total Hg	Methylated (%)
Poland marine, Baltic sea	164 ±250	0.66 ±0.74
Poland fresh water	21.1 ±13.8	1.52 ±0.38
Malaysia, China Sea	61 ±47	0.11 ±0.12
Russia, Bering Sea	3 339 ±711	0.22 ±0.22

Table 13.2. Percentages of total mercury as methyl mercury in marine sediments

Location	Methylated Hg (%)	Reference
Baltic Sea, Sweden	0.1 – 3.52	Jernelöv et al. 1975
Baltic Sea, Poland	0.02 – 2.27	Kannan and Falandysz 1998
Irish Sea , UK	<0.01 – 1.35	Bartlett et al. 1987
Elbe Estuary, Germany	2 – 8	Wilken and Hintlemann 1991
Mulde (Elbe tributary), Germany	10	Wilken and Hintlemann 1991
Scheldt Estuary, Belgium	0.7	Muhaya et al. 1997
Adriatic Sea, Croatia	0.43 – 0.54	Mikac et al. 1989
South China Sea, Malaysia	0.02 – 0.27	Kannan and Falandysz 1998
Bering Sea, Russia	0.02 – 0.7	Kannan and Falandysz 1998
San Francisco Bay, USA	0.03 – 1.0	Olson and Cooper 1974
Everglades, USA	0.03 – 0.07	Andren and Harris 1973
Patuxent River Estuary, USA	0.1 – 0.5	Benoit et al. 1998

The authors suggest that many factors need to be taken into account, including oxygen, temperature, pH, organic matter and sulfate, before a complete understanding of the biogeochemical cycle of mercury can be understood.

Concentrations of up to 10 000 $ng\,g^{-1}$ of mercury were found in the very polluted Saguenay Fjord in Canada (Gagon et al. 1997). The mercury, which originated from a chlor-alkali plant that closed in 1976, was found to be bound to organic matter and with Mn/Fe oxides. There was also binding to anomalously abundant acid volatile sulfides. Remobilisation of Hg from deep layers was slow. The reservoirs and fluxes of mercury in the Saguenay Fjord are modelled in Fig. 13.3.

In the much less polluted Patuxent Estuary in the USA (Table 13.2), Benoit et al. (1998), found that Hg_T content of the sediments appeared to be controlled by organic matter. While methyl mercury in the sediments was positively correlated with Hg_T and organic matter, it was negatively correlated with sulfide. They concluded that in this system sulfide limits the production and accumulation of methyl mercury. Similarly, in the Scheldt Estuary in Belgium (Muhaya et al. 1997), both Hg_T and methyl mercury increased with increase of organic matter in the sediments. Whereas in the sediments (Table 13.2) methyl mercury was 0.7% of Hg_T, in the polychaete worm, *Neries diversicolor*, methyl mercury was 18% of Hg_T.

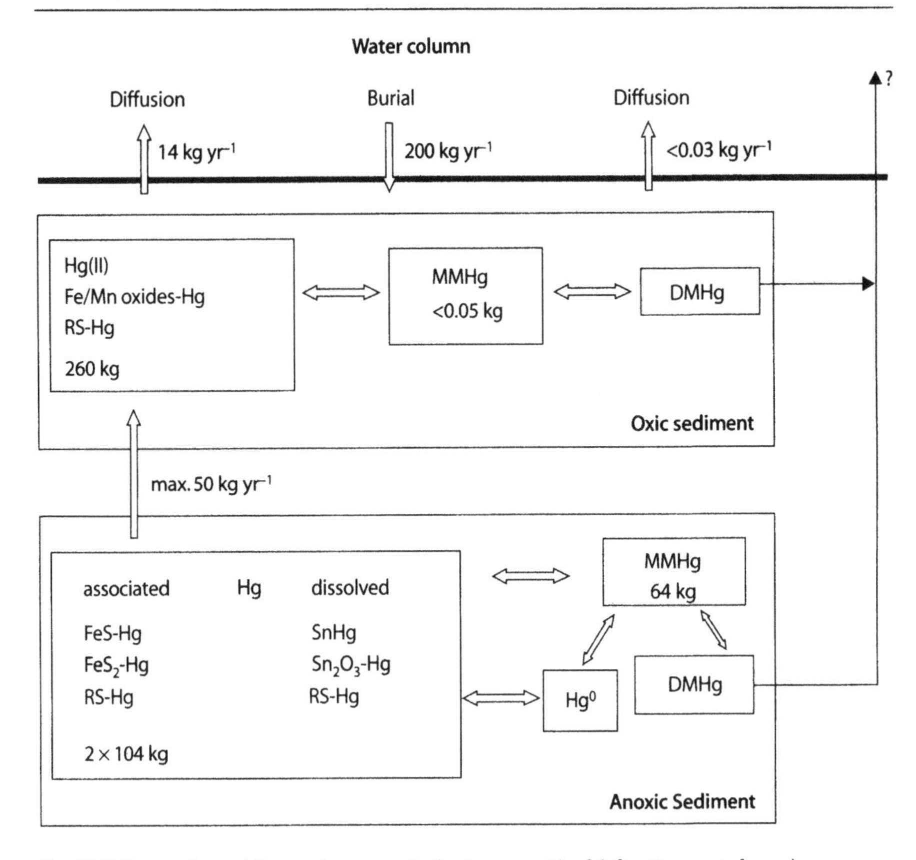

Fig. 13.3. Reservoirs and fluxes of mercury in the Saguenay Fjord (after Gagon et al. 1997)

13.4 Mercury in Waters

Values for concentrations in waters are reported in the literature in a variety of units, these have been normalised to ng l^{-1}.

13.4.1 Baltic and North Seas

It has been estimated, from long term pollution measurements, that 50% of the mercury that enters North Sea and Baltic waters is from atmospheric deposition, with 80% of the deposition in the vapour (metallic, Hg^0) form (Pacyna 1996).

In a survey of mercury in Baltic and North Sea waters Schmidt (1992) reported that Hg_T concentrations in surface waters from the Baltic Sea were in the range 1.4–3.6 ng l^{-1}. This contrasted with values from the German Bight where the distribution showed most values between 5 and 25 ng l^{-1}, with some values as high as 100 ng l^{-1}. This location showed particularly high values in the Elbe Estuary (Section 13.3), where the concentration in unfiltered waters rose to more than 1 000 ng l^{-1} in the extremely polluted

Port of Hamburg. Schmidt also reported values of mercury concentrations, in samples from 10 m depth, carried out in two surveys of the entire North Sea in May/June 1986 and January/March 1987. The first survey showed values of around 1 ng l^{-1} in the open regions, with higher values in the German Bight (due to the Elbe) and also in some areas east of Scotland and in the Skagerrak, where values were between 1 and 10 ng l^{-1}. The second winter survey resulted in a different distribution of Hg concentrations, where most areas were below the detection limit of 0.5 ng l^{-1}, the bight was again around 10 ng l^{-1}, but 10 stations east of Scotland showed values in excess of 200 ng l^{-1}. The author speculated that there might be some connection between these high levels and gas production.

The concentrations of elemental mercury Hg^0 in surface waters from the Scheldt Estuary and the North Sea have been reported (Baeyens and Leermarkers 1998). As with total mercury, a seasonal variation was noticed with concentrations of Hg^0 reported to be 0.02–0.76 ng l^{-1} in winter and 0.48–0.13 ng l^{-1} in summer. In the North Sea, Hg^0 concentrations ranged from 0.012 to 0.16 ng l^{-1} with the concentrations being higher in coastal stations. Positive correlation with phytoplankton concentration was observed, and the possibility of the involvement of phytoplankton in the production of Hg^0 was suggested. The authors stressed the need for studies on photocatalytic reduction of mercury species and on bacterial demethylation.

13.4.2 Mediterranean Sea

It was pointed out some years ago that many marine organisms living in the Mediterranean basin have higher mercury body burdens than do the same species living in the Atlantic or other Oceans (Stoeppler et al. 1979; Bernhard 1985, 1988; Renzoni et al. 1998). Using the measured concentrations of mercury in sardines from both the Mediterranean and Atlantic Ocean, Bernhard (1985) calculated the mercury levels in Mediterranean and Atlantic plankton, and then assuming that the mercury concentration in plankton is 5 000 times that of sea water, he concluded that the Mediterranean should contain 5 ng l^{-1} and the Atlantic 1 ng l^{-1} of mercury. Values for the mercury concentration in the Mediterranean have reported to be in the range 0.5–1.5 ng l^{-1} (May and Stoeppler 1983; Kniewald et al. 1987; Dorten et al. 1991). The high levels of mercury in Mediterranean biota have been discussed for many years. It has been pointed out (Bernhard 1985; Renzoni et al. 1998) that about 55% of the world's mercury resources are in the Mediterranean basin (an area about 1% of the total area of the planet), and that the turnover of water through the straits of Gibraltar is low, thus Mediterranean biota have high levels of mercury from natural sources. Riverine inputs of Hg to the Mediterranean Sea, from the Rhone, Ebro, Arno, Krka, Nile and Tiber have been estimated to be relatively small and lower than previously suggested (Dorten et al. 1991).

13.4.3 The Pacific and Atlantic Oceans

The measurement of chemical species of mercury at low concentrations in the open oceans has been made possible by recent improvements in analytical techniques (Gill and Fitzgerald 1985; Bloom and Fitzgerald 1988). Mason and Fitzgerald (1990, 1991, 1993, 1994) reported the presence of both MMHg and DMHg in the low oxygen waters of the equatorial Pacific Ocean and thus demonstrated that a pathway exists for the accumulation of methylated mercury in marine pelagic fish. Results from sampling and analysis during a cruise in January/February 1990 between the Panama Canal and American Samoa lead to a number of conclusions:

- The substrate for methylation is *labile inorganic mercury*, which is composed of labile inorganic and organic complexes of Hg(II) and of labile particulate associations.
- Labile inorganic mercury is supplied primarily by atmospheric deposition; it then reaches the subthermocline waters by vertical mixing and dissolution of particles.
- DMHg and MMHg are produced in the O_2 minimum regions and the concentrations of methylated species decrease as the minimum oxygen concentration rises.
- Biological methylation in anoxic environments is the dominant methylating process.
- DMHg is the dominant methylated mercury compound, whereas MMHg dominates in fresh waters.
- Demethylation is the principal source of Hg^0 in low oxygen waters and direct reduction of Hg(II) is the main source of Hg^0 in the mixed layer.
- The main processes consuming Hg(II) in natural waters are methylation, reduction and particulate scavenging.

From these conclusions Mason and Fitzgerald (1990, 1991, 1993, 1994) proposed a model showing the principal cycles and processes that affect the distributions of various mercury species in the equatorial Pacific Ocean (Fig. 13.4)

Mason et al. (1995, 1998) investigated species of mercury in the deep ocean waters of the North Atlantic Ocean. Their investigations confirmed the processes that had been found in the equatorial Pacific Ocean. The results also suggested the presence of organic complexes of mercury in the surface water. Some concentrations of the mercury species are shown in Table 13.3. The speciation of mercury in the open ocean is dominated by Hg^0 and interconversion between the species accounts for the complex profiles. It was suggested (Mason et al. 1998) that Hg^0 formation is constrained more by rate of supply than by rate of conversion. There is a general increase in the concentrations of reactive mercury, and of DMHg in the colder, deeper waters of the North Atlantic, suggesting that the latter compound is relatively stable under these conditions. Winter mixing releases DMHg to the atmosphere. There is also evidence of MMHg throughout the water column.

13.5 Toxicity of Mercury to Marine Life

Mance (1987) has collected data on toxicity studies on marine biota. Exposure tests did not reveal any differences between mercury chemical species: all compounds tested (organic or inorganic Hg(II)) had adverse effects at "low" concentrations. For the embryos of the fish *Fundulus heteroclitus*, the 4-day LC_{50} was 0.067 mg l^{-1}. However, exposure for one day followed by 3 days in clean water gave and LC_{50} of 0.09 mg l^{-1}, indicating that the effects were rapid and irreversible (Sharp and Neff 1980). Crustaceans and mollusks (Martin et al. 1981; McClurg 1984) were also found to be sensitive to Hg(II).

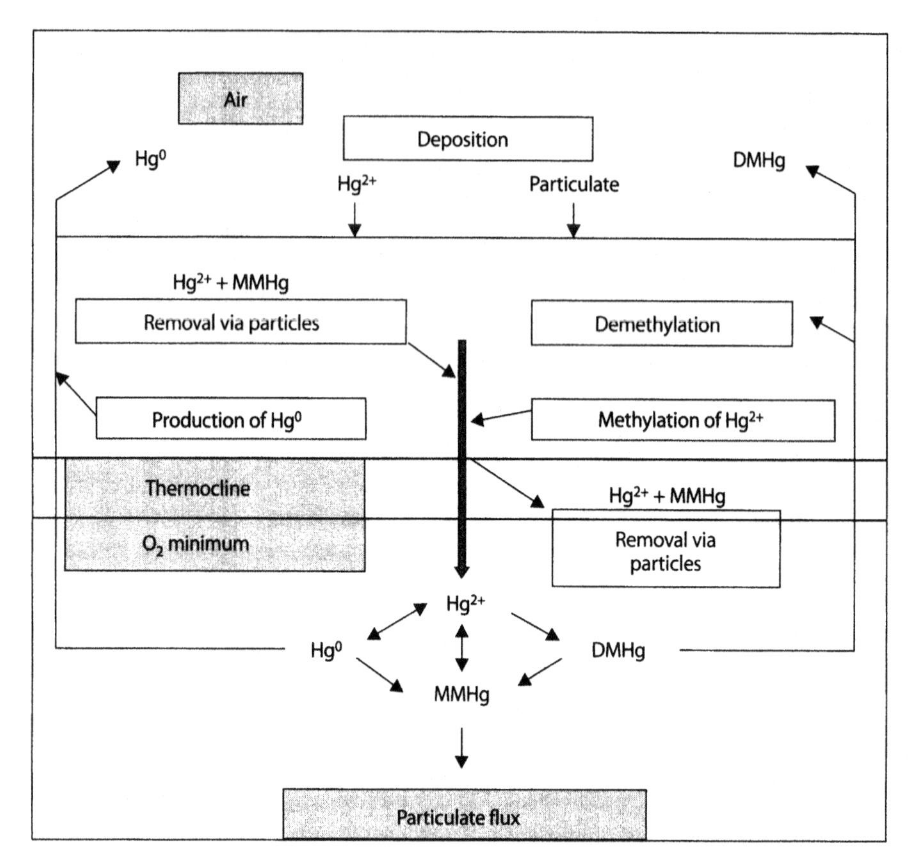

Fig. 13.4. Principal cycles and processes that affect the distributions of various mercury species in the equatorial Pacific Ocean (from Mason and Fitzgerald 1991)

Table 13.3. Mean concentrations of mercury species found in different water masses on deep ocean waters of the North Atlantic (56.6 °N, 25.5 °W); data from Mason et al. 1995

		Mixed layer	Thermocline	Intermed. water	Deep water
Hg_T	(pM)	1.55	4.25	3.15	1.7
	($ng\ l^{-1}$)	0.31	0.85	0.66	0.34
Hg^0	(pM)	0.89	0.07	0.3	0.56
	($ng\ l^{-1}$)	0.18	0.014	0.06	0.11
Reactive Hg	(pM)	0.95	0.94	0.75	1.03
	($ng\ l^{-1}$)	0.19	0.19	0.15	0.21
DMHg	(pM)	<0.01	0.05	0.15	0.2
	($ng\ l^{-1}$)	–	0.01	0.03	0.04

13.6 Mercury in the Marine Food Chain

13.6.1 Bioaccumulation

The production of methyl mercury in aquatic systems is the first stage in the aquatic bioaccumulation process (WHO/IPCS 1990). Bioaccumulation is distinct from biomagnification (Mance 1987). Bioaccumulation involves the uptake from water independently by each trophic level and species, whereas biomagnification involves the transfer from one trophic level to another by ingestion. Förstner and Wittmann (1979) concluded that because of its high affinity for organic substances, mercury (together with arsenic) was likely to show biomagnification of the food chain. There is, however, conflicting evidence in the literature. Some studies have reported an increase along the food chain (e.g. Knauer and Martin 1972), whereas others have shown no consistent increase (e.g. Leatherland and Burton 1975). It has been shown, however, that Hg_T concentrations and the percentage of methyl mercury increase in older organisms and those at the top of the food chain (Bernhard 1985, 1988). It was suggested, from models, that this is brought about because the inorganic mercury reaches a constant concentration early in the lifespan, while that for methyl mercury increases with the size of the specimen and the trophic level of the species.

Minganti et al. (1996) studied the accumulation of mercury along the web of seafood species in the Ligurian Sea. They investigated five crustaceans, whose trophic positions in the food web from herbivore/omnivore in the second level to species in high trophic levels. For most species Hg_T was positively correlated with size, and this increase in total mercury was brought about by an increase in methyl mercury thus validating the earlier model. Hg_T also increased with trophic level.

More recently, Lawson and Mason (1998) demonstrated that different ligands influenced the uptake of inorganic mercury/methyl mercury by phytoplankton (*Thalassiosira weissflogii*) which are at the base of the estuarine food chain. Copepods and amphipods feeding on the algal cells assimilated methyl mercury more efficiently than inorganic mercury. Finally, fish feeding on the copepods assimilated methyl mercury more efficiently than inorganic mercury because of the larger fraction of methyl mercury found in the tissues of the copepods.

13.6.2 Human Exposure to Methyl Mercury

The major source of mercury in the human diet is from the consumption of fish. Most edible fish contain low amounts of methyl mercury in their tissues (up to 200 ng g^{-1}) (WHO/PCS 1990). Predatory species (ocean tuna, shark, swordfish and fresh-water species such as pike and walleye) may contain methyl mercury in excess of 1 000 ng g^{-1}. Mediterranean tuna are known to reach 4 500 ng g^{-1} (Bernhard 1985). Poisonings in Japan by the consumption of fish containing methyl mercury demonstrated the neurological damage and teratogenic effects of mercury on human populations. Damage is now known to be mainly psychomotor disturbances and mental retardation (Skerfving 1991). Chromosome damage has also been reported (Skerfving 1970).

Communities who eat large amounts of contaminated seafood are at risk from neurological damage and harm to the unborn foetus. The effects of methyl mercury on humans and dose response relationships have been discussed (WHO/IPCS 1990). The biomarkers of exposure are blood and hair for methyl mercury and urine for inorganic mercury. Mean reference values are (WHO/IPCS 1990): blood 8 $\mu g\ l^{-1}$; hair, 2 $\mu g\ g^{-1}$; urine 4 $\mu g\ l^{-1}$. Although hair is the preferred indicator, the concentrations in hair can be affected by a number of factors (Suzuki 1988). Until recently Hg_T concentrations in the hair of pregnant women up to 10–20 $\mu g\ g^{-1}$ have been considered to be safe (WHO/IPCS 1990), however recent data suggests that this value may be too high (Grandjean et al. 1997, 1998).

Methyl mercury is well absorbed and the biological half life in fish has been reported to be in the range 300–1 000 days, and that for inorganic mercury of about 100 days (Bernhard 1985). In humans mercury can reach high levels (Berlin 1986). Ingested methyl mercury is absorbed into the blood stream (Fig. 13.5) and distributed to the tissues. Because methyl mercury equilibrates among the tissues rapidly, the ratio of that in the blood to that in the whole body is essentially constant and approximates to the average concentration in the body (Farris and Smith 1997). The most important metabolite of methyl mercury is inorganic mercury, which accumulates within the body, and at steady state (for example during chronic exposure), where it represents a significant proportion of the total mercury within the body. Farris and Smith point out that earlier workers did not consider the metabolite and reported an erroneously long biological half-life for methyl mercury. They propose that for an accurate description of the pharmacokinetics of methyl mercury, the model must contain at least two compartments. Faecal mercury derives primarily from the methylmercury compartment, whereas urinary mercury derives from the inorganic mercury compartment.

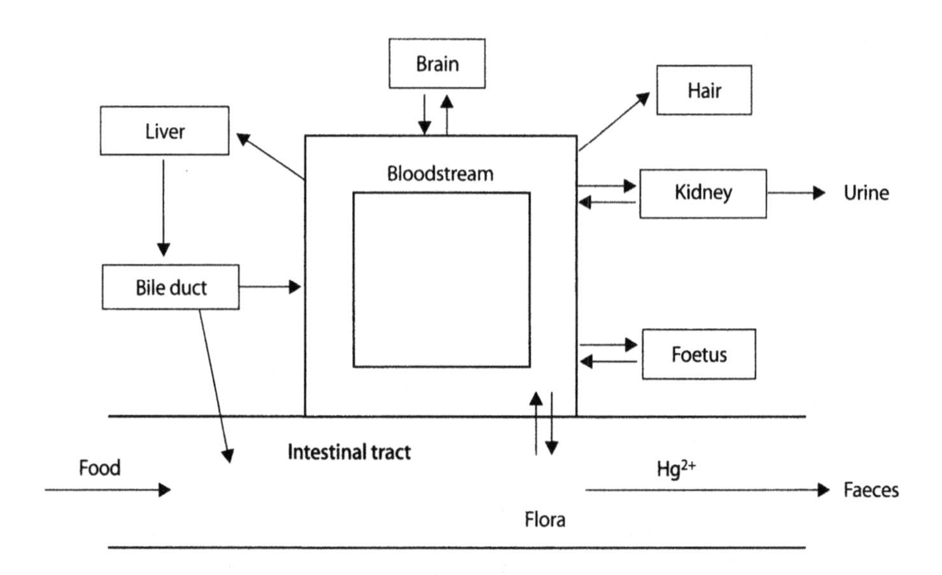

Fig. 13.5. The distribution of methyl mercury in human tissues (from Clarkson 1994)

13.6.3 Human Exposure to Mercury in some Fish Eating Communities

The best known of the poisoning incidences was at Minamata Bay on the south-west coast of Kyushu (Takeuchi 1972; Smith and Smith 1973; D'Itri 1991), where residents showing symptoms relating to the peripheral and central nervous systems were noted in the 1950s. By 1956 it was established that the disease was caused by the consumption of contaminated shell fish and fish from Minamata Bay. The source of the contamination was a plastics factory using mercuric chloride as catalyst. Evaluation of the incident makes it appear most likely that the contamination was not only inorganic mercury but also methyl or other alkylated mercury compounds.

In a number of marine fish eating coastal and island communities, studies have been carried out to assess levels in seafood, human exposure and consequences for health. These include Greenland (Hansen et al. 1976), UK (Haxton et al. 1979) Iceland (Jòhannesson et al. 1981), Seychelles (Matthews 1982), Singapore (Foo et al. 1988), Maldives (Renzoni 1989), Canada (WHO/IPCS 1990), the Mediterranean (Renzoni 1992; Renzoni et al. 1998; Franchi et al. 1994), Minamata (Harada et al. 1998), the Faroe Islands (Weihe et al.1997; Grandjean et al. 1997, 1998) and Madeira (Renzoni 1992; Renzoni et al. 1998; Gaggi et al. 1996). Some of the more recent case studies are discussed briefly below.

13.6.3.1 *The Mediterranean*

As discussed above (Section 13.4.2), many marine organisms living in the Mediterranean Sea have higher mercury body burdens than do the same species living in the Atlantic or other Oceans (Stoeppler et al. 1979; Bernhard 1985, 1988). Species from the North Tyrrhenian Sea show particularly enhanced mercury in most species. The correlation between body weight and total mercury concentrations in the tissues (wet weight) of the Norwegian lobster from this area, is shown in Fig. 13.6. Positive correlations between mercury levels in benthic crustaceans and mollusks and those in sediments have been demonstrated for the eastern Mediterranean (Hornung et al. 1984). Because of the high levels of mercury in fish and seafood tissue, there are risks of high mercury intake, since many residents in the coastal areas consume, on average, more than 4 seafood meals per week. An investigation of mercury in hair from residents of fishing villages along the Tyrrhenian Sea (Table 13.4) demonstrated a positive correlation with the number of meals consumed (Renzoni 1992; Renzoni et al. 1998), in addition, a few values over 50 $\mu g\ g^{-1}$ were found. These individuals were investigated further and positive correlations were found between mercury in blood and chromosomal aberrations (Franchi et al. 1994).

13.6.3.2 *Present Day Exposure in the Minamata Area*

A study of present day inhabitants (ages 32–82 years) of the Minamata area (Harada et al. 1998) showed that of the 185 subjects only 6 had total mercury in scalp hair >10 $\mu g\ g^{-1}$. There was an upward trend correlated with the intake of seafood (Table 13.5).

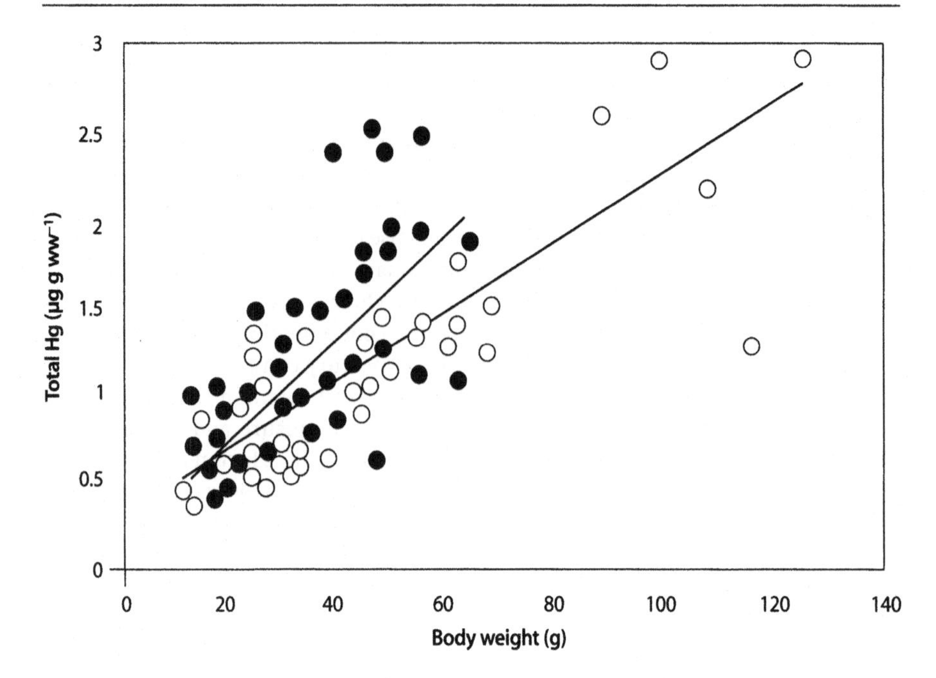

Fig. 13.6. Total mercury in muscle tissue of the Norwegian lobster, *Nephros norvegicus*, from the Tyrrhenian Sea (from Renzoni et al. 1998). *Open circles*, males; *closed circles*, females

Table 13.4. Total mercury concentrations ($\mu g\ g^{-1}$) in scalp hair samples from two fishing villages along the Tyrrhenian coast (from Renzoni 1992, 1998)

Number of seafood meals	Mean ±SD	*n*
One or fewer per month	1.03 ±0.21	40
Two to four per week	8.89 ±5.74	88
More than four per week	36.36 ±21.62	39

Table 13.5. Total mercury concentrations ($\mu g\ g^{-1}$) in scalp hair samples from present day residents of the Minamata area (Harada et al. 1998)

Seafood meals eaten	Mean ±SD	*n*
On less than 5 days per week	2.0 ±1	14
On 5 days per week	2.6 ±1.2	39
On 6 days per week	3.5 ±2.6	17
Every day	4.0 ±3.5	105

Some of the inhabitants had eaten the contaminated fish in the original incident, and it appeared that, in spite of the low levels of total mercury in the hair samples, the subjects showed various neurological symptoms, and the authors concluded that there are many people with slight atypical Minamata disease in the area in addition to those officially recognised as Minamata patients.

13.6.3.3
The Faroe Islands

In addition to seafood, which constitutes a large proportion of the diet of inhabitants of the Faroe Islands, the meat and blubber from the pilot whale are also consumed. The concentration of mercury in the muscle of the Faroese pilot whale is about 3.3 $\mu g\ g^{-1}$, half of which is methyl mercury (Weihe et al. 1997). Grandjean et al. (1997, 1998) have studied a cohort of children borne in the Faroe Islands, and prenatal exposure was assessed from the concentration of mercury in the maternal hair. At the age of seven a detailed neurobehavioural examination was carried out. A case group of 112 children, where the prenatal maternal hair concentrations were in the range 10–20 $\mu g\ g^{-1}$ (considered to be "safe"), were matched with a control group of children where the concentration was <3 $\mu g\ g^{-1}$. The case group showed mild decrements, relative to controls, in motor function, language and memory. Thus, effects are being noticed at levels of prenatal exposure previously considered to be safe. Effects are not severe clinical symptoms as observed in Minamata disease, but resemble the effects of early lead exposure.

13.6.3.4
The Island of Madeira

Volcanic sources enhance mercury in the waters off the Atlantic Island of Madeira. High mercury levels in hair were found in coastal fisherman and their families (Renzoni 1992; Renzoni et al. 1998; Gaggi et al. 1996). In the village of Camara de Lobos, mean mercury concentrations in scalp hair were for males: 38.9 $\mu g\ g^{-1}$, maximum 114.3 $\mu g\ g^{-1}$ ($n = 66$); and for females 16.2 $\mu g\ g^{-1}$, maximum 56.1 $\mu g\ g^{-1}$ ($n = 22$). A more detailed study was carried out on women visiting the health centre for pregnancy tests (Table 13.6 and 13.7). Concentrations of mercury in hair of greater than 20 $\mu g\ g^{-1}$ have been associated with risk to foetal brain development. Even lower values which were thought to be safe have recently given cause for concern (Grandjean et al. 1997; 1998).

Table 13.6. Total mercury in hair ($\mu g\ g^{-1}$) and blood ($\mu g\ l^{-1}$) from pregnant women from Camara de Lobos, Madeira (Renzoni et al. 1998)

	Total mercury in hair	Total mercury in blood
N	181	183
Mean ±SD	10.30 ±6.85	31.54 ±29.91
Median	8.62	26.43
Range	1.93–42.61	4.55–142.38

Table 13.7. Frequency distribution of hair mercury concentrations ($\mu g\ g^{-1}$) among 181 from pregnant women from Camara de Lobos, Madeira (Renzoni et al. 1998)

Mercury concentration	No. of pregnant women	%
0 – 6	49	27
6.01 – 10	65	36
10.01 – 20	54	30
> 20	13	7

13.7 Conclusions

Our understanding of the part played by the oceans in the global cycle of mercury is increasing, largely due to the analytical techniques which allow very low levels of mercury species to be determined in marine samples. The part played by biota in the cycle, including biomethylation, is receiving study. It is now known that mercury is a pollutant of importance, and adverse health effects at relatively low exposure levels have been documented from the ingestion of marine seafood.

References

Ahrland S (1985) Inorganic chemistry of the ocean. In: Irgolic KJ, Martell AE (eds) Environmental inorganic chemistry. VCH Publishers Inc., Deerfield Beach FL, pp 65–88
Andren AW, Harris CR (1973) Methyl mercury in estuarine sediments. Nature 245:256–257
Baeyens W, Leermarkers M (1998) Elemental mercury concentrations and formation rates in the Scheldt Estuary and the North Sea. Mar Chem 60:257–266
Bartlett PD, Craig, PJ, Morton, SF (1987) Sci Tot Env 10:24–249
Beijer J, Jernelöv A (1979) Methylation of mercury in aquatic environments. In: Nriagu JO (ed) The biogeochemistry of mercury in the aquatic environment. Elsevier, Amsterdam, Oxford, New York, pp 203–210
Benoit JM, Gilmour CC, Mason RP, Riedel GS, Riedel GF (1998) Behaviour of mercury in the Patuxent River estuary. Biogeochemistry 40:249–265
Berdowski JJM, Bloos JJ, Visschedijk AJH (1998) The European emission inventory of heavy metals for 1990. In: Skeaff JM (ed) Metals and the environment. Canadian Institute of Mining, Metallurgy and Petroleum, Montreal (Canada), pp85–93
Berlin M (1986) Mercury. In: Friberg L, Nordberg GF, Voulk V (eds) Handbook on the toxicology of metals, 2nd edn. Elsevier, Amsterdam, Oxford, New York
Bernhard M (1985) Mercury accumulation in a pelagic food chain. In: Irgolic KJ, Martell AE (eds) Environmental inorganic chemistry. VCH Publishers Inc., Deerfield Beach FL pp 349–358
Bernhard M (1988) Mercury in the Mediterranean. UNEP Regional Seas Reports and Studies No. 98
Bloom NS, Fitzgerald WF (1988) Determination of volatile mercury species at the picogram level by low temperature gas chromatography with cold vapour atomic fluorescence detector. Anal Chimica Acta 208:151–161
Capone DG, Kiene RP (1988) Comparison of microbial dynamics in marine and fresh-water sediments. Contrasts in anaerobic carbon metabolism. Limnol Oceanogr 33:725–749
Carpi A (1997) Mercury from combustion sources: A review of chemical species emitted and their transport in the atmosphere. Water Air Soil Pollut 98:241–254
Choi SC, Bartha R (1994) Environmental factors affecting mercury methylation in estuarine sediments. Bull Environmental Contam Toxicol 53:805–812
Clarkson TW (1994) The toxicology of mercury and its compounds. In: Watras CJ, Huckerbee JW (eds) Mercury pollution integration and synthesis. Lewis Publishers, Boca Raton, London, pp 631–641
Compeau G, Bartha R (1983) Effects of sea salt anions on the formation and stability of methyl mercury. Bull Environmental Contam Toxicol 31:486–493
Compeau G, Bartha R (1985) Sulphate reducing bacteria – principal methylators of mercury in anoxic estuarine sediments. App Environ Microbiol 50:498–502
Craig PJ (1980) Metal cycles and biological methylation In: Hutzinger (ed) The handbook of environmental chemistry, vol I. Part A: The natural environment and the biogeochemical cycles. pp 169–227
Craig PJ (1986) Organomercury compounds in the environment. In: Craig PJ (ed) Organometallic compounds in the environment. J Wiley and Sons New York, Chichester, Toronto
D'Itri FM (1991) Mercury contamination: What have we learned since Minamata?. Environmental Monitoring and Assessment 19:165–182
Dorten WS, Elbaz-Poulichet F, Mart LR, Martin J-M (1991) Reassessment of the river input of trace metals into the Mediterranean Sea. Ambio 20:2–6
Fagerstrom T, J, Jernelöv A (1972) Wat Res 6:1193. Cited in Mason and Fitzgerald (1991)
Farris FF, Smith JC (1997) Pharmacokinetics of methyl mercury. Abstracts of the International Conference on Human Health and Mercury Exposure, Faroe Islands, June 1997

Fitzgerald W (1989) Atmospheric and oceanic cycling of mercury. In: Riley JP, Chester R (eds) Chemical oceanography. Academic Press, London
Foo SC, Ngim CH, Phon WO, Lee J (1988) Mercury in scalp hair of healthy Singapore residents. Science of the Total Environment 72:113–122
Förstner U, Wittmann GTW (1979) Metal pollution in the aquatic environment. Springer-Verlag, Berlin, Heidelberg, New York, pp 486
Franchi E, Loprieno G, Ballardin M, Petrozzi L, Migliore L (1994) Cytogenic monitoring of fishermen with environmental mercury exposure. Mutat Res 320: 23–29
Gaggi C, Zino F, Duccini D, Renzoni A (1996) Levels of mercury in scalp hair of fishermen and their families from Camara de Lobos, Madeira (Portugal). Bull Environ Contam Toxicol 56:860–865
Gagon C, Pelletier É, Mucci A (1997) Behaviour of anthropogenic mercury in marine coastal sediments. Mar Chem 59: 159–176
Gill GA, Fitzgerald WF (1985) Mercury sampling of open ocean waters at the picomolar level. Deep Sea Research 32:287–290
Gilmore CG, Henry EA (1991) Mercury methylation in aquatic systems affected by acid deposition. Environ Pollut 71:131–136
Grandjean P, Weihe P, White RF., Debes F, Araki S, Yokoyama K, Murata K, Sørensen N, Dahl R, Jørgensen PJ (1997) Cognitive deficit in 7-year-old children with prenatal exposure to methyl mercury. Neurotoxicol Teratol 19:417–428
Grandjean P, Weihe P, White RF, Debes F (1998) Cognitive performance of children exposed prenatally to "safe" levels of methyl mercury. Environ Res A 77:165–172
Hansen JC, Wulf HC, Kromann N, Alboge K (1976) Human exposure to heavy metals in East Greenland. 1. Mercury. Sci Total Environ 26:233–243
Harada M, Nakanishi J, Ohno K, Kimura T, Yamaguchi H, Tsuruta K, Kizaki T, Ookawara T, Ohno H (1998) The present mercury contents of scalp hair and clinical symptoms in inhabitants of the Minamata area. Environ Res A 77:160–164
Haxton J, Lidsay DG, Hislop JS, Salmon L, Dixon EJ, Evans WH, Reid JR, Hewitt CJ, Jeffries DF (1979) Duplicate diet study of fishing communities in the United Kingdom: Mercury exposure in a critical group. Environ Res 18:351–386
Herbert MT, Acres GJK, Hughes JE (1980) Platinum, gold, silver and mercury. In: Hatfull D (ed) Future metal strategy. The Metals Society, London
Holm HW, Cox MF (1975) Transformation of elemental mercury by bacteria. Appl Microbiol 29: 491–495
Hornung H, Krumholz BS, Cohen Y (1984) Mercury pollution in benthic animals, fish and sediments in Haifa Bay, Israel. Mar Environ Res 12, 191–208
Jensen S. Jernelöv A (1969) Biological methylation of mercury in aquatic organisms. Nature 223: 753–754
Jernelöv A, Ramel C (1994) Mercury in the environment. Ambio 23:166
Jernelöv A, Landner L, Larsson T (1975) Swedish perspectives on mercury pollution. J Wat Pollut Control Fed 47:810–822
Jòhannesson T, Lunde G, Steinnes E (1981) Mercury, arsenic, cadmium selenium and zinc in human hair and salmon fries in Iceland. Acta Pharmacol Toxicol 48:185–189
Kannan K, Falandysz J (1998) Speciation and concentrations of mercury in certain coastal marine sediments. Water Air Soil Pollut 103:129–136
Knauer GA, Martin GH (1972) Mercury in a marine pelagic food chain. Limnol Oceanogr 17:868–876
Kniewald C, Kwokal, Z, Branica M (1987) Marine sampling by scuba diving. 3 sampling procedures for measurement of mercury concentrations in estuarine waters and seawater. Mar Chem 22:2–4, 342–352
Lawson NM, Mason RP (1998) Accumulation of mercury in estuarine food chains. Biogeochemistry 40:235–247
Leatherland TM, Burton JD (1974) The occurrence of trace metals in coastal organisms with particular reference to the Solent Region. J Mar Bio Ass (UK) 54:547–568
Leermakers M, Baeyens W, Ebinghaus R, Kuballa J, Kock HH (1997) Determination of atmospheric mercury during the North Sea Experiment. Water Air Soil Pollut 97:257–263
Lindberg, Stokes P, Goldberg E, Wren C (1987) Mercury. In: Hutchinson TW, Meemz KM (eds) Lead, mercury, cadmium and arsenic in the environment. John Wiley and Sons, New York, Chichester, Toronto, pp 17–34
Lindqvist O, Jernelöv A, Johansson K, Rodhe R (1984) Mercury in the Swedish environment: Global and local sources. SOLNA, Swedish National Environmental Protection Board Report No 1816
Mance G, (1987) Pollution threat of heavy metals in aqueous systems. Elsevier Applied Science, London, New York
Martin M, Osborne KE, Billig P, Glockstein N (1981) Toxicities of ten metals to *Crassostrea gigas* and *Mytilus edulis* embryos and *Cancer magister* Larvae. Mar Pollut 12:305–308
Mason RP, Fitzgerald WF (1990) Mercury species in the equatorial Pacific. Nature 347:457–459

Mason RP, Fitzgerald WF (1991) Mercury speciation in open ocean waters. Water Air Soil Pollut 56: 779–789
Mason RP, Fitzgerald WF (1993) The distribution and biogeochemical cycling of mercury in the equatorial Pacific Ocean, Deep-Sea Ocean Research 40: 1897–2001
Mason RP, Fitzgerald WF (1994) Elemental mercury cycling within the mixed layer of the equatorial Pacific Ocean. In: Watras CJ, Huckerbee JW (eds) Mercury pollution integration and synthesis. Lewis Publishers, Boca Raton, London, Tokyo, pp 83–97
Mason RP, Fitzgerald WF, Morel FMM (1994) The biogeochemical cycling of elemental mercury: Anthropogenic influences. Geochim Cosmochim Acta 58:3192–3198
Mason RP, Rolfhus KR, Fitzgerald WF (1995) Methylated and elemental mercury cycling in surface and deep ocean waters of the North Atlantic. Water Air Soil Pollut 80:665–677
Mason RP, Rolfhus KR, Fitzgerald WF (1998) Mercury in the North Atlantic. Mar Chem 61:37–53
Matthews AD (1982) Mercury content of commercially important of Seychelles and hair mercury levels of a selected part of the population. Environ Res 30:305–312
May K, Stoeppler M (1983) Studies on the biogeochemical cycle of mercury. I Mercury in inland waters and food products. Proc. Int. Conf. Heavy Metals in the Environment. Sept 1983, vol I, pp 241–244
McClurg TP (1984) Effects of fluoride, cadmium and mercury on the estuarine prawn *Penaeus indicus*. Water SA10:40–45
McGrath SP (1997) Behaviour of trace elements in terrestrial ecosystems. In: Prost R (ed) Contaminated soils. 3rd International Conference on the Biogeochemistry of Trace Elements Paris, May 1995, Institut National de la Researche Agronomique, Paris (France), pp 35–56
Miganti V, Capelli R, De Pellegrini R, Orsi-Relini L, Relini G (1996) Total and organic mercury concentrations in offshore crustaceans of the Ligurian Sea and their relations to the trophic levels. Sci Tot Env 184:149–162
Mikac N, Kwokal Z, May K, Branica M (1989) Mercury distribution in the Krka River estuary (E. Adriatic coast). Mar Chem 28:109–126
Miskimmin BM, Rudd JWM, Kelly CA (1992) Influence of dissolved oxygen, carbon, pH and microbial respiration rates on mercury methylation and demethylation in lake water. Can J Fish Aquat Sci 49:17–22
Mitch WJ, Gosselink JG (1986) Wetlands. Van Nostrand Reinhold, New York
Muhaya BBM, Leermakers M, Baeyens W(1997) Total mercury and methyl mercury in sediments and in the polychaete *Nereis diversicolor* at Groot Buitenschoor (Scheldt Estuary, Belgium). Water Air Soil Pollut 94:109–123
Nriagu JO, Pacyna JM (1988) Quantitative assessment of world-wide contamination of air, water and soils by trace metals. Nature 333:134–139
Olson BH, Cooper RC (1974) In situ methylation of mercury in estuarine sediments, Nature 252:682–683
Pacyna JM (1987) Atmospheric emissions of arsenic, cadmium, lead and mercury from high temperature processes in power generation and industry. In: Hutchinson TW, Meemz KM (eds) Lead, mercury, cadmium and arsenic in the environment. John Wiley and Sons, New York, Chichester, Toronto, pp 69–88
Pacyna JM (1996) Monitoring and assessment of metal contaminants in the air. In: Chang LW, Magos L, Suzuki T (eds) Toxicology of metals. CRC, Lewis Publishers, New York, London, Boca Raton, pp 9–26
Rasmussen PE, Edwards GC, Kemp RJ, Fitzgerald-Hubble CR, Schroeder WH (1998) Towards an improved natural sources inventory for mercury. In: Skeaff JM (ed) Metals and the environment. Canadian Institute of Mining, Metallurgy and Petroleum, Montreal (Canada), pp 73–81
Regnell O (1994) The effect of pH and dissolved oxygen levels on methylation and partitioning of mercury in fresh water model systems. Environ Pollut 84:7–13
Renzoni A (1989) Mercury in scalp hair of Maldivians. Mar Pollut Bull 20:93–94
Renzoni A (1992) Comparative observations on levels of mercury in scalp hair of humans from different islands. Environ Management 16:597–602
Renzoni A, Zino F, Franchi E, (1998) Mercury levels along the food chain and risks for exposed populations. Environ Res A 77:68–72
Ridley WP, Dizikes LJ, Wood JM (1977) Biomethylation of toxic elements in the environment. Science 197:329–332
Schlesinger WH (1997) Biogeochemistry: An analysis of global change. Academic Press, London, San Diego
Schmidt D (1992) Mercury in Baltic and North Sea waters. Water Air Soil Pollut 62:43–55
Sharp JR and Neff JM (1980) Effects of the duration of exposure to mercuric chloride on the embryogenisis of the estuarine teleost *Fundulus heteroclitus*. Mar Environ Res 3:195–213
Skerfving S (1970) Chromosome breakage in humans exposed to methyl mercury through fish consumption. Arch Environ Health 21: 133–139
Skerfving S (1991) Exposure to mercury in the population. In: Suzuki T, Imura N, Clarkson T (eds) Advances in mercury toxicology. Plenum Press, New York, pp 411–425

Slemr F, Langer E (1992) Increase in global atmospheric concentrations of mercury inferred from measurements over the Atlantic Ocean. Nature 355: 434–437
Smith WE, Smith AM (1973) Minamata. Holt, Rinehart and Winston, New York
Stoeppler M, Bernhard M, Backhaus F, Schulte E (1979) Comparative studies on trace metal levels in marine biota. I Mercury in marine organisms from the western Italian coast, the Strait of Gibraltar and the North Sea. Sci Total Environ 209–212
Summers AO, Silver S (1978) Microbial transformation of metals. Ann Rev Microbiol 32: 637–642
Suzuki, T (1988) Hair and nails: Advantages and pitfalls when used in biological monitoring. In: Clarkson TW, Friberg L, Nordberg GF, Sager PR (eds) Biological monitoring of toxic metals. Plenum Press, New York, pp 623–640
Takeuchi T (1972) Distribution of mercury in the environment of Minamata Bay and the Inland Ariake Sea. In: Hartung R, Dinaman BD (eds) Environmental mercury contamination. Ann Arbor Science, Ann Arbor, pp79–81
Topping G, Davies IM (1981) Methyl mercury production in the marine water column. Nature: 290: 243–244
Weihe P, Grandjean P, Jørgensen (1997) Methyl mercury exposure in the Faroes. Abstracts of the International Conference on Human Health and Mercury Exposure, Faroe Islands, June 1997, pp 18
Weiss HV, Koide M, Goldberg ED (1971) Mercury in a Greenland ice sheet: Evidence of recent input by man. Science 174:692–694
WHO/IPCS (1990) Environmental health criteria, 101. Methyl mercury. International Programme on Chemical Safety, World Health Organisation, Geneva
WHO/IPCS (1991) Environmental Health Criteria, 118. Inorganic Mercury. International Programme on Chemical Safety, World Health Organisation, Geneva
Wilken R-D, Hintlemann H (1991) Mercury and methyl mercury in sediments and suspended particles from the River Elbe, North Germany. Water Air Soil Pollut 56:427–4370
Wood JM, Kennedy FS, Rose CG (1968) Synthesis of methyl mercury compounds by extracts of a methogenic bacterium. Nature 220:173–174

Occurrence, Formation and Fate of Organoantimony Compounds in Marine and Terrestrial Environments

P.J. Craig · S.N. Forster · R.O. Jenkins · D.P. Miller · N. Ostah · L.M. Smith · T.-A. Morris

14.1 Occurrence of Organoantimony Compounds in Marine and Terrestrial Environments

Antimony is a relatively common group 15 metalloid, with world industrial production around 80 000 tonnes. It occurs at 0.2 ppm crustal abundance (arsenic at 1.8 ppm, Carmalt and Norman 1997). The majority of antimony metal usage is in metallurgy where it is used as an alloy to impart hardness to lead and other metals (Maeda 1994). The main use of antimony compounds is as a flame retardant, with Sb_2O_3 being the most commonly used compound (ca. 90% of the annual usage of Sb_2O_3 in the US and Japan is as a flame retardant (Maeda 1994). The ubiquitous nature of such compounds in consumer products can lead to worrying amounts of antimony in landfill sites and in emissions from municipal incinerators. Organometallic antimony compounds – unlike those of As, Sn and Pb – are not exposed to the natural environment during usage. The only common use of organometallic antimony compounds is as precursors in the manufacture of semiconductors. This use is restricted to very controlled environments, hence there is no obvious route for organometallic antimony compounds to enter the environment. Anthropogenic emissions of antimony to the atmosphere have been calculated at 6 tonnes per year, while that for arsenic is 31 tonnes; natural emissions are 3 and 12 tonnes respectively.

Antimony is a heavy element and its toxicity is comparable to arsenic. The nature of its toxicity, however, has not been fully investigated. The toxic nature of antimony depends on the oxidation state, with the trivalent state being more toxic than the pentavalent state(Oehme 1979). The poor solubility of most antimony compounds slows down excretion, thereby causing accumulation of antimony in the body. Antimony compounds react with thiol groups in enzymes and other cellular constituents and have particular affinity for the liver, kidney and thyroid (Maeda 1994). The toxicity of most antimony compounds has, as mentioned, been little investigated and for organometallic compounds there is little information available regarding toxicity. Antimony oxide has been shown to cause cancer in some studies (Jones 1994).

Relatively little is known about the chemistry of antimony, compared to that of arsenic. As mentioned above, the natural abundance of antimony is low compared to its congener arsenic (Carmalt and Norman 1997). The natural chemistry of arsenic seems to be much more diverse than that of other related heavy elements (e.g. Hg, Sn, Se, Te) and encompasses large organic structures such as the arsenosugars, arsenobetaine and arsenocholine (Cullen and Reimer 1989). These involve methyl arsenic moieties being bonded to complex organic counter-ions. This may occur for other metals, including antimony, but nothing is known yet. The chemistry of organometallic antimony compounds has been significantly expanded over the last 30 years. Previously, only aromatic compounds of antimony were known and well characterized; this is due to the greater stability of these species (Doak and Freedman 1970). Antimony(V) organometallic compounds are generally more stable than antimony(III) species and antimony(V) compounds were the first to be synthesized and characterized (Doak and Freedman 1970). Trimethylantimony(V) dihalides were characterized in 1968 (Doak et al. 1967), although a synthesis had been previously published (Morgan and Davies 1926). Formation of organometallic antimony compounds in the environment, rather than their use, is the most logical explanation for their presence. Present information suggests that methyl to metal bonds are the only organometallic products of any natural environmental organometallic synthesis. This process is called biomethylation and is well established for arsenic and a number of other heavy elements, such as Hg, Se, Sn and Te (Fatoki 1997). One example of an environmental ethyl to metal linkage is known (C_2H_5Hg-) (Cai et al. 1997), but this may have been caused by a transalkylation. Antimony biomethylation was reported in a Ph.D. thesis (Barnard 1947), where it was noted that an inorganic antimony compound, when incubated with mould cultures, appeared to be mobilised (possibly as $(CH_3)_3Sb$) to a distant part of the apparatus. The techniques at that time (1947) did not allow any conclusive identification. Very recently, similar work has demonstrated that methylated antimony is actually formed in such experiments (Jenkins et al. 1998a, 1998b). Oxidation of this compound is reported to be rapid and to cause some breakage of the antimony carbon bonds (Parris and Brinckman 1976). This suggests an overview indicating which organometallic species we might expect to find in the natural environment i.e. antimony(V) compounds with between one and three antimony carbon bonds. Oxidation would also give rise to oxide hydroxide linkages that would increase solubility of the antimony compounds.

The first modern report of organometallic antimony compounds in the aquatic environment was by Andreae et al. in 1981. They reported that methyl and dimethyl stibonic acids were present in polluted water samples (Fig. 14.1). The stibonic acids reported were acid analogues, dimethyl stibonic acid $(CH_3)_2SbOOH$ (DMSA) and methylstibinic acid $CH_3SbO(OH)_2$ (MSA).The analysis was performed using a hydride generation technique with element specific detection. This technique preserves and gives information on antimony carbon linkages, but destroys the inorganic portion of the compound. The species produced are methylantimony hydrides, and to identify these species standard compounds are required. In this case (Andreae et al. 1981) they were obtained from a group which had reportedly synthesized the compounds (Meinema and Noltes 1972). The amounts of methylantimony species found varied between sites, two polluted German rivers had 1.2 and 1.8 ngl^{-1}Sb as MSA and no DMSA. A group of rivers which drain into the Gulf of Mexico each had less than 1 ngl^{-1}Sb MSA and no DMSA, except the Mississippi which contained 2.3 ngl^{-1}Sb MSA. The more in-

Fig. 14.1. Stibonic and stibinic acid

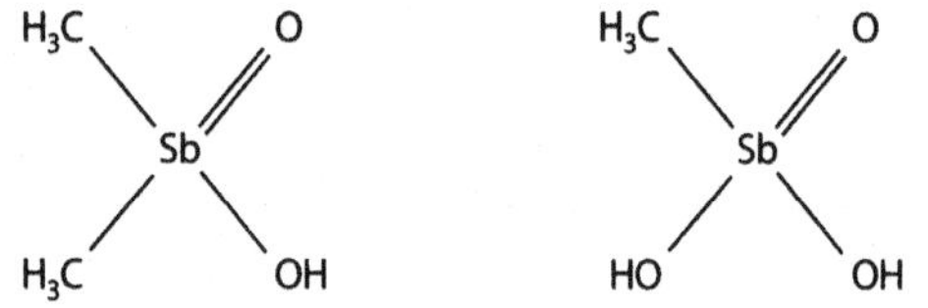

H_3C O Sb H_3C OH

H_3C O Sb HO OH

teresting results were found in the estuary and bay samples. The Ochlockonee Bay Estuary was tested along the salinity gradient, and the concentration of MSA increased with increasing salinity. At the lowest salinity (4.3%) MSA concentration was 0.8 $ngl^{-1}Sb$ with no DMSA detected. DMSA was detected when salinity reached 23.8% and in subsequent samples at 30.2 and 33.4% salinity at 1.1 and 1.5 $ngl^{-1}Sb$ respectively. The highest detected levels of MSA were also in these samples (10.9 and 12.6 $ngl^{-1}Sb$). The Gulf of Mexico, Apalachee Bay was the only other location where DMSA was detected; the samples contained the highest detected level of DMSA at 3.2 $ngl^{-1}Sb$, while only 5.3 $ngl^{-1}Sb$ as MSA was detected.

Further reports from this group confirmed the presence of methylantimony species in disparate marine locations. Polluted estuarine water from Portugal and sea water from the Baltic were analysed and found to contain low levels of methylantimony species (Andreae et al. 1983; Andreae and Froelich 1984). Since these reports, the nature of the standard compounds used to identify MSA and DMSA has been questioned and further attempts to synthesize these compounds have been unsuccessful (Dodd et al. 1992). The ability of these compounds to form in nature has therefore been questioned. Nevertheless, the reports of Andreae et al. did constitute the first knowledge of methylantimony compounds found in the environment and the existence of an antimony carbon linkage is not in question. The molecular structure of the compounds remains unresolved; MSA and DMSA were not necessarily the actual methylantimony species present in the environments tested. A further report of organoantimony compounds in the environment came from a group in Germany investigating the production of volatile metalloid species from sewage and landfill gases (Hirner et al. 1994; Feldmann and Hirner 1995). Trimethylantimony was found in these emissions along with many other organometallic species: the anoxic nature of these environments allowed this species to be detected where it would normally oxidize to antimony(V) species in an oxic environment. Dodd et al. (1996) have recently reported organometallic antimony compounds in pond weed (*Potamogeton pectinatus*) extracts from a polluted region in Canada. The method used was hydride generation coupled with gas chromatography-mass spectrometry (GC-MS). Trimethylantimony, dimethylantimony hydride and methylantimony dihydride were all identified in this sample. There was no study of the inorganic components. This work, however, does confirm that all three antimony carbon linkages can survive the oxidation process and that methylantimony species exist in the natural environment. Total antimony concentration in the pond weed was 48 $\mu g\ g^{-1}$; quantification of the methylantimony species was not reported.

That there are few reports of organoantimony species in the environment may be due to the low crustal abundance of the element in relation to detection limits of available analysis methods. Detection limits of analytical systems generally are falling, which will undoubtedly increase the reports of organometallic antimony species in the environment. Another factor influencing the number of reports is that few groups so far have studied antimony chemistry in the environment. Increasing emissions and uses of antimony may increase concentrations of this material, which will lead to further study of antimony and its compounds. This is likely to enhance our understanding of environmental antimony chemistry generally. Very recently methylantimony compounds have been found in biota associated with an antimony-rich terrestrial region (Craig et al. 1999).

14.2 Analysis of Organoantimony Compounds

The advent of highly sensitive analytical instruments has resulted in a variety of techniques that can be employed for the speciation of organoantimony compounds. Many workers have simply adapted techniques already developed for similar elements, such as arsenic, and applied the knowledge gained for use with antimony. Although this has not always been straightforward, our understanding of these organoantimony compounds is continually increasing.

14.2.1 Sample Preparation

The type of analytical instrument(s) used for the detection and speciation of antimony compounds will dictate the necessary sample preparation needed. Often samples to be analysed are of a biological or environmental nature, and the vast array of auxiliary chemical species present usually necessitates a cleanup stage to remove potential interferents. Also, in these complex matrices there is sometimes a need to derive the antimony species into a form that is suitable for analysis (e.g. volatile). Processes such as freeze-drying, homogenisation and acid digestion are common examples of sample preparation. Samples to be analysed by inductively-coupled plasma mass spectrometry (ICP-MS) must not contain solid material larger than 20 microns or the risk of blocking the nebulizer is introduced and hence acid digestion followed by filtration is required. Once in an acceptable physical form, the sample can be analysed using one or a combination of sensitive techniques. A selection of the most commonly utilized techniques is outlined below.

14.2.2 Hydride Generation

Derivatisation of organoantimony species by hydride generation with sodium tetraborohydride ($NaBH_4$) has proved to be a useful technique for their speciation. Gaseous hydride species can be produced directly from environmental samples, thus providing a relatively simple cleanup and matrix separation process and avoiding any pretreatment steps. Generally hydride generation only provides information about the oxidation state of an inorganic compound, by inference from previous knowledge of the reaction conditions under which a metal in a known oxidation state is successfully derivatised. With respect to alkylated compounds, hydride generation can only provide information about the degree of organic substitution of the compound, and this only in conjunction with gas chromatographic methods. It cannot provide information about the full molecular nature of an environmental compound, even in conjunction with techniques such as GC-MS. For this liquid chromatographic techniques and standards are required. There is a total lack of environmental organoantimony standards. This may be due to the instability of the methylhydrides to oxygen (reported as $10^3\ m^{-1}\ s^{-1}$) (Parris and Brinckman 1976) and partially due to the difficulty of synthesizing pure mono- and dimethyl compounds. It is not even certain which methyl-Sb species should be synthesized! Recent attempts to prepare both mono- and dimethyl compounds has resulted in the formation of insoluble polymers (Dodd et al. 1992). Only trimethylantimony compounds have been successfully synthesized and isolated as standards by a number of researchers (Gürleyük et al. 1997; Dodd et al. 1992; Craig et al. 1999; Gates et al. 1997). A further problem has been that on hydride generation of these trimethylantimony standards (i.e. $(CH_3)_3SbCl_2$), many researchers have detected four peaks relating to the one standard compound, irrespective of the detector used. The structures of the compounds corresponding to each of the four peaks have been elucidated, using GC-MS, as $(CH_3)_3Sb$ and the expected dismutation products SbH_3, $(CH_3)SbH_2$ and $(CH_3)_2H$. These are eluted in order of their boiling points on non-polar chromatographic columns. Some researchers have attributed this dismutation to the influence of pH on the derivatization system (Koch et al. 1998), while others have thought that it may be due to a very fast oxidation rate of trimethylantimony in the gaseous phase (Gürleyük et al. 1997). Conversely, this rearrangement has been used to good effect in headspace analysis (Jenkins et al. 1998a, 1998b, 1998c), with hydride generation of a single standard under conditions that will deliberately cause rearrangement being utilized. The identification of inorganic, monomethyl-, dimethyl-, and trimethylantimony compounds by comparison with retention times, achieved through chromatographic separation, can therefore be utilized for standards work. It is clear from a review of the literature (see for example Andreae et al. 1981; Dodd et al. 1996) that great care must be taken in performing hydride generation of methylantimony compounds from environmental matrices. For both quantitative and qualitative analysis, consideration must be given to reaction conditions. Koch et al. (1998) have shown that environmental matrices directly affect molecular rearrangement, and they recommend that only the standard additions method be used as a calibration technique in the hydride generation of organoantimony compounds from environmental samples. Table 14.1 illustrates the various hydride generation conditions utilized in the literature and the subsequent rearrangement observed.

It should be noted that hydride generation precludes analysis as such of large organoantimony molecules, such as the stibnolipid suggested by Benson (1988). These (from the As precedent) are not expected to be hydride derivatised under normal conditions. Only if techniques such as microwave digestion or ultraviolet photolysis are included on-line before the borohydride step (in order to breakdown these molecules), can these compounds be detected. Even so, the organic counter-ion species would be destroyed and only the methylantimony portions detected, i.e. there would not be a full molecular speciation.

14.2.3 Gas Chromatographic Separation

Both gas chromatography and cold trap methods have been utilized in conjunction with hydride generation to ensure the gaseous species produced are adequately separated. Essential to the problem of partially alkylated compounds is the requirement that measurement be capable of both rapid and non-destructive molecular separation. To reach this goal the preconcentrated sample must be separated without strong interactions between the volatile metal compound and the stationary phase of a chromatographic system. This is generally achieved at relatively low temperatures, in comparison with more conventional GC, on a non-polar column with elution according to

Table 14.1. Conditions used for hydride generation

Method	Batch/S-C	Purging gas	Reagents	Conditions	Re-arrangement, DL	References
HG-CT-QF-AAS	B	Helium 100 ml min^{-1}	6 N HCl (0.5ml), di- and monomethyl standards (0.05–20ng of Sb) made up with water (100ml), 4% w/w $NaBH_4$ (1 ml) stabilised with NaOH	Degassing (3 min) Reaction and purge (6 min)	Rearrangement not reported Standards only 52.5 and 42% pure DL = 30–60 pg	Andreae et al. 1981, Andreae and Froelich 1984
HG-GC-AAS	B	Helium	pH 1.5–6 range, in water sol. (60 ml), ±KI, tris-(hydroxymethyl)-amino-methane, HCl, tri- and dimethylantimony standards, 4% KBH_4 (6 ml)	Volatile hydrides were passed through lead acetate trap and a CO_2/isopropanol trap before reaching the column	Rearrangements occurred under all conditions	Dodd et al. 1992
HG-GC-MS	S-C	Helium	i) 1 M HCl, ii) 4 M acetic acid, trimethylantimony standards, sample, 2% m/v $NaBH_4$ in 0.1% m/v NaOH	Stibines swept through gas-liquid separator, through CO_2/acetone trap to GC	No rearrangement, when acetic acid, $NaBH_4$ and distilled water rinsed apparatus, occurs when only rinsed with distilled water, DL = 15 ng of Sb	Dodd et al. 1996
a) HG-GC-AAS b) HG-CT-ICP-MS	a) S-C b) B	a) Helium b) Helium 80 ml min^{-1}	$(CH_3)_3SbCl_2$ (100–200 ng), sample (0.1–3 ml), pH 0.51 to 6.76 using various acids and buffers, 2% w/v $NaBH_4$ b) de-ionised water (10 ml), $(CH_3)_3SbCl_2$ (1 ng), 1 M HCl (re-distilled), 6% w/v $NaBH_4$ (0.8 ml)	a) HG and trapping on column = 3 min b) Reaction mixture purged = 6 min CO_2/acetone trap used	a) No rearrangement at pH 6.10–6.76, rearrangement at pH 0.51–3.30 (max = 24%), sample matrix affects rearrangement, DL = 4 ng; b) pH similar results to a) max. rearr. 68%, DL = 0.01 ng	Koch et al. 1998
HG-CT-AAS	B	Helium 40 ml min^{-1}	In millipore-Q water (53 ml),acetic acid (3 ml), trimethylantimony standard (4.9 µl), $NaBH_4$ (10 mg)	Purged and collected on cold trap		Jenkins et al. 1998b,c
HG-CT-ICP-MS	B	Helium 400 ml min^{-1}	Sample (1 g), bidistilled H_2O (5 ml), HCl, pH 2 5% $NaBH_4$ (1.5 ml)	Degassing(1 min), purging (10 min)	Rearrangement not reported	Krupp et al. 1996
HG, a) GC-ICP-MS b) GC-MS-MS	B	Helium	6% m/m $NaBH_4$ (purged with H_2 for 30 min) (0.5 ml) $(CH_3)_3SbCl_2$ (30 ng as Sb), deionised water (1 ml), pH 7	Conducted in a sealed vial (15 ml) injected onto column using a gas-tight syringe	No rearrangement reported, only trimethylantimony detected	Feldmann et al. 1998
a) HG-GC-AAS b) HG-GC-ICP-MS	a) S-C b) B	Helium 70 ml min^{-1}	a) Sample (4 ml), 0.05 M citrate buffer pH 6, 2% $NaBH_4$ b) water (10 ml), 6% $NaBH_4$ **(0.8 ml)**	b) Degassing (1 min), purge and trapping (6 min) column in liq. N_2	a) and b) No rearrangement seen	Andrewes et al. 1998
HG-CT-GC-ICP-MS	B	Helium 500 ml min^{-1} degassing 100 ml min^{-1} purging	Sample (condensed water) (10 ml), 5% $NaBH_4$ (1 ml), pH 2 acidified with 1 M HCl (1.5 ml)	Degassing (10 min)	No rearrangement reported	Feldmann et al. 1994

Table 14.2. Conditions used for separation of organoantimony compounds

Method	Stationary phase	Heating	Dimensions	References
HG-CT-QF-AAS	15% OV-3 on Chromosorb W 60/80 mesh	Chromel wire (3 Ω)(1 m) (ca. 13 V, 4 A)	30 cm × 6 mm o.d. Pyrex U-tube, 2/3 packed with stationary phase	Andreae et al. 1981 Andreae a. Froelich 1984
HG-GC-AAS	Silanised porapak-PS (Mesh 80/100)	50 °C (1 min) then 25 °C min^{-1} til 150 °C		Dodd et al. 1992
HG-GC-MS	Porapak-PS, 80–100 mesh (Waters Scientific)	70 °C (1 min) then 25 °C min^{-1} til 150 °C	50 cm × 0.4 cm i.d., Teflon column	Dodd et al. 1996
HG-GC-ICP-MS (method B)	10% SP2100 on Chromosorb (Supelco) 45/60 mesh	−196 °C to 150 °C	22 cm × 6 mm o.d.	Koch et al. 1998
a) GC-MS b) Chemiluminescence	1 µm 5% phenyl, 95% methyl polysiloxane (a and b)	−20 °C (1 min) then 20°C min^{-1} til 250 °C (a & b)	a) 30 m × 0.25 mm i.d. capillary column b) 30 m × 0.32 mm i.d. capillary	Gürleyük et al. 1997
a) GC-ET-QFAAS(1) b) GC-MS(2)	3% OV101 (a) OV1 column (b)	c) Electrothermically heated at 125 °C over 5 min (6) 40 to 100 °C at a rate of 10°C min^{-1}	c) 50 cm, glass (1) × 4 mm i.d. d) 30 m × 0.32 mm i.d.	Jenkins et al. 1998a,b,c
HG-GT-ICP-MS	Supelcoport 10% SP-2100	−100 to 160 °C	22 cm × 6 mm i.d.	Krupp et al. 1996
Trap-ICP-MS	Glass beads (DMCS treated 60/80 mesh, Alltech)	Nichrome wire (5.5 Ω m^{-1})(0.5 mm, o.d.) heated from −80 to 120 °C	12 mm o.d. × 22 cm long	Hirner et al. 1994
(HG) GC-ICP-MS)	10% SP-2100 on Supelcoport (60/80 mesh)	−100 to 180 °C	22 cm, i.d. 6 mm	Feldmann and Hirner (1995) Feldmann et al. 1994
a) GC-MS-MS b) GC-ICP-MS	a) PTE-5, 0.25 µm Supelco 2-4143 poly (5% diphenyl–95% dimethylsiloxane) b) 10% SP2100 on Chromosorb	a) injector 200 °C, column 40 to 150 °C at 15°C min^{-1} b) −196 to 150 °C	a) 30 m × 0.32 mm i.d. capillary column b) 22 cm × 6 mm o.d.	Feldmann et al. 1998
GC-ICP-MS	10% SP-2100 on Chromosorb	−196 to 150 °C within 3 min	22 cm × 6 mm o.d.	Feldmann and Cullen (1997)
a) HG-GC-AAS b) HG-GC-ICP-MS	a) Porapak PS b) 10% SP-2100 on Chromosorb	a) 70 to 150 °C at 30 °C min^{-1} b) −196 to 200 °C	b) 22 cm × 6 mm o.d.	Andrewes et al. 1998

compounds boiling points. Table 14.2 illustrates conditions which have been utilized for successful separation of organoantimony compounds.

14.2.4
Detection Systems

14.2.4.1
Atomic Absorption Spectrometry

Atomic absorption spectrometry (AAS) is the most commonly utilized detection technique, owing to its selectivity, sensitivity and relatively low running costs, together with the speed and ease of operating the instrument. The characteristic source emission is generated by one of two types of lamp. Hollow cathode lamps are commonly used, however Andreae et al. (1981) reported an improvement by a factor of three when using an electrode discharge lamp. This power source also needs to be replaced less frequently than the former, although the cost of initial installation is higher. Principal resonance lines for antimony are found at 217.6, 206.8, and 231.2 nm. The 217.6 nm line, which is the most sensitive, should be used with a spectral band-pass of less than 0.2 nm to isolate this line from two non-absorbing lines at 217.0 nm and 217.9 nm. Without this, reductions in sensitivity and linearity will occur. An electrically heated quartz furnace offers the most sensitive method of introduction of the organoantimony compounds into the light beam, owing to its long path length. The quartz furnace also facilities the use of a hydrogen-air flame, which improves sensitivity, owing to a complex series of reactions between hydrogen and oxygen radicals as the sample passes through. Using an AAS quartz furnace, detection limits as low as 30–60 pg have been observed (Andreae et al. 1981).

14.2.4.2
Mass Spectrometry

The use of mass spectrometry (MS) in conjunction with hydride generation (HG) and gas chromatography (GC) allows for the characterization of the antimony compounds based on their retention times and mass spectral data. However, it should be noted that this detection system is less sensitive than atomic absorption spectroscopy. Antimony has two stable isotopes: ^{121}Sb (relative abundance 57%) and ^{123}Sb (43%), hence, the mass spectra will be more complex than those of arsenic, for example, which has only one stable isotope. The occurrence of these two stable isotopes provides a useful signature for the detection of antimony compounds by MS in the natural environment. Dodd et al. (1996) used a semi-continuous HG-GC-MS with a detection limit of ~15 ng of antimony in the narrow scan mode (m/z 117–170). Jenkins et al. (1998 a,b,c) used GC-MS to analyse the headspace of cultures for organoantimony compounds (Fig. 14.2). Here, the culture headspaces were initially preconcentrated onto Tenax-TA tubes and a thermal-desorption unit was used to introduce sample into the GC-MS instrument. Gürleyük et al. (1997) also used GC-MS to sample headspaces from cultures, but this was by direct gas injection, undertaken by sampling the headspace using a gas-tight syringe

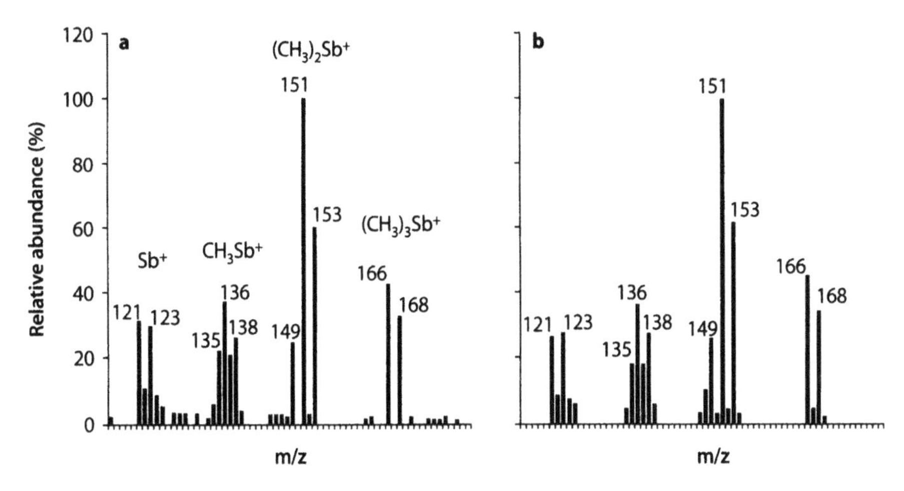

Fig. 14.2. MS identification of trimethylantimony; **a** Typical mass spectrum of trimethylantimony from culture headspace gases from methanogen enrichment culture of pond sediment, supplemented with potassium antimony tartrate; **b** Mass spectrum (reference) of TMA from NIST library (from Jenkins et al. 1998c, reproduced with author's permission)

14.2.4.3
Inductively-Coupled-Plasma-Mass Spectrometry (ICP-MS)

ICP-MS is a very sensitive analytical technique (<0.01 ppb) which has limited matrix effects, although sample preparation is a necessity as samples introduced into the instrument must be particulate free. This technique can be used as an element-specific detector for GC in order to identify volatile metal or metalloid species, although it is more commonly used as a 'stand-alone' technique for analysing total concentrations of soluble antimony species. In this latter regard the use of ICP-MS is non-species specific. Krupp et al. (1996) have used an integrated series of instruments consisting of HG-GC-ICP-MS to analyse for organoantimony compounds in sediment samples. The analytical apparatus consists of a low-temperature GC, coupled on-line with an ICP-MS as a multi-element detector. The chosen antimony isotope was ^{121}Sb, with ^{103}Rh and ^{203}Tl as selected internal standards. Trimethylantimony, dimethylantimony hydride and monomethylantimony dihydride and triethylantimony speciation was determined from a boiling point/retention time correlation. ICP-MS has also frequently been used to analyse for organoantimony compounds in gaseous samples (Hirner et al. 1994; Feldmann and Hirner 1995; Feldmann et al. 1994; Feldmann and Cullen 1997).

14.2.4.4
Chemiluminescence

Chemiluminescence detection following capillary GC separation has been proved to be highly selective and sensitive to many elements which are biomethylated in the environment. More recently, fluorine-induced chemiluminescence has been successfully applied to the analysis of organoantimony compounds by Gürleyük et al. (1997). Detection limits were found to be in the order of 15 pg Sb.

14.2.4.5
Liquid Chromatography

Lintschinger et al. (1997) have produced effective separation of antimony(V) and organoantimony [as $(CH_3)_3SbCl_2$ and $(CH_3)_3Sb(OH)_2$] with an anion exchange high performance liquid chromatography (HPLC) column under alkaline conditions and a complexing reagent in the mobile phase. Element specific detection was obtained using a flame atomic absorption system which produced a detection limit in the mg l^{-1} range, or alternatively when using ICP-MS as the detector system, the detection limit was in the μg l^{-1} range. A more recent paper published by Lintschinger et al. (1999) reported improvements on the anion exchange HPLC-ICP-MS technique, with a detection limit of 5 ng l^{-1}, achieved mainly by avoiding contamination from chromatographic devices. Andrewes et al. (1998), have used Alumina-B SPE cartridges rinsed with ammonium carbonate buffer (50 mM, pH 12) to remove most of the inorganic antimony(III) from a medium. This allowed detection with HG-GC-AAS and HG-GC-ICP-MS of the organoantimony compounds present in much smaller quantities, which previously would have been hidden by a broad tail of SbH_3 in the chromatograms.

14.2.5
Conclusion

All of the analytical instruments described above are reproducible and reliable techniques used for the analysis and speciation of antimony compounds. The availability of this wide range of techniques leaves the choice of systems utilized at the discretion of the analytical laboratory concerned, whereby, the most appropriate integrated series of techniques to be used is dependent upon cost, availability, efficiency and/or sensitivity.

14.3
Microbial Biotransformation of Antimony Compounds

14.3.1
Biomethylation of Antimony

The term biomethylation is used to describe the process mediated by living organisms in which methyl groups are transferred to various metals or metalloids. Many elements of environmental importance undergo biomethylation, including As, Hg, Sn, Se, Te, Ge, Sb, P, S, Cl and Pb. Several biological methylating agents are known, the most important of which are S-adenosylmethionine and methylcobalamine. S-adenosylmethionine has been described as 'the ubiquitous methylating agent' and has been the most extensively studied of all methylating agents. It has been established that S-adenosylmethionine is a methylating agent for arsenic (Cullen et al. 1977) and selenium (Takahashi et al. 1990), and it is thought to be involved in the methylation of tellurium (Fatoki 1997). Methylcobalamin is a derivative of vitamin B_{12}, and has been established as the methylating agent for mercury (Choi et al. 1994). In contrast to S-adenosylmethionine, methyl group transfer by methylcobalamin requires that the acceptor atoms be electrophilic (e.g. Hg^{2+}). Other methylating agents reported in the literature include selenoadenosylmethionine (Kajander et al. 1991) (the selenium analogue of S-adenosylmethionine), coenzyme F-430 (Jaun 1993) and dimethyl-ß-propiothetin (Brinckman et al. 1985). The attatchment of a methyl group to a heavy metal atom can profoundly influence the chemical, physical, and toxicological properties of the element. The general effect is an increase in the lipophilic nature and volatility. Biomethylation can therefore lead to increased mobility of metals and metalloids in the environment and enhance their accumulation in biological systems, including higher organisms. The toxicity of that element in its compound may increase (e.g. mercury) or decrease (e.g. arsenic). Antimony and arsenic are two related elements belonging to group 15 of the periodic table. The biomethylation of arsenic is well established and has been extensively reviewed (Cullen and Reimer 1989; Pratt 1993). A range of microorganisms can convert this metalloid to methylated arsenic species. Challenger in the 1930s resolved the phenomenon of the evolution of Gosio gas, trimethylarsine, by moulds present on wallpaper containing Scheele's green pigment [Cu $(AsO_2)_2$]. *Scopulariopsis brevicaulis*, a filamentous fungus, growing on breadcrumbs produced trimethylarsine when supplied with As_2O_3. The mechanism proposed by Challenger (1978) for biomethylation involved alternating reduction and methylation steps (Scheme in Fig. 14.3). To identify the methyl donor, Challenger added ^{14}C-labelled methionine to cultures and detected labelled trimethylarsine. S-adenosylmethionine was subsequently identified as the methyl donor, with the methyl group being transferred as a carbocation to the nucleophilic arsenic(III). It has been suggested (Thayer 1984, 1995) that the Challenger mechanism for biomethylation of arsenic (and selenium) might also apply to antimony on the basis of similarity of reduction potentials of these metalloids.

Challenger and Barnard also studied the interaction of several antimony compounds with various fungi. In a thesis published in 1947 (Barnard 1947), they report traces of colouration in a wet chemical test (Gutzeit test), and suggested that *Penicillium notatum* incubated aerobically in the presence of $KSbO_3$ produced a volatile form of antimony, which was detectable in a remote part of the apparatus. Molecular identification of the volatile product(s) was not possible at that time. Numerous workers over the past half-century have speculated on the possibility of antimony biomethylation. Considering the similarities between antimony and many of the elements that are known to be biomethylated, there is no obvious chemical, thermodynamic or kinetic barrier to biomethylation of antimony. In the absence of any widespread industrial use of me-

Fig. 14.3. The Challenger mechanism for the biomethylation of arsenic (an analogous scheme may apply to the biomethylation of antimony)

$$H_3AsO_4 \xrightarrow{2e^-} H_3AsO_3 \xrightarrow{[CH_3]^+} CH_3AsO(OH)_2 \xrightarrow{2e^-}$$

$$CH_3As(OH)_2 \xrightarrow{[CH_3]^+} (CH_3)_2AsO(OH) \xrightarrow{2e^-} (CH_3)_2As(OH) \xrightarrow{[CH_3]^+}$$

$$(CH_3)_3AsO \xrightarrow{2e^-} (CH_3)_3As$$

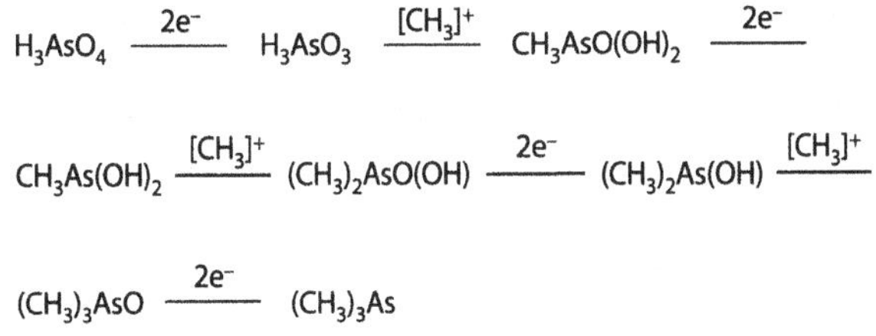

thyl antimony compounds, their detection in the natural environment (see Section 14.1) suggests that a biomethylation pathway for antimony must exist. Over recent years, concern regarding the postulated involvement of antimony in the toxic gas hypothesis for sudden infant death syndrome (SIDS) has fuelled interest in the possibility of antimony biovolatilisation. Sb_2O_3 had been incorporated as a fire retardant into PVC cot mattress covers (Turner-Report 1995) and it has been hypothesised that antimony trihydride (SbH_3) is generated from cot mattresses by microbial action (Richardson 1994). *S. brevicaulis*, a known methylator of inorganic arsenic, was thought to be the fungus involved. The toxic gas hypothesis for SIDS has been disputed by several researchers (see for example Jenkins et al. 1998a, Gates et al. 1997) and there is no evidence to support toxic gases being formed in a cot environment (Anon. 1998). Jenkins et al. (1998b) reported the formation of trimethylantimony by *S. brevicaulis*, which was the first report of antimony methylation by a characterized microorganism. The fungus was grown aerobically in the presence of inorganic antimony (e.g. potassium antimony tartrate, Sb_2O_3, Sb_2O_5), and volatile antimony evolved into the headspace above the fungal cultures was quantified by remote trapping and analysis by ICP-MS. Antimony was mobilised from both the (III) and (V) oxidation states of the element, but occurred far less readily from the latter state. The most productive phase of antimony volatilisation in the liquid cultures (malt extract medium) was at the end of the linear growth phase. Identification of trimethylantimony as the biogenic antimony species involved was established by remote trapping onto Tenax-TA, and analysis by GC-MS and by gas chromatography-electrothermal-atomic absorption spectrometry. Jenkins et al. (1998b) reported that exclusion of oxygen in the culture experiments was necessary for the identification to avoid trimethylantimony being oxidized to less volatile forms. No other volatile compound containing antimony was detected in *S. brevicaulis* headspace gases by these researchers, even though other methylated forms of antimony and antimony trihydride could be detected in derivatized standards. A subsequent paper by Jenkins et al. (1998a), reported that the yields of volatilised antimony in relation to fungal biomass were around two-fold higher on a solid medium (ca. 6 μg antimony (g dry weight biomass)$^{-1}$), when compared to a liquid culture. The order of antimony substrates in relation to ease of biovolatilisation was reported as: PAT >>> Sb_2O_3 >>> Sb_2O_5 > $KSb(OH)_6$. These researchers found no evidence of antimony volatilisation by other fungi (*Penicillium* spp., *Aspergillus* spp., *Alternaria* sp.) or bacteria (*Bacillus* spp.) tested. The ability of *S. brevicaulis* to methylate inorganic antimony has recently been confirmed by other workers (Andrewes et al. 1998) who detected methylantimony compounds, principally $(CH_3)_3SbO$, at concentrations up to 7.1 μg Sb l^{-1} in culture media containing inorganic antimony(III) compounds. The other volatile antimony compounds detected – SbH_3, CH_3SbH_2, $(CH_3)_2SbH$ and $(CH_3)Sb$ – were considered to be very minor products of *S. brevicaulis* metabolism. However, these authors did not discount the possible oxidation of trimethylantimony to produce other methylated forms of antimony in solution. Indeed the oxidation of a dimethylantimony precursor could produce the reported environmental methyl stibonic acids. The production of only low quantities of methylantimony compounds by *S. brevicaulis*, and the ability of this fungus to grow well at high concentrations of either antimony(III) or (V) (Jenkins et al. 1998b, Andrewes et al. 1998), suggests that antimony biomethylation does not function as an antimony detoxification mechanism for the organism. In contrast, biomethylation of arsenic is thought to have evolved as a process for converting the highly toxic inorganic arsenic compounds found in the environment to methylated species that are much less toxic. It remains to be seen whether trimethylantimony formation by *S. brevicaulis* is a consequence of antimony substituting in the biomethylation process for arsenic. Trimethylantimony production by anaerobic soil/sediment enrichment culture has also been described recently (Gürleyük et al. 1997; Gates et al. 1997; Jenkins et al. 1998c). Gürleyük et al. (1997) reported the detection of trimethylantimony in the culture headspace of soil bacteria enrichment cultures, when supplied with inorganic antimony in the (III) or (V) oxidation states. The incubation conditions were designed to promote bacterial growth through the use of nitrate as terminal electron acceptor in anaerobic respiration. This report represents the first characterisation of a volatile antimony compound thought to arise biogenically from an inorganic antimony substrate, although the identity of microorganisms in soils responsible for the biomethylation was unknown. Gates et al. (1997) have also reported the detection of trimethylantimony in the culture headspace of undefined mixed cultures. In this work, pond sediment samples were incubated under conditions (highly proteinaceous medium) that promote growth of fermentative bacteria and supplied with potassium antimony tartrate. Jenkins et al. (1998c), have recently reported positive results for trimethylantimony with variable frequency for four of six soils tested and for three types of enrichment culture, designed to encourage growth of fermentative, nitrate-reducing or methane-producing bacteria (Fig. 14.2). These authors proposed that, as for arsenic, different metabolic categories of prokaryotic organisms are able to methylate antimony and that this capability is widely distributed in the natural environment. Unlike arsenic, however, all reports so far of antimony biovolatilisation by soil bacteria suggest that the trimethyl form is the sole or principal species produced. We are not aware of reports in the literature of antimony biomethylation by monoseptic cultures of bacteria. There is evidence, however, that monoseptic cultures of the methane-producing bacteria *Methanobacterium formicicum* and *Methanosarcina barkeri* are capable of methylating $SbCl_3$ with the formation of trimethylantimony (A.V. Hirner, personal communication), which is consistent with reports of detection of methyl antimony species in landfill and sewage gases (Feldmann and Hirner 1995; Feldmann et al. 1994).

14.3.2 Bioreduction and Bio-Oxidation of Antimony

Microbial reduction of inorganic antimony compounds has not been demonstrated unequivocally. Detection of antimony trihydride (SbH_3) in landfill gas (Feldmann and Hirner 1995; Feldmann et al. 1994) could reflect the capability for microbial reduction of antimony in highly reducing environments. SbH_3 has also been detected in picogram quantities in aerobic cultures of *S. brevicaulis* supplemented with potassium antimony tartrate (Andrewes et al. 1998), although the biological formation of this trihydride is yet to be confirmed. SbH_3 has never been detected in controlled laboratory experiments in which trimethylantimony has been generated from inorganic antimony compounds by cultures of anaerobic bacteria (Gürleyük et al. 1997; Gates et al. 1997; Jenkins et al. 1998c). In contrast, certain soil bacteria have been shown to produce arsine (AsH_3) as the sole As containing product when incubated anaerobically in the presence of arsenicals (Cheng and Focht 1979). Bioreduction of trimethylantimonydibromid to trimethylantimony by *Pseudomonas fluorescens* growing anaerobically in sealed vessels has been reported (Gürleyük et al. 1997).

Many microorganisms are able to able to reduce arsenic(V) to arsenic(III). This reduction is the first step in the arsenic biomethylation pathway of microorganisms. Other bacteria are able to derive energy and grow by reduction of arsenic in the process of anaerobic respiration (Laverman et al. 1995; Ahmann et al. 1994). In the case of antimony – assuming the Challenger mechanism (Cai et al. 1997) for biomethylation – the recent direct (Gürleyük et al. 1997) and indirect (Jenkins 1998b) evidence for trimethylantimony formation from antimony(V) substrates suggests that certain microorganisms are capable of reducing antimony. There are no reports of microorganisms able to derive energy and grow by reduction of antimony. It is worthy of note, however, that the reduction potentials for arsenic(V) and antimony(V) are similar and there appears to be no thermodynamic reason why dissimilatory antimony reduction could not provide sufficient energy to sustain bacterial growth. There is some evidence in the literature for microbial bio-oxidation of antimony. The bacterium *Stibiobacter senarmontii* has been reported to oxidize antimony trioxide (Lyalikova and Korbutaev 1989), while there is mention (supporting data not provided) of the fungus *S. brevicaulis* being able to oxidize antimony(III) to antimony(V) (Andrewes et al. 1998). In contrast, a wide range of microorganisms are known to oxidize arsenic(III) to arsenic(V). It is evident from the literature that there are many significant differences in the interaction of microorganisms with arsenic and antimony compounds. Many of these apparent differences may be caused by selectivity of cellular uptake, rather than distinct intracellular processing of these metalloids. There is a need therefore for a comparative study of antimony and arsenic biotransformation capability in systems where differences in cell uptake selectivity for the various forms of the two metalloids are avoided, such as in cell-free extracts or in permeabilised cells.

References

Ahmann D, Roberts AL, Krumholz LR, Morel FMM (1994) Microbe grows by reducing arsenic. Nature 371: 750

Andreae MO, Froelich PN (1984) Arsenic antimony and germanium biogeochemistry in the Baltic Sea. Tellus, Series B – Chemical and Physical Meteorology 36:101–117

Andreae MO, Asmodé JF, Foster P, Van't Dack L (1981) Determination of antimony(III), antimony(V) and methylantimony species in natural waters by atomic absorption spectrometry with hydride generation. Anal Chem 53:1766–71

Andreae MO, Byrd JT, Froelich PN (1983) Arsenic, antimony, germanium, and tin in the Tejo estuary, Portugal – Modeling a polluted estuary. Environ Sci Technol 17:731–737

Andrewes P, Cullen WR, Feldmann J, Koch I, Polishchuk E, Reimer E (1998) The production of methylated organoantimony compounds by *Scopulariopsis brevicaulis*. Applied Organometallic Chemistry, 12:827–842

Anonymous (1998), Expert group to investigate cot death theories: Toxic gas hypothesis (Chairman: Lady Limerick), Final Report. Department of Health, London

Barnard K (1947) *Title unknown*. Ph.D. Thesis, University of Leeds (UK), pp 9–29

Benson AA (1988) Antimony metabolites in marine algae. In: Craig P, Glockling F (eds) The biological alkylation of heavy metals. Royal Society of Chemistry, London, p 135

Brinckman FE, Olson GJ, Thayer JS (1985) Biological mediation of marine metal cycles; the case of methyl iodide. In: Sigleo AC, Hattori A (eds) Marine and estuarine geochemistry. Lewis, Chelsea, Mich., p 227–238

Cai Y, Jaffe R, Jones R (1997) Ethylmercury in the soils and sediments of the Florida Everglades. Environ Sci Technol 31:302–305

Carmalt CJ, Norman NC (1997) Arsenic, antimony and bismuth: Some general properties and aspects of periodicity. In: Norman NC (ed) Chemistry of arsenic antimony and bismuth. Blackie, London

Challenger F (1978) Biosynthesis of organometallic and organometalloidal compounds. In: Brinckman FE, Bellama JM (eds) Orgnaometals and organometalloids: Occurrence and fate in the environment. American Chemistry Society, Washington, p 1–22
Chen CN, Focht DD (1979) Production of arsine and methylarsines in soil and in culture. Appl Environ Microbiol 38:494–498
Choi SC, Chase T, Bartha R (1994) Metabolic pathways leading to mercury methylation in *Desulfovibrio desulfuricans* LS. Appl Environ Microbiol 60:1342–1346
Craig PJ, Forster SN, Jenkins RO, Miller D, Ostah N (1999) An analytic method for the detection of methyl antimony species in environmental matrices - methyl antimony levels in some UK plant material. Analyst, in press
Cullen WR, Reimer KJ (1989) Arsenic speciation in the environment. Chem Rev 89:713–764
Cullen WR, Froese CL, Lui A, McBride BC, Patmore DJ, Reimer M (1977) The aerobic methylation of arsenic by microorganisms in the presence of L-methionine-methyl-d. Journal of Organometallic Chemistry, 139:61–69
Doak GO, Freedman LD (1970) Organometallic compounds of arsenic antimony and bismuth. Wiley Interscience, London
Doak GO, Long GG, Key ME (1967) Trimethylantimony dihalides. Inorganic Syntheses, 9:92–97
Dodd M, Grundy SL, Reimer KJ, Cullen WR (1992) Methylated antimony(V) compounds - synthesis, hydride generation properties and implications for aquatic speciation. Applied Organometallic Chemistry 6:207–211
Dodd M, Pergantis SA, Cullen WR, Li H, Eigendorf GK, Reimer KJ (1996) Antimony speciation in freshwater plant extracts by using hydride generation-gas chromatography-mass spectrometry. Analyst, 121:223–228
Fatoki OS (1997) Biomethylation in the nature environment: A review. South African Journal of Science 93:366–370
Feldmann J, Cullen WR (1997) Occurrence of volatile transition metal compounds in landfill gas: Synthesis of molybdenum and Tungsten carbonyls in the environment. Environ Sci Technol 31:2125–2129
Feldmann J, Hirner AV (1995) Occurrence of volatile metal and metalloid species in landfill and sewage gases. International Journal of Environmental Analytical Chemistry 60:339–359
Feldmann J, Grümping R, Hirner AV (1994) Determination of volatile metal and metalloid compounds in gases from domestic waste deposits with GC ICP-MS. Fresenius Journal of Analytical Chemistry 350:228–234
Feldmann J, Koch I, Cullen WR (1998) Complementary use of capillary gas chromatography mass spectrometry (ion trap) and gas chromatography inductively-coupled plasma mass spectrometry for the speciation of volatile antimony, tin and bismuth compounds in landfill and fermentation gases. Analyst 123:815–820
Gates PN, Harrop HA, Pridham JB, Smethurst B (1997) Can microorganisms convert antimony trioxide or potassium antimony tartrate to methylated stibines? Sci Total Environ 205:215–221
Gürleyük H, Van Fleet-Stalder V, Chasteen TG (1997) Confirmation of the biomethylation of antimony compounds. Applied Organometallic Chemistry 11:471–483
Hirner AV, Feldmann J, Goguel R, Rapsomanikis S, Fischer R, Andreae MO (1994) Volatile metal and metalloid species in gases from municipal waste deposits. Applied Organometallic Chemistry 8:65–69
Jaun B (1993) Methane formation by methanogenic bacteria: Redox chemistry of co-enzyme F430. In: Sigel H, Sigel A (eds) Metal ions in biological systems: Biological properties of metal alkyl derivatives. Marcel Dekker, New York, 29:287–332
Jenkins RO, Craig PJ, Goessler W, Irgolic KJ (1998a) Biovolatilizationof antimony and sudden infant death syndrome (SIDS). Hum Exp Toxicol 12:231–238
Jenkins RO, Craig PJ, Goessler W, Miller D, Ostah N, Irgolic KJ (1998b) Biomethylation of inorganic antimony compounds by an aerobic fungus: *Scopulariopsis brevicaulis*. Environ Sci Technol 32: 882–885
Jenkins RO, Craig PJ, Miller DP, Stoop LCAM, Ostah N, Morris TA (1998c) Antimony biomethylation by mixed cultures of microorganisms under anaerobic conditions. Applied Organometallic Chemistry 12:449–455
Jones RD (1994) Survey of antimony workers-mortality 1961–1992. Occup Environ Med 51:772–776
Kajander EO, Harvima RJ, Eloranta TO, Martikainen M, Kantola M, Karenlampi SO, Akerman K (1991) Metabolism, cellular actions, and cytotoxicity of selenomethionine in cultured-cells. Biol Trace Elem Res 28:39–45
Koch I, Feldmann J, Lintschinger J, Serves SV, Cullen WR, Reimer KJ (1998) Demethylation of trimethylantimony species in aqueous solution during analysis by hydride generation gas chromatography with AAS and ICP MS detection. Applied Organometallic Chemistry 12:129–136
Krupp EM, Grümping R, Furchtbar UR, Hirner AV (1996) Speciation of metals and metalloids in sediments with LTGC/ICP-MS. Fresenius Journal of Analytical Chemistry 354:546–549

Laverman AN, Switzer Blum J, Schaeffer JK, Phillips EJP, Lovley DR, Oremland RS (1995) Growth of strain SES-3 with arsenate and other diverse electron acceptors. Appl Environ Microbiol 61:3556–3561
Lintschinger J, Koch I, Serves S, Feldmann J, Cullen WR (1997) Determination of antimony species with high-performance liquid chromatography using element specific detection. Fresenius Journal of Analytical Chemistry 359:484–491
Lintschinger J, Schamel O, Kettrop A (1999) The analysis of antimony species by using ESI-MS, and HPLC-ICP-MS. Fresenius Journal of Analytical Chemistry 361:96–102
Lyalikova NN, Norbutaev AK (1989) Role of bacteria in Sb and Mo oxidation. In: Anke M et al. (eds) 6th International Trace Element Symposium, Leipzig, pp 246–250
Maeda S (1994) Safety and environmental effects. In: Patai S (ed) The chemistry of organic arsenic, antimony and bismuth compounds. John Wiley & Sons, Chichester (UK), p 725
Meinema HA, Noltes JG (1972) Investigations on organoantimony compounds VI. Preparation and properties of thermally stable dialkylantimony(V) compounds of the types $R_2Sb(OR')_3$, $R_2Sb(OAc)_3$ and $R_2Sb(O)OH$. Journal of Organometallic Chemistry, 36:313–322
Morgan GT, Davies GR (1926). Antimonyl analogues of the cacodyl series. Proceeding of the Royal Society of Chemistry, Series A, p 523
Oehme FM (1979) Toxicity of heavy metals in the environment, part 2. Marcel Dekker Inc, New York
Parris GE, Brinckman FE (1976) Reactions which relate to environmental mobility of arsenic and antimony. II. Oxidation of trimethylarsine and trimethylstibine. Environ Sci Technol 10:1128–1134
Pratt HM (1993) Making and breaking the co-alkyl bond in B12 derivatives. In: Sigel H, Sigel A (eds) Metal ions in biological systems: Biological properties of metal alkyl derivatives, vol XXIX. Marcel Dekker, New York, pp 229–280
Richardson BA (1994) Sudden infant death syndrome: A possible primary cause. J Forensic Sci 34: 199–204
Takahashi K, Yamauchi K, Mashiko M, Yamamura Y (1990) Effect of s-adenosylmethionine on methylation of inorganic arsenic. Nippon Koshu Eisei Zasshi 45:613; Chemical Abstracts 114:96350k
Thayer JS (1984) Organometallic compounds and living organisms. Academic Press,New York
Thayer JS (1995) Environmental chemistry of the heavy elements: Hydrido and organo compounds. VCH, New York
Turner-Report (1991) Sudden infant syndrome (SIDS). HMSO, London

Chapter 15

Redox Processes of Chromium in Sea Water

M. Pettine

15.1 Introduction

Chromium exists in natural waters in two oxidation states, Cr(III) and Cr(VI), that are characterized by different chemical behaviour, bioavailability and toxicity. Dissolved Cr(III) has a strong tendency to adsorb to surfaces (Cranston and Murray 1978) and is characterized by a low solubility. Candidates for the solubility control of chromium(III) in natural waters are not fully understood. They include $Cr(OH)_{3(s)}$ (Baes and Mesmer 1976), mixed hydroxides of the type $(Cr_xFe_{1-x})(OH)_{3(s)}$ (Sass and Rai 1987; Rai et al. 1989) and chromite ($FeCr_2O_4$) (Murray et al. 1983) and give variable concentrations from 10^{-7} to 10^{-10} M. The substitution of waters of hydration in aquated Cr(III) ions is extremely slow and the stability of the hydration sphere imposes limitations on the mechanism of electron transfer reactions. Cr(III) has a low toxicity toward organisms since its bioavailability is controlled by solubility limitations and the formation of strong complexes. The very low assimilation of Cr(III) has led to its use as an inert tracer of ingested particulate matter to measure carbon assimilation in marine invertebrates (Wang et al. 1997). At low levels Cr(III) is an essential element for mammals. Cr(VI) forms anionic oxy-compounds that also have an adsorption affinity for certain proton-specific mineral surfaces (Zachara et al. 1987). However, their adsorption is limited in sea water where both Cr(VI) species and particle surfaces have negative charges, and high concentrations of sulfate (0.028 M) compete with chromate for adsorption sites. Cr(VI) is much more bioavailable to aquatic organisms than Cr(III) and it is thought that this oxidized species is transported into the cells as a sulfate analogue (Riedel 1984). The competition between Cr(VI) and sulfate is responsible for significant changes in the bioavailability and toxicity of Cr(VI) in estuarine waters where sulfate concentrations increase markedly with salinity (Riedel 1984; Riedel 1985). Cr(VI) compounds are toxic due to their oxidation of intracellular compounds and are also known as a human carcinogen and mutagen.

The bioavailability of Cr and the rates and routes of its accumulation in aquatic organisms are critical for understanding Cr toxicity and its risk assessment for aquatic organisms and for public health (Wang et al. 1997). To this end, the speciation and redox behaviour of Cr deserve a high priority, since they control the interactions of this metal with biota and the possible environmental effects.

15.2 Speciation of Cr(III) and Cr(VI)

Cr(III) speciation in sea water is dominated by Cr(III) hydrolysis products. Elderfield (1970), using available association constants, calculated that $Cr(OH)_2^+$ was the most important hydroxyl complex in seawater. Recently, Rai et al. (1987) have revised the hydrolysis constants of Cr(III) and suggested that $Cr(OH)_2^+$ contributes only marginally to the total soluble chromium (III), while $Cr(OH)_3$ is the dominant complex in aquatic systems at the most common pH. The hydrolysis constants of Cr^{3+} resulting from these recent findings (Rai et al. 1987; Rai et al. 1989) have been compared with those proposed by Baes and Mesmer (1976) (see appendix). The Cr(III) speciation at seawater ionic strength (0.75 M), according to the revised constants, is shown in Fig. 15.1a: $Cr(OH)_3$, $Cr(OH)^{2+}$ and $Cr(OH)_2^+$ give approximately similar contributions at a pH of about 6.4, $Cr(OH)_3$ is dominant in the pH range 7–11, while $Cr(OH)_4^-$ becomes dominant above pH 11.3. Ion pairing with specific ions other than OH^- (F^-, Cl^-, Br^- and SO_4^{2-}) is not considered to be important in seawater (Elderfield 1970; Turner et al. 1981). Unfortunately, it is not possible to make reliable speciation calculations for ligands such as CO_3^{2-} or $B(OH)_4^-$ because the stability constants are not available. The association constants of some polynuclear species ($Cr_2(OH)_2^{4+}$, $Cr_3(OH)_4^{5+}$, $Cr_4(OH)_6^{6+}$) are (Rai et al. 1987) many orders of magnitude lower than previously reported, and, at the very low levels of chromium(III) in sea water, their formation does not occur (Van Der Weijden and Reith 1982; Rai et al. 1989). Information on organic Cr(III) species is limited and opinions on the role of organic chromium species in aquatic systems are conflicting (Nakayama et al. 1981a; Osaki et al. 1983).

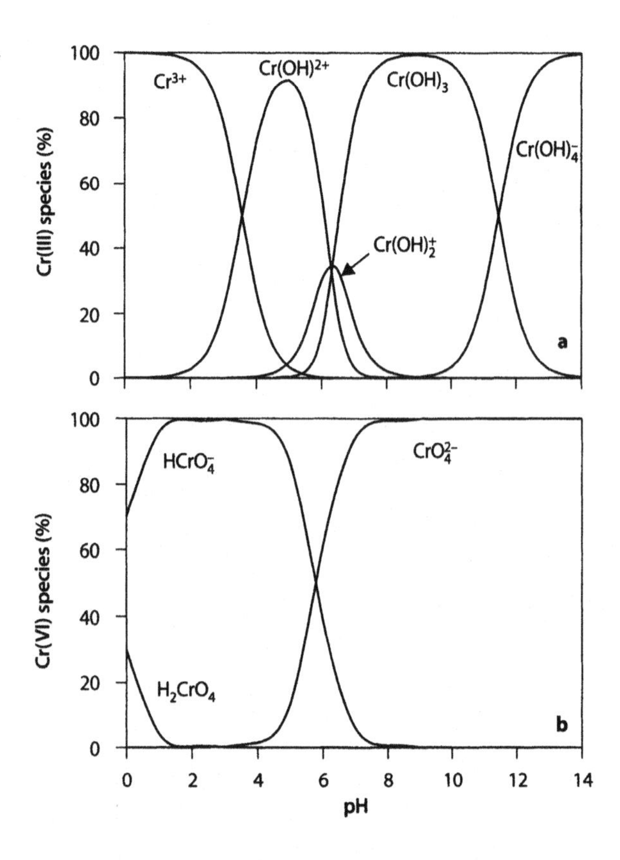

Fig. 15.1. Speciation as a function of pH; **a** Cr(III); **b** Cr(VI)

Cr(VI) can exist in a number of different forms as a function of pH, including H_2CrO_4, $HCrO_4^-$, CrO_4^{2-} and the dimer $Cr_2O_7^{2-}$ (see Fig. 15.1b and Appendix). At micromolar or lower levels of Cr(VI), typical of marine environments, the concentration of the dimer is negligible and the predominant Cr(VI) species at pH > 7 is chromate (Fig. 15.1b).

The reduction of CrO_4^{2-} to $Cr(OH)_2^+$ has been described by Elderfield (1970) as

$$CrO_4^{2-} + 6\,H^+ + 2\,H_2O + 3\,e^- \longrightarrow Cr(H_2O)_4(OH)_2^+ \tag{15.1}$$

The equilibrium constant for Reaction 15.1 from free energies of formation given by Hem (1977) is $\log K = 66.19$ at 25 °C and $I = 0$ M. This gives $\log K = 65.31$ at the ionic strength of seawater by assuming $\gamma_{Cr(OH)_2^+} = 0.82$ (Elderfield 1970) and $\gamma_{H^+} = 0.97$ and $\gamma_{CrO_4^{2-}} = 0.13$ (Millero and Schreiber 1982). According to the above reaction, the equilibrium ratio of dissolved Cr is given by (Murray et al. 1983)

$$\log([Cr(VI)]\,/\,[Cr(III)]) = 6\,pH + 3\,pe - \log K \tag{15.2}$$

At pH 8.1 and pe 12.5, the ratio of Cr(VI) to Cr(III) should be about 10^{20} in both diluted and saline solutions; Cr(VI) should predominate. If pe is 6.5, consistent with control of sea water pe by the O_2/H_2O_2 couple and $H_2O_2 = 0.1$ µM (Moffett and Zika 1983), the ratio falls to about $10^{2.8}$, again suggesting a large dominance of Cr(VI).

By using the Cr(III) hydrolysis constants revised by Rai et al. (1987), $Cr(OH)_3$ is the dominant species in seawater (Fig. 15.1 and 15.2) and the equilibrium ratio of dissolved Cr is given by

$$\log([Cr(VI)]\,/\,[Cr(III)]) = 5\,pH + 3\,pe - \log K \tag{15.3}$$

where $\log K$ is 59.84 and 58.86 at $I = 0$ and 0.75 M.

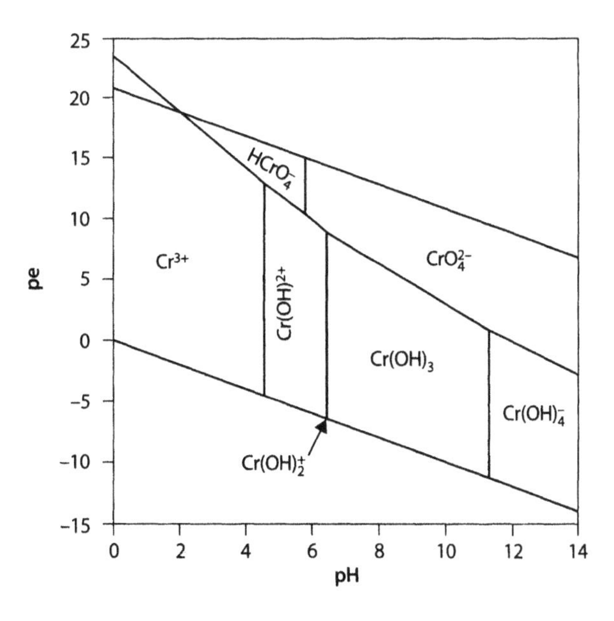

Fig. 15.2. pe-pH diagram of Cr species at 25 °C and $I = 0.75$ M (Equilibria data from Rai et al. 1987, see appendix)

The Cr(VI)/Cr(III) ratios at seawater ionic strength become in this case $10^{19.4}$ and $10^{1.1}$ with pe of 12.5 and 6.5, respectively. According to Reaction 15.3, Cr(VI) should be about a factor of 10 higher than Cr(III) at a pe of 6.5. This Cr(VI)/Cr(III) ratio is about an order of magnitude lower than that resulting from Relation 15.2 at the same pe. In anoxic regions the ratio Cr(VI)/Cr(III) should be about 10^{-20} or less, suggesting analytically insignificant concentrations of Cr(VI).

15.3 Environmental Concentrations

The chromium content of natural solids varies widely from tens of nanomoles to a few micromoles per gram of solid (Richard and Bourg 1991); while the dissolved concentrations range from one to ten nanomoles per liter in unpolluted or slightly polluted aquatic systems. Typical concentrations recorded in oceanic, coastal and estuarine waters are given in Table 15.1. These available data indicate that Cr(VI)/Cr(III) ratios range from <1 to >100 for oxic conditions (Chuecas and Riley 1966; Fukai 1967; Fukai and Vas 1969; Kuwamoto and Murai 1970; Grimaud and Michard 1974; Emerson et al. 1979; Cranston and Murray 1980; Nakayama et al. 1981b; Nakayama et al. 1981c; Cranston 1983; Murray et al. 1983; Campbell and Yeats 1984; Ahern et al. 1985; Pettine et al. 1992; Pettine et al. 1997). Earlier results (Chuecas and Riley 1966; Fukai 1967; Fukai and Vas 1969; Kuwamoto and Murai 1970; Grimaud and Michard 1974) giving values <1 were probably affected by the alteration due to Cr(VI) reduction on storage at low pH (Fukai and Vas 1969; Kuwamoto and Murai 1970; Grimaud and Michard 1974; Ahern et al. 1985; Pettine et al. 1988) or reflected variable extraction yields of organically bound Cr(III) (Nakayama et al. 1981a; Jeandel and Minster 1984). Ratios of about 0.5 and 2 were found in Saanich Inlet below the O_2/H_2S interface (Emerson et al. 1979) and at stations along a transect from Hawaii to the coast of Baja California at the top of oxygen minimum zone (Murray et al. 1983), respectively.

Thus, field distributions of Cr(III) and Cr(VI) appear in many cases to be far from thermodynamic equilibria in both oxic and anoxic environments. However, the revision of the hydrolysis constants of Cr(III) (Rai et al. 1987) makes the extent of the discrepancy of the field results with thermodynamic expectations lower than was previously highlighted by Elderfield (1970) for oxic environments.

Many factors, including unreliable analytical data, existence of unknown redox species, and kinetic control by purely chemical and biologically mediated reactions may be invoked to explain the deviations of the field concentrations with respect to thermodynamic predictions. This discrepancy has stimulated studies on the kinetics and speciation of chromium aimed at elucidating processes responsible for Cr(III)-Cr(VI) interconversions in environmental matrices. Data recently gathered on the kinetics of the main redox reactions affecting the cycle of chromium in natural waters will be reviewed in the next section.

15.4 Kinetic Studies

Until recently, geochemical and environmental models have considered the fate of metals in the environment as the result of equilibrium processes only, while it has now

Table 15.1. Examples of chromium concentrations in estuarine, coastal and oceanic systems

Locations	Cr(VI) (nM)	Cr(III) (nM)	References
British coastal	0	8.8	Chuecas and Riley 1966
Ligurian sea	0.96 – 6.9	0.38 – 4.8	Fukai 1967; Fukai and Vas 1969
Pacific Ocean	1.3 – 5.2	2.5 – 17.3	Kuwamoto and Murai 1970
Equatorial Pacific Ocean	0.58 – 2.1	4.6 – 10	Grimaud and Michard 1974
Northeast Pacific Ocean	2.9	0.01	Cranston and Murray 1978
Saanich Inlet			Emerson et al. 1979
Surface waters	1.5 – 2	< 0.2	
O_2/H_2S interface	0.75 – 0.25	0.2 – 0.8	
H_2S zone	0.4 – 0.5	0.8 – 0.9	
Columbia River Estuary			Cranston and Murray 1980
River	3.14	0.06	
Sea	3.5	0.28	
Cascadia Basin	1.6 – 3.3	0.01 – 0.30	Cranston 1983
Hawai-Baja California			Murray et al. 1983
Surface waters (O_2 = 200–300 μM)	3	≤ 0.2	
Top oxygen minimum (O_2 ≅ 10 μM)	2	0.5 – 1	
Deep waters (O_2 ≅ 130 μM)	4.5	0.2	
St. Lawrence Estuary			Campbell and Yeats 1984
River	13.1 ± 1.1[a]		
Sea	4.4 – 4.8[a]		
Pacific Ocean	2.3 – 3.8	0.9 – 1.8[a]	Nakayama et al. 1981b[b]
Sea of Japan	2 – 3.2	0.8 – 1.6[b]	Nakayama et al. 1981c[b]
North of Sydney	3.1 – 6.0	1.9 – 8.3	Ahern et al. 1985
Po River Estuary			Pettine et al. 1992, 1997
River	18.3 ± 8.6	3.8 ± 3.1	
Sea	6 – 13	5.8 – 8.8	

[a] Values refer to Cr_{TD}.
[b] The authors report also data on organic Cr: 4.2–6.1 (Pacific Ocean); 3.6–5.2 (Japan Sea).

become obvious that the kinetics of reactions play an important role as well (Saleh et al. 1989). Environmental concentrations of Cr(III) and Cr(VI) reflect the inputs and redox rates of chromium species. The redox transformations of Cr(III) into Cr(VI) or vice versa involve the exchange of three electrons with another redox couple which may accept or donate them. In natural aquatic environments, the main significant redox couples are: H_2O/O_2, H_2O_2/O_2, Mn(II)/Mn(IV), N_2/NO_3, NH_4/NO_3, Fe(II)/Fe(III), HS^-/SO_4^{2-} and CH_4/CO_2 (Schroeder and Lee 1975; Eary and Rai 1987; Saleh et al. 1989; Richard and Bourg 1991).

15.4.1 Oxidation Processes

Because the redox potential of the Cr(VI)/Cr(III) couple is so high (Fig. 15.3), there are few oxidants capable of oxidizing Cr(III) to Cr(VI). Dissolved oxygen and manganese oxides were first considered as the most important species responsible for the oxidation of Cr(III) in aquatic environments (Eary and Rai 1987).

The kinetics of the oxidation of Cr(III) to Cr(VI) with O_2 have been investigated in laboratory and field studies, and the reaction was found to be slow with half lives ranging from 2 to 20 months (Schroeder and Lee 1975; Cranston and Murray 1978; Emerson et al. 1979; Eary and Rai 1987).

The oxidation of Cr(III) by manganese oxides is reported to be more rapid than by dissolved oxygen. These oxides are present in aquatic environments as grain coatings, crack deposits or finely disseminated grains, related or not with bacterial activities (Richard and Bourg 1991). The oxidation reaction occurs through three steps, including adsorption of Cr(III) onto MnO_2 surface sites, oxidation of Cr(III) to Cr(VI) by surface Mn(IV) and desorption of the reaction products, Cr(VI) and Mn(II). The Cr(III) oxidation rates were found to be strongly related to the amount and the surface area of Mn oxides, changing from initially rapid values to significantly lower rates after 20–60 min (Schroeder and Lee 1975; Nakayama et al. 1981d; Eary and Rai 1987; Fendorf and Zasoski 1992). Different theoretical stoichiometry and rates have been suggested for different Mn oxides (Eary and Rai 1987; Richard and Bourg 1991)

$$Cr^{3+} + 1.5\ \delta\text{-}MnO_2(s) + H_2O \longrightarrow HCrO_4^- + 1.5\ Mn^{2+} + H^+ \tag{15.4}$$

$$Cr(OH)^{2+} + 3\ \beta\text{-}MnO_{2(s)} + 3\ H_2O \longrightarrow HCrO_4^- + 3\ MnOOH_{(s)} + 3\ H^+ \tag{15.5}$$

Eary and Rai (1987) have reported that Cr(III) is readily oxidized to Cr(VI) by β-$MnO_2(s)$ over the pH range from 3.0 to 10.1 and calculated a half life of 95 days for the oxidation of a 10^{-5} M Cr(III) solution for 1 kg of a 20% porosity soil containing 0.05 % wt β-$MnO_{2(s)}$ with a surface area of 5.0 $m^2\,g^{-1}$. Other forms of soil manganese oxides such as birnessite (δ-MnO_2) and cryptomalene (α-MnO_2) that are likely to have higher surface energies than β-$MnO_{2(s)}$ may cause a more rapid oxidation of Cr(III) than β-$MnO_{2(s)}$. Therefore, manganese oxides are likely to be the most important oxidants for Cr(III) in ground-water porous systems and at the sediment-water interface of oxic aquatic systems. Under different field conditions where concentrations of Mn particles are lower, and competition for surface sites and inorganic/organic coating of reactive sites increase, the role of Mn oxide should be minor.

Recent findings (Cooper and Zika 1983; Zika et al. 1985) on the photochemical production of H_2O_2 in surface sea water have stimulated our interest in studying its role in Cr speciation through the oxidation of Cr(III). The rates of oxidation of Cr(III) to Cr(VI) with H_2O_2 have been measured under pseudo first order conditions in borate buffered $NaClO_4$ and NaCl solutions (Pettine and Millero 1990; Pettine et al. 1991) as a function of pH, temperature, ionic strength and H_2O_2 concentration. The oxidation rates were found to be first order with respect to both $1/H^+$ or OH^- in the pH range 7–9 and H_2O_2 in the range 112 to 2 621 μM. Fig. 15.4 shows the values of the logarithm of the pseudo-first-order constant as a function of pH. Results showed a marked de-

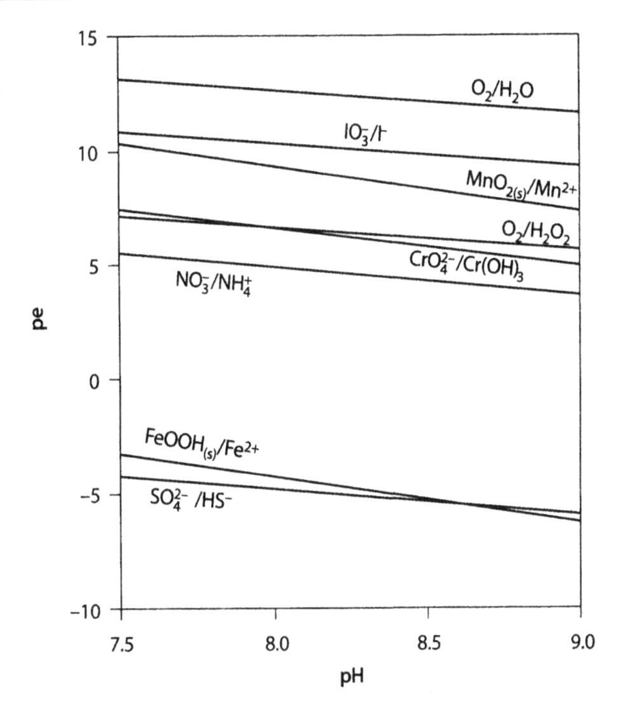

Fig. 15.3. Calculated values of pe for the most important redox couples and Cr in the most usual pH range of sea water. The following conditions were assumed: $[Mn^{2+}] = [Fe^{2+}] = 10^{-7}$ M; $O_2 = 0.2$ atm; $[H_2O_2] = 10^{-7}$ M; for the redox couples Cr(VI)/Cr(III), NO_3^-/NH_4^+, O_4^{2-}/HS^-, Fe(III)/Fe(II) a concentration ratio of 1/1 was chosen

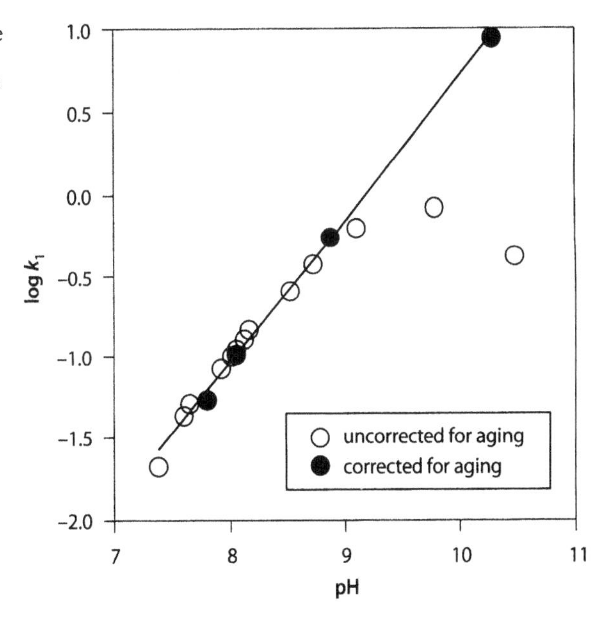

Fig. 15.4. Values of $\log k_1$ for the oxidation of Cr(III) (1.9 μM) with H_2O_2 (447 μM) at 25 °C as a function of pH ($NaClO_4$ 1 M) (Pettine and Millero 1990)

viation from the linearity above pH 9. This deviation was caused by aging Cr(III) solutions over a short time before the addition of H_2O_2. The effect was noted at pH > 9 where the kinetics of the aging process increased markedly (Fig. 15.5) and was due to the precipitation of Cr(III) inert to oxidation. The kinetics of the aging at any pH was found to be much slower than that of the oxidation of Cr(III) with H_2O_2, thus exerting an influence on the slope of ln [Cr(III)] vs. time, but not on the linearity of the curve. The aging was strongly influenced by the temperature of the medium and the concentrations of magnesium, borate and carbonate ions other than pH. Borate caused a reduction of the aging effect, while carbonate and magnesium produced an increase of the aging effect. The Mg^{2+} influence (Fig. 15.6) was much stronger than CO_3^{2-}, when these ions were present at their seawater level in individual Na-Mg-Cl and Na-Cl-CO_3^{2-} solutions. When CO_3^{2-} and Mg^{2+} were present at the same time, the combined aging effect was less than that found in simple Na-Mg-Cl solutions (Pettine et al. 1991). The effect of CO_3^{2-} and Mg^{2+} on the aging process involves the formation of new species (probably mixed species such as $CrCO_3OH$ and $Cr_xMg_{(1-x)1.5}(OH)_3$) which increase the rates of the precipitation. The values of the overall rate constant ($k = k_1 / [H_2O_2]$, $M^{-1}\,min^{-1}$) for the oxidation of Cr(III) with H_2O_2 resulting from kinetic runs in $NaClO_4$ and NaCl media fit the equation (SD = ±0.06 in log k)

$$\log k = -4.60 - 5.13 / T + 0.87\,\mathrm{pH} \qquad (15.6)$$

from 5 to 40 °C, pH 7–9 and I = 0 to 1 M. Rates were independent of ionic strength and gave an energy of activation for the reaction of 9.9 ±0.8 kcal mol^{-1} (Pettine and Millero 1990). In the presence of borate, the rates of the oxidation of Cr(III) with H_2O_2 increased (Pettine et al. 1991), suggesting that the rate-determining steps involve two chromium(III) species according to

$$Cr(H_2O)_5OH^{2+} + H_2O_2 \longrightarrow \text{products} \qquad (15.7)$$

$$Cr(H_2O)_4(OH)[B(OH)_4]^{+} + H_2O_2 \longrightarrow \text{products} \qquad (15.8)$$

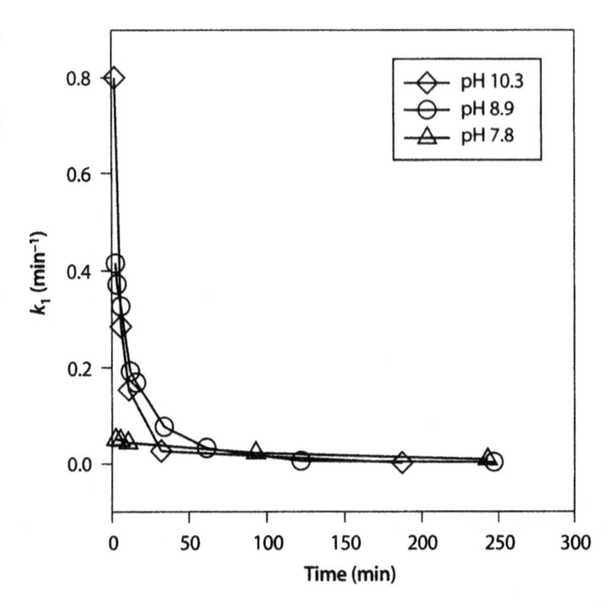

Fig. 15.5. Effect of aging on the values of k_1 (min^{-1}) obtained for the oxidation of Cr(III) (1.9 μM) with H_2O_2 (447 μM) at 25 °C (Pettine and Millero 1990)

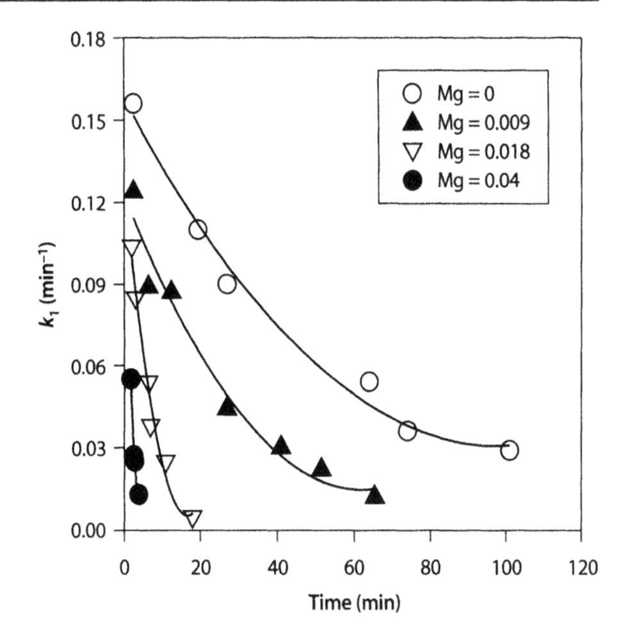

Fig. 15.6. Values of k_1 for the oxidation of 1.9 µM Cr(III) with 447 µM H_2O_2 as a function of aging time at pH 8.00 ±0.05 and 25 °C in NaCl solutions buffered with 1.85 mM $B(OH)_4^-$ and spiked with different molar Mg^{2+} concentrations at 0.5 M ionic strength (Pettine et al. 1991)

The rate equation is in this case given by

$$d[\text{Cr(III)}] / dt = -k_3[\text{Cr(III)}][H_2O_2][OH^-][B(OH)_4^-]^{0.3} \tag{15.9}$$

By assuming a concentration of 0.1 µM H_2O_2, the half life of Cr(III) in NaCl or $NaClO_4$ solutions (0.009 M borax) is 24 days according to Eq. 15.6, while direct measurements in artificial and real sea water at natural boron levels gave 45 days. Since estimates of the half life for the oxidation of Cr(III) with O_2 under similar conditions are about 500 days (Schroeder and Lee 1975), our calculations, although crude, indicate that the levels of H_2O_2 in natural waters are 20 times more effective in oxidizing Cr(III) than is O_2. The kinetics of the oxidation of Cr(III) by H_2O_2 are also faster than those by Mn oxides suggesting that H_2O_2 is the most efficient oxidant of Cr(III) in natural surface waters.

15.4.2 Reduction Processes

Many chemicals may reduce Cr(VI) to Cr(III) in water or inside the cells of organisms. Nakayama et al. (1981d) studied the pH dependence of Cr(VI) reduction by several compounds, including ascorbic acid, hydroxylamine, humic acid and formaldehyde, but they did not measure the rates of reduction. Ascorbic acid and hydroxylamine are able to reduce Cr(VI) at the pH of sea water, but the concentrations expected for these compounds in natural waters are too low to really affect the distribution of chromium. Humic or fulvic acid are naturally present at effective levels, but experimental results (Nakayam et al. 1981d; Eckert et al. 1990; Wittbrodt and Palmer 1995) indicate that these reducing agents do not affect chromium distribution at a pH higher than 6.5.

A number of low molecular weight organic compounds including substituted carboxylic acids and phenols were also found to be able to reduce Cr(VI), and reactions were found to be catalyzed by oxide surfaces (Deng and Stone 1996). Fe(II) and H_2S appear to be the most important candidates in the reduction of Cr(VI) in natural waters (Saleh et al. 1989). Both these chemicals are widely used in treatment processes of industrial wastes for Cr(VI) removal through its reduction to Cr(III) and the subsequent precipitation of Cr(III) ions (Eary and Rai 1988).

Reduced forms of iron and sulfur can be produced in the pore waters of sediments and in anoxic basin waters due to chemical or biological processes as well as in hydrothermal waters due to geochemical processes. Fe(II) and sulfides may also be produced in surface sea waters by photooxidation of organic matter (Collienne 1983; Miller et al. 1995; Voelker et al. 1997) and hydrolysis of carbonyl sulfides. Recently, we have studied the kinetics of the Cr(VI) reduction by H_2S (Pettine et al. 1994). The overall rate constant (k, $M^{-1}\ min^{-1}$) in

$$d\mathrm{Cr(VI)} / dt = -k[\mathrm{Cr(VI)}][\mathrm{H_2S}] \tag{15.10}$$

was given by (SD = 0.07 in log k)

$$\log k = 16.19 - 1.06\ \mathrm{pH} - 2301 / T \tag{15.11}$$

and was shown to be independent of ionic strength. Based on these kinetic measurements, the half times of Cr(VI) in NaCl media range from 0.5 to 500 ays at 1 mM and 1 μM, respectively (Pettine et al. 1994). The reduction of Cr(VI) with H_2S was found to be catalyzed by heavy metals (Pb^{2+}, Cu^{2+}, Cd^{2+}, Ni^{2+}), which caused large increases in the reduction rates at micromolar concentrations (Pettine et al. 1998a). The effect was attributed to the formation of $MeCrO_4$ complexes which react faster with sulfide than free chromate and follow the sequence $NiCrO_4 > PbCrO_4 > CuCrO_4 > CdCrO_4$. Fe(III) was also found to exert a catalytic effect on the reduction of Cr(VI) with H_2S (Pettine et al. 1998a), but the effect was, in this case, attributed to the substitution of the slow one step reaction of Cr(VI) with H_2S with a fast two-step process consisting of Fe(III)+H_2S, followed by the reaction of Fe(II)+Cr(VI). The latter process involves cyclic reduction and oxidation of Fe with electron transfer from H_2S to Cr(VI).

The role of Fe(II) in the reduction of Cr(VI) in aqueous systems has been recently investigated by various workers (Fendorf and Li 1996; Buerge and Hug 1997; Sedlak and Chan 1997; Pettine et al. 1998b). Our recent paper (Pettine et al. 1998b) extended the previous findings by describing the reaction rate dependence on chromate and ferrous ion speciation, pH, temperature and ionic strength, and by allowing one to predict chromium reactions under most naturally occurring conditions from fresh to salt water environments.

The rate of Cr(VI) reduction is given by the general expression

$$-d[\mathrm{Cr(VI)}] / dt = k[\mathrm{Cr(VI)}][\mathrm{Fe(II)}] \tag{15.12}$$

where k ($M^{-1}\ min^{-1}$) can be determined from the

$$\log k = 6.74 - 1.01\ \mathrm{pH} - 188.5 / T \tag{15.13}$$

$$\log k = 11.93 + 0.95\,\text{pH} - 4\,260 / T - 1.06\, I^{0.5} \tag{15.14}$$

which are valid for the pH ranges 1.5–4.5 ($\sigma = 0.2$) and 5–8.7 ($\sigma = 0.2$), respectively, from 5 to 40 °C and 0.01 to 2 M ionic strength. The logarithm of the overall constant ($\log k$) shows a typical parabolic dependence on pH, decreasing from pH 1.5 to 4.5 and increasing from pH 5 to 8.7 (Fig. 15.7). The effect of pH, temperature and ionic strength on the reaction led to the reactions at low pH being due to

$$H_2CrO_4 + Fe^{2+} \xrightarrow{k_{H_2A\text{-}Fe}} \text{products} \tag{15.15}$$

while the reactions at high pH are due to

$$HCrO_4^- + FeOH^+ \xrightarrow{k_{HA\text{-}FeOH}} \text{products} \tag{15.16}$$

and

$$HCrO_4^- + Fe(OH)_2 \xrightarrow{k_{HA\text{-}Fe(OH)_2}} \text{products} \tag{15.17}$$

The overall rate expression over the entire pH range can be determined from ($H_2A = H_2CrO_4$)

$$k = k_{H_2A\text{-}Fe}\alpha_{H_2A}\alpha_{Fe^{2+}}) + k_{HA\text{-}FeOH}\alpha_{(HA^-)}\alpha_{(FeOH^+)} + k_{HA\text{-}Fe(OH)_2}\alpha_{HA^-}\alpha_{Fe(OH)_2} \tag{15.18}$$

where $k_{H_2A\text{-}Fe} = 5 \times 10^6$, $k_{HA\text{-}FeOH} = 1 \times 10^6$, $k_{HA\text{-}Fe(OH)_2} = 5 \times 10^{11}$ and $\alpha(i)$ are the molar fractions of Cr(VI) and Fe(II) which can be determined from the dissociation constants of chromic acid (Shen-Yang and Ke-An 1986) and the hydrolysis constants of

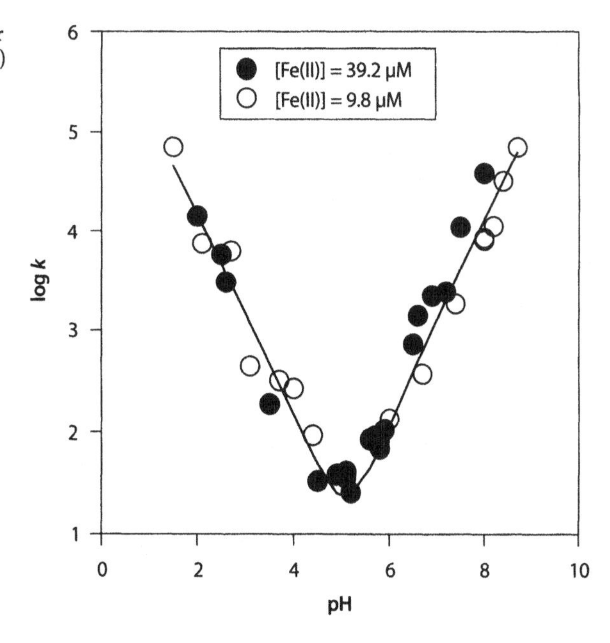

Fig. 15.7. Linear fits of $\log k$ for the reduction of 0.95 µM Cr(VI) with Fe(II) as a function of pH in 0.1 M NaCl at 10 °C (Pettine et al. 1998b)

iron (Millero et al. 1995). Half times for the reduction of Cr(VI) with Fe(II) are much faster than those estimated for the reduction with sulfide: at pH 8.0 and 25 °C they are about 0.5 and 58 h in sea water in the presence of 1 and 0.01 μM Fe(II).

15.5 The Chromium Cycle in Sea Water

Total dissolved chromium concentrations in the oceans show a slight depletion at the surface (2–3 nM) compared to deep waters (4–5 nM) and a general increase toward the bottom (up to 15.8 nM) (Jeandel and Minster 1984). Surface depletion suggests that Cr is affected by biochemical activity (Richard and Bourg 1991), and significant correlations of Cr with Si and P support the involvement of chromium in both a deep and shallow regeneration cycle, consisting in the release of Cr from siliceous parts and soft tissues of organisms, respectively (Cranston 1983).

Our recent studies improve upon existing models for the circulation of chromium in aquatic systems (Nakayama et al. 1981d; Richard and Bourg 1991; Kieber and Helz 1992) according to which Cr(III) is oxidized by Mn oxides and Cr(VI) is reduced by biologically mediated processes and other unspecified compounds (Fe(II) and organics).

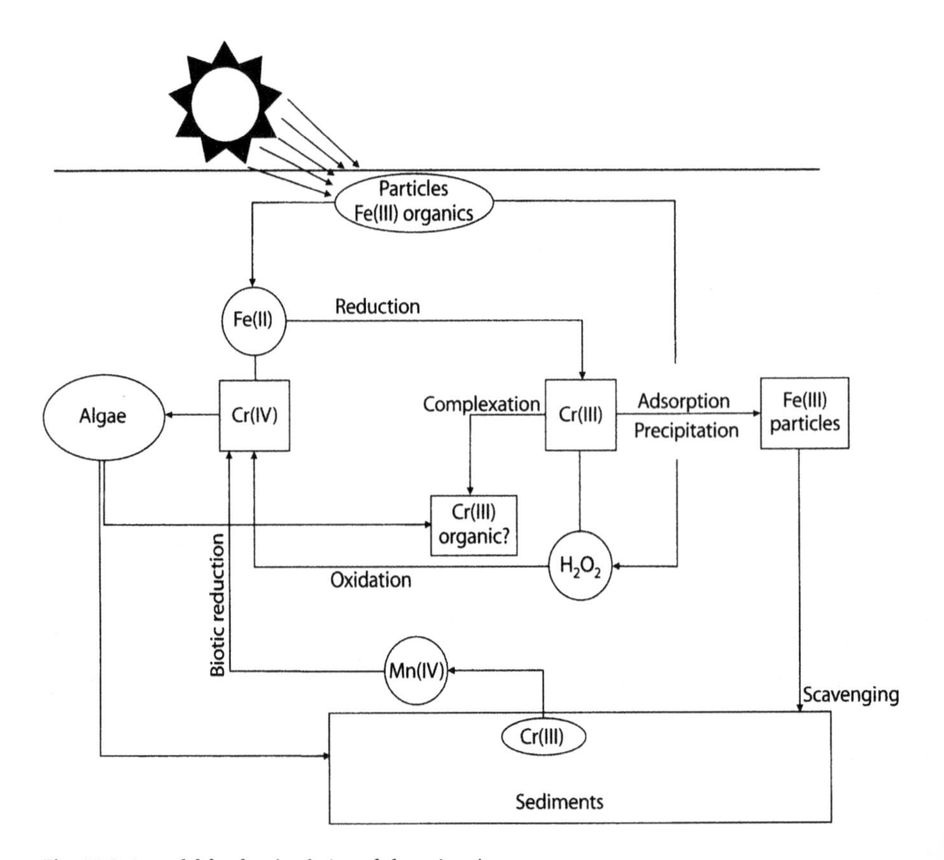

Fig. 15.8. A model for the circulation of chromium in sea water

Cr(III)-Cr(VI) interconversions in the photic layer of aquatic systems appear to be strongly controlled by the concentrations of photochemically produced Fe(II) and H_2O_2 (Fig. 15.8). By assuming pH = 8.2, $I = 0.74$, Fe(II) = 1 nM and H_2O_2 = 100 nM, which are conditions typical of surface sea water (Zika et al. 1985; Voelker et al. 1997), the kinetic balance between the reduction of Cr(VI) by Fe(II) and the oxidation of Cr(III) with H_2O_2 is reached when Cr(III) has approximately the same concentration as Cr(VI). Complexation of Cr(III) by organic ligands and sorption to surfaces also occur in the water column. Cr(III) scavenged from the water column may be reoxidized by Mn oxides at the sea/sediment interface of oxic systems producing Cr(VI) species which diffuse into the bottom waters.

Precipitation and sorption processes of Cr(III) explain the relatively short residence time of chromium ($\sim 10^4$ years) in oceans compared to sulfur ($\sim 10^7$ years) and molybdenum ($\sim 10^6$ years). These three elements behave similarly at their highest oxidation state: they form tetracoordinate oxyanions of identical charge and nearly identical size in oxygenated surface waters, are actively acquired by marine plankton by using the same anion channel and neither strongly sorb onto inorganic surfaces nor give rise to strong bioaccumulation processes (Kieber and Helz 1992). However, the kinetic control of the Cr(VI)/Cr(III) ratios in seawater, which provides Cr(III) concentrations higher than those expected on a thermodynamic basis, favours the removal of chromium from the water column through sorption and precipitation processes and lowers its residence time.

Further data on organic Cr species are needed for understanding their role in the cycle of chromium in natural waters.

Acknowledgements

The author wishes to thank Prof. F.J. Millero of the University of Miami for the revision of the English text. I wish also express my gratitude to Frank Millero for his continuous scientific stimulation during a twenty-year friendship.

References

Ahern F, Eckert JR, Payne NC (1985) Speciation of chromium in seawater. Anal Chim Acta 175: 147–151

Baes CF, Mesmer RE (1976) The hydrolysis of cations. Wiley, London

Buerge IJ, Hug SJ (1997) Kinetics and pH dependence of chromium(VI) reduction by iron(II). Environ Sci Technol 31:1426–1432

Campbell JA, Yeats P (1984) Dissolved chromium in the St. Lawrence Estuary. Estuar Coastal Shelf Sci 19:513–522

Chuecas L, Riley JP (1966) The spectrophotometric determination of chromium in seawater. Anal Chim Acta 35:240

Collienne RH (1983) Photoreduction of iron in epilimnion lakes. Limnol Oceanogr 28:83–100

Cooper WJ, Zika RG (1983) Photochemical formation of hydrogen peroxide in surface and ground waters exposed to sunlight. Science 20:711–712

Cranston RE (1983) Chromium in Cascadian basin, northeast Pacific Ocean. Mar Chem 13:109–125

Cranston RE, Murray JW (1978) The determination of chromium species in natural waters. Anal Chim Acta 99:275–282

Cranston RE, Murray JW (1980) Chromium species in the Columbia River and Estuary. Limnol Oceanol 25(6):1104–1112

Deng B, Stone AT (1996) Surface catalyzed chromium(VI) reduction: Reactivity comparisons of different organic reductants and different oxide surfaces. Environ Sci Technol 30:2484–2494

Eary LE, Rai D (1987) Kinetics of chromium(III) oxidation to chromium(VI) by reaction with manganese dioxide. Envir Sci Technol 21:1187–1193
Eary LE, Rai D (1988) Chromate removal from aqueous wastes by reduction with ferrous ion. Environ Sci Technol 22:972–977
Eckert JM, Stewart JJ, Waite TD, Szymczak R, Williams KL (1990) Reduction of chromium(VI) at sub-$\mu g\ l^{-1}$ levels by fulvic acid. Anal Chim Acta 236:357–362
Elderfield H (1970) Chromium speciation in seawater. Earth Planet Sci Lett 9:10–16
Emerson S, Cranston RE, Liss PS (1979) Redox species in a reducing fjord: Equilibrium and kinetic considerations. Deep-Sea Res 26:859–878
Fendorf SE, Li G (1996) Kinetics of chromate reduction by ferrous iron. Environ Sci Technol 30:3328–3333
Fendorf SE, Zasoski RJ (1992) Chromium(III) oxidation by δ-MnO_2 1. Characterization. Environ Sci Technol 26:79–85
Fukai RM (1967) Valency state of chromium in sea water. Nature 213:901
Fukai RM, Vas D (1969) Changes in the chemical forms of chromium on the standing of seawater samples. J Oceanogr Soc Jpn 25:47–49
Grimaud D, Michard G (1974) Concentration du chrome dans deux profils de l'ocean Pacifique. Mar Chem 2:229–237
Jeandel C, Minster JF (1984) Isotope dilution measurement of inorganic chromium(III) and total chromium in seawater. Mar Chem 14:347–364
Hem J (1977) Reactions of metal ions at surfaces of hydrous iron oxide. Geochim Cosmochim Acta 41:527–538
Kieber RJ, Helz GR (1992) Indirect photoreduction of aqueous chromium(VI). Environ Sci Technol 26:307–312
Kuwamoto T, Murai S (1970) Preliminary report of the Hakuho-maru cruise KH-68-4. Ocean Research Institute, University of Tokyo, p 72
Miller W, King DW, Lin J, Kester DR (1995) Photochemical redox cycling of iron in coastal seawater. Mar Chem 50:63–77
Millero FJ, Schreiber DR (1982) Use of the ion pairing model to estimate activity coefficients of the ionic components of natural waters. Am J Sci 282:1508–1540
Millero FJ, Yao W, Aicher J (1995) The speciation of Fe(II) and Fe(III) in natural waters. Mar Chem 50:21–39
Moffett JW, Zika RG (1983) Oxidation kinetics of Cu(I) in seawater. Implications for its existence in the marine environment. Mar Chem 13:239–251
Murray JW, Spell B, Paul B (1983) The contrasting geochemistry of manganese and chromium in the eastern Pacific Ocean. In: Wong CJ, Boyle E, Bruland KW, Burton JD, Goldberg ED (eds) Trace metals in sea water. NATO Conf. Ser. 4, Plenum, pp 643–669
Nakayama E, Kuwamoto T, Tsurubo S, Kohoro H, Fujinaga T (1981a) Chemical speciation of chromium in sea water. Part 1. Effect of naturally occurring organic materials on the complex formation of chromium(III). Anal Chim Acta 130:289–294
Nakayama E, Kohoro H, Kuwamoto T, Fujinaga T (1981b) Dissolved state of chromium in seawater. Nature 290:768–770
Nakayama E, Kuwamoto T, Kohoro H, Fujinaga T (1981c) Chemical speciation of chromium in seawater. Part 3. The determination of chromium species. Anal Chim Acta 131:247–254
Nakayama E, Kuwamoto T, Tsurubo S, Fujinaga T (1981d) Chemical speciation of chromium in sea water. Part 2. Effects of manganese oxides and reducible organic materials on the redox processes of chromium. Anal Chim Acta 130:401–404
Osaki S, Osaki T, Hirashima N, Takashima Y (1983) The effect of organic matter and colloidal particles on the determination of chromium(VI) in natural waters. Talanta 30:523–526
Pettine M, Millero FJ (1990) Chromium speciation in seawater: The probable role of hydrogen peroxide. Limnol Oceanogr 35:730–736
Pettine M, La Noce T, Liberatori A, Loreti L (1988) Hydrogen peroxide interference in the determination of chromium(VI) by the diphenylcarbazide method. Anal Chim Acta 209:315–319
Pettine M, Millero FJ, La Noce T (1991) Chromium(III) intercations in seawater through its oxidation kinetics. Mar Chem 34:29–46
Pettine M, Camusso M, Martinotti W (1992) Dissolved and particulate transport of arsenic and chromium in the Po River (Italy). Sci Total Environ 119:253–280
Pettine M, Millero FJ, Passino R (1994) Reduction of chromium(VI) with hydrogen sulfide in NaCl media. Mar Chem 46:335–344
Pettine M, Mastroianni D, Camusso M, Guzzi L, Martinotti W (1997) Distribution of As, Cr and V species in the Po-Adriatic mixing area, (Italy). Mar Chem 58:335–349
Pettine M, Barra I, Campanella L, Millero FJ (1998a) Effect of metals on the reduction of chromium(VI) with hydrogen sulfide. Wat Res 32(9):2807–2813

Pettine M, D'Ottone L, Campanella L, Millero FJ, Passino R (1998b) The reduction of chromium(VI) by iron(II) in aqueous solutions. Geochim Cosmochim Acta 62(9):1509–1519
Rai D, Sass BM, Moore DA (1987) Chromium(III) hydrolysis constants and solubility of chromium(III) hydroxide. Inorg Chem 26:345–349
Rai D, Eary LE, Zachara TM (1989) Environmental chemistry of chromium. Sci Total Environ 86:15–23
Richard FC, Bourg ACM (1991) Aqueous geochemistry of chromium: A review. Wat Res 25(7):807–816
Riedel GF (1984) Influence of salinity and sulphate on the toxicity of chromium(VI) to the estuarine diatom *Thalassiosira pseudonana*. J Phycol 20:496–500
Riedel GF (1985) The relationship between chromium(VI) uptake, sulphate and chromium(VI) toxicity in the estuarine diatom *Thalassiosira pseudonana*. Aquat Toxicol 7:191–204
Saleh FY, Parkerton TF, Lewis RV, Huang JH and Dickson KL (1989) Kinetics of chromium transformations in the environment. Sci Total Environ 86:25–41
Sass BM, Rai D (1987) Solubility of amorphous chromium(III)-iron(III) hydroxide solid solutions. Inorg Chem 26:2228–2232
Schroeder DC, Lee GF (1975) Potential transformations of chromium in natural waters. Wat Air Soil Pollut 4:355–365
Sedlak D, Chan PG (1997) Reduction of hexavalent chromium [Cr(VI)] by ferrous iron [Fe(II)]. Geochim Cosmochim Acta 61:2185–2192
Shen-Yang T, Ke-An L (1986) The distribution of chromium(VI) species in solution as a function of pH and concentration. Talanta 33:775–777
Turner DR, Whitfeld M, Dickson AG (1981) The equilibrium speciation of dissolved components in fresh water and seawater at 25 °C and 1 atm pressure. Geochim Cosmochim Acta 45:855–881
Van Der Weijden CH, Reith M (1982) Chromium(III)-chromium(VI) interconversions in seawater. Mar Chem 11:565–572
Voelker BM, Morel FMM, Sulzberger B (1997) Iron redox cycling in surface waters: Effect of humic substances and light. Environ Sci Technol 31:1004–1011
Wang WX, Griscom SB, Fischer NS (1997) Bioavailability of Cr(III) and Cr(VI) to marine mussels from solute and particulate pathways. Environ Sci Technol 31:603–611
Wittbrodt PR, Palmer CD (1995) Reduction of Cr(VI) in the presence of excess soil fulvic acid. Environ Sci Technol 29:255–263
Zachara JM, Girvin DC, Schmidt RL, Resch CT (1987) Chromate adsorption on amorphous iron oxyhydroxide in the presence of major ground-water ions. Environ Sci Technol 21:589–594
Zika RG, Moffett JW, Petasne RG, Copper WJ, Saltzman ES (1985) Spatial and temporal variations of hydrogen peroxide in Gulf of Mexico water. Geochim Cosmochim Acta 49:1173–1184

Appendix

Table 15.A1. Hydrolysis reactions

Hydrolysis reactions	log *K* Baes and Mesmer 1976		log *K* Rai et al. 1987	
	I = 0.00 M	*I* = 0.75 M[a]	*I* = 0.00 M	*I* = 0.75 M[b]
$Cr^{3+} + H_2O \longrightarrow Cr(OH)^{2+} + H^+$	-4.0	-4.4	-3.57	-4.56
$Cr^{3+} + 2\,H_2O \longrightarrow Cr(OH)_2^+ + 2\,H^+$	-9.7	-10.6	-9.84	-11.0
$Cr^{3+} + 3\,H_2O \longrightarrow Cr(OH)_3 + 3\,H^+$	-18.0	-19.2	-16.19	-17.45
$Cr^{3+} + 4\,H_2O \longrightarrow Cr(OH)_4^- + 4\,H^+$	-27.4	-28.2	-27.65	-28.74
$H_2CrO_4 \longrightarrow H^+ + HCrO_4^-$	0.20	0.37		
$HCrO_4^- \longrightarrow H^+ + CrO_4^{2-}$	-6.51	-5.80		
$2\,HCrO_4^- \longrightarrow H^+ + Cr_2O_7^{2-}$	1.523	1.91		

[a] Values calculated according to the ionic strength dependences given by Baes and Mesmer 1976.
[b] Values calculated by assuming $\gamma_{H^+} = 0.97$; $\gamma_{Cr(OH)^{2+}} = 0.5$; $\gamma_{Cr(OH)_2^+} = 0.82$; $\gamma_{Cr(OH)3} = 1$; $\gamma_{Cr(OH)_4^-} = 0.7$; $\gamma_{CrO_4^{2-}} = 0.13$ and $\gamma_{Cr^{3+}} = 0.05$.

Table 15.A2. Redox reactions

Redox reactions	log*K* $I = 0.0$ M	$I = 0.75$ M
$1/3\ HCrO_4^- + 7/3\ H^+ + e^- \longrightarrow 1/3\ Cr^{3+} + 4/3\ H_2O$	23.18	23.50
$1/3\ HCrO_4^- + 2\ H^+ + e^- \longrightarrow 1/3\ Cr(OH)^{2+} + H_2O$	21.99	21.98
$1/3\ CrO_4^- + 7/3\ H^+ + e^- \longrightarrow 1/3\ Cr(OH)^{2+} + H_2O$	24.15	23.92
$1/3\ CrO_4^- + 2\ H^+ + e^- \longrightarrow 1/3\ Cr(OH)_2^+ + 2/3\ H_2O$	22.06	21.77
$1/3\ CrO_4^- + 5/3\ H^+ + e^- \longrightarrow 1/3\ Cr(OH)_3 + 1/3\ H_2O$	19.95	19.62
$1/3\ CrO_4^- + 4/3\ H^+ + e^- \longrightarrow 1/3\ Cr(OH)_4^-$	16.13	15.86

Part IV

Analytical Methodologies and Chemometrics for Sea Water

Chapter 16

Oceanic *DOC* Measurements

C. Minero · E. Pelizzetti · M.R. Preston

16.1 Introduction

Dissolved organic carbon (*DOC*) is by far the largest of the organic carbon reservoirs in the oceans with a carbon load that for many years was accepted to be approximately 200 Gt C (Hedges 1987). It is made up of a very complex mixture of chemicals, which are generally known collectively as 'dissolved organic matter' or *DOM*, and which derive either directly from biological processes or from the enzymatic or bacterial degradation of defunct organisms. The heterogeneous nature of the composition of *DOM* is also reflected in its molecular size distribution which ranges from that of the simplest organic molecules up to complex macromolecules whose properties lie at an indistinct boundary between truly dissolved and colloidal or particulate matter (Gough and Mantoura 1991).

This *DOM* has a major influence in determining the nature of a wide variety of physical, geochemical and biological processes in the oceans (Gagosian and Steurmer 1977; Jackson 1988). For example, the accumulation of *DOM* at the air-sea interface alters the surface tension of the water and may modify the transfer of wind energy, air-sea gas exchange rates and the incursion of light (Mopper et al. 1991). The adsorption of *DOM* onto the surfaces of inorganic particles radically modifies their physicochemical properties with profound consequences for other adsorption reactions (e.g. involving trace metals) and interactions with the biota (Stumm 1992). Degradation of *DOM*, normally by bacteria, may alter both dissolved oxygen concentrations and redox potentials with significant consequences for other biogeochemical processes.

The first studies of *DOM* in the oceans occurred in the 1930s (Krough 1934; Datzko 1939) and to some extent difficulties over the selection of the most appropriate analytical procedures were evident from the outset. Early efforts using dichromate or chromic acid were abandoned fairly soon, but discrepancies emerged between dry combustion methods (used by e.g. Skopintsev et al. 1966,1968 and Starikova 1970; MacKinnon 1978) and wet chemical/UV oxidation methods (used by others, e.g. Menzel 1970; Banoub and Williams 1972; Gershey et al. 1979).

As a general rule, dry combustion methods gave dissolved organic carbon (*DOC*) values that were roughly twice those of wet chemical methods. The debate about the merits of these various figures has been reviewed by, for example, Wangersky (1993) and was conducted along the competing charges of incomplete reaction (levelled at adherents of the wet chemical oxidation procedure) and contamination of the reaction (levelled at the dry combustion enthusiasts). By the end of the 1970s, and well into the 1980s, the view of the proponents of wet chemical/UV oxidation methods had largely prevailed. As a result some sort of consensus that marine *DOC* concentration should fall in the approximate range of 15–100 μM C had been reached. Methods such as that published by Collins and Williams (1977) or Mantoura and Woodward (1983) had become the norm. However, there was still no real understanding of either the vertical or seasonal variability in *DOC* other than the observation that concentrations were highest in productive surface waters with featureless profiles in deeper waters (Toggweiler 1988, 1989; Williams and Druffel 1988). Nor was there any real insight into the cycle of *DOC* where mass balance calculations into the fate of terrigenous organic matter in the oceans revealed considerable discrepancies, leading to hypotheses about unknown remineralisation processes (Hedges 1987).

Seasonal variability exists in the surface *DOC* concentrations: *DOC* accumulates due to the spring bloom, it is partially consumed in summer and autumn, and it is transported under the euphotic layer by convection in winter. The *DOC* concentration in deep water is almost constant from the North Atlantic to the Pacific (Martin and Fitzwalter 1992). Above a depth of 400 m, *DOM* plays an important role in the biogeochemical cycles. *DOM* is carried by the Ekman transport from the equatorial to the subtropical region. This transport reduces the nutrient trapping effect in the high production areas and supplies nutrient to the low production areas. Geochemical modelling suggests that above a depth of 400 m exists the so-called semilabile *DOM*, with half-life of half a year, and that its vertical and horizontal transport plays an important role for the marine biogeochemical cycle. Below that depth only the inert refractory *DOM* exists, of minor importance for the biogeochemical cycle (Yamanaka and Tajika 1997).

In the late 1980s and early 1990s a number of reports, deriving in considerable part from Suzuki and co-workers (Sugimura and Suzuki 1988; Suzuki et al. 1990; Suzuki and Tanoue 1991; Suzuki et al. 1992 but see also Martin and Fitzwalter 1992 and Kumar et al. 1990), reported non-volatile *DOC* concentrations that were considerably higher than previously accepted as the norm by a factor of between about 2 and 4. These measurements were conducted using a new high temperature catalytic oxidation method (HTCO) and the data were given credence by an apparent link between measured *DOC* values and an independent parameter of geochemical significance, the apparent oxygen utilization (*AOU*) parameter (Sugimura and Suzuki 1988).

The impact of these papers on the scientific community was considerable because it necessitated an upward revision of the global ocean inventory of *DOC* up to around 800 Gt. The 600 Gt difference between this new value and the 200 Gt previously estimated is extremely large and equivalent to the entire pool of atmospheric carbon (Hedges 1987). A major change in the accepted analytical procedures and historical data had profound implications for the Joint Global Ocean Fluxes Study Programm (JGOFS) to examine carbon fluxes (Sharp et al. 1995) and on the modelling of biogeochemical general circulation (Yamanaka and Tajika 1997).

As can readily be imagined a considerable degree of controversy surrounded all of these *DOC* determinations. Whilst some scientists accepted the new data and speculated about its significance, others recognised that the key issue lay not in the interpretation of the data but rather, the quality of the analytical method that produced it. Recent data validating the utilization of several different techniques indicate that the equatorial Pacific oceanic *DOC* values in near surface waters are on the order of 60–70 μM C and deep water values on the order of 35–40 μM C (Sharp et al. 1995). Peltzer and Hayward (1996) reported that the total organic carbon concentration in the Pacific surface water along 140° W is 60 μM C at the equator and 80 μM C at 10° N and 10° S, but in deep water it is almost constant at about 36 μM C.

The new data also necessitated a re-evaluation of the extent of 'characterized' (meaning of known molecular structure) vs. 'uncharacterized' organic matter by a similar multiple. The implications of the new study were that this 'new' (i.e. newly discovered) material consisted of biologically derived macromolecules which were biologically reactive, particularly to bacteria (Kirchman et al. 1991; Kepkay and Wells 1992; Coffin et al. 1993) *but* resistant to chemical and photochemical oxidation (Sugimura and Suzuki 1988; Suzuki and Tanoue 1991). However, other studies pointed out that the bulk of the oceanic *DOM* comprises small molecules, some of which are relativley unavailable to microorganisms (Amon and Benner 1994), and that the carbon that is missed by wet chemical oxidation methods is in the low-, rather than high-molecular mass fraction (Ogawa and Ogura 1992). Measurements on ^{14}C showed that the *DOC* in deep water is very old (Bauer et al. 1992). The application of ultrafiltration techniques to the study of oceanic samples has allowed further characterization of *DOC*. Spanning the size range from 0.4–0.2 μm to 1 nm, ultrafiltered dissolved organic matter (*UDOM*) is composed of macromolecular components and true colloids. Depth-weighted results of the size distribution of organic matter in sea water indicated that about 75% of marine organic carbon was low-molecular-weight *DOC* (<1 nm), ~24% was high-molecular weight *DOC* (1–100 nm) and ~1% was particulate organic carbon (>100 nm) (Benner et al. 1997). The re-evaluated distribution of carbon in surface water was therefore shifted to a greater relative abundance of larger fractions, suggesting a diagenetic sequence from macromolecular material to small refractory molecules. Surface *UDOM* is greater than deep *UDOM* by a factor of 2–3. Elemental as well as molecular-level aldose and amino acid compositions were recently determined (McCarthy et al. 1996). *UDOM* has a characteristic organic composition, clearly distinct from other marine materials such as fresh plankton, sinking particles or humic substances. It is rich in galactose and deoxy sugars, and ubiquitous regardless of depth or location. In contrast the amino acid content of *UDOM* is a relatively minor component.

Various methods for the measurement of *DOC*, as well as the various problems common to all the techniques and those that are method specific, have been reviewed by Wangersky (1993). This paper is intended to bring the reader up to date with the developments on the *DOC* measurements and to provide a survey of new concepts in such analytical procedures.

16.2 *DOC* Measurement Techniques

The sheer complexity of *DOM* rules out any kind of rigorous analytical approach designed to individually quantify all of the contributory organic chemicals. Instead, the strategy that has generally been adopted is to convert the organic carbon into CO_2. *OC* may be classified according to the particle size. Since the distribution of molecular components of sea water covers a continuous size spectrum, ranging from small molecules to colloids and larger particles, operational criteria have been developed for the classification of *OM*. Accordingly, filtration may be performed using nominal pore size of 0.45 μm, to get the so-called dissolved organic carbon (*DOC*). In the filtrate most of the colloids are recovered. The colloidal fraction contributes to 10–30%

of the *DOC* in sea water (Wells and Goldberg 1991, 1992, 1993; Ogawa and Ogura 1992). The particulate retained on the filter is less than 5% of *OC* in the ocean and may be usually neglected. However, the particulate peaks around 10% of *OC* during phytoplankton blooms (Wangersky 1993).

The high ratio of dissolved inorganic carbon (*DIC*) to *DOC* in sea water (~54/1) means that it is first necessary to remove the *DIC* by acidification and purging. Volatile organic carbon (*VOC*) belongs to *TOC*, and part of it is lost during the stripping of inorganic carbon. Most commercial analysers neglect the amount of lost *VOC*. Marine *VOC* concentrations may rise locally to 10% of *DOC*, either due to anthropogenic input or to intense biogenic production (Juttner and Henatsch 1986; Bianchi and Varney 1992). Normally *VOC* is a minor contribution to *TOC* (≤5 μM C) and comparable to the precision of contemporary *TOC* analyses (±1–2 μM C). Inorganic compounds like cyanates, isocyanates and thiocyanates are determined in the *DOC* fraction as *TOC*.

The CO_2 that is evolved can then be directly measured using a suitable detector (e.g. a non-dispersive infrared gas analyser, NDIR) or converted to methane which may then be determined using a conventional flame ionisation detector (FID). The NDIR has proved to be the more popular and easily applicable unit, though its usage has not been without problems.

The analytical steps may then be broken down into the following stages:

i. sample collection and storage
ii. pretreatment and removal of *DIC*
iii. the oxidation of *DOC*
iv. the removal of interfering species
v. the detection and quantitation

The issue of instrumental and methodological blank (*vi.*) pertains to all of the steps and strongly influences analytical precision and accuracy.

16.2.1
Sample Collection and Storage

An examination of much of the early literature regarding *DOC* measurements reveals a considerable paucity of details regarding sample collection and storage. Many of the major improvements of analytical data quality in recent years have come about because of a greatly increased awareness of the importance of these aspects. Inappropriate sampling procedures can result in sample contamination by the ship's gear, the sampling bottles themselves and, perhaps most importantly, by the surface microlayer. All lead to erroneously elevated values. Detailed sampling procedures have been described by Statham and Williams (1983), Peltzer and Brewer (1993) and Sharp and Peltzer (1993).

Collected samples have often had to be stored rather than subjected to immediate analysis. The main problem appears to derive from microbial activity, and this is a particular problem in highly productive waters (Chen et al. 1996). Studies of different storage procedures conducted in recent years have shown that even the sign of the change in measured *DOC* concentrations may not be consistent when different methods are used. Concentrations may be elevated by contamination or decreased by degradation or adsorption. Early methods which involved the use of filtration (precombusted glass fibre filters) followed by poisoning with $HgCl_2$ or simply freezing in precombusted glass ampoules have been shown to be as satisfactory as any more modern alternatives (Tupas et al. 1994). The last researchers suggested that samples can be conveniently stored at −20 °C in acid-cleaned polypropylene tubes, high-density polypropylene bottles or combusted glass ampoules even for extended periods (5 months after collection). Preservation at 4 °C needs the addition of 0.025% of H_3PO_4.

16.2.2
Pretreatment and Removal of *DIC*

The procedure for removing *DIC* has not altered much during the history of *DOC* measurements. The sample of sea water is acidified to a pH < 3 using H_3PO_4 or HCl. It is then purged with a CO_2-free gas such as N_2 or O_2. This purging needs to be maintained for a period of time that is proportional to the gas flow rate and the geometry of the purging vessel. It might typically range from about 5 minutes for a flow rate of 500 ml min^{-1} up to 20 minutes for a flow rate of 175 ml min^{-1} (Peltzer and Brewer 1993). No effect of the purging was reported for *DOC* abundance measurements (Tupas et al. 1994).

16.2.3
The Oxidation of *DOC*

It is the oxidation step in *DOC* analysis that has probably provoked the greatest difficulties over the years. The various approaches to this problem are briefly reviewed in this section.

16.2.3.1
Dry Combustion

Dry combustion, often called sealed-tube combustion (STC), methods generally involve the combustion of a dried sample over a catalyst in a sealed ampoule. A catalyst such as copper oxide and H_3PO_4 (Alperin and Martens 1993; Fry et al. 1993; Chen et al. 1996; Peltzer et al. 1996) or sulfuric acid (Fry et al. 1996) is added. The second catalyst avoids the problem of CO_2 absorption present with copper basic salts (carbonates) formed during the later combustion. The evaporation of the water is obtained directly in the ampoule by vacuum distillation, freeze drying or oven drying. After sealing and a suitable heating period under controlled conditions (typically ~580 °C), the ampoule is opened in such conditions as to quantitatively recover the evolved CO_2 which is then collected and quantified. A tube cracker under a closed atmosphere is needed. Depending on the measurement technique (such as linear response capacitance manometer), the CO_2 generated in combustion may necessitate purification from traces of water and other contaminant gases, using vacuum manifolds and sequential sublimation through liquid nitrogen and dry-ice/ethanol cold traps. Dry combustion has been identified as a reference method (Peltzer et al. 1996) because analyses can also be performed in commercial elemental analysers (MacKinnon 1978).

The method is handicapped by the difficulties in handling the large quantities of water involved in the analysis of low-*TOC* samples and is it is more suitable for the analysis of fresh water, because the salt present in sea water makes this approach difficult (Cauwet 1994). Troubles with blanks when drying low carbon offshore *DOC* samples using a freeze-dryer has been reported (Fry et al. 1993; Peterson et al. 1996), mainly due to back-streaming of vacuum pump oil vapours. Oven drying is therefore preferred (Fry et al. 1996) not only because of this problem but also because the salty brine solution formed during freeze-drying is increasingly resistant to the removal of the last traces of water.

16.2.3.2
Wet Chemical Oxidation (WCO) and UV-Assisted WCO

Wet chemical methods employ an oxidizing agent (typically $K_2S_2O_8$ (PS)) and heating in a closed vessel, followed by a similar quantitative step as for the dry combustion (Van Hall et al. 1965; Miller et al. 1993b). Similarly, the use of UV photooxidation has sometimes relied on irradiation in a sealed (UV transparent) container followed by release and quantitation (Armstrong et al. 1966). The efficiency of the UV oxidation procedure depends to a considerable degree on both the type and the age of the source lamp. The type of the lamp influences the spectral characteristics of the emitted radiation and the intensity of the irradiation decreases with age. Most systems use high power (1 kW) medium pressure mercury lamps as suggested by Collins and Williams (1977) for 20 min to 1 h treatment time.

The advantages of both the chemical and UV oxidation procedures are that they lend themselves to applications in continuous flow systems. A typical, combined PS/UV system has been described by Mantoura and Woodward (1983) and is shown diagrammatically in Fig. 16.1. As well as the potassium persulfate reagent, this unit uses a borate buffer to stabilize the pH and hydroxylamine hydrochloride to inhibit corrosion of the peristaltic pump tubes by the free chlorine generated during the oxidation step. Such systems give a total system blank of around 20 μM C with a precision for real samples of ±6 μM C at a concentration of 145 μM C (Miller 1996), or about 5% on marine samples (Sharp et al. 1995).

The systems are robust enough for use at sea though some early designs of detectors were more susceptible to vibration than others, and care needs to be taken to ensure that vibration does not cause analytical problems.

The drawbacks are in the possibility that the compound may not be transformed into CO_2. For example, ^{14}C-labelled *DOC* produced by some species of phytoplankton grown in the presence of ^{14}C bicarbonate resists both persulfate and UV oxidation (Ridal and Moore 1993). Resistance to oxidation was also reported recently for compounds in which the carbon is in its higher oxidation state (like CCl_4 and cyanuric acid for which oxidation is not allowed), and thermal hydrolysis is slow (Calza et al. 1997b). To overcame the difficulties associated with the possible incomplete oxidation by persulfate of the original material (or volatile compounds produced in the oxidation process), the wet oxidation step may be combined with a catalytic oxidation in a furnace at 900 °C over CuO (Fung et al. 1996). The volatile compounds are distilled from the acidified sample containing persulfate under a flux of oxygen at about 2–3 ml min^{-1}. Finally the CO_2 formed in both steps is collected in alkaline solution and

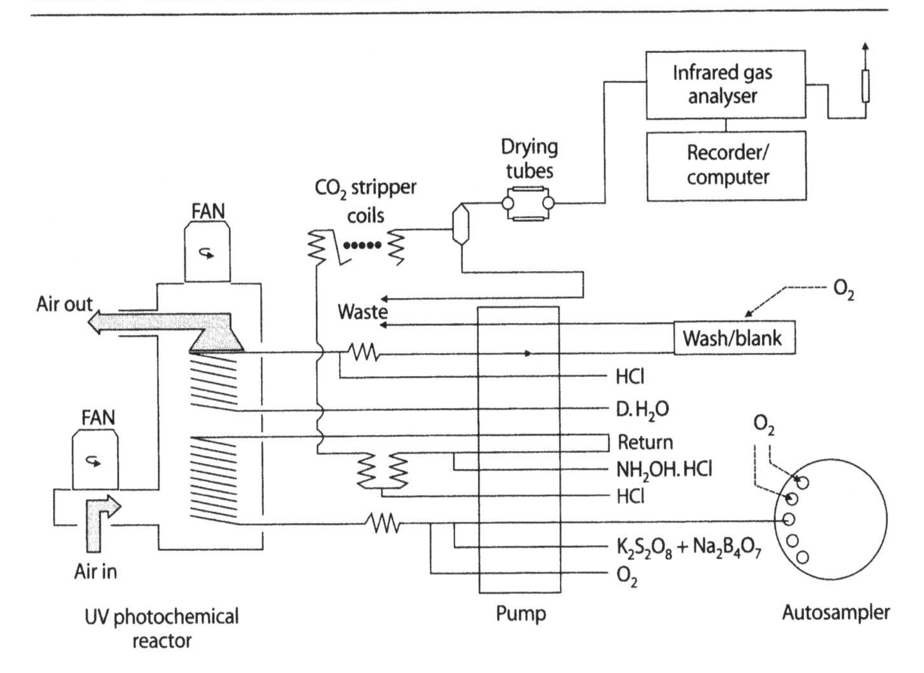

Fig. 16.1. A combined PS-UV oxidation unit (from Mantoura and Woodward 1983)

determined by ion chromatography. Since the flask for sample oxidation can be chosen as large as wanted, the detection limit may be very low. It is critical to reach a low blank value, mainly at the startup.

Another drawback is related to the chloride ion content of sea water. Aiken (1992), and McKenna and Doering (1995) have addressed this issue experimentally and reported that high chloride concentration or low persulfate to chloride ratios can lead to incomplete yields of oxidation.

16.2.3.3
High Temperature Combustion (HTC)

A schematic diagram of the Sugimura and Suzuki (1988) high temperature catalytic oxidation system (HTCO) is shown in Fig. 16.2. They claimed that HTCO method was able to more completely oxidize organic matter. Further design details were given in a later paper, and as became apparent later, the design of the system is critical to its oxidation efficiency (Suzuki et al. 1992). In this original system small volumes (200 µl) of acidified (pH < 2, H_3PO_4), purged (O_2) sea water were directly injected onto a column containing a catalyst (3% Pt-Al_2O_3) held at 680 °C. Detection was carried out by NDIR spectrometry. Glucose was used for the calibration of the instrument and a variety of reference materials were recovered quantitatively (with the exception of sulfathiazole).

The new generation of HTC instruments gives reproducible data with a high sensitivity. The average precision is 1.5–3% for commercial instruments (Sharp et al. 1995). They are capable of providing high quality measurements of large series of samples (Benner and Strom 1993).

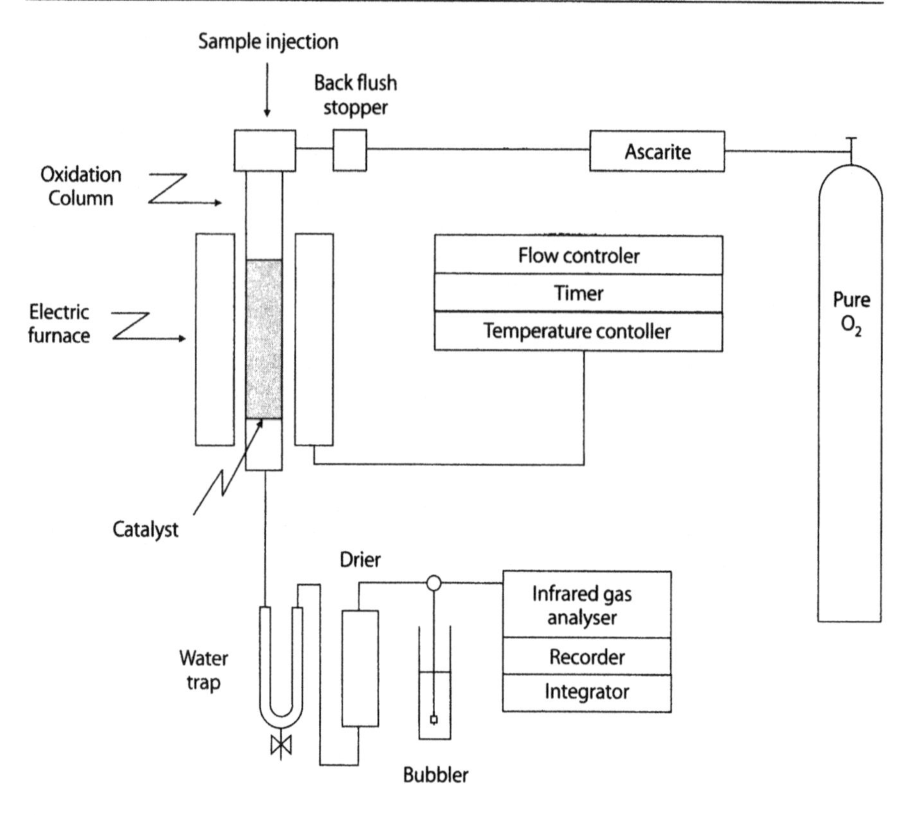

Fig. 16.2. The Sugimura and Suzuki (1988) *DOC* analyser

In many investigations the catalyst was perceived to be the key element to the new technique. The original catalyst used by Sugimura and Suzuki had a very restricted availability, so a number of alternative formulations were tested with mixed results (see e.g. Bauer et al. 1990; Cauwet 1992; Miller et al. 1993a). Some researchers found that the original highly platinised alumina gave higher *DOC* results (Cauwet et al. 1990) whilst others questioned whether platinum was involved in the process at all (Williams et al. 1993). It is not clear what function, if any, the catalyst plays in the oxidation process (Perdue and Mantoura 1993; Bauer et al. 1993). Thus, the more correct term is HTC (Sharp et al. 1995). Also the role of oxygen is questionable. Instruments operated with carrier gases not containing oxygen produced signals equivalent to those with air (Perdue and Mantoura 1993; Perdue et al. 1994; Skoog et al. 1997). Water is likely to be the source for reacted oxygen also when molecular oxygen is present in the carrier gas. The water contained in a 100 μl injection expands rapidly to a volume >400 ml, assuming the gas law, 1 atm pressure and T > 650 °C. Because the void volume of the combustion tube is generally around 50 ml, the main portion of the sample carbon is not exposed to any oxidation agent other than the sample water and the small amount of oxygen dissolved in it.

Some recent papers outlined the key role of the oxidation unit and the performance of instrumental units. From an instrumental point of view, poor reproducibility and poor accuracy can arise from the sample injection system, from memory effects and from sensitivity to instability of instrumental parameters (temperature, vibrations etc.). Due to the rapid expansion after injection, the stability of the catalyst bed becomes an important issue. After long use with marine water a salt crust is formed in the top part of the catalyst and, surprisingly, improved peak sensitivity and shape is achieved. This was attributed to the formation of a funnel shape of the precipitate that could help guide the sample into the catalyst (Skoog et al. 1997). The use of a loop-type injector, allowing precise injections under closed atmosphere, improves the precision (0.6% RSD for $n = 12$ on Antarctic seawater samples with 85 μM C level, 4.0% RSD for $n = 4$ on distilled water with 7 μM C level) and eliminates the possibility of contamination during injection (Qian and Mopper 1996).

The memory effect arises from void volume at the top of the oxidation column (Qian and Mopper 1996) and from adsorptive properties of the catalyst (Cauwet 1994). It is evident when injection of sea water or high *OC* standards are followed by injections of distilled water (DI): the first injection of DI gives signals always larger than the following replicate injections of DI. This effect originates from incomplete combustion, and especially from flaking off of uncombusted *OC* in the cold zone at the top of the column. It can be reduced by eliminating the void volume by direct injection into the high temperature zone through a platinum tube, or accurately direct the injected flow on the hot zone. As far as the type of catalyst is involved, quartz or silica have lower memory effects than alumina, which, being an amphoteric oxide, may adsorb CO_2 depending on the partial pressure of CO_2 and column temperature. In this case the memory effect is proportional to the injected amount of C. However, the results from injection of ^{14}C-labelled organic material showed that the oxidation of organic carbon to gaseous compounds on Pt/alumina beads is complete, and that the catalyst is not a source of carry over signal between injections. The blank signal from further injection of water was attributed to carbonaceous compounds contained in the catalyst particles (Skoog et al. 1997).

As for column packing, the recovery (based on peak area) is independent on oxidation column temperature in the range 650–900 °C (Sharp et al. 1995; Qian and Mopper 1996) and the flow rates of oxygen, air or even nitrogen (Skoog et al. 1997). Low recoveries at lower temperatures may be due to memory effects, which may become worse with decreasing temperature. However these parameters influence the sensitivity, that is the slope of the calibration curve, and the peak broadening. The peak broadening changes significantly with temperature and the nature of the analyte (see Fig. 16.3). Often the peak originated by natural samples contains two shoulders. Suzuki et al. (1992) suggested that the two shoulders are an effect of compounds having different combustion rates. However, due to the flash evaporation of the sample, the two shoulders could also be caused by pressure effects on the detector, or to the cooling down of the catalyst enough to combust only part of the sample, leaving the remaining carbon to combust after reheating (Skoog et al. 1997).

The decrease in sensitivity at low temperatures or high flow rates is related to the decrease in oxidation efficiency due to the lower rates of oxidation and shorter residence time in the column. Conversely, the increase in precision at lower flow rates is somewhat offset by peak broadening, which may adversely affect the integration precision. For high precision a very stable gas flow, especially during injection, is needed.

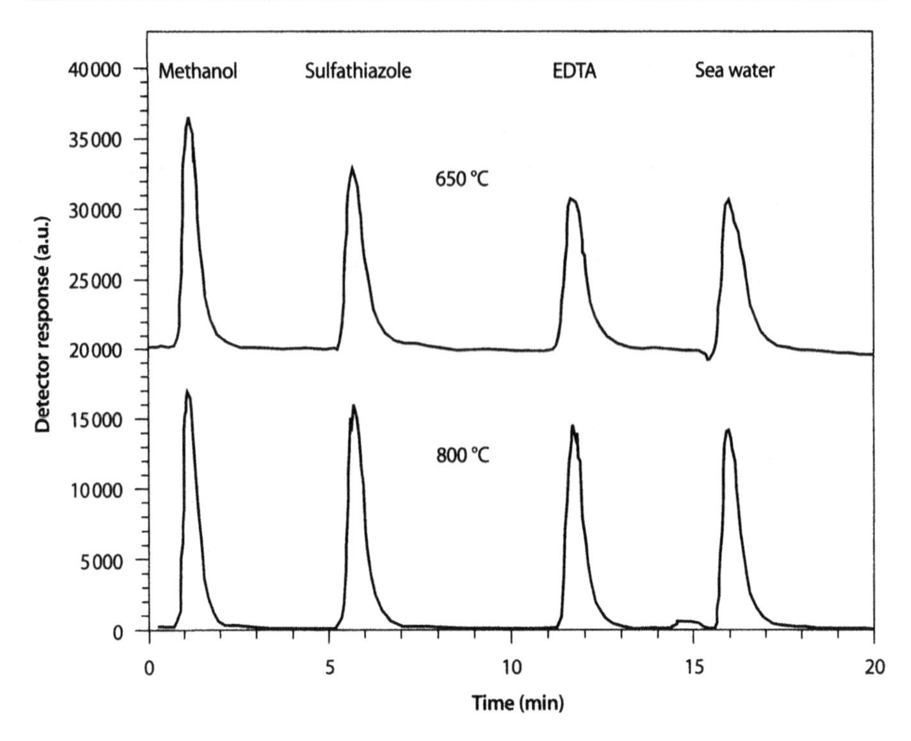

Fig. 16.3. Effect of column temperature on peak broadening for different analytes (from Qian and Mopper 1996, with permission)

16.2.4 The Detection and Quantitation Step

The formed carbon dioxide can be determined by several methods. Several detectors have been proposed, as previously outlined. Among these NDIR, conductivity, and coulometric detectors, carbon dioxide sensors, and FID after reduction to methane are the more common. Depending on the chosen detector, the removal of interfering species follows. The NDIR detector is very sensitive, almost calibration-free and accurate. It suffers little interference from other compounds that are eliminated by proper scrubbers. Usually an electronic or chemical ($Mg(ClO_4)$) dehumidifier is connected in series with acidic (H_3PO_4) traps, containing $AgNO_3$ (either in solution or as beads), or 4,4'-methylenebis(N,N-dimethylaniline) for removal of halides, and metallic Cu grains (Sharp et al. 1995). However, the NDIR detector is expensive and some designs are not robust enough to shocks for onboard measurements (Tupas et al. 1994). They also require additional gas for CO_2 stripping, and it may not be easily miniaturized for portable or remote systems. Conductivity and coulometric detectors are inexpensive, easily miniaturized, needing few calibrations. They are utilized on several commercial instruments. The principles of coulometric detection are described in standard textbooks (Skoog and West 1976), and have been applied to marine measurements of TIC since 1985 (Johnson et al. 1985; 1987). FID detection of methane formed after reduction of CO_2 was marketed until a few years ago. This method of detection requires additional hydrogen, an additional reduction chamber maintained at high temperature, and a stable electrometer, all of which increase the costs. Although its sensitivity is very high and it is almost interference-free, frequent calibrations are needed and unattended operation is not allowed.

The quantitation step implies the calibration of the system and the transformation of the obtained signal in the needed chemical information. The influence of the analytical system itself on the overall signal is strongly related to the quality of the analytical results. If there is a consistent, but unrecognised blank signal, this leads to a bias in the results. Much more dangerously, an inconsistent, unrecognised signal in the analysis leads to data corruption with, at best, decreased precision and at worst, seriously misleading results. The sources of blank signals are:

i. *The system/instrument blank*, which is a composite signal resulting from the sum of the contributions of the components of the analytical system itself, e.g. carrier gas, catalyst, reagents, pipework etc. It is due to anything that adds or subtracts CO_2 to that produced, or which adds non-CO_2 compounds, which interfere with the detection.
ii. *The water blank* (a signal that derives from the presence of residual organic carbon in the low organic carbon water commonly used for the preparation of standard solutions).
iii. *The sample blank*, due to production of non-CO_2 components during oxidation or combustion which can give spurious signals at the detector.
iv. *The sample memory blank*, due to memory effects observed when the signal generated is to some extent influenced by the previous samples that have been analysed. Another term for this phenomenon might be a carry-over blank.
v. *The procedural blank*, which is the signal that derives from the three components identified above plus any additional factors, which may arise because of organic carbon contamination or removal during sample collection and storage procedures.

Considerable efforts have been made to quantify all of the possible influences of the various blank contributions identified above (Benner and Strom 1993), but the variety of systems in use in different laboratories has made a consistent approach difficult. These problems are particularly evident in the analysis of sea water, for which the low *TOC* concentration is associated with high salt content. A number of authors have published uncorrected data together with their determined blank values to try and circumvent this problem (e.g. Ogawa and Ogura 1992; Kaplan 1992; de Baar et al. 1993 and Zweifel et al. 1993, 1995).

The contribution of the different blanks in the WCO and HTCO methods was analysed by Cauwet (1994) using MilliQ water. In WCO method the blank includes the reagent blank and the instrument blank, giving, after conditioning of the automated system, a stable base line. When samples are processed the water blank is absent, giving rise to a lower base line by 12 ±1 µM C with respect to carrier gas. In the HTCO method the blank was estimated by injecting volumes of water ranging from 50 to 200 µl. From the linear correlation obtained, the blank was estimated (2.5 µM C with 100 µl injection). Some instruments (e.g. Shimadzu-*TOC* 5 000) have the possibility of collecting the pyrolized water and using it for blank checking. This blank check made from carbon-free water gives a good estimation of the instrument blank. Cauwet (1994) reported that this blank depended strongly on the catalyst used. It was high for 3% Pt/Al_2O_3 (42 µM C) and lower when using Pt impregnated quartz wool (5 µM C).

The recommendations for analytical work are that the analyst should:

1. Calibrate the instrument daily. The heterogeneous nature of marine *DOM* means that selection of a suitable material (or materials) for the calibration of *DOC* analyses is not straightforward. Glucose has been a common choice, but some fears have been expressed that glucose solutions are not stable in unsterile solutions (Miller et al. 1993a). Phthalate standards are frequently recommended, because it is readily soluble in both seawater and distilled water and is commercially available in a high purity form. Wangersky (1993) has published a summary of compounds that show virtually complete oxidation by UV or wet chemical oxidation methods. However, a procedure for calibrating seawater *DOC* analysers has not yet been universally accepted.
2. Make the appropriate blank correction using carbon-free distilled water. This can be obtained by a high quality water purifier, such as MilliPore MilliQ Water system, collection of HTC processed water, prepared by H_2O_2/UV oxidation or distilled over persulfate (Benner and Strom 1993; Cauwet 1994; Peltzer et al. 1996). Instrument blanks measured with ultrapure water in different laboratories had average values 10.9 $\pm$1.2 μM C. Conditioning of the catalyst bed in HTC methods is beneficial, as well as the careful preparation and handling for reducing the potential for errors from contaminated samples (Sharp et al. 1995).
3. At least once, determine the combustion efficiency for the method either by direct comparison with a referee technique or by oxidation test performed on recalcitrant compounds. It is impossible to prove the completeness of oxidation for a method dealing with natural samples owing to the unknown nature of the material under analysis. Since the comparisons with STC may be impracticable for routine use, a suite of recalcitrant compounds to be added to marine water samples has to be identified: leucine and proline are resistant to WCO, as is urea for UV oxidation, and several organic compounds under HTCO conditions (see Fig. 16.3).
4. Certify the accuracy of the method by the regular measurement of reference samples. This is a difficult task to achieve, also because of storage problems. Gershey et al. (1979) first called attention to deep oceanic waters as a standard for low concentration reference standard. Such a standard has not yet been identified for high-*DOC* reference standard.

16.2.5 Comparative Performance of the Different Techniques

The very high level of interest in the Sugimura and Suzuki system and the problems that it posed, about analytical procedures in general and marine *DOC* analysis in particular, led to a major international exercise held in Seattle in 1991 that was reported in a special edition of "Marine Chemistry" in 1993 (Hedges and Lee 1993). Suzuki later found that there had been a variety of problems with his published data and published a retraction of the 1988 paper (and an earlier paper on dissolved organic nitrogen, Suzuki et al. 1993). However, Hedges et al. (1993) strongly resisted any inclinations to disregard all of the previous work because, by this time, a number of independent workers had also reported elevated *DOC* levels. It was clear that although the original data may have been flawed, Sugimura and Suzuki's work had raised a number of major analytical issues of international importance. The careful utilization of HTCO methods has demonstrated the importance of blank measurements and the control of several parameters. One of the major outcomes of the Seattle meeting was a realization that the determination of appropriate blanks for the analytical system is a non-trivial exercise.

A variety of both home made and commercially produced clones of the original system were produced but little consensus could be reached about the better performance of some methods. As stated by Peltzer et al. (1996), there is no generic persulfate, HTC, commercial or home-made methods, but rather there are various techniques practiced by individual analysts, each of which requires calibration and verification.

Various checks on oxidation efficiency have been made by using comparisons of high temperature catalytic oxidation and sealed tube combustion of the same samples (Hedges and Farrington 1993; Sharp et al. 1995). Some earlier intercomparision and field studies were inconclusive (Miller et al. 1993a,b), but later comparison showed that comparable results (±7.5%) were produced using five different HTC instruments and WCO methods (Sharp et al. 1995). Measurements carried out by Peltzer et al. (1996) on a limited number of samples showed that there was agreement among HTC (three methods), PS and STC methods, but yields for all methods decreased compared to the STC technique at concentrations higher than 400 µM C. However, these researchers outlined that the intercomparison of various methods needs a proper evaluation of instrument blanks, which can be reduced by proper conditioning of the catalyst bed in HTC methods. Contamination problems were also reported. As recommended by Gershey et al. (1979), a regression analysis would provide important insights on the importance of blanks. When comparing two methods, if the intercept of the linear regression analysis is different from zero, the difference between the pair of methods might be due to a blank problem. In a similar manner, the slope of the fitted line gives information about the relative oxidation efficiency. For the best statistical approach for doing this regression analysis, see Peltzer et al.(1996).

Ogawa and Ogura (1992) subjected to replicate *DOC* determinations using both WCO and HTCO methods and marine dissolved organic matter fractionated by ultrafiltration. Three fractions have been collected, $f_1 < 1\,000$, $1\,000 < f_2 < 10\,000$, $f_3 > 10\,000$, respectively. There was a close agreement between WCO and HTCO results for fraction f_3 (accounting for 3–22% of the total *DOC*) and fraction f_2 (accounting for 32–56% of the total *DOC*). HTCO results were somewhat higher than those from WCO for fraction f_1. The mean difference was reported in the range 15–25%, supporting the conclusion that a portion of *DOC* is missed in WCO methods. Other experiments agreed that the difference between HTCO and WCO represents no more than 10–15% (Cauwet 1994). Since naturally occurring volatile substances are a minor part of *DOC*, this portion has to be attributed to organic species with low boiling points, formed during the PS oxidation. The WCO system proposed by Fung et al. (1996) mentioned above would overcome this effect.

16.3 New Concepts for *DOC* Determination

The detection of *TOC* in a sample can be achieved theoretically with all the techniques able to sense the amount of carbon in the sample, independent of the type of substances present. If the C signal is independent on the sample composition, quantification of *DOC* is possible. Other than the detection of CO_2 formed, the signal can be made independent of the sample composition using C^+ signal obtained by ICP-MS. The inductively-coupled plasma ionization technique ensures an effective matrix decomposition. Using isotope dilution techniques and MS, the quantification (De Bièvre and Peiser 1997) of *DOC* is possible without need of absolute intensity determination, as required by atomic emission spectroscopy, which can suffer from some matrix effects. This principle was recently proposed for the determination of *TOC* content of HPLC fractions of heavy metal complexes with humic substances (Vogl and Heumann 1998).

However, the search for efficient oxidation systems which can be operated unattended, remotely or in restricted and mobile environments, with limited energy consumption and without the use of chemical reagents and bottles of gases, led in recent years to the development of *TOC* measuring systems based on the photocatalytic oxidation capabilities of organics to CO_2 in the presence of the anatase form of titanium dioxide.

It has long been recognized that titanium dioxide illuminated with band gap energy of greater than 3.2 eV (<380 nm) efficiently mineralizes most organic compounds (Pelizzetti et al. 1993). Semiconductor-assisted photocatalysis as a method for destroying pollutants has been the subject of extensive investigations in the last 20 years. Books, reviews and exhaustive collections are available, and dedicated conferences and workshops have been held (Schiavello 1985; Pelizzetti and Serpone 1986; Serpone and Pelizzetti 1989; Anpo 1989; Pelizzetti and Minero 1994; Blake 1994, 1995, 1997; Hoffmann et al. 1995; Bolton 1996; Al-Ekabi 1997).

The key of the process is the absorption of light from the semiconductor creating valence band holes (strong oxidants) and conduction band electrons (mild reductants) (see Fig. 16.4). Heterogeneous photocatalytic processes involve reactions at the surface-solution interface, where the oxidizing species can be holes (more oxidizing than

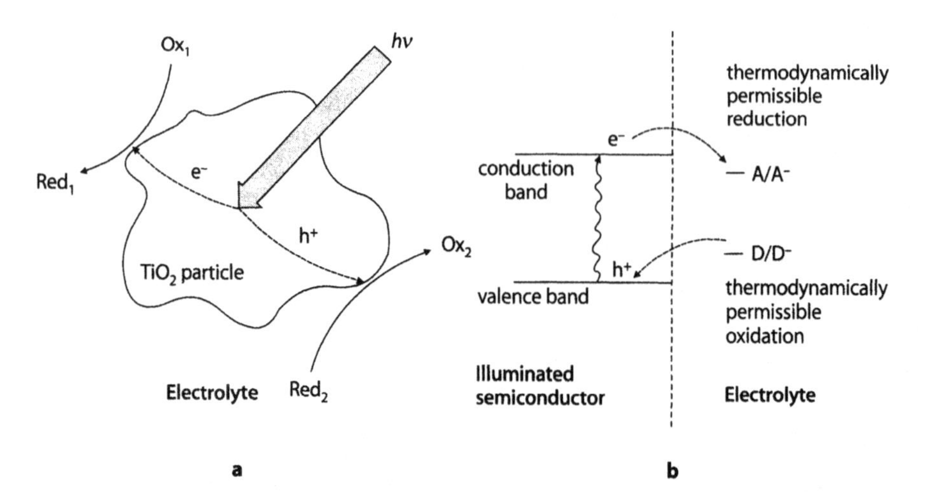

Fig. 16.4. a Schematic representation of electron-hole formation and electron transfer reaction at the semiconductor particle; **b** Thermodynamic constraints for electron exchange at illuminated semiconductor-electrolyte interfaces (*A* = electron acceptor, *D* = electron donor)

aqueous hydroxyl radical) or trapped holes (less oxidizing than aqueous hydroxyl radical), and where the reducing species can be conduction band electrons or trapped electrons, both being much less reducing than aqueous hydrated electrons (Ollis and Al-Ekabi 1993; Pelizzetti and Schivallo 1991).

A series of events lead, in the presence of oxygen, to the conversion of the organic species into CO_2 and water, the other heteroatoms being transformed into inorganic species (Pelizzetti and Minero 1993). Besides the solvent, other species are present at the interface or in solution (e.g. $HO_2\cdot$ / $O_2^{-}\cdot$, O_2, H_2O_2) which can contribute to the complex degradation scheme leading to the final mineralization of the organic compounds. Several degradation mechanisms have been reported for hydrocarbons, halohaliphatic, haloaromatics, surfactants, herbicides and several other classes of organic compounds, respectively.

The process may operate either with artificial or with solar light. This has stimulated very active research (Schiavello 1988). The impressive amount of scientific contributions and reports has led to pilot-scale work and testing of solar driven plants for water decontamination (Minero et al. 1996) and also for marine oil spill removal (Minero et al. 1997a).

By contrast only a few analytical applications of this process have been developed (Low and McEvoy 1996) and only recently detectors for *TOC* monitoring of ultrapure water and a benchtop *TOC* analyser based on this principle have been marketed. For the last, the organic species are photocatalytically decomposed using a TiO_2 slurry to carbon dioxide, with measurement as carbonate by conductivity. The system is depicted in Fig. 16.5. The graph of the logarithm of the conductivity against the logarithm of the amount of organic carbon present is essentially linear for amounts of carbon greater than 100 μM C and can be further linearized to less than 5 μM C by subtracting a constant linearizing factor (Matthews et al. 1990). Conductivity measurements are almost incompatible with waters of high salinity. However, the photocatalytic oxidation has been coupled also with a NDIR analyser (Abdullah and Eek 1996).

Until now the performance of the new oxidation technique has not been verified by the scientific community, and the discussion of the benefits and limits of the new concept can be based only on the work of developers and the knowledge of the technology for pollution abatement (i.e. organic compound oxidation).

Titanium dioxide is preferred among all other possible photocatalyst because it is stable to photocorrosion, insoluble in a wide pH range, non-toxic and cheap. Among the several inherent advantages of photocatalytic oxidation with respect to UV or PS and PS/UV, the following are probably the most important. The destruction of organic compounds is achieved in the absence of dissolved oxidants other than oxygen (from air), either by oxidation or reduction processes, with the possibility of transforming into CO_2, also of organics in which carbon is at the highest oxidation number (Pelizzetti and Minero 1994, 1999). The oxidation of quite recalcitrant organics, stable even to other oxidation processes, as well as hydrophilic solutes and semivolatile compounds (e.g. 1,4-dioxane (Maurino et al. 1997), ethylene glycol (Parent et al. 1996), halocarbons (Calza et al. 1997a, 1997b)) is possible at rates comparable to other organic compounds. The reported conversions for a *TOC* system are on average 99% for a variety of compounds (Matthews et al. 1990). Being a heterogeneous process, the catalyst surface may adsorb organic species (Crittenden et al. 1997) favouring their degradation. The oxidation process is also very effective for ppb levels of water contaminants (Pelizzetti

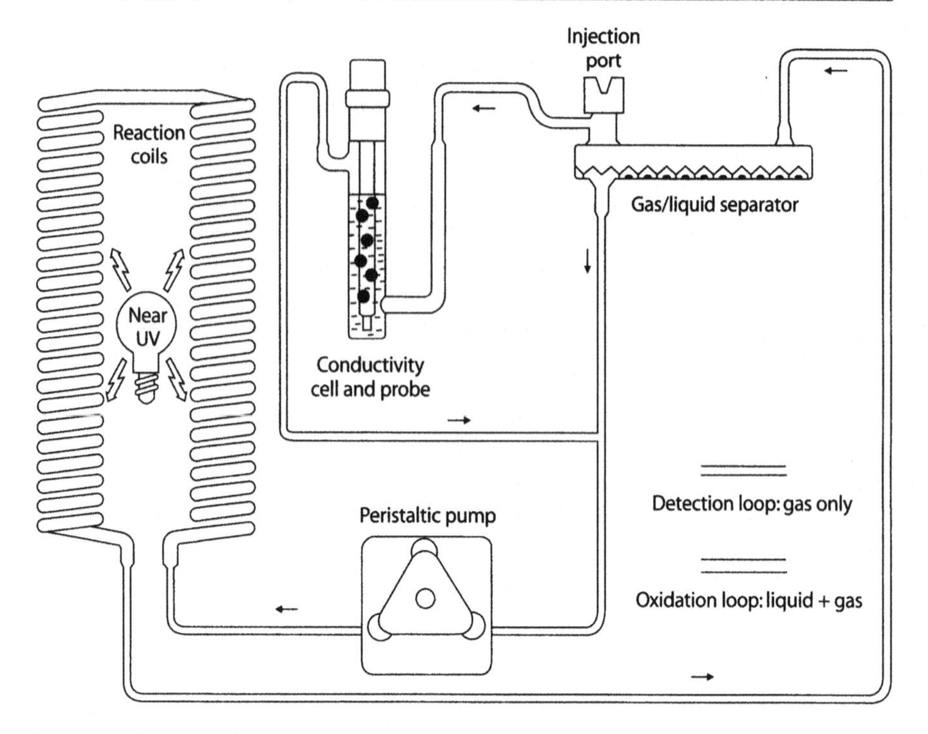

Fig. 16.5. Schematics of the photocatalytic oxidation based *TOC* analyser (from Low and McEvoy 1996, with permission)

et al. 1990), thus promising very low detection limits. Matthews et al. (1990) reported a relative standard deviation of 0.7% for ten determination of *TOC* at 160 μM C, and Abdullah and Eek (1996) showed that replicate analysis of fresh water gave 1.6–3.0% reproducibility at about 600 μM C. No adjustment of pH may be required, but best conditions of pH are required to avoid incomplete oxidation and favour CO_2 removal.

In spite of the above-mentioned advantages, several factors may limit the performance of the photocatalytic mineralization. Some of these factors, like the relatively slow overall rates and relatively low quantum yields, and the possibility of low-order dependence of rates on radiation intensity, are of major interest for pollution abatement and to the related engineering aspects. The intensity at which the rate goes from linear dependence to low-order seems to depend largely on the nature and concentration of the organic matter. Owing to the dependence of the oxidation rate on light intensity, for *DOC* oxidation various illumination systems have been tested, from low power fluorescent lamps (Matthews et al. 1990) to high power medium-pressure xenon lamps (Abdullah and Eek 1996). For the *TOC* measurements, the sensitivity to the nature of the organic composition of the sample is of major concern, since this affects the reaction times, the operational parameters, and eventually the possibility of obtaining complete conversion to CO_2. Experiments at various concentrations of formic acid, and for different solutes, showed that, under the experimental conditions used, *TOC* conversion to CO_2 is quantitative for 7–15 minutes of illumination, permitting a rapid sample turnaround (Matthews et al. 1990).

Specific areas in which it is possible to envisage improvements in the efficiency of the photocatalytic method for *TOC* measurements are the improvement of the catalyst performance (Anpo 1989; Hoffmann et al. 1995; Ollis and Al-Ekabi 1993) and catalyst immobilisation on solid support. Suspended catalysts require a gas-liquid separator for CO_2 measure, adding to the cost and complexity of the overall process, with the possibility of incomplete equilibrium distribution (Matthews et al. 1990). Immobilised materials (Zeltner et al. 1993) would facilitate the engineering of the system, although these materials will probably be more costly and subject to fouling. The suspension is in general more suited for the analysis of dirty water. Titanium dioxide has been attached to a variety of supporting materials (glass (Matthews et al. 1990), Teflon® (Low and Matthews 1990), silica, quartz). The performance of the supported catalyst is generally less than in suspension, by a factor of two or more. The decrease has been assigned to a diminution in the number of active sites and to the mass transfer limitation (Bideau et al. 1995). Attention has to be paid also to the extent of mineralization when porous supports are adopted (Matthews 1991). The pumping speed has thus a significant effect on the rate of oxidation (Matthews et al. 1990).

Oxygen is essential for the complete mineralization of organic compounds, so it is generally crucial that oxygen depletion not occur during the photooxidation process. However, when the *DOC* is highly oxidated (e.g. heavily halogenated hydrocarbons, carbonyl compounds, polyhydroxo- or polycarboxyl compounds), the reductive pathways may initially predominate and, since oxygen is a competitor for the conduction band electrons, a detrimental effect of oxygen concentration on the degradation rate is observed (Pelizzetti and Minero 1999).

Other electron scavengers, such as peroxydisulfate, generally, but not always, show a beneficial effect on the photooxidation processes. For example, whereas peroxydisulfate shows an astonishing increase of the rate of formation of cyanuric acid as the end product of atrazine degradation (Pelizzetti et al. 1991), a detrimental effect is reported for ethylene glycol (Parent et al. 1996). The need of PS (1.5 mM) for photocatalytic *TOC* measurement of urea was examined, suggesting a little beneficial effect (Abdullah and Eek 1996). Hydrogen peroxide shows an even more complex behavior dependent on several operational parameters (Pichat et al. 1995). The use of these additives may be envisaged when *DOC* amount is high to reach rapidly the total conversion to CO_2.

The fouling and poisoning of the catalyst, as well as the catalyst regeneration procedure, are under current research, and severely depend on the real composition of *DOC*. It was reported (Parent et al. 1996) that humic acids reduce the activity of titanium dioxide toward the target substances, but it was recently shown that the mineralization process still occurs efficiently and chloride evolution ensures a complete dehalogenation of the chlorinated compounds (Sega et al. 1999). The photocatalytic *DOC* oxidation seems effective on macromolecular components isolated by ultrafiltration of sea water (Abdullah and Eek 1996).

Simple inorganic ions, such as chloride, bromide, phosphate and sulfate have been found to reduce the photocatalytic performances (Abdullah et al. 1990). Recent studies have shown the formation of halogenated (Cl and Br) organics if photocatalytic degradation occurs in the presence of the correspondent halides (Minero et al. 1997b). This is of major importance for marine *DOC* measurements. However, Matthews et al. (1990) reported that the inhibition is not dramatic when working with sea water at acidic pH.

Acknowledgements

C. M. and E. P. are grateful to MURST, and CEE MASTOC project# MAS3-CT97-0138 for their financial support.

References

Abdullah MI, Eek E (1996) Automatic photocatalytic method for the determination of dissolved organic carbon (*DOC*) in natural waters. Wat Res 30:1813–1822
Abdullah MI, Low G, Matthews RW (1990) Effect of common inorganic anions on rates of photocatalytic oxidation of organic carbon over illuminated titanium dioxide. J Phys Chem 94:6820–6826
Aiken GR (1992) Chloride interference in the analysis of dissolved organic carbon by the wet oxidation method. Environ Sci Technol 26:2435–2439
Al-Ekabi H (1997) Proceedings of the Third International Conference on TiO_2 Photocatalytic Purification and Treatment of Water and Air, Orlando, FL, September 23–26, 1997
Alperin MJ, Martens CS (1993) Dissolved organic carbon in marine pore waters: A comparison of three oxidation methods. Mar Chem 41:135–143
Amon RMW, Benner R (1994) Rapid cycling of high-molecular-weight dissolved organic matter in the ocean. Nature 369:549–551
Anpo M (1989) Photocatalysis on small particles TiO_2 catalysts. Reaction intermediates and reaction mechanisms. Res Chem Intermed 11:67–106
Armstrong FAJ, Williams PM, Strickland, JDH (1966) Photo-oxidation of organic matter in sea water by ultraviolet radiation. Analytical and other applications. Nature 221:481–483
Banoub, MW, Williams PJLeB (1972) Measurements of microbial activity and organic material in the western Mediterranean Sea. Deep-Sea Res 19:433–443
Bauer JE, Williams PM, Druffel ERM, Suzuki Y (1990) Deep profiles of dissolved organic carbon in the Sargasso Sea south of Bermuda. EOS, Trans AGU 71:154
Bauer JE, Williams PM, Druffel ERM (1992) ^{14}C activity of dissolved organic carbon fractions in the north-central Pacific and Sargasso Sea. Nature 357:667–670
Bauer JE, Ocelli ML, Williams PM, McCaslin PC (1993) Heterogeneous catalyst structure and function: Review and implications for the analysis of the dissolved organic carbon and nitrogen in natural waters. Mar Chem 41:75–89
Benner R, Strom M (1993) A critical evaluation of the analytical blank associated with *DOC* measurements by high temperature catalytic oxidation. Mar Chem 41:153–160
Benner R, Biddanda B, Black B, McCarthy M (1997) Abundance, size distribution, and stable carbon and nitrogen isotopic compositions of marine organic matter isolated by tangential-flow ultrafiltration. Mar Chem 57:243–263
Bianchi A, Varney MS (1992) Sampling and analysis of volatile organic compounds in estuarine air by gas chromatography and mass spectrometry. J Chromatogr 643:11–23
Bideau M, Claudel B, Dubien C, Faure L, Kazouan H (1995) On the "immobilization" of titanium dioxide in the photocatalytic oxidation of spent waters. J Photochem Photobiol A Chem 91:137–144
Blake DM (1994) Bibliography of work on the photocatalytic removal of hazardous compounds from water and air. National Renewable Energy Laboratory, Golden (CO), NREL/TP-430-6084
Blake DM (1995) Bibliography of work on the photocatalytic removal of hazardous compounds from water and air. National Renewable Energy Laboratory, Golden (CO), NREL/TP-473-20300
Blake DM (1997) Bibliography of work on the photocatalytic removal of hazardous compounds from water and air. National Renewable Energy Laboratory, Golden (CO), NREL/TP-430-22197
Bolton JR (ed) (1996) Special issue: Solar detoxification. Solar Energy 56:375–477
Calza P, Minero C, Pelizzetti E (1997a) Photocatalytic assisted hydrolysis of chlorinated methanes under anaerobic conditions. Environ Sci Technol 31:2198–2203
Calza P, Minero C, Pelizzetti E (1997b) Photocatalytic transformations of chlorinated methanes in the presence of electron and hole scavengers. J Chem Soc Faraday Trans 93:3765–3771
Cauwet G (1992) The relevance of *DOC*-HTCO analysis to estuaries and coastal waters. Wat Poll Res Rep 28:183–196
Cauwet G (1994) HTCO method for dissolved organic carbon analysis in seawater: Influence of catalyst on blank estimation. Mar Chem 47:55–64
Cauwet G, Sempere R, Saliot A (1990) Carbone organique dissous dans l'eau de mer: Confirmation de la sou-estimation antérieure. C R Acad Sci Paris 311:1061–1066
Chen RF, Fry B, Hopkinson CS, Repeta DJ, Peltzer ETP (1996) Dissolved organic carbon on Georges Bank. Cont Shelf Res 16:409–420

Coffin RB, Connolly JP, Harris PS (1993) Availablity of dissolved organic carbon to bacterioplankton examined by oxygen utilization. Mar Ecol Prog Ser 101:9–22
Collins JJ, Williams PJLeB (1977) An automated photochemical method for the determination of dissolved organic carbon in sea and estuarine waters. Mar Chem 5:123–141
Crittenden JC, Sri RPS, Perram DL, Hand DW (1997) Decontamination of water using adsorption and photocatalysis. Wat Res 31:411–418
Datzko VG (1939) Organic matter in water of certain seas. Comptes Rendues Dokl Akad Nauk S.S.S.R. 24:294–297
De Baar HJW, Brussard J, Hegeman S, Stoll MHC (1993) Sea trials of three different methods for measuring non-volatile dissolved organic carbon in seawater during the JGOFS North Atlantic pilot study. Mar Chem 41:145–152
De Bièvre P, Peiser HS (1997) Basic equations and uncertainties in isotope-dilution mass spectrometry for traceability to SI of values obtained by this primary method. Fresenius J Anal Chem 359:523–525
Fry B, Saupe S, Huller M, Peterson BJ (1993) Platinum-catalyzed combustion of *DOC* in sealed tubes for stable isotopic analysis. Mar Chem 41:187–193
Fry B, Peltzer ET, Hopkinson CS jr, Nolin A, Redmond L (1996) Analysis of marine *DOC* using a dry combustion method. Mar Chem 54:191–201
Fung YS, Wu Z, Dao KL (1996) Determination of total organic carbon in water by thermal combustion-ion chromatography. Anal Chem 68:2186–2190
Gagosian RB, Steurmer DH (1977) The cycling of biogenic compounds and their diagenetically transformed products in seawater. Mar Chem 5:605–632
Gershey RM, MacKinnon MD, Williams PJLeB, Moore RM (1979) Comparison of three oxidation methods for the analysis of the dissolved organic carbon in seawater. Mar Chem 7:289–306
Gough MA, Mantoura RFC (1991) Advanced methods for the characterisation of macromolecular organic matter. In: Hilfe ER, Tuszynski W (eds) Proceedings of the particle desorption mass spectrometry for marine organic geochemistry. 3rd International Workshop on the Physics of Small Systems, pp 114–130
Hedges JI (1987) Organic matter in sea water. Nature 330:205–206
Hedges JI, Farrington J (1993) Measurement of dissolved organic carbon and nitrogen in natural waters: Workshop report. Mar Chem 41:5–10
Hedges JI, Lee C (eds) (1993) Measurement of dissolved organic carbon and nitrogen in natural waters. Proceedings of NSF/NOAA/DOE Workshop, Seattle, WA, 15–19 July 1991. Mar Chem 41(1–2): pp 290
Hedges JI, Bergamaschi BA, Benner R (1993) Comparative analyses of *DOC* and *DON* in natural water. Mar Chem 41:121–134
Hoffmann MR, Martin ST, Choi W, Bahnemann DW (1995) Environmental applications of semiconductor photocatalysis. Chem Rev 95:69–96
Jackson GA (1988) Implications of high dissolved organic matter concentrations for oceanic properties and processes. Oceanogr 1:28–33
Johnson KM, King AE, Sieburth JMcN (1985) Coulometric TCO_2 analysis for marine studies: An introduction. Mar Chem 16:61–82
Johnson, K.M., Sieburth J.McN., Williams P.J and Brandstrom, L. (1987), Coulometric total carbon dioxide analysis for marine studies: Automation and calibration, Mar. Chem., 21, 117–133
Juttner F, Henatsch JJ (1986) Anoxic hypolimnion is a significant source of biogenic toluene. Nature 323:797–798
Kaplan LA (1992) Comparison of high-temperature and persulfate oxidation methods for determination of dissolved organic carbon in fresh waters. Limnol Oceanogr 37:1119–1125
Kepkay PE, Wells ML (1992) Dissolved organic carbon in North Atlantic surface waters. Mar Ecol, Prog Ser 80:275–283
Kirchman DL, Suzuki Y, Garside C, Ducklow HW (1991) High turnover rates of dissolved organic carbon during a spring phytoplankton bloom. Nature 352:612–614
Krough A (1934) Conditions of life at great depths in the oceans. Ecol Monogr 4:430–439
Kumar MD, Rajendran A, Somasundar B, Jenisch A, Shou Z, Ittekkot V, Desai BN (1990) Dissolved organic carbon in the northwestern Indian Ocean. Mar Chem 31:299–316
Low GCK, Matthews RW (1990) Flow injection determination of organic contaminants in water using an ultraviolet-mediated titanium dioxide film reactor. Anal Chim Acta 231:13–20
Low GCK, McEvoy S (1996) Analytical monitoring systems based on photocatalytic oxidation principles. Trends Anal Chem 15:151–156
MacKinnon MD (1978) A dry oxidation method for the analysis of *TOC* in seawater. Mar Chem 7:17–37
Mantoura RFC, Woodward EMS (1983) Conservative behaviour of riverin dissolved organic carbon in the Severn Estuary: Chemical and geochemical implications. Geochim Cosmochim Acta 47:1293–1309
Martin JH, Fitzwalter SE (1992) Dissolved organic carbon in the Atlantic, Southern and Pacific Oceans. Nature 356:699–700

Matthews RW (1991) Environment: Photochemical and photocatalytic processes. Degradation of organic compounds. In: Pelizzetti E, Schiavello M (eds) Photochemical conversion and storage of solar energy. Kluwer, Dordrecht, pp 427–450
Matthews RW, Abdullah M, Low GK-C (1990) Photocatalytic oxidation for total organic carbon analysis. Anal Chim Acta 233:171–179
Maurino V, Calza P, Minero C, Pelizzetti E, Vincenti M (1997) Light assisted 1,4-dioxane degradation. Chemosphere 35:2675–2688
McCarthy M, Hedges J, Benner R (1996) Major biochemical composition of dissolved high molecular weight organic matter in seawater. Mar Chem 55:281–297
McKenna JH, Doering PH (1995) Measurements of dissolved organic carbon by wet chemical oxidation with persulfate: Influence of chloride concentration and oxidant volume. Mar Chem 48:109–114
Menzel DW (1970) The role of in situ decomposition of organic matter on the concentration of non-conservative properties in the sea. Deep-Sea Res 17:751–764
Miller AEJ (1996) A reassessment of the determination of dissolved organic carbon in natural waters using high temperature catalytic oxidation. Ph.D. thesis, University of Liverpool
Miller AEJ, Mantoura RFC, Preston MR (1993a) Shipboard investigation of *DOC* in the NE Atlantic using platinum-based catalysts in a Shimadzu TOC-5000 HTCO analyser. Mar Chem 41: 215–221
Miller AEJ, Mantoura RFC, Suzuki Y, Preston MR (1993b) Preliminary study of *DOC* in the Tamar Estuary, UK, using UV-persulphate and HTCO techniques. Mar Chem 41:223–228
Minero C, Pelizzetti E, Malato S, Blanco J (1996) Large solar plant photocatalytic water decontamination: Effect of operational parameters. Solar Energy 56:421–428
Minero C, Maurino V, Pelizzetti E (1997a) Solar-assisted photocatalytic transformations of hydrocarbons at the marine water-air interface. Mar Chem 58:361–372
Minero C, Maurino V, Calza P, Pelizzetti E (1997b) Photocatalytic formation of tetrachloromethane from chloroform and chloride ions. New J Chem 21:841–842
Mopper K, Zhou X, Kieber RJ, Kieber DJ, Sikorski RJ, Jones RD (1991) Photochemical degradation of dissolved organic carbon and its impact on the oceanic carbon cycle. Nature 353:60–62
Ogawa H, Ogura N (1992) Comparison of two methods for measuring dissolved organic carbon in sea water. Nature 356:696–698
Ollis DF, Al-Ekabi H (1993) Photocatalytic purification and treatment of water and air. Elsevier, Amsterdam
Parent Y, Blake D, Magrini-Bair K, Lyons C, Turchi C, Watt A, Wolfrum E, Prairie M (1996) Solar photocatalytic processes for the purification of water: State of development and barriers to commercialization. Solar Energy 56:429–437
Pelizzetti E, Minero C (1993) Mechanism of the photo-oxidative degradation of organic pollutants over TiO_2 particles. Electrochimica Acta 38:47–55
Pelizzetti E, Minero C (1994) Metal oxides as photocatalysts for environmental detoxification. Comments Inorg Chem 15:297–337
Pelizzetti E, Minero C (1999) Role of the oxidative and reductive pathways in the photocatalytic degradation of organic compounds. Colloids Surf A 151: 329–338
Pelizzetti E, Schiavello M (eds) (1991) Photochemical conversion and storage of solar energy. Kluwer, Dordrecht
Pelizzetti E, Serpone N (eds) (1986) Homogeneous and heterogeneous photocatalysis. Reidel, Dordrecht
Pelizzetti E, Maurino V, Minero C, Carlin V, Pramauro E, Zerbinati O, Tosato ML (1990) Photocatalytic degradation of atrazine and other s-triazine herbicides. Environ Sci Technol 24:1559–1565
Pelizzetti E, Carlin V, Minero C, Graetzel M (1991) Enhancement of the rate of photocatalytic degradation of organic pollutants on TiO_2 by inorganic oxidizing species. New J Chem 15:351–359
Pelizzetti E, Carlin V, Minero C, Gratzel M (1993) Photoinduced degradation of atrazine over different metal oxides. New J Chem 17:315–322
Peltzer ET, Brewer P (1993) Some practical aspects of measuring *DOC* – sampling artifacts and analytical problems with marine samples. Mar Chem 41:243–252
Peltzer ET, Hayward NA (1996) Spatial and temporal variability of total organic carbon along 140° W in the equatorial Pacific Ocean in 1992. Deep Sea Res II 43:1155–1180
Peltzer ET, Fry B, Doering PH, McKenna JH, Norrman B, Zweifel UL (1996) A comparison of methods for the measurements of dissolved organic carbon in natural waters. Mar Chem 54:85–96
Perdue EM, Mantoura RFC (1993) Mechanism subgroup report. Mar Chem 41:51–60
Perdue EM, Rodgers MO, Garland L, Wang W (1994) A mechanistic study of high temperature oxidation of organic carbon in carbon analyzers. Abstr Am Chem Soc Div Geochem, No 95
Peterson BJ, Fry B, Saupe S, Ullar MH (1996) The distribution and stable carbon isotopic composition of dissolved organic carbon in estuaries. Estuaries 17:111–121

Pichat P, Guillard C, Maillard C, Amalric L, D'Oliveira J-C (1995) Assessment of the importance of the role of H_2O_2 and $O_2^{-}\cdot$ in the photocatalytic degradation of 1,2-methoxybenzene. Sol En Mat Sol Cells 38:391–399
Qian J, Mopper K (1996) Automated high-performance, high temperature combustion total organic carbon analyzer. Anal Chem 68:3090–3097
Ridal JJ, Moore RM (1993) Resistance to UV and persulfate oxidation of dissolved organic carbon produced by selected marine phytoplankton. Mar Chem 42:167–188
Schiavello M (ed) (1985) Photoelectrochemistry, photocatalysis and photoreactors: Fundamentals and development. Reidel, Dordrecht
Schiavello, M (ed) (1988) Photocatalysis and environment. Kluwer, Dordrecht
Sega M, Minero C, Pelizzetti E, Friberg SE, Sjoblom J (1999) The role of humic substances in the photocatalytic degradation of water contaminants. J Disper Sci Technol 20:643–661
Serpone N, Pelizzetti E (eds) (1989) Photocatalysis. Fundamentals and applications. Wiley, New York
Sharp JH, Peltzer ET (1993) Procedures subgroup report. Mar Chem 41:37–49
Sharp JH, Benner R, Bennet L, Carlson CA, Fitzwater SE, Peltzer ET, Tupas LM (1995) Analyses of dissolved organic carbon in seawater: The JGOFS EqPac methods comparison. Mar Chem 48:91–108
Skoog DA, West DM (1976) Fundamental of analytical chemistry. Saunders College, Philadelphia, PA
Skoog A, Thomas D, Lara R, Richter K-U (1997) Methodological investigations on *DOC* determinations by the HTCO method. Mar Chem 56:39–44
Skopintsev BA, Timofeyeva SN, Vershinna OA (1966) Organic carbon in the near-equatorial and southern Atlantic and in the Mediterranean. Oceanol 6:201–210
Skopintsev BA, Romenskaya NN, Sokolova MV (1968) Organic carbon in the waters of the Norwegian Sea and of the northeast Atlantic. Oceanol 8:178–186
Starikova ND (1970) Vertical distribution patterns of dissolved organic carbion in sea water and interstitial solutions. Oceanol 10:796–807
Statham PJ, Williams PJ Le B (1983) The automatic determination of dissolved organic carbon. In: Grasshof K, Erhardt MM, Kremling K (eds) Methods of seawater analysis, 2nd edn. Verlag Chemie GmbH, Weinheim, pp 380–395
Stumm W (1992) Chemistry of the solid-water interface. John Wiley & Sons, London
Sugimura Y, Suzuki Y (1988) A high-temperature catalytic oxidation method for the determination of non-volatile dissolved organic carbon in seawater by direct injection of a liquid sample. Mar Chem 24:105–131
Suzuki Y (1993) On the measurement of *DOC* and *DON* in seawater. Mar Chem 41:287–288
Suzuki Y, Tanoue E (1991) Dissolved organic carbon enigma: implications for ocean margins. In: Mantoura RFC, Martin J-M, Wollast R (eds) Ocean margin processes in global change. Report of the Dahlem Workshop on Ocean Margin Processes in Global Change, Berlin, March 18–23, 1990. Wiley, Chichester, pp 197–209
Suzuki Y, Peltzer ET, Brewer PG (1990) *DOC* analyses during the North Atlantic spring bloom. EOS, Trans AGU 71:1550
Suzuki Y, Tanoue E, Iroshi I (1992) A high-temperature catalytic oxidation method for the determination of dissolved organic carbon in seawater: Analysis and improvement. Deep-Sea Res 39:185–198
Toggweiler JR (1988) Deep-sea carbon, a burning issue. Nature 334:468
Toggweiler JR (1989) Is downward *DOM* flux important in carbon transport? In: Berger WH, Smetacek VS, Wefer G (eds) Productivity of the ocean: present and past. John Wiley & Sons, Chichester, pp 65–83
Tupas LM, Popp BN, Karl DM (1994) Dissolved organic carbon in oligotrophic waters: Experiments on sample preservation, storage and analysis. Mar Chem 45:207–216
Van Hall CE, Barth D, Stenger VA (1965) Elimination of carbonates from aqueous solutions prior to organic carbon determination. Anal Chem 37:769–771
Vogl J, Heumann K (1998) Development of an ICP-IDMS method for dissolved organic carbon determination and its application to chromatographic fractions of heavy metal complexes with humic acid substances. Anal Chem 70:2038–2043
Wangersky PJ (1993) Dissolved organic carbon methods: A critical review. Mar Chem 41:61–74
Wells M, Goldberg ED (1991) Occurrence of small colloids in seawater. Nature 353:342–344
Wells M, Goldberg ED (1992) Marine sub-micron particles. Mar Chem 40:5–18
Wells M, Goldberg ED (1993) Colloid aggregation in seawater. Mar Chem 41:353–358
Williams PM, Druffel ERM (1988) Radiocarbon in dissolved organic matter in the central North Pacific Ocean. Nature 330:246–248
Williams PM, Bauer JE, Robertson KJ, Wolgast DM, Ocelli ML (1993) Report on *DOC* and *DON* measurements made at Scripps Institution of Oceanography, 1988–1991. Mar Chem 41:271–281
Yamanaka Y, Tajika E (1997) Role of dissolved organic matter in the marine biogeochemical cycle: Studies using an ocean biogeochemical general circulation model. Glob Biogeochem Cycles 11:599–612

Zeltner WA, Hill CG Jr, Anderson MA (1993) Reactor concepts for photodegradation using supported titania. Chemtech 23(5):21–29
Zweifel UL,Norrman B, Hagström A (1993) Consumption of dissolved organic carbon by marine bacteria and demand for inorganic nutrients. Mar Ecol Prog Ser 101:23–32
Zweifel UL, Wikner J, Hagström A, Lundberg E, Norrman B (1995) Dynamics of dissolved organic carbon in a coastal ecosystem. Limnol Oceanogr 40:299–305

Characterization of Marine Toxins by Means of Liquid Chromatography – Electrospray Ionization – Mass Spectrometry

M. Vincenti · A. Irico

17.1 Introduction

In a previous review on application of mass spectrometric methods to the analysis of natural and anthropogenic marine contaminants (Vincenti 1997), it was advanced that rapid development and diffusion of electrospray ionization (ESI) interface for combining liquid chromatography with mass spectrometry (LC-MS) was going to have strong impact on research and routine analysis of the biological substances produced by marine microorganisms, among which marine toxins. This trend has been confirmed in the last two years. It is today clear that ESI represents one of the decisive steps in the development of mass spectrometry. The dramatic improvement of LC-MS effectiveness, as a consequence of ESI outbreak, has made the interfacing techniques previously developed obsolete. The present review will only consider the applications of marine toxin LC-MS determination in which an ESI interface has been used.

The principles (Kebarle and Ho 1997) and instrumental design (Bruins 1997) of ESI has been reviewed in several texts. In ESI, the ions initially present or formed in an electrolytic solution are transferred to the gas phase, by means of an electrical field applied to the tip of a capillary through which the solution is flowed. Upon formation of a double layer on the meniscus of the solution by electrolyte ion separation, a spray of charged droplets is released in the gas phase. The efficiency of this process depends on a number of parameters, including the composition and electrolyte concentration in solution, the flow rate, the voltage and polarity of the electrical field. To assist the charged spray formation and expand the conditions (expecially the solution flow rate) feasible to ESI, various devices have been alternatively added to the capillary including a coaxial sheath liquid, a nebulizing gas (pneumatically-assisted ESI or Ionspray), an ultrasonic transducer (ultrasonic-assisted ESI or Ultraspray), and some form of heating. Once the charged droplets are formed, solvent evaporation occurs, leading to cycles of droplet shrinkage, increasing coulombic repulsion and droplet explosion. At the end of this cyclic process, desolvated ions are transferred into the mass spectrometer and analysed. During spray formation and desolvation, several solution and gas-phase ion-molecule reactions occur, at atmospheric pressure, possibly leading the formation of charged analyte species. In this sense, ESI is one of the ways leading to atmospheric pressure ionization (API). To assist the desolvation process, heating of the interface walls or a counter current flow of dry nitrogen are frequently employed. Ion transfer from the atmospheric pressure region into the high vacuum of the mass spectrometer takes place through a series of orifices and/or capillaries and a final skimmer plate located before the mass analyser. By regulating the skimmer voltage, incoming ions can be accelerated and possibly fragmented by collisional activation with an inert gas or solvent molecules, in a region of relatively high pressure (about 1 torr). This fragmentation, intentionally induced in ESI spectra, is often useful for the structural characterization of analytes. Without this "in-source" collision induced dissociation (CID), ESI spectra generally exhibit only the protonated or deprotonated molecular ions of solution components, as well as cationized and polymeric adduct ions.

Alternatively, fragmentation can be induced by CID in a tandem mass spectrometry (MS-MS) experiment. The experiment most frequently encountered in marine toxin LC-ESI-MS-MS analysis consists of selecting the toxin molecular ion by the first mass analyser, then forcing it into a collision cell where it experiences multiple collisions with an inert gas. These collisions deposit an excess of internal energy into the impinging ions, which eventually fragment. The daughter ions are determined by the second mass analyser, operating either under continuous scanning (MS-MS or daughter ion spectra) or under mass-hopping among the fragment ion of larger abundance. The latter operating mode, called selected reaction monitoring (SRM) represents the MS-MS equivalent of selected ion monitoring (SIM) in single-stage MS.

As mentioned before, ESI operates with constant flow of an electrolyte solution. This can either be a suitable matrix, in which a solution of the pure toxin sample is flow-injected (FIA), the liquid flow of a capillary electrophoresys (CE) device, or, most frequently, the eluate of a liquid chromatograph (LC). In the latter case, some restrictions are imposed on the eluant composition by ESI, which does not tolerate some buffer components and high concentrations of ionic species. In general, the LC separation should be programmed keeping in mind such restrictions. The efficiency of ESI is maximized at low LC solvent flow rates. When the flow rate is increased, the increased sample throughput does not determine an increased ion signal, but this remains approximately constant. Consequently, the sensitivity remains constant in terms of concentration, but decreases in terms of mass consumption, and the instrument contamination from solvent impurities increases considerably, thus affecting the long-term stability of the instrument. Therefore, LC-ESI-MS is often performed with small-bore or microbore columns. Alternatively, most parts of the eluate are split and either sent to an UV detector or thrown away.

17.2 Dinophysistoxins and Related Toxins

One of the applications within the field of marine toxins, in which LC-ESI-MS has been most extensively applied is the detection of diarrhetic shellfish poisoning toxins. Diarrhetic shellfish poisoning (DSP) is a serious gastroenteritis, resulting from eating contaminated shellfish. Besides acute toxic effects, these toxins are also reported to be potent carcinogens. DSP toxins accumulate in shellfish, such as mussels and clams, but are produced by toxic strains of certain dinoflagellates (phytoplankton). Upon ingestion of these dinoflagellates shellfish becomes toxic. After the first identification in Japan, the cases of DSP have markedly increased in many parts of the world, especially in Japan and Europe, posing a serious problem to human health (Ragelis 1984; Yasumoto et al. 1989; Yasumoto 1990; Hall and Strichartz 1990; Falconer 1993; Smayda and Shimizu 1993).

Three different classes of polyether toxins have been found to cause DSP, namely

i. okadaic acid (OA) and its derivatives dinophysistoxins (DTXs),
ii. pectenotoxins (PTXs), and
iii. yessotoxins (YTXs).

whose structures are shown in Fig. 17.1.

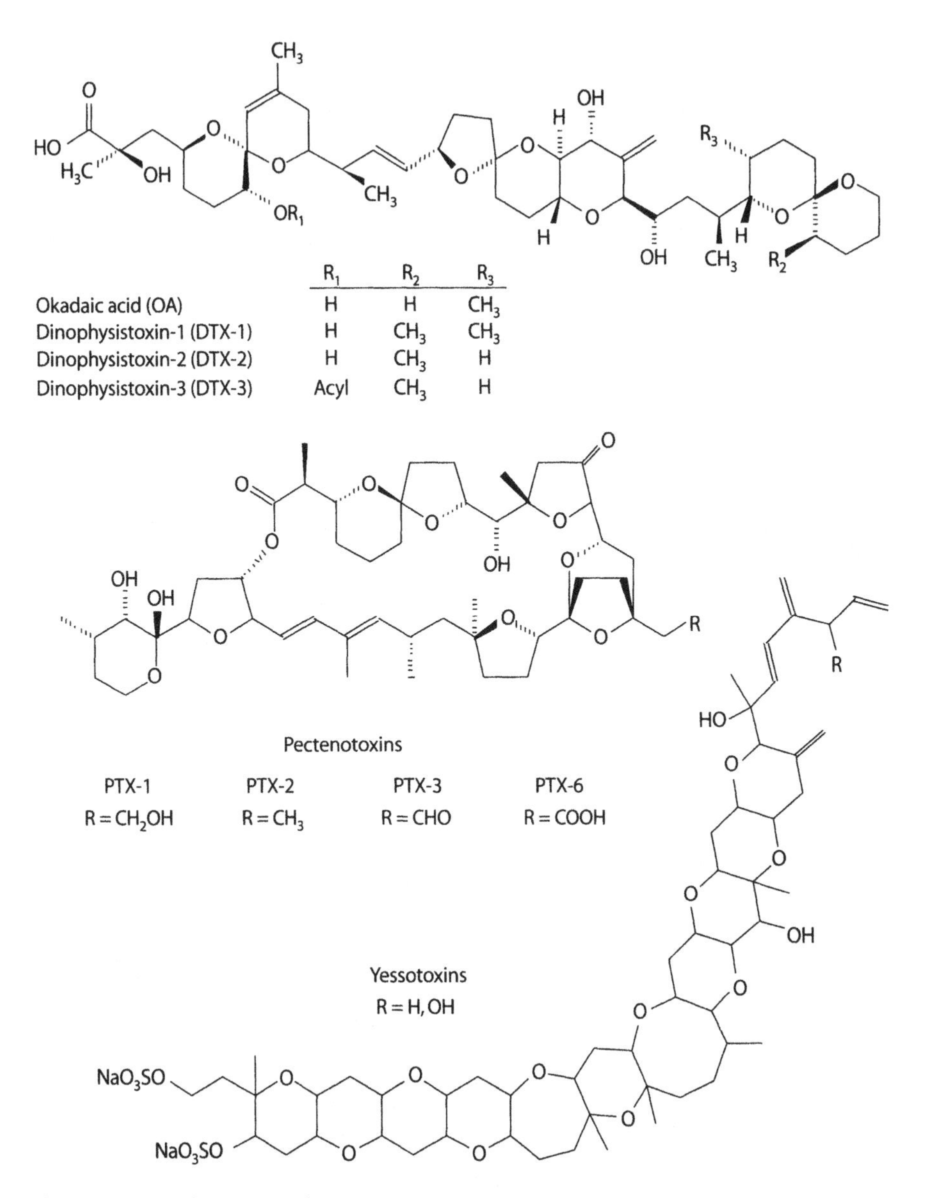

Fig. 17.1. Chemical structures of some common DSP toxins

Sometimes, only one compound is responsible for the toxicity, while in other cases several molecules with distinct structures contribute to the overall poisoning effect, making toxicological evaluations problematic. Therefore, the traditional bioassay tests based on acute effects in mice and rats proved to be scarcely sensitive, unspecific and occasionally unreliable. The analytical methods initially developed were based on the derivatization of toxins by fluorescent markers followed by LC with fluorimetric detection (Quilliam 1995; Shen et al. 1997). The refinement and increased availability of LC-ESI-MS made this intrumentation the analytical technique of choice for the identification and detection of DSP toxins, since it is more sensitive and specific than fluorimetric detection, does not require derivatization pretreatments and allows much higher preconcentration factors (Pleasance et al. 1992b). Figure 17.2 provides a clear picture of the superior MS selectivity, which is reflected in an improved S/N ratio and, consequently, in extremely high sensitivity.

Most experimental work has been addressed to the detection of OA and its isomers, homologues and derivatives collectively defined DTXs, since most of DSP cases proved to be related to the presence of these toxins. The kind of samples considered were either mussels (Marr et al. 1992; Pleasance et al. 1992b; Draisci et al. 1995; Quilliam 1995; James et al. 1997b; Draisci et al. 1998b) or marine phytoplankton specimen (James et al. 1997a; Draisci et al. 1998a; Morton et al. 1998). Since the mussels are generally available in large amounts and tend to accumulate the toxins, the analysis of their hepatopancreas apparatus is relatively simple and does not require extensive cleanup, provided that LC separation is combined with a selective detection method, such as ESI-MS (Draisci et al. 1995; Draisci et al. 1998b). In this case, a conventional 4.6 mm LC column has been generally used. By means of a rather complex sample treatment (including up to five different chromatographic steps (Draisci et al. 1998a)), the single toxins were isolated and subsequently used as analytical standards, whenever not commercially available. Since the analysis of single compounds did not require LC separation, they were generally flow-injected into the ESI-MS apparatus in order to optimize the experimental parameters (solvent flow rate, nebulizer gas pressure, tip voltage, skimmer voltage) (Quilliam 1995; Draisci et al. 1995). In contrast to mussels, pure phytoplankton specimen are generally available in limited amount, since various phytoplankton species are generally mixed together, especially when they can not be cultured in a laboratory. Draisci, James and co-workers were able to analyse single phytoplankton species by picking individual cells from a microscope slide (James et al. 1997a; Draisci et al. 1998a). The resulting extract, available in a very small amount, was subsequently analysed by micro-LC-MS-MS, requiring only 0.2 μl injection.

It should be noted that the decreased amount of sample loaded in micro-LC (approximately 100 times lower than in conventional LC) is compensated for by a largely improved ionization and ion-trasmission efficiency associated with the low solvent flow rate of micro-LC, resulting in comparable sensitivities, in terms of analyte concentration (Pleasance et al. 1992b; Quilliam 1995; James et al. 1997a).

The ESI positive ion mass spectra of OA and DTXs are characterized by an abundant protonated molecular ion and a series of fragments corresponding to consecutive losses of water molecules, due to the presence of several labile hydroxyl groups in the toxin structures. The relative abundance of such fragment ions strongly depends on the experimental parameters adopted, particularly the skimmer voltage, which can induce the molecular ion dissociation (Marr et al. 1992). The selective detection of these

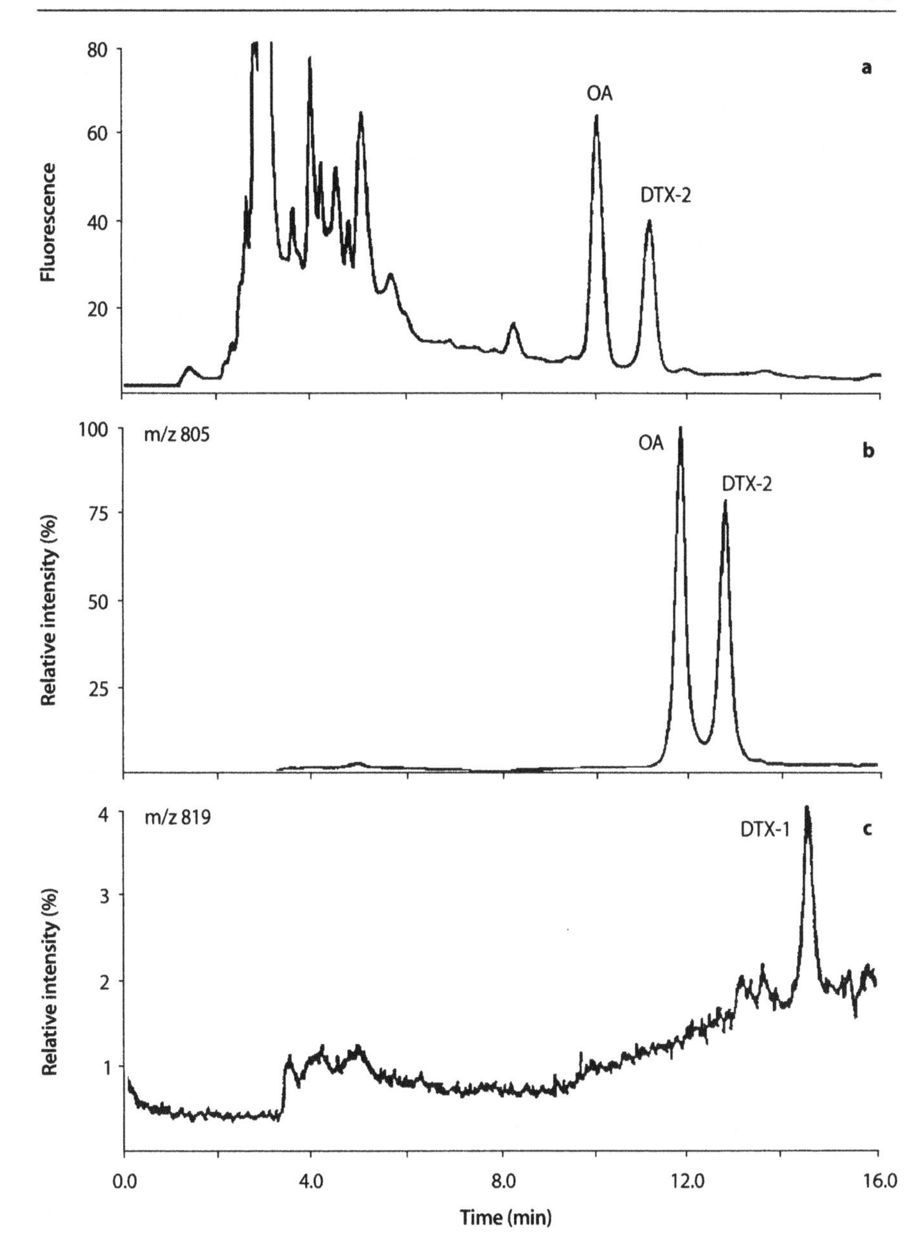

Fig. 17.2. Analysis of digestive-gland extract of DSP-toxin-contaminated Irish mussels; **a** by LC with fluorescence detection of 9-anthryldiazomethane-derivatized toxins; **b,c** by LC-ESI-MS single ion chromatograms, obtained from SIM operating mode (reprinted with permission from Pleasance et al. 1992b)

toxins is executed in selected ion monitoring (SIM) by hopping the analyser among the masses of the various protonated molecular ions, respectively m/z 805 for OA, DTX-2, DTX-2B, DTX-2C (all positional isomers of OA), m/z 819 for DTX-1 (Draisci et al. 1995; Draisci et al. 1998a) and m/z 1015, 1041, 1043, 1057, 1069 for DTX-3 (a mixture of fatty acid esters of OA and DTX-1, possibly produced by shellfish as toxin metabolites) (Marr et al. 1992). Various degrees of fragmentation could be obtained by collision induced dissociation either by raising the voltage of the skimmer without preliminary parent ion selection (Quilliam 1995) or by introducing an inert gas in the central quadrupole of a triple quadrupole instrument, running in the MS-MS mode. In the first case, the masses of some fragment ions could be added to the SIM mass sequence in order to confirm the toxin determination, ruling out the chance of false positive identification. A higher degree of selectivity was obtained by tandem mass spectrometric experiments, run in the SRM mode, where the transition between protonated molecular ions and their fragments, corresponding to the neutral loss of two, three and four molecules of water, was selectively monitored. For example, the first mass analyser was set on m/z 805, while the second was hopping between m/z 769, 751 and 733 (James et al. 1997a). Figure 17.3 illustrates the outcome of this experiment, in terms of chromatograms of single fragmentation transitions. While Quilliam (Quilliam 1995) advanced the hypothesis that some sensitivity was lost in the double mass selection inherent to MS-MS experiments, it is more likely that this effect was due to

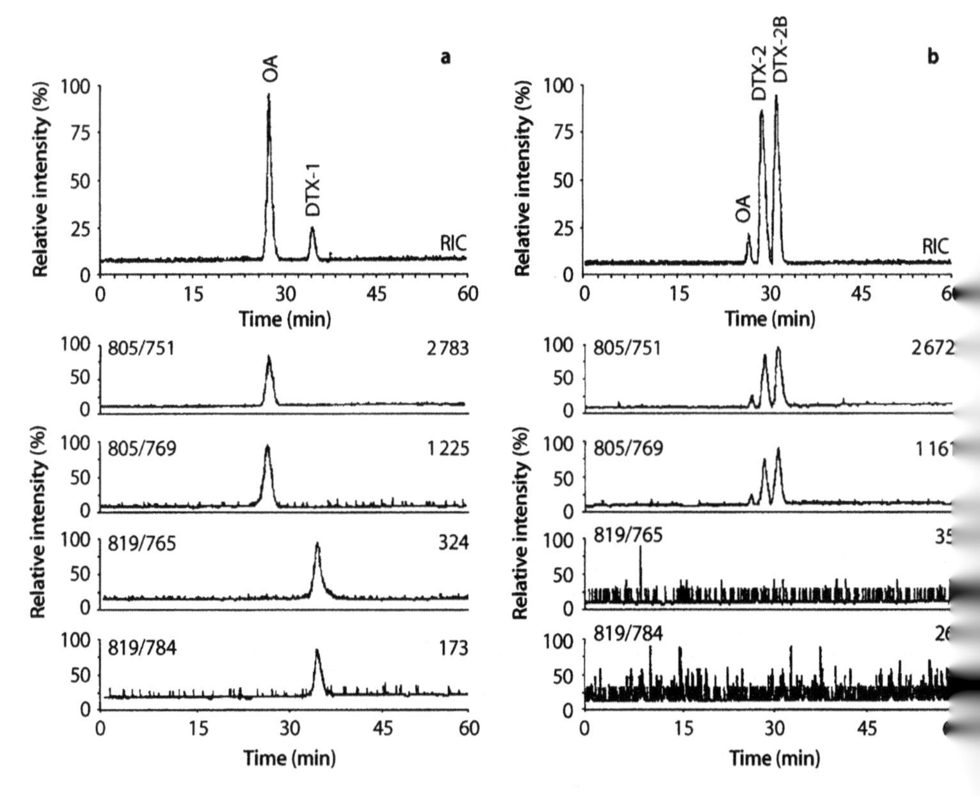

Fig. 17.3. Selected reaction monitoring (SRM) LC-ESI-MS-MS chromatograms of extracts from hepatopancreas from Italian and Irish toxic mussels; **a** Italian mussel: OA (1.51 $\mu g\,g^{-1}$), DTX-1 (0.29 $\mu g\,g^{-1}$); **b** Irish mussel extract fraction after cleanup on silica: DTX-2B (0.080 $\mu g\,g^{-1}$) (reprinted with permission from Draisci et al. 1998b)

poor instrument performance, since MS-MS experiments generally allow higher S/N ratio than single MS mode, due to dramatic reduction of the chemical noise. Satisfactory performance of the MS-MS approach was confirmed by the lower detection limits obtained from this method (15 ppb on the extract) (Draisci et al. 1998b) than in the single MS mode (Pleasance et al. 1992b). Quantitation of DSP toxins was executed either by external calibration (Draisci et al. 1998b) or with a synthesized internal standard (7-O-acetyl-OA) (Quilliam 1995).

It is clear that the fragments arising from consecutive losses of water molecules are somehow distinctive of the DTX class of toxins, but they do not allow the discrimination among isomers. The same conclusion was drawn from other fragmentation processes involving the cleavage of the toxin skeleton, because all left intact the part of the structure containing the isomer specific groups (Draisci et al. 1998a,b; Curtis et al. 1996), as can be assumed from Fig. 17.4. It is unfortunate that a study on the fragmentations induced by EI and CI was restricted to OA and not to its isomers, because some of them appear to be isomer specific (i.e. m/z 111) (Bencsath and Dickey 1991). The same information could in theory be provided also by extensive ESI-MS-MS investigation. Negative ion ESI-MS and ESI-MS-MS spectra, incidentally reported in two recent studies (Draisci et al. 1998a; Moeller et al. 1998), also appear to be promising, both in terms of absolute sensitivity and for structural identification; they are likely to be more extensively investigated in the future.

Pectenotoxins have a distinct structure from DTXs, as they are cyclic and do not carry acidic groups, allowing their separation on alumina column in the "neutral" fraction, but they are still polyether DSPs with similar mass spectral features. These include the consecutive losses of water molecules as the main fragmentation process (Draisci et al. 1996). Unlike DTXs, PTXs can be directly analysed by LC, as they are strong UV adsorbers (235 nm). However, also in the case of PTX, tandem mass spec-

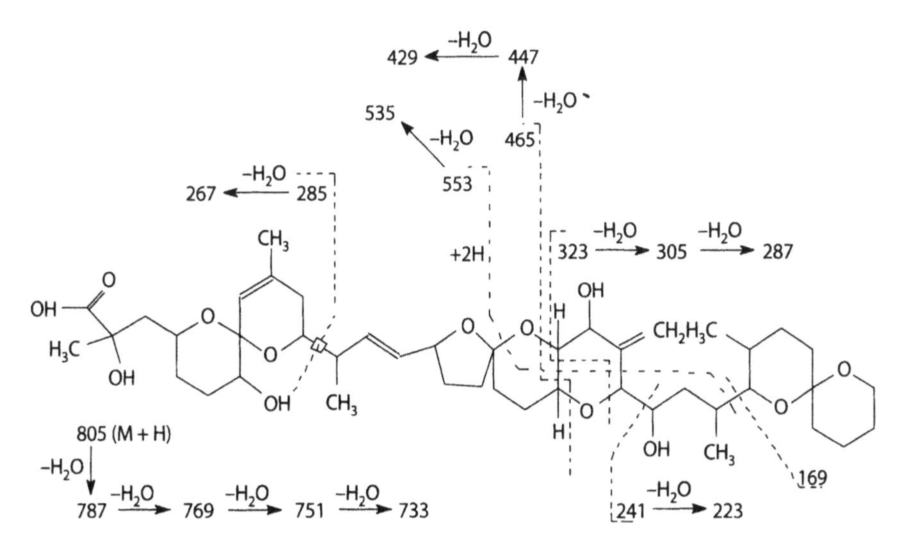

Fig. 17.4. Proposed origin of fragment ions in the ESI mass spectra of OA produced by CID of the protonated molecular ion (reprinted with permission from Quilliam 1995)

trometry proved much superior to UV-spectrophotometry in terms of sensitivity and selectivity.

17.3 Brevetoxins

Other cyclic polyether toxins are brevetoxins, which are produced by the marine dinoflagellate *Gymnodinium breve*, a toxicogenic single-cell organism whose uncontrolled growth (or bloom) produces a calamitous phenomenon called "red tide" (Yasumoto 1990). Brevetoxins are the toxic agents of these red tides, which cause massive fish kills, shellfish contamination and various health effects on mammals and humans, including severe respiratory problems and eye and skin irritation. At least nine brevetoxins have been identified up to now, with structures that can be collected into two types, as shown in Fig. 17.5. Type-1 toxins are the ones with the highest toxicity. The proposed toxicological mechanism implies their specific binding to the proteins modulating sodium channels in cell membranes. As several other cyclic polyethers, brevetoxins are strong ligands for alkali metal ions.

The analysis of brevetoxins is performed by *(i)* bioassay methods, mainly on mice; *(ii)* immunoassays and *(iii)* chromatographic methods (Hua et al. 1995). While immunoassay methods are very sensitive, chromatographic methods are the only ones capable of determining single brevetoxins. LC with UV detection (215 nm) has been

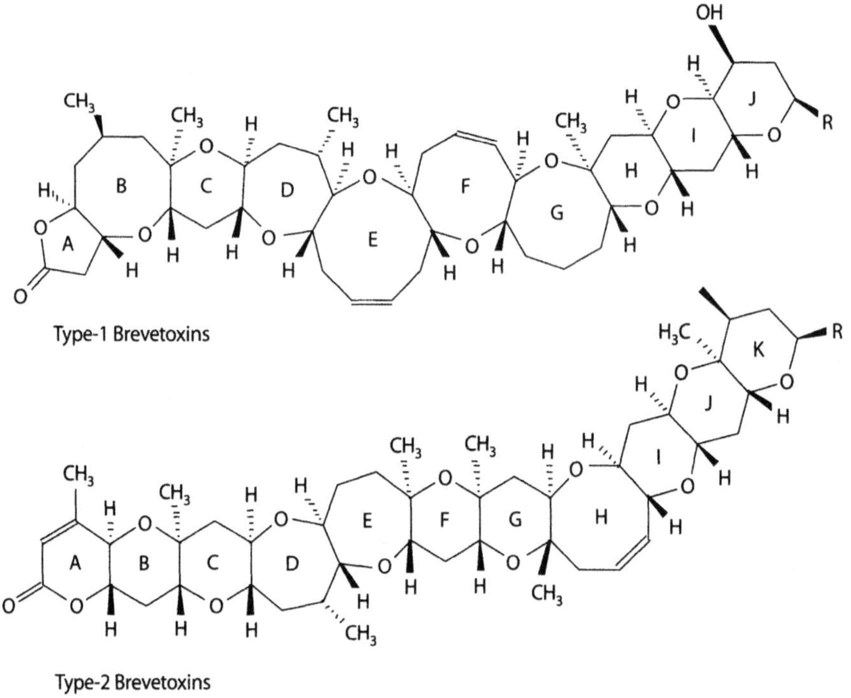

Type-1 Brevetoxins

Type-2 Brevetoxins

Fig. 17.5. Chemical structures of two classes of brevetoxins

widely used to detect them. Recently, Cole and co-workers developed an LC-ESI-MS method that proved to be 3 to 100 times more sensitive than LC-UV, depending on the specific toxin to be determined (Hua et al.1995 1996). They used a microbore column, 100×1 mm i.d., with an eluent flow rate of 4–8 μl min^{-1}. The eluent was split into a 3/1 volume ratio and the two portions were conveyed respectively into a UV detector and a mass spectrometer with electrospray ionization, so that the corresponding chromatograms could be compared to one another. When the mass analyser was scanned, the total ion chromatogram profile was similar to that arising from UV detection, even if some peak resolution was lost in the longer and more complex ESI interface.

Information of higher quality is gained from single ion chromatograms and ESI mass spectra, examples of which are provided in Fig. 17.6, relative to a "red tide" seawater extract. On the right side, a series of ESI mass spectra are reported, corresponding to seven different brevetoxins. The main peak in each spectrum is provided by the adduct ion formed between the brevetoxin and one sodium ion, present as ubiquitary contaminant even after sample cleanup. This finding confirms the large affinity of brevetoxins toward alkali metal ions. Other mass peaks typical of ESI correspond to the sodiated brevetoxin dimer $[2 M + Na]^+$ and a doubly-charged adduct formed by three brevetoxin molecules and two sodium ions $[3 M + 2 Na]^{2+}$. Obviously, these doubly charged ions, frequently encountered in ESI, appear at an m/z value equal to one half of the adduct ion total mass. On the left side of Fig. 17.6, the single ion chromatograms relative to the most abundant ion of the brevetoxin spectra are plotted. The peaks are very neat and unaffected by interferences. Their quantitation, by external calibration with authentic standards, proved accurate. In contrast, some quantitative results determined by LC-UV were significantly higher (for example, 65 ng μl^{-1} instead of 18 ng μl^{-1}) than expected (Hua et al. 1996). These higher concentrations were attributed to interfering substances coeluting with brevetoxins. In LC-ESI-MS analysis not only the component peaks are separated in different diagrams, but also these interferences were removed.

17.4 Saxitoxins

Saxitoxins are potent neurotoxins causing paralytic shellfish poisoning (PSP), a syndrome of variable gravity, with occasional cases of human death. They are also related to toxic red algal blooms, from a particular species of dinoflagellates. The family of saxitoxins comprises more than thirty closely related compounds, and new structures are continuously identified by the use of NMR and ESI-MS techniques (Arakawa et al. 1995; Onodera et al. 1997). Most important from an analytical point of view is that, while the basic structure of saxitoxin carries two positive charges, most analogs carry one or more easily-ionizable hydroxysulfate groups, which compensate for the positive charge (see Fig. 17.7). As the ESI efficiency strongly depends on the ionic state of desovated analyte ions, setting up experimental conditions (both chromatographic and mass spectrometric) suitable for all saxitoxin analogs is extremely difficult. Traditionally, the analytical problem was solved by LC with post-column oxidation of saxitoxins and fluorescence detection. More recently, Pleasance and co-workers compared several mass spectrometric approaches (Pleasance et al. 1992a), finding out that FIA-ESI-MS and FIA-ESI-MS-MS were particularly suited for the structural identification

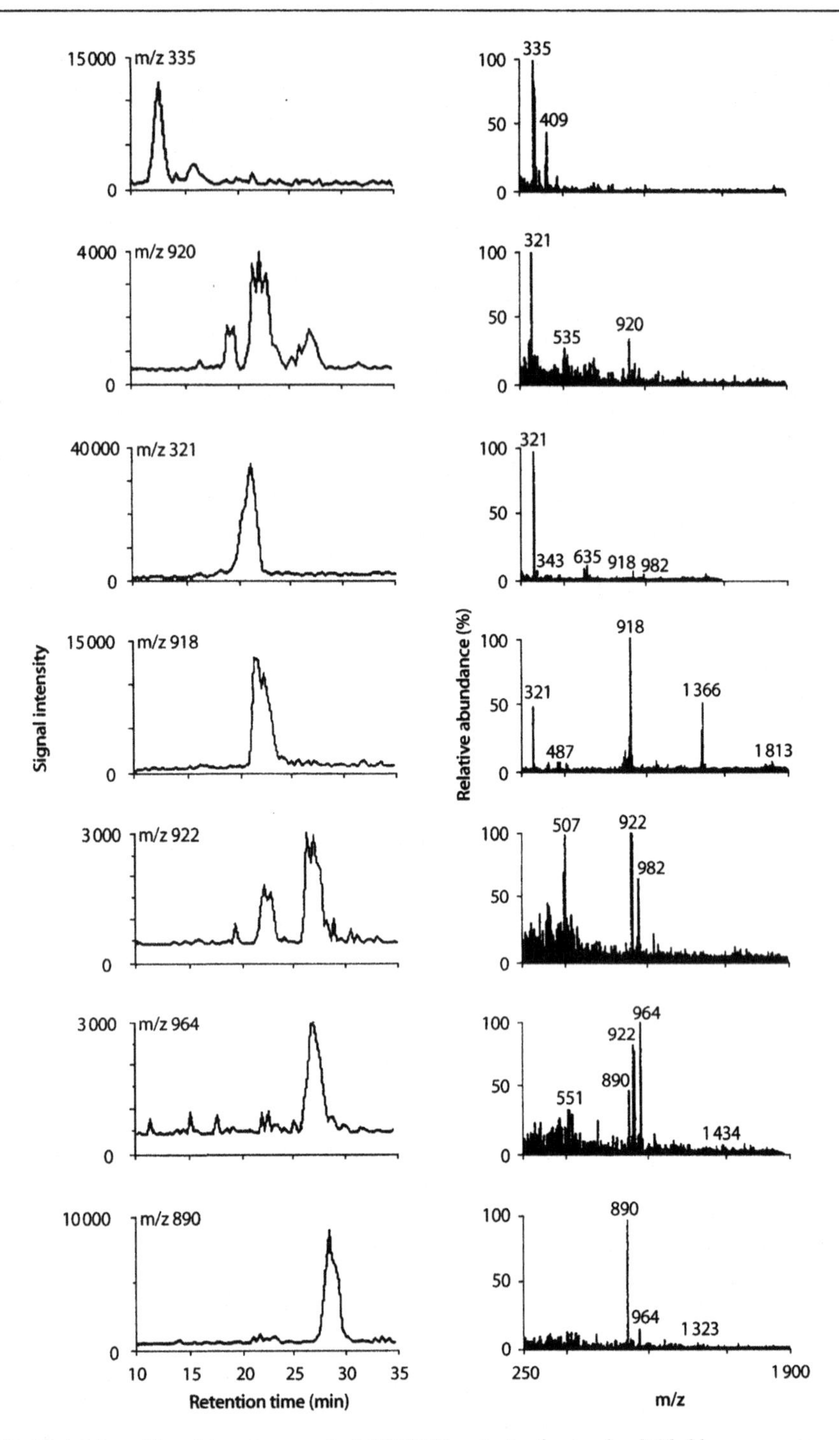

Fig. 17.6. Selected ion chromatograms for LC-ESI-MS analysis of natural red tide bloom seawater extract (*left*); ESI mass spectra obtained as an average of four scans at peak maxima from the corresponding selected ion chromatograms on the left (*right*) (reprinted with permission from Hua et al. 1996)

Fig. 17.7. General chemical structure of saxitoxins

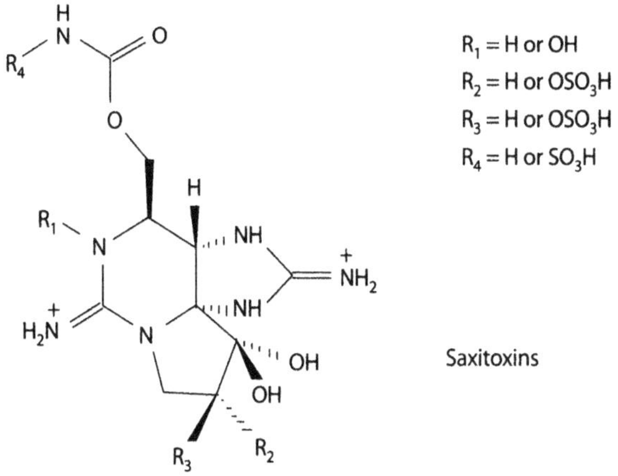

R_1 = H or OH
R_2 = H or OSO_3H
R_3 = H or OSO_3H
R_4 = H or SO_3H

Saxitoxins

of separated saxitoxin analogs, whereas capillary electrophoresis (CE)-ESI-MS can conveniently replace LC methods, including LC-ESI-MS, in which the composition of the eluent necessary to separate the saxitoxin analogs tended to suppress the ESI signal. Determination of saxitoxins by LC-ESI-MS is made easier by preliminary oxidation of the analytes (Quilliam et al. 1993). A comparison between CE with UV detection and CE-ESI-MS, in the analysis of a mixture of four saxitoxins, is provided in Fig. 17.8. It is evident that CE-MS provides much higher signal-to-noise ratios for all saxitoxins.

In structural analysis of saxitoxin analogs, a potentially powerful MS-MS technique was used: besides the molecular ions, also the fragment ions originated in ESI-MS by collisional activation in the declustering region (depending on the skimmer voltage) were mass-selected and fragmented further. In this way, the structure of fragments is investigated in detail, often resulting in isomeric discrimination. The presence of hydroxysulfate groups in saxitoxin analogs is revealed by facile loss of SO_3 molecules upon collisional activation, as well as from abundant ESI-MS signal in the negative ion mode (Pleasance et al. 1992a). In contrast, the positive charge on the saxitoxin precursor precludes the formation of negative ions by ESI.

17.5
Ciguatoxins and Maitotoxins

Ciguatoxins and maitotoxins are also dinoflagellate cyclic polyether toxins, which interfere with the cellular sodium channel, causing gastrointestinal and neurological disorders of variable gravity. This intoxication, called ciguatera, has been known for hundreds of years and is caused by ingestion of contaminated fish, captured in tropical seas, mostly inside coral reefs. The chemistry and the methods used to determine ciguatoxins and maitotoxins have been recently reviewed (Yasumoto and Satake 1996). Despite the similar origin and mechanism of action, the chemical structures of maitotoxin-1 and -2 are much more complex than those of ciguatoxins, as their molecular weights are as high as 3422 and 3298 Dalton. ESI mass spectra have nevertheless been obtained for these large toxins (Lewis et al. 1994) even if their molecular

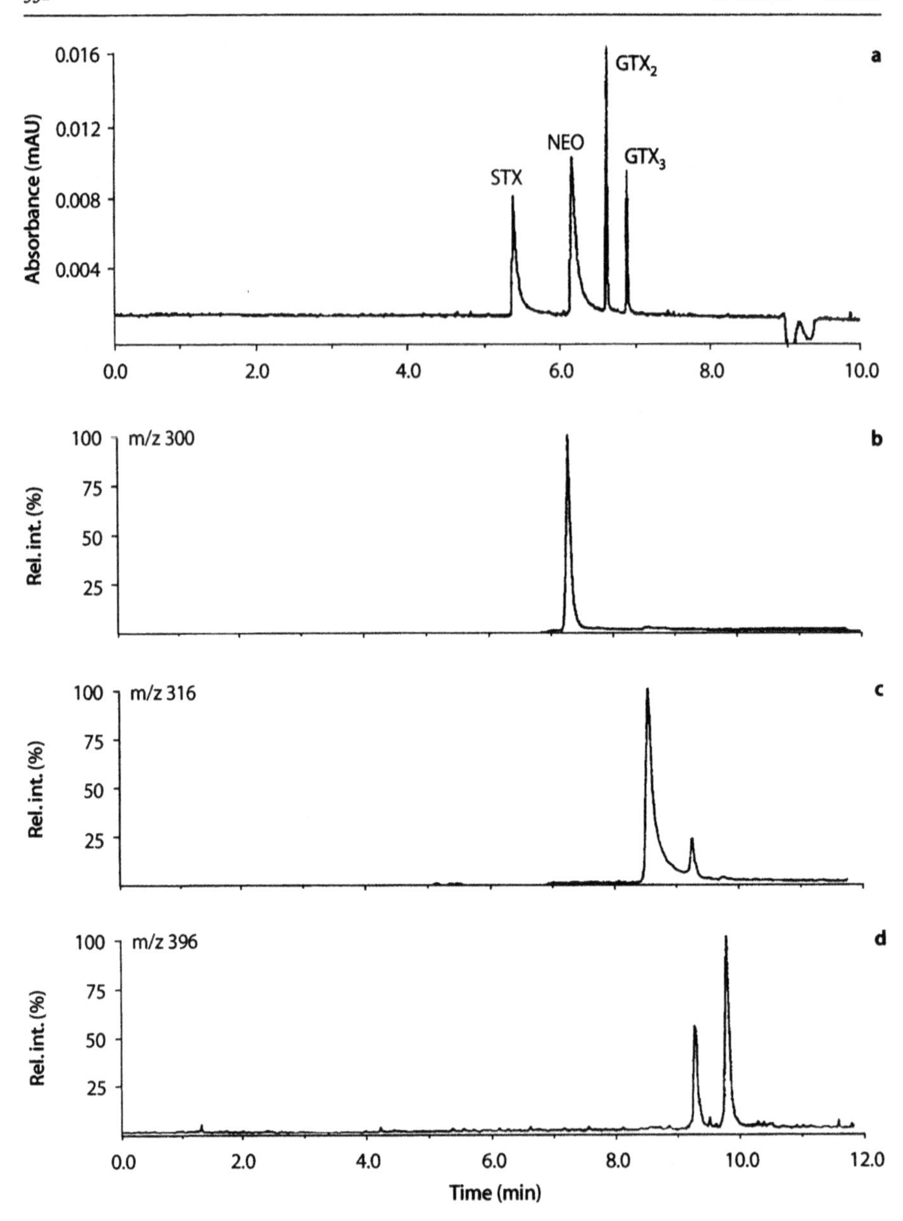

Fig. 17.8. Analysis of saxitoxin mixture isolated from a culture of the dinoflagellate *Alexandrium excavatum* by capillary electrophoresis (CE); **a** CE-UV at 200 nm; **b,c,d** selected ion mass chromatograms from CE-ESI-MS (reprinted with permission from Pleasance et al. 1992a)

weight fell beyond the mass range of the instrument used. The detection of several doubly and triply charged ionic species, such as $[M + 2H + Na]^{3+}$, $[M + 2H + K]^{3+}$, $[M + H + Na]^{2+}$, $[M + 2H + K]^{2+}$, giving signals at m/z values equal to one third and one half of their mass, allowed easy deduction of maitotoxin molecular weights (Lewis et al. 1994).

In a more recent study (Lewis and Jones 1997), the application of LC-ESI-MS allowed *(i)* the determination of up to fourteen ciguatoxin congeners in highly purified extracts of Pacific Ocean moray eel viscera and *(ii)* the determination of the major congener (P-CTX-1) in crude extracts. It is noteworthy that eleven congeners were present at trace levels and could not be detected by mouse bioassay. In another LC-ESI-MS study (Vernoux and Lewis 1997), Lewis and co-workers could demonstrate that the ciguatoxin congeners present in contaminated fish from the Caribbean Sea (C-CTXs) are different from those arising from fish of the Pacific Ocean (P-CTXs), thus justifying the different symptoms described for ciguatera intoxication in the two regions. The major component of the Caribian mixture C-CTX-1 was found to be 30 Da larger than the major component of Pacific Ocean ciguatoxin mixture P-CTX-1. C-CTX-1 is also more hydrophobic and 10 times less toxic than P-CTX-1. The NMR and ESI mass spectral features for C-CTX-1 and P-CTX-1 are very close to one another, confirming the similarity of their structures. However, the precise chemical structure of C-CTX-1, and its differences from P-CTX-1, have not been elucidated yet (Poli et al. 1997). The ESI mass spectra of C-CTX-1 and -2 at, respectively, 120 and 80 V skimmer voltage are presented in Fig. 17.9. Signals are evident at m/z 1179.9 $[M+K]^+$, 1163.8 $[M+Na]^+$, 1158.8 $[M+NH_4]^+$, 1141.8 $[MH]^+$, 1123.8 $[MH-H_2O]^+$, 1105.8 $[MH-2\,H_2O]^+$, 1087.9 $[MH-3\,H_2O]^+$, 1069.7 $[MH-4\,H_2O]^+$, 1051.6 $[MH-5\,H_2O]^+$. These spectral features confirm both the high alkali metal ion affinity and the polyhydroxylated nature of ciguatoxins. It is noteworthy that high orifice (skimmer) voltages favor the formation of both alkali cationized ions and extensively dissociated fragments. While extensive fragmentation is expected, according to an energetic "in-source" CID, the reasons for enhancement of alkali cationized species are unclear.

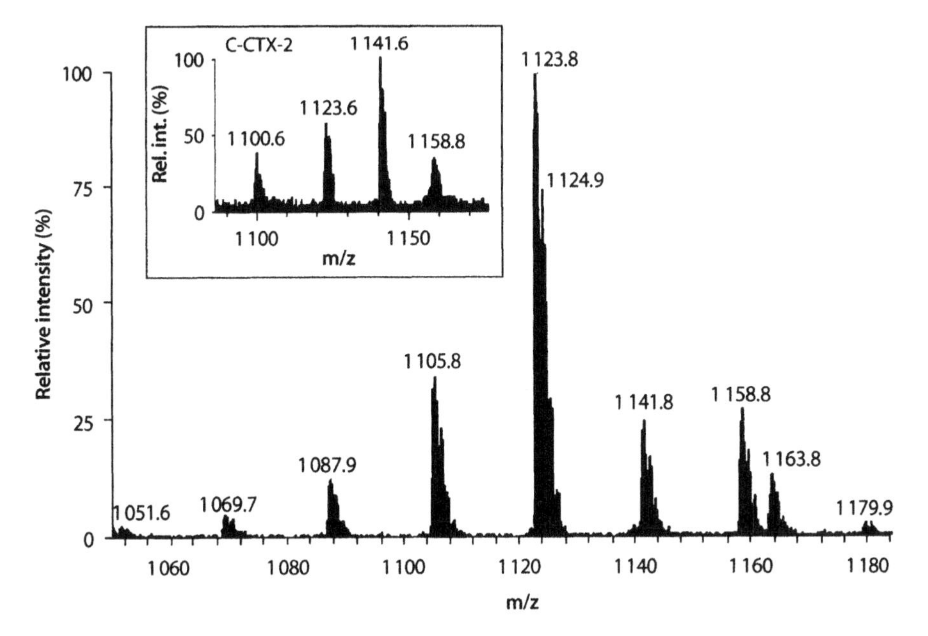

Fig. 17.9. ESI mass spectra of C-CTX-1 and C-CTX-2 (*inset*). The skimmer voltage was set to respectively 120 V and 80 V (reprinted with permission from Vernoux and Lewis 1997)

As the most refined analysis of dinophysistoxins took advantage of the great selectivity allowed by MS-MS, the same is occurring for recent ciguatoxin determinations. Good linearity of response was observed in MS-MS quantitative determinations, and detection limits in the range of hundreds of picograms (40–200 ppt in fish flesh) were obtained (Jones et al. 1997). As for dinophysistoxins, the specific MS-MS experiment adopted was SRM, in which multiple loss of water molecules was followed.

17.6 Polypeptidic Toxins: Conotoxins

Conotoxins are biochemically active peptides present in the venom of some tropical marine snails. These toxins are active on sodium and calcium channels as well as on nicotinic acetylcholine receptors. Although relatively small, quite specific receptor properties are imparted by the constrained and rigid tridimensional conformation of conotoxins. This is mainly due to the presence of several cysteine residues in the aminoacidic sequence, forming multiple disulfide bonds. Also γ-carboxyglutaric residues contribute significantly to the tridimensional conformation. The structural elucidation of conotoxins is complicated by the presence of several post-translational modifications, including C-terminal amidation, proline hydroxylation, tryptophan halogenation and glutaric acid carboxylation.

Mass spectrometric methods have been extensively used for protein sequence determination, and proved particularly useful in the identification of post-translational modifications, which are not feasible by Edman degradation techniques (Hunt et al. 1986). In the general procedure, the protonated molecular ion of the polypeptide, either produced by ESI or by other soft-ionization techniques, is mass-selected and subsequently fragmented by CID in an MS-MS apparatus. The secondary product ions are detected by scanning the second mass analyser. The resulting daughter-ion spectrum is interpreted on the basis of regular and rather predictable sequences of bond cleavages in the peptide backbone, with charge retention either on the N-terminal (sequences a_n, b_n, c_n) or on the C-terminal (sequences x_n, y_n, z_n) fragment (Hunt et al. 1986). Each amino acid (unmodified or post-translationally modified) is identified by the mass gap between consecutive fragments in each sequence, possibly with the support of interpreting software programs. Single amino acids can also be mass-analysed after peptide degradation, and disulfide bonds can be detected by comparing ESI-MS and ESI-MS-MS spectra before and after cysteine derivatization.

The whole set of procedures mentioned and the availability of an extremely wide set of literature on peptide analysis by MS methods, make the structural identification of an unknown polypeptide a more standardized task than that of a cyclic polyether. Nevertheless, investigation of conotoxins is rather complicated, since the large number of disulfide bonds, post-translational modifications and individual variations (interspecies, intraspecies and within-individual) put the interpreting procedure out of traditional schemes. A general overview of such problems and ESI-MS tools to solve them has been recently published (Bingham 1996). The same group, in a research paper (Jones et al. 1996), used an LC-ESI-MS approach to identify several conopeptides and to demonstrate the large compositional variability of venom mixtures. Disulfide bonds were determined by stepwise reduction and alkylation of an isolated conotoxin. The derivatization mixture was subsequently analysed by LC-ESI-MS (see Fig. 17.10a)

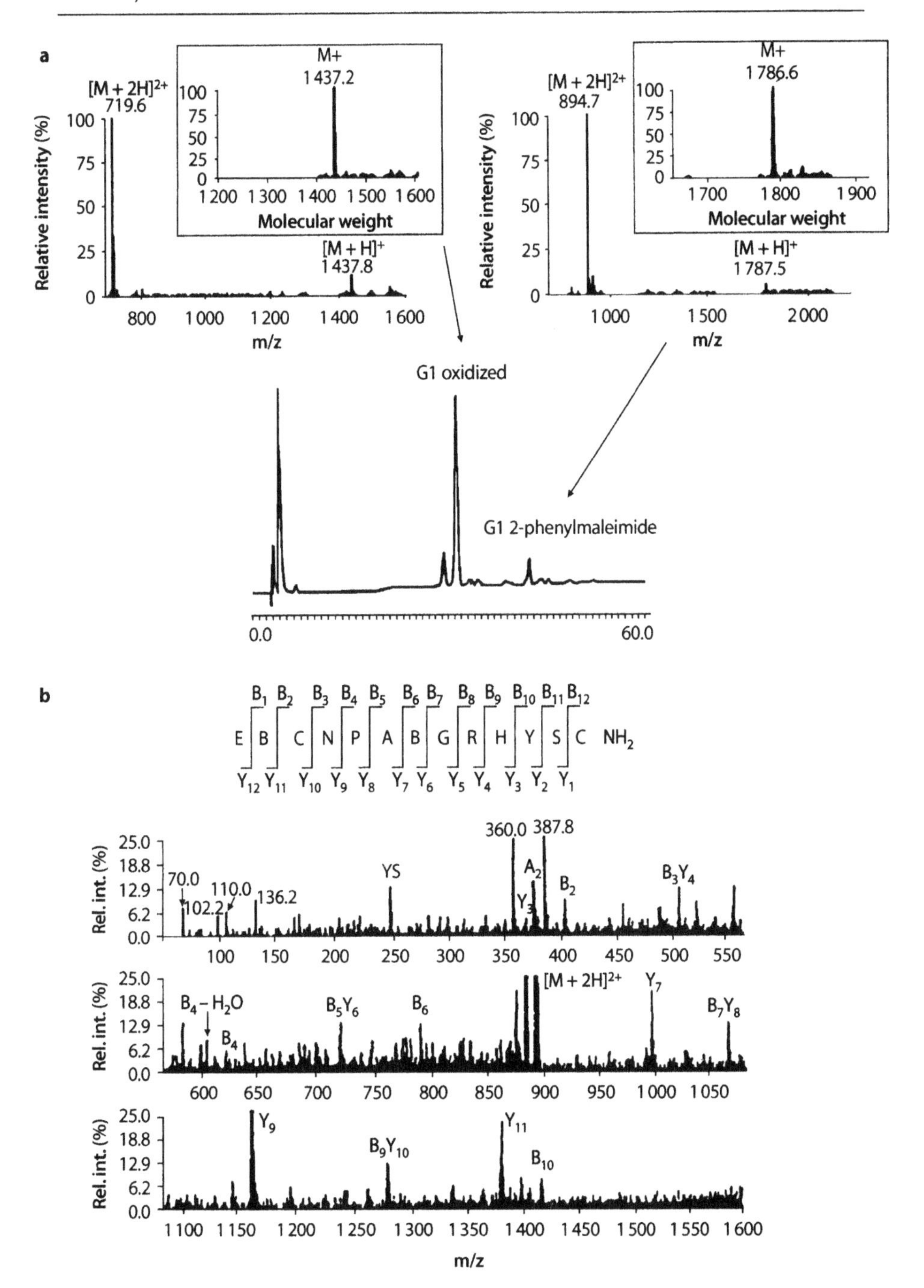

Fig. 17.10. a LC-ESI-MS chromatogram of partially reduced and alkylated G1 α-conotoxin. Stepwise reduction and alkylation were carried out with tris(2-carboxyethyl)phosphine and N-phenylmaleimide, respectively. *Inset*: ESI mass spectra of original and derivatized conotoxin; **b** MS-MS daughter ion spectrum of reduced and alkylated G1 α-conotoxin molecular ion. In the interpretation scheme, amino acids are represented by conventional one-letter symbols; *B* indicates N-phenylmaleimide-substituted cysteine (reprinted with permission from Jones et al. 1996

and the daughter-ion MS-MS spectra of both original and alkylated conopeptides revealed the position of disulfide-bound cysteines. Figure 17.10b shows the ESI-MS-MS spectrum of alkylated conotoxin and the corresponding interpretation scheme. A different group also used ESI-MS-MS spectra of singly- and multiply-charged conotoxin molecular ions to characterize the amino acidic sequence of *Conus* venom components, without LC separation (Krishnamurthy et al. 1996).

In a more recent study (Eluan Kalume et al. 1998), eleven conotoxins have been investigated in detail, some of which show a large number of post-translationally modified amino acids. Among these, the presence of glycosylic groups has been detected for the first time in conotoxins. From the isotopic pattern of both conotoxin molecular ions obtained by ESI-MS, and ESI-MS-MS fragment ions obtained upon CID in an ion-trap mass spectrometer, it has been possible to distinguish the number and position of brominated triptophane residues in six conotoxins. In general, mass spectrometric data were used to complement the gaps present in amino acid sequence data obtained by Edman degradation, and arising from post-translational modifications.

17.7 Conclusions

The examples of LC-ESI-MS characterization of marine toxins reported so far clearly indicate the rapidly increasing involvement of such instrumentation in a field previously dominated by biological and biochemical assays. ESI-MS is extremely sensitive and selective in the same time, so that its on-line combination with a powerful separation method such as LC allows one to undertake the complex task of attributing the toxicological properties of a toxin mixture to each individual component. From this analysis, revealing the complex nature and variability of most seafood intoxications, it is possible to set the basis for understanding and preventing these syndromes.

However, the potentiality of chromatographic and MS techniques does not appear to be fully exploited in toxin analysis at the moment. For example, the combination of capillary electrophoresis with ESI-MS is likely to be more extensively utilized in the future, especially for polar and intrinsically-charged toxins. More striking is the relatively infrequent application of MS-MS methods on a routine basis, i.e. to achieve a higher degree of selectivity in quantitative determinations. Tandem mass spectrometric techniques are likely to have an even higher impact in the structural characterization of unknown toxins. The availability of LC-ESI-ion-trap mass spectrometers from at least two suppliers will favor the diffusion of these instruments capable of performing multiple MS (MS^n). Thus, the molecular ion of an unknown toxin could be mass-selected and then fragmented, producing a secondary spectrum. If this secondary spectrum is still not sufficient for complete structural elucidation, then the fragment ions could be mass-selected and fragmented further until structural features and isomeric distinctions are singled out.

In conclusion, it can be foreseen that LC-ESI-MS and LC-ESI-MS-MS will increasingly complement NMR techniques in the structural elucidation of marine toxins and will substitute them in those cases when limited pure toxin availability and presence of impurities prevents the application of NMR. For the routine determination of known toxins in raw seafood materials, LC-ESI-MS will probably not replace the fast and sensitive immunochemical assays for screening protocols, but will provide more precise and reliable information in the cases turning out positive to the screening.

Acknowledgements

Financial support from the project "Sistema Lagunare Veneziano", M.U.R.S.T. and C.N.R. is gratefully acknowledged.

References

Arakawa O, Nishio S, Noguchi T, Shida Y, Onoue Y (1995) A new saxitoxin analog from a xanthid crab *Atergatis floridus*. Toxicon 33:1577–1584
Bencsath FA, Dickey RW (1991) Mass spectral characteristics of okadaic acid and simple derivatives. Rapid Comm Mass Spectrom 5:283–290
Bingham J-P, Jones A, Lewis RJ, Andrews PR, Alewood PF (1996) *Conus* venom peptides (conopeptides): Interspecies, intraspecies and within-individual variation revealed by ion-spray mass spectrometry. Biochem Aspects Mar Pharmacol 13–27
Bruins AP (1997) ESI source design and dynamic range considerations. In: Cole RB (ed) Electrospray ionization mass spectrometry. John Wiley & Sons, New York, pp 107–136
Curtis JM, Hu T, Wright JLC (1996) Studies of sulfated marine toxins by MS and MS/MS using a hybrid sector/orthogonal acceletation time-of-flight mass spectrometer. Proceedings of the 44th ASMS Conference on Mass Spectrometry and Allied Topics, Portland, Oregon, May 12–16, p 884
Draisci R, Lucentini L, Giannetti, L, Boria P, Stacchini A (1995) Detection of diarrohetic shellfish toxins in mussels from Italy by ionspray liquid chromatography-mass spectrometry. Toxicon 33: 1591–1603
Draisci R, Lucentini L, Giannetti L, Boria P, Poletti R (1996) First report of pectenotoxin-2 (PTX-2) in algae (*Dinophysis fortii*) related to seafood poisoning in Europe. Toxicon 34:923–935
Draisci R, Giannetti L, Lucentini L, Marchiafava C, James KJ, Bishop AG, Healy BM, Kelly SS (1998a) Isolation of a new okadaic acid analogue from phytoplankton implicated in diarrhetic shellfish poisoning. J Chrom A 798:137–145
Draisci R, Lucentini L, Giannetti L, Boria P, James KJ, Furey A, Gillman M, Kelly SS (1998b) Determination of diarrhetic shellfish toxins in mussels by microliquid chromatography-tandem mass spectrometry. J AOAC Int 81:441–447
Eluan Kalume D, Roepstorff P, Furie B, Furie B, Czervjec E, Hambe B, Stenflo J (1998) Characterization of heavily modified peptide toxins from sea snail (*Conus textile*) by MALDI-TOF and ESI-ion-trap mass spectrometry. Presented at the 46th ASMS Conference on Mass Spectrometry and Allied Topics, Orlando, May 31–June 4
Falconer IR (1993) Algal toxins in seafood and drinking water. Academic Press, London
Hall S, Strichartz G (1990) Marine toxins: Origin, structure and molecular pharmacology. American Chemical Society, Washington DC
Hua Y, Lu W, Henry MS, Pierce RH, Cole RB (1995) On-line high-performance liquid chromatography-electrospray ionization mass spectrometry for the determination of brevetoxins in "red tide" algae. Anal Chem 67:1815–1823
Hua Y, Lu W, Henry MS, Pierce RH, Cole RB (1996) On-line liquid chromatography-electrospray ionization mass spectrometry for determination of the brevetoxin profile in natural "red tide" algae blooms. J Chrom A 750:115–125
Hunt DF, Yates RJ III, Shabanowitz J, Winston S, Hauer CR (1986) Protein sequence by mass spectrometry. Proc Natl Acad Sci 83:6233–6237
James KJ, Bishop AG, Gillman M, Kelly SS, Roden C, Draisci R, Lucentini L, Giannetti L, Boria P (1997a) Liquid chromatography with fluorimetric, mass spectrometric and tandem mass spectrometric detection for investigation of the seafood-toxin-producing phytoplankton, *Dinophysis acuta*. J Chrom A 777:213–221
James KJ, Carmody EP, Gillman M, Kelly SS, Draisci R, Lucentini L, Giannetti L (1997b) Identification of a new diarrohetic toxin in shellfish using liquid chromatography with fluorimetric and mass spectrometric detection. Toxicon 35:973–978
Jones A, Bingham J-P, Gehrmann J, Bond T, Loughnan M, Atkins A, Lewis RJ, Alewood PF (1996) Isolation and characterization of conopeptides by high-performance liquid chromatography combined with mass spectrometry and tandem mass spectrometry. Rapid Comm Mass Spectrom 10: 138–143

Jones A, Lewis RJ, Vernoux J-P (1997) Application of HPLC/MS and HPLC/MS/MS in the detection of multiple ciguatoxins in fish. Proceedings of the 45th ASMS Conference on Mass Spectrometry and Allied Topics, Palm Springs, California, June 1–5, p 350
Kebarle P, Ho Y (1997) On the mechanism of electrospray mass spectrometry. In: Cole RB (ed) Electrospray ionization mass spectrometry. John Wiley & Sons, New York, pp 3–64
Krishnamurthy T, Prabhakaran M, Long SR (1996) Mass spectrometric investigations on *Conus* peptides. Toxicon 34:1345–1359
Lewis RJ, Jones A (1997) Characterization of ciguatoxins and ciguatoxin congeners present in giguateric fish by gradient reverse-phase high-performance liquid chromatography/mass spectrometry. Toxicon 35:159–168
Lewis RJ, Holmes MJ, Alewood PF, Jones A (1994) Ionspray mass spectrometry of ciguatoxin-1, maitotoxin-2 and -3, and related marine polyether toxins. Natural Toxins 2:56–63
Marr JC, Hu T, Pleasance S, Quilliam MA, Wright JLC (1992) Detection of new 7-O-acyl derivatives of diarrheic shellfish poisoning toxins by liquid chromatography-mass spectrometry. Toxicon 30:1621–1630
Moeller PDR, Lanoue B, Busman M (1998) Determination of marine biotoxins by negative mode ESI and APCI. Presented at the 46th ASMS Conference on Mass Spectrometry and Allied Topics, Orlando, May 31–June 4
Morton SL, Moeller PDR, Young KA, Lanoue B (1998) Okadaic acid production from the marine dinoflagellate *Prorocentrum belizeanum* Faust isolated from the Belizean coral reef ecosystem. Toxicon 36:201–206
Onodera H, Satake M, Oshima Y, Yasumoto T, Carmichael WW (1997) New saxitoxin analogs from freshwater filamentous cyanobacterium *Lyngbya wollei*. Nat. Toxins 5:146–151
Pleasance S, Ayer SW, Laycock MV, Thibault P (1992a) Ionspray mass spectrometry of marine toxins. III. Analysis of paralytic shellfish poisoning toxins by flow-injection analysis, liquid chromatography/mass spectrometry and capillary electrophoresis/mass spectrometry. Rapid Comm Mass Spectrom 6:14–24
Pleasance S, Quilliam MA, Marr JC (1992b) Ionspray mass spectrometry of marine toxins. IV. Determination of diarrhetic shellfish poisoning toxins in mussel tissue by liquid chromatography/mass spectrometry. Rapid Comm Mass Spectrom 6:121–127
Poli MA, Lewis RJ, Dickey RW, Musser SM, Buckner CA, Carpenter LG (1997) Identification of Caribbean ciguatoxins as the cause of an outbreak of fish poisoning among U.S. soldiers in Haiti. Toxicon 35:733–741
Quilliam MA (1995) Analysis of diarrhetic shellfish poisoning toxins in shellfish tissue by liquid chromatography with fluorometric and mass spectrometric detection. J. AOAC Int 78:555–570
Quilliam MA, Janecek M, Lawrence JF (1993) Characterization of the oxidation products of paralytic shellfish poisoning toxins by liquid chromatography/mass spectrometry. Rapid Comm Mass Spectrom 7:482–487
Ragelis EP (1984) Seafood toxins. American Chemical Society, Washington, DC (ACS Symp. Series, No. 262)
Shen J-L, Hummert C, Luckas B (1997) Sensitive HPLC fluorometric and HPLC-MS determination of diarrhetic shellfish poisoning (DSP)-toxins as 4-bromomethyl-7-methoxycoumarin esters. Fres J Anal Chem 357:101–104
Smayda TJ, Shimizu I (1993) Toxic phytoplankton blooms in the sea. Elsevier, Amsterdam
Vernoux J-P, Lewis RJ (1997) Isolation and characterization of Caribbean ciguatoxins from the horse-eye jack (*Caranx latus*). Toxicon 35:889–900
Yasumoto T (1990) Marine microorganism toxins – an overview. In: Granéli E, Sundström B, Edler L, Anderson DM (eds) Toxic marine phytoplankton. Elsevier, New York
Yasumoto T, Satake M (1996) Chemistry, ethiology and determination methods of ciguatera toxins. J Toxicol, Toxin Rev 15:91–107
Yasumoto T, Murata M, Lee JS, Torigoe K (1989) Polyether toxins produced by dinoflagellates. In: Natori S, Hashimoto K, Ueno T (eds) Mycotoxins and phycotoxins '88. Elsevier, Amsterdam
Vincenti M (1997) Application of mass spectrometric techniques to the detection of natural and anthropogenic substances in the sea. In: Gianguzza A, Pelizzetti E, Sammartano S (eds) Marine chemistry. An environmental analytical chemistry approach. Kluwer Academic Publishers, Dordrecht, pp 189–209

Chapter 18

Metals Analysis by High Performance Liquid Chromatography

C. Sarzanini

18.1 Introduction

Metal ion determination and speciation by liquid chromatography has received particular attention during the last years and some reviews (Robards et al. 1991; Sarzanini and Mentasti 1997), with particular regard to ion chromatography (Dasgupta 1992) and complexation ion chromatography (Timerbaev and Bonn 1993; Jones and Nesterenko 1997), summarize its potentialities.

Why high performance liquid chromatography (HPLC) for metals analysis? The instruments used most widely for metal determination (e.g. atomic absorption and inductively-coupled plasma atomic emission spectrometers) are subject to both spectral and chemical interferences and, in some cases, are unsuitable for trace analyis in a complex matrix or for studies on metal speciation. High performance liquid chromatography has become in recent years the most flexible tool for separating different species, removal of matrix intereferences, and may be coupled with different kinds of detectors enabling the determination of specific or group elements. Several devices including electrochemical detectors (ED), inductively-coupled plasma (ICP) and graphite furnace atomic absorption (GFAA) spectrometers have been used, as well as inductively-coupled plasma-mass spectrometry (ICP-MS) coupled to liquid chromatography (Hill et al. 1993; Tomlinson et al. 1994; Sutton et al. 1997).

The approaches for metal determinations by HPLC are:

- normal phase chromatography
- reversed phase chromatography
- ion chromatography
  - ion exchange
  - ion exclusion
- ion interaction chromatography
- chelation ion chromatography
- multidimensional and multimode chromatography

The basic principles and retention mechanisms (processes involving solute interactions in both the mobile and the stationary phases) are summarized hereafter.

18.2
Retention Mechanisms in Liquid Chromatography

18.2.1
Normal Phase Chromatography

Adsorption or straight phase or normal phase chromatography (NPC) is characterized by the use of an inorganic adsorbent or chemically bonded stationary phase with polar functional groups (e.g. silica, cyanoalkyl-silica) and a non aqueous mobile phase (one or more polar solvents diluted to the desired eluting power). A layer of adsorbed mobile phase molecules is formed at the surface of the stationary phase. The simplest model for analyte separation assumes the stationary phase covered by a monolayer of mobile phase and solute molecules; a competition between analytes and eluent for the surface is the driving force. Secondary solvent effects must also be taken into account for polar compounds (e.g. hydrogen bonding) and conversely to the competition model, a solvent bilayer adsorbed onto the stationary phase is assumed.

Retention and selectivity in NPC are dramatically influenced by the presence of polar additives (water) in the mobile phase.

18.2.2
Reversed Phase Chromatography

Reversed phase chromatography (RPC) employs as a stationary phase non polar solids of high-surface area (usually alkyl-bonded silica packing, e.g. C_8 or C_{18} groups grafted to the silica surface), and an aqueous-organic solvent mixture is used as the mobile phase.

Retention in RPC occurs by non-specific hydrophobic interactions of the solute with the stationary phase. Secondary chemical equilibria (see below) optimize separation selectivity varying the mobile phase composition. Solute retention is attributed to both adsorption and partition phenomena; neutral and ionic solutes can be separated simultaneously.

18.2.3
Ion Chromatography

The term ion chromatography (IC) in the past included only ion-exchange chromatography. This technique is based on a reversible interchange of ions or ionized substances (e.g. by pH manipulation or complexation) between a solution and a solid, inorganic or polymeric insoluble material, containing fixed ions and exchangeable counter-ions. Analytes are separated on the basis of their different relative affinity for the ionic centers of the stationary phase. Ion exchange chromatography is based on electrostatic interactions between the ions to be exchanged but other kinds of reactions may occur, e.g. hydrophobic interactions between the sample and non-ionic regions of the stationary phase.

18.2.3.1
Ion Exclusion Chromatography

This kind of IC is based on the use of both cation and anion exchangers as stationary phases. The separation mechanism is governed by the Donnan effect, that is, strongly ionized species are repelled by the charge on the surface of the sationary phase and neutral or weakly dissociated analytes are retained by interaction with the matrix. Charge and solute size are the dominant factors affecting the separation, but hydrophobic adsorption also plays a role for the retention mechanism of ion-exclusion chromatography (IEC).

The main advantage of IEC is that the degree of ionization of the solute can be modified, as can its retention, by modulating the pH value of the eluent.

18.2.3.2
Ion Interaction Chromatography

The above mentioned secondary chemical equilibria (SCE) refer to all other equilibria (the primary equilibrium is the distribution of the solute between the mobile phase and the stationary phase), providing additional mechanism(s) for control of retention and selectivity during a separation, e.g. ionization, metal complexation, solute-micelle association and ion pairing.

Ion interaction chromatography (IIC), also called soap chromatography, ion pair chromatography and dynamic ion exchange chromatography, is a typical example of a chromatographic process based on secondary equilibria. Eluents containing an ion interaction reagent (IIR) (e.g. alkylammonium salts, alkylsulfates or alkylsulfonates) are used, and stationary phases (conventional RP or polymers) are dynamically modified into low-capacity ion-exchangers. During the separation, the retention of neutral analytes, analytes having the same or opposite charge with respect to the IIR, will not be affected, decreased or increased. Elution of cations is achieved by their complexation with the eluent ligand and ion-pairing of the negatively charged complex formed with IIR or their cation exchange with the counter-ion of IIR. Research in this field is devoted to the evaluation of the nature and concentration of proper ligands and IIR, as well as the organic modifier and eluent pH. The pH of the mobile phase affects retention and resolution of chelates since it can alter their stoichiometry and over-all charge.

Another approach to IIC is the use of common reversed phase stationary phases permanently coated with suitable hydrophobic agents such as alkylsulfonates or alkylsulfates with a sufficiently long alkyl group. The mechanism of elution is governed by the mobile phase in the following two ways:

i. eluents containing a strong driving cation and a small amount of complexing agent ("push-pull" method, e.g. mobile phase containing ethylenediammonium cation and tartaric acid)
ii. eluents containing a very weak driving cation and higher concentrations of complexing agent ("pure pull" mechanism)

A recent study of the chromatographic behaviour of metal ions in IIC when the stationary phase is modified with various alkanesulfonates (1–10 carbon atoms in the alkyl chain) must be mentioned; the paper shows that a very good resolution is achieved even if in the ion interaction mode the number of theoretical plates of the column is lower than that obtained in the reversed-phase mode (Zappoli et al. 1996).

18.2.4 Chelation Ion Chromatography

Chelation ion chromatography, or, more correctly, high performance chelation ion chromatography (HPCIC) is based on the use of high-performance substrates for trace metal separation and determination. Chelation exchange involves an ion-exchange process and the formation of coordinate bonds. The different kinds of stationary phases, available in the market or *laboratory made*, have recently been reviewed (Jones and Nesterenko 1997) and are silica or polymer based materials with chelating agents grafted onto the surface. Alternatively, chelating ligands can be immobilized by adsorption onto a styrene-divinylbenzene copolymer, silica gel or other synthetic polymers. These precoated columns use for the separation of metal ions an aqueous mobile phase at a relatively high concentration of an inorganic salt, e.g. 0.5–1.0 M KNO_3. More recently a graphitic porous carbon reversed phase column has been coupled with a mobile phase containing a selective metallochromic ligand for the separation of alkaline earth metals (Paull et al. 1996) and a polymeric reversed phase column with a mobile phase containing methylthymol blue for the separation of Mn, Zn, Cd and Pb (Paull et al. 1998). Since the ligand is a part of the eluent, we can say that, in analogy with the dynamic ion exchange, a dynamic chelating chromatography has been developed. For the HPCIC the acting separation mechanisms, as a function of pH, ionic strength and organic modifier content of the eluent, are *chelation* and *ion exchange*.

Each of the techniques mentioned is suitable for metal separations and determination in different kinds of samples, and the main problems encountered are due to matrix composition and analyte concentration. In some cases, as detailed below, more than one mechanism or column is coupled in order to enhance chromatographic resolution and so called multimode or multidimensional techniques are used for complicated samples.

We will consider hereafter some of the procedures that have been developed, focusing our attention on the methods dedicated to seawater analysis.

18.3 Procedures for Metal Separation

18.3.1 Normal Phase Chromatography

The current approach for normal phase chromatography (NPC) is based on the formation of metal chelates (e.g. diacetylbisthiobenzohydrazones, dithizone, diethyldithiocarbamate), their extraction from the sample, injection and their elution, performed with organic solvent mixtures of n-heptane/benzene, toluene, diethyl ether/acetonitrile or similar. A typical reaction for heavy metal ions is with ammonium tetramethylenedithiocarbammate (ATDC), which gives neutral complexes. Since this kind of chelate is usually uncharged, it can be eluted in the same manner as neutral compounds do on normal phase silica columns (e.g. 1% propanol in hexane).

The main problems encountered in NPC are connected with metal complex solubility, the need of extraction and the risk of pollution of the sample when traces must be detected. Among the mechanisms mentioned, normal phase chromatography is not, at present, the most widely used.

18.3.2 Reversed Phase Chromatography

This technique has been widely used for the separation of neutral or weakly charged metal complexes, but the more extensive applications are based on the ion pairing mechanism, so related procedures will be detailed hereafter.

Analysis for trace metals is carried out by the formation of metal chelates with separation by RPC on C_{18} columns and the use of organic-based mobile phases. Dithiocarbamates are the most frequently reported complexing agents due to the strong chelating ability of their sulfur groups and their ability to form nearly water-insoluble metal salts with all metals except sodium and other alkali and alkali earth metals. A comprehensive study on reversed-phase HPLC behaviour of diethyldithiocarbamate (DEDTC) complexes of Cu, Co, Cr, Ni and Hg with a variety of columns and mobile phases was completed (Dilli et al. 1990). In this case DEDTC complexes were preformed off-column (60 °C, 15 min), extracted into chloroform and finally dissolved in CH_3OH and injected for the separation onto a C_{18}-colum (μBondapak, Waters). The study showed that the ligand must also be present in the mobile phase for low concentration of chelates, to avoid their dissociation.

Various azo dyes have also been considered for the chromatographic separation of metal chelates on a reversed phase RP-18 column, and the study focused on the separation and determination of V(V) at trace levels in natural waters (Miura 1990). The originality of this investigation is due to optimization of the RP column selectivity by introducing a tetraalkylammonium salt into the system. The metal chelates considered are neutral or cationic and ion-paired complexes are not involved, whereas other metal ions (e.g. Fe, Al) do not interfere in the determination.

8-Quinolinol (HQ) is another extensively used ligand for the separation of metal ions by HPLC. For this ligand also, methods are based on metal ion complexation, usually by heating the sample in the presence of HQ, one or two step extraction with a compatible eluent solvent and injection of complexes into the chromatographic system. To give an example, the simultaneous determination of Mo(VI), V(V), Cu(II) and Fe(III) at ppb level in sea water can be mentioned (Ohashi et al. 1991).

18.3.3 Ion Chromatography

Ion exchange chromatography (IEC) of metal ions is performed by using both cation and anion exchangers. In the cation-exchange technique the metal ions are normally reacted with an anion of a weak acid to reduce their charge density in the eluent solution before entering the separation column, where they are separated according to their respective affinities toward the active sites of the separating resin. Ligands are also required to avoid precipitation when an acidic eluent is not suitable for the columns selected. A low-capacity silica-based cation-exchange column was used for the separation of transition metals (Co, Cu, Fe) and coupled with post-column chemiluminescence detection to enhance sensitivity (Yan and Worsfold 1990). In this case attention has been paid to eluent composition not only to improve separation but also for its compatibility with the post-column reaction.

Ion chromatographic separation of metal ions based on anionic exchange offers the potential for a different selectivity, reduced problems for metal ion hydrolysis and for application to complex sample matrices. Notwithstanding the fact that many organic acids from mono-, di-, tricarboxylic acids to chelating agents such as α-hydroxyisobutyric acid (HIBAα), tartaric, citric, oxalic, pyridine-2,6-dicarboxylic acid (PDCA), 1,2-diaminocyclohexanetetraacetic acid (DCTA), and diethylenetriaminopentaacetic acid (DTPA) have been evaluated for simultaneous ion-chromatography of anions, alkali, alkaline earth and heavy metals (Yan and Schwedt 1990; Cardellicchio et al. 1997), ethylenediaminetetraacetic acid (EDTA) plays a fundamental role. EDTA has also been used as a masking agent, to avoid metal ion interference arising from possible precipitation due to eluent pH. Since EDTA forms, at the proper pH, negatively charged complexes with divalent or trivalent metal ions, the possibility for simultaneous separation of anions from metal ions as well as the speciation of metal ions is also feasible. In these procedures complexes can be obtained in two ways: the first one is through their formation before the chromatographic separation (precolumn complexation, complexes must be stable enough to avoid decomposition during separation or ligand must be added to the eluent); the second way is based on the complexation in the chromatographic column itself. Some examples of applications of an EDTA eluent have been reported, and experiments were also performed with binary eluent systems comprising EDTA as complexing agent. For seawater samples (LeGras 1993), silica based anion exchange analytical columns have enhanced sensitivity and enabled detection limits to be reached from 20 $\mu g\,l^{-1}$ for Mg^{2+} to 0.4 $mg\,l^{-1}$ for Ca^{2+} with UV and conductivity detection and eluent pH at 4.8.

18.3.4 Ion Interaction Chromatography

As mentioned above, the elution of cations is achieved by their complexation with an eluent ligand and ion-pairing of a negatively charged complex formed with IIR or their cation exchange with the counter-ion of IIR. Research in this field is devoted to the evaluation of the nature and concentration of proper ligands and IIR as well as an organic modifier and eluent pH. Ion-interaction chromatography is one of the most suitable techniques for metal ion determination in complicated samples. The main approaches involve metal complex formation by adding the ligand to the sample or in situ complexation by reaction of metal ions with the ligand added to the eluent. Naturally the eluent must contain a proper ion pairing reagent and an organic modifier, if required, enabling the separation of complexes. The sequence mentioned is useful but it is not sufficient when trace metals have to be determined. In this case a preconcentration step is required (see below).

An extensive study on IIC of metal complexes of nitrosonaphthol sulfonates ion-paired, with liquid-liquid extraction and on-line derivatization, has been carried out by Sirén (1991). A modification of the methods proposed consists of an on-line derivatisation of metal ions; in this procedure metals are injected into a methanol-water eluent containing a quaternary ammonium bromide (e.g. cetyltrimethyl-ammonium, CTA) and, after the column, they are mixed with a ligand solution (1-nitroso-2-naphthol-6-sulfonate) (Sirén and Riekkola 1992). The metal ion separation by this procedure is governed by the kinetics of formation of complexes and ion-pairs and retention onto the post-column mixer-reactor system. The chromatographic behaviour of 3-(5-chloro-2-hydroxyphenylazo)-4,5-dihydroxy-naphthalene-2,7-disulfonic acid (Plasmocorinth B) and its metal ion complexes has been studied in ion-pairing reactions for metal preconcentration and separation by HPLC (Sarzanini et al. 1993a,b). The separation of analytes was optimized with a flow-gradient elution and the method, successfully applied to river water samples, enabled analyte metals to be separated from alkaline and alkaline earth elements.

Ohtsuka et al. have separated several ion pairs of anionic metal chelates with pyridyl-azosulfoaminophenol derivatives (PAPS) on a C18 stationary phase (Ohtsuka et al. 1991,1992). They elucidated the retention behaviour of metal chelates (PAPS) in IIC as a function of mobile phase composition (Ohtsuka et al. 1994) with respect to the significant differences found in methanol-water and acetonitrile-water systems as a function of the volume fraction of water (Alvarez-Zepeda et al. 1992).

Octadecyl-bonded silica permanently coated with sodium dodecylsulfate in the presence of complexing agents was considered for the separation of transition metals (Janos and Broul 1992). In the work mentioned an ion-exchange mechanism similar to that of fixed sites exchangers seems to occur; both the pushing effect of the eluting cation and the pulling effect of the complexing anion take place but the latter plays a dominant role in the process of elution. A significant example of the approach mentioned, is the detailed study by Cassidy and Sun (1993). They compared the performance of an anion separation with a cation separation both based on an ion-interaction system that used cetylpyridinium chloride or n-octanesulfonate to modify a reversed stationary phase. In the first case transition metals (Mn, Co, Ni, Cu and Zn) were eluted with an oxalate eluent. The anion-exchange system provided column efficiences comparable with that for the cation system. This approach may be attractive for solving analytical problems taking into account the considerably different order of separation obtained by the two systems.

Reversed-phase ion-pair procedures involving EDTA have also been considered in optimizing separation and detection of metal species. Different techniques like precolumn derivatization without a complexing agent in the eluent or on-column derivatization may be less efficient and give rise to peak broadening. Ion-pair reversed phase high performance chromatography has been investigated by coupling EDTA with tetraethylammonium (TEA), tetrapropylammonium (TPA) (Marina et al. 1993) and tetrabutylammonium (TBA) (Sacchero et al. 1991; Jen and Chen 1992; Marina et al. 1993) bromide ion pairing agents. TBA proved to be the most suitable ion pairing agent in all cases, and the use of EDTA in the eluent (Sacchero et al. 1991; Marina et al. 1993), together with high complexation constants, shifted the equilibrium in favour of chelate formation attaining lower detection limits. The data obtained (Marina et al. 1993) clarify some aspects of the separation mechanism of ion-interaction chromatography for different oxidation states of metal ions and confirm that the retention of divalent and trivalent metal ions complexed with EDTA takes place through an ion exchange mechanism in which the ion exchanger is dynamically generated by the retention of the counterion in the stationary phase (Sacchero et al. 1991).

Studies for the separation and determination of metal ions (Cu, Co, Fe, Ni, V(V), Pd) as 5-Br-PADAP chelates (5-Br-2-pyridylazo-5-diethylaminophenol) by RP-HPLC showed that only the retention of the Co(III)-5-Br-PADAP complex is affected by varying the concentration of surfactant added to the eluent (Zhao and Fu 1990). In addition a stronger interaction resulted for TBA in respect to CTA and cetylpyridinium (CP). An alkyl group, such as that of CTA or TBA, may interact with the C18 chain on the stationary phase by molecular interaction so that the charged part of surfactant is exposed on the surface, increasing its polarity so that the Co chelate is eluted earlier.

18.3.5 Chelation Ion Chromatography

For samples of high ionic strength such as sea waters or concentrated brines, the selectivity of the chromatographic separation can be enhanced by the use of chelating chemically bonded phases. Complexation reactions in the stationary phase, ion exchange due to free or protonated chelating groups which act as ion exchange sites, and in some cases complexation in the eluent, are responsible for the separation. Two main approaches can be followed to obtain proper stationary phases:

i. chemical bonding of the chelating group to the substrate
ii. coating of a substrate with a ligand which is permanently trapped onto the substrate

Voloschik et al. (1994a) used a silica gel-based sorbent with chemically bonded amidoxime functional groups for the selective determination of metals in waters (Mn, Cd, Co, Cu, Ni, Pb, Zn and Hg). Since this kind of resin showed a weak affinity for Mg and Ca and a very strong affinity for Fe it was possible to eliminate their interference in the determination. The separation of mono- and divalent cations on a polybutadiene maleic acid (PDMA) stationary phase coated on silica materials has been performed with different eluents, and a detailed study on this material (Nair et al. 1994) enabled the determination of some transition metals by conductivity detection within detection limits from 35 to 410 $\mu g\ l^{-1}$ for Cd and Pb respectively.

The most widely used resin for preconcentrating and for separating elements and groups of elements in seawater, applied for the first time almost three decades ago (Riley and Taylor 1968), is an iminodiacetate resin (i.e. Chelex-100). The low degree of cross-linking of the polystyrene-divinylbenzene (PS-DVB) supporting polymer made the Chelex-100 unsuitable for high pressure liquid chromatography applications. A more highly cross-linked macroporous PS-DVB resin containing the iminodiacetate functional group that allows operation at high pressure without physical degradation (MetPac CC-1 column, Dionex) has been more recently developed (Siriraks et al. 1990).

The alternative approach to ligands chemically bound to silica or polymer phases refers to a technique, applied for a long time in metal ions preconcentration and matrix removal before spectroscopic determinations, based on the permanent loading of sorbents by chelating agents (Abollino et al. 1990,1993; Porta et al. 1992; Sarzanini et al. 1992). Jones and co-workers widely investigated this field in order to improve detectability of both alkaline-earth and heavy metals in brine samples. Using high performance chelation ion chromatography (HPCIC) the order of retention of metal ions, including earth metals, is reversed with respect to IIC and so barium is eluted first as a sharp peak followed by Sr, Mg and Ca. The barium separation and determination at mg l^{-1} concentration in 1600 mg l^{-1} Ca samples with neutral hydrophobic resin (PS-DVB), precedingly impregnated with methylthymol blue (3,3'-bis[N,N-di(carboxymethyl)aminomethyl]thymol-sulfonephthalein) coupled with an acid elution (0.5 M KNO_3 + 0.5 M lactic acid) and UV detection by post-column reaction (4-(2-pyridylazo)-resorcinol + ZnEDTA) (Paull et al. 1994a) was successfully achieved showing a detection limit of 3 μg l^{-1} for Ba and Sr, as well as Sr and Ca separation was optimized in milk powder analysis with a phthalein purple (o-cresolphthalein-3',3"-bis-methyleneiminodiacetic acid) impregnated column (Jones et al. 1994). Similar studies devoted to transition metal determination (Jones et al. 1992; Challenger et al. 1992,1993) resulted in an interesting procedure using a 10 μm particle size, 100 Å pore size PS-DVB resin impregnated with Xylenol Orange, stable from pH 0.5 to 11.5 (Paull et al. 1994b). The stationary phase enables Ca and Mg removal during the sample on-column preconcentration at pH 6 and the separation and determination of Zn, Pb, Ni and Cu with a step-gradient pH elution. The only drawback of the method is Cd-Mn coelution, but satisfactory results are obtained for seawater samples (CASS-2 certified sample). This is a good example showing how small variations in chelating ability between dyes can be very useful in allowing specific separations. A recent paper provides ample details on an investigation into the parameters involved in the production of a range of dye impregnated chelating columns (10 chelating dyes, mainly based on triphenylmethane or azo based dyes); for preconcentration and separation of alkaline earth, transition and heavy metals at trace levels within an application devoted to Al determination in sea water (Paull and Jones 1996). Finally Haddad et al. developed the *dynamic chelating chromatography* (Paull et al. 1998) by studying a procedure where a polymeric reversed-phase was dynamically coated with the methylthymol blue (MTB) ligand. The new system, compared to MTB precoated chelating columns enabling only the separation of alkaline earth metals, allows the separation of transition and heavy metals. pH and MTB concentration are the key parameters for the retention, and the method is suitable for samples at high ionic strength values (1.0 M NaCl).

18.3.6
Multimode and Multidimensional Liquid Chromatography

Multimode and multidimensional chromatographic techniques respectively involve:

a. the use of two different mechanisms, e.g. ion exchange coupled with ion exclusion chromatography
b. the use of two or more columns "switching" the total flux or a portion of eluate to each other

The multimode technique is not usually employed for liquid chromatography of metal ions but, in this field, there are some applications of liquid-gas chromatography.

The multidimensional technique is also known as "heart cut" column switching. The term "column switching" includes all techniques by which the direction of the flow of the mobile phase is changed by valves so that the effluent (or a portion of it) from the primary column is passed to a secondary column for a defined period of time. The two following examples can summarize the application of multidimensional chromatography.

Transition and heavy metals have been determined in seawater samples by multidimensional liquid chromatography (Voloschik et al. 1994b). The separation was achieved by coupling a dynamically coated sorbent column, a preconcentrator and a chelating column. The metal concentrations (Cu, Ni, Co, Mg, Ca, Sr, Fe and Pb, Zn, Mn, Cd) in the seawater sample were 5.0–50 $\mu g\ l^{-1}$.

Samples containing alkali metals, alkaline earth cations and ammonium ion are difficult to analyse; environmental samples, at low levels of ammonium in matrices with a high concentration of sodium, are a typical case. This is mainly due to the similar selectivities, of ammonium and sodium ions, for the common stationary phases containing sulfonate or carboxylate cation-exchange functional groups. This problem has been solved by a column-switching technique which enables the determination of trace concentrations of the common inorganic cations (Li, Na, K, Mg, Ca) and ammonium in the presence of large concentrations of either sodium or ammonium (Rey et al. 1997).

It must be mentioned that a great number of papers referring to column switching or multidimensional liquid chromatography concern methods not actually using two chromatographic columns, but simply a short column (preconcentrator) coupled with an analytical column.

18.3.7 Preconcentration

To overcome the problems arising from off-line complexation and preconcentration (e.g. sample poisoning, extraction procedures) great efforts have been devoted to the development of on-line preconcentration procedures. The main approaches, for metal ion determinations, are based on the retention of metals on a preconcentrator as such (e.g. on a chelating microcolumn which does not retain alkali and alkaline earths) or after their complexation, usually with negatively charged sulfonated dyes. In this case, metal anionic complexes can be retained as such on an anion exchanger or on a reversed phase material preloaded with an ion pairing agent, enabling a specific retention through ion pairing and their hydrophobic interaction with the network of the stationary phase.

In all cases attention must be paid to ensure the compatibility of the preconcentration step (strength of the retention of analytes on the preconcentrator) with the composition of the eluent used for the subsequent recovery of the analytes and their separation.

18.4 Analytical Applications

To give an example, a column switching technique has been proposed (Ryan and Meaney 1992) where two compatible eluents of different eluotropic strengths were selected, one (CH_3CN/H_2O) to concentrate the metal-8-quinolinol (HQ) derivative complexes on a precolumn (Nucleosil C18), and the other ($CH_3CN/H_2O/HQ$) to elute the analytes from the precolumn onto the analytical column (C18). The linear dynamic range is from 5 ppb to 10 ppm for Al and from 40 ppb to 5 ppm for Cu and Fe.

DEDTC chelate formation-preconcentration and RP-HPLC separation have been used, on-column, for the determination of Cd, Cu and Ni in seawater samples (Comber

1993). A C2-bonded silica microcolumn, loaded with a dithiocarbamate-cetyltrimethylammonium ion pair, enabled the retention of metal ions and their complexation. Elution on an ODS analytical column was optimized by adding cetyltrimethylammonium (CTA) bromide to a CH_3CN-H_2O mixture, owing to the relative instability of the Cd-DEDTC anionic complex which was eluted as a neutral ion pair.

On the basis of previous studies (Zhao and Fu 1990), a selective preconcentration method with a cation exchange resin for RP-HPLC of the Co-5-Br-PADAP has been more recently developed (Uehara et al. 1994). Co complex, in aqueous solution, is readily oxidized to the Co(III)-5-Br-PADAP inert cationic complex which is retained on a sulfonated XAD-4 resin. Co is detected spectrophotometrically (588 nm) after elution onto a C18 analytical column (Capcell SG-120) with a methanol-water eluent to which EDTA and TBA have been added and without 5-Br-PADAP. The absence of 5-Br-PADAP favours the dissociation, e.g. of Cu and Zn chelates, and other metal ions eluted later, like Fe and Ni, do not interfere. The detection limit for Co in water samples is reported to be 5.9 ng l^{-1}.

A feature of these techniques is the high selectivity obtainable by a proper choice of ligand and chromatographic parameters. The method developed for Fe(II) determination in aerosol, rainwater and seawater (remote marine aerosol) is an example (Yi et al. 1992). A solid phase extraction on Sep-Pak C18 cartridges loaded with 3-(2-pyridyl)-5,6-diphenyl-1,2,4-triazine-p,p'-disulfonic acid (ferrozine, FZ) enabled the separation from matrix and preconcentration of Fe(II) as Fe(II)-FZ cation complex. The complex and FZ were eluted with methanol, and excess ligand was separated by ion pair-reversed phase LC. In this way Fe(III), Ni(II), Co(II) and Cu(II) interferences were removed, and a detection limit of 5.6 ng l^{-1} was obtained.

The detection limits of the whole IIC procedure, based on metal ion precomplexation with Plasmocorinth B (Sarzanini et al. 1993a), were lowered by coupling a preconcentration step. This last was performed by eluting samples through a hydrophobic microcolumn (C18) after carrying out the ion-pair reaction directly on the sample. The procedure gave enrichment factors up to 900 with detection limits between 15–90 ng l^{-1} (Sarzanini et al. 1993b).

The above mentioned highly cross-linked macroporous PS-DVB resin containing the iminodiacetate functional group that allows operation at high pressure without physical degradation (MetPac CC-1 column, Dionex) (Siriraks et al. 1990) seems a good example of the column switching technique involving a preconcentration step and selective matrix removal. At pH 5.2–5.6 polyvalent metal ions are selectively concentrated into MetPac CC-1, alkali metals and anions are not retained, and a selective elution of alkali earth metals can be achieved using ammonium acetate. Lanthanides and heavy metals, with the exception of chromium, are eluted with acid into a cation exchange column, acting as interface before the analytical column, to which the above are successively driven with a PDCA or an oxalate complexing eluent. UV-VIS spectrophotometric detection is accomplished after post-column derivatization (PAR), and detection limits (Fe, Cu, Ni, Zn, Co, Mn, Cd, Pb) range from 0.2 to 1 µg l^{-1} (Siriraks et al. 1990; Dionex 1990). The procedure was successfully used for metal ion determination in seawater from the Venice Lagoon, and with 60 ml sample preconcentration the detection limits for Cu, Ni, Zn, Co and Mn were lowered to 0.05–0.1 µg l^{-1} (Caprioli and Torcini 1993). A modification of the above mentioned procedure performed metal recovery from MetPac CC-1 resin with the same eluent used for the ion chromatographic separation (75 mM H_2SO – 100 mM HCl – 100 mM KCl) and a cation exchange column with higher capacity. In this way detection limits of 10 and 30 ng l^{-1} are achieved for Cd and Pb respectively for seawater samples (200 ml) (Cardellicchio et al. 1993).

18.5 Detection and Hyphenated Techniques

Refractive index and spectrophotometric and electrochemical detectors were traditionally used in liquid chromatography, while IC introduced the use of suppressed eluent conductivity detection. Analysis of complex matrices (e.g. foods, new materials, pharmaceutical and environmental samples) and speciation studies, in the field of metal analysis, are fundamental topics and have stimulated studies for the improvement of detection sensitivity and selectivity. For instance, spectrophotometric or electrochemical detection were coupled with post-column reactions, but a new approach, *hyphenation*, has become the emerging field of research. Hyphenation concerns the coupling of *unconventional* detectors with LC. This means that atomic spectroscopies (absorption and emission: flame, graphite furnace, hydride generation, cold vapour, plasma) were applied to LC.

In the field of atomic emission techniques, the majority of applications have been based on inductively-coupled plasma (ICP). This source is more suitable for LC since the chromatographic flow rate is compatible with conventional, i.e. pneumatic chamber, ICP interfaces. The use of an ultrasonic nebulizer or direct injection nebulization (DIN), with microbore columns, increased transport efficiency to the ICP interface. Finally, ICP has been used as a source for mass spectrometry (MS) and ICP-MS has become one of the most powerful techniques for speciation analysis when coupled with separation procedures.

For more details both the advances in detection techniques for IC (Buchberger and Haddad 1997), and the ICP-MS detection modes for chromatography and capillary electrophoresis (Sutton et al. 1997), have recently been reviewed. Some examples, concerning difficult sample analysis (e.g. rare earth) or metal speciation (As, Se, Hg and Cr), will be given hereafter.

The determination of rare earth elements is critical due to the nature of the samples, e.g. nuclear-power waste, rock samples. An ion-exchange chromatographic separation of rare earth elements in seven geological reference materials to remove matrix interferences in ICP-AES determination (Farinas et al. 1995) as well as a review of their determination (Kumar 1994) could summarize the classic approaches to rare earth element separation and determination. Reversed phase columns with both isocratic and gradient separation were coupled with ICP-MS detection. Isocratic conditions allowed rare earth separation into groups but was preferred for the powerful selectivity of ICP-MS, enabling 0.4–5.0 μg l^{-1} detection limits, and for providing reduced time for analysis (Braverman 1992). Another example is the ICP-MS determination of fission product isotopes. U or U + Pu are preeluted with 1 M HCl or 0.4 M HNO_3, and isobaric overlaps, present in direct mass spectrometric determinations, are removed as shown by measurements of isotopic composition of Nd in a high U and Pu matrix (Röllin et al. 1996). Finally the separation of some lanthanides with chelating chromatography must be mentioned (Kobayashi et al. 1992), where an HCl solution, instead of a ligand, was used as a mobile phase coupled with a selective chelating resin, namely a porous polymer impregnated with 2-ethylhexyl hydrogen 2-ethylhexyl phosphonate

(PC-88A). The main advantages of this approach, in addition to high selectivity, are that sample solutions do not contain concentrated salts or chelating agents, and clogging of ICP torch and spectral interferences due to molecular bands are removed; nebulization efficiency is also optimized in respect to more viscous eluents.

The molecular forms of arsenic subject to speciation analysis are anions, i.e. arsenite As(III), arsenate As(V), monomethylarsonate (MMA) dimethylarsinate (DMA) or cations, e.g. arsenobetaine (AsB), arsenocholine (AsC) and tetramethylarsonium ions (TMAs) or uncharged compounds at neutral pH, e.g. arsenous acid. An IC-HPLC procedure for As(III), As(V), MMA and DMA separation and ICP-AES determination have been developed (Rauret et al. 1991) by coupling the systems with the hydride generation sample introduction technique. The procedure was improved (Rubio et al. 1992) by checking two different kinds of columns (Nucleosil-5SB and Hamilton PRP X-100) and by comparing isocratic and gradient elution. The peak profile was improved by filtering the data corresponding to low concentration with the Fourier transform. With such a procedure detection limits between 2.7 As(III) and 11.4 As(V) $\mu g\ l^{-1}$ were obtained. Ion chromatographic methods for elemental speciation (As, Se and Cr) using microbore columns with direct injection nebulization by ICPAES have been described (Gjerde et al. 1993). Arsenite, arsenate, MMA, DMA, AsB and AsC have been separated by an anion HPLC procedure with a phosphate eluent and the analytes were determined by hydride generation atomic absorption spectrometry (HG-AAS) (López-Gonzálvez et al. 1996). The HPLC-UV-HG-AFS (AFS, atomic fluorescence spectrometry) method has also been applied to investigate the stability of arsenic species in relation to food (seafoods and mushrooms) treatment procedures (van Elteren and Slejkovec 1997). Reversed-phase microbore columns and eluents containing ion-pairing agents could be coupled with mass spectrometric detection for arsenic speciation by the use of a direct injection nebulizer interface (Shum et al. 1992a,b). Good efficiency was also obtained using interfaces based on hydride generation manifolds. Hydride generation was tried to avoid the poor efficiency of conventional pneumatic nebulizers in LC-ICP-MS studies of arsenic speciation (Branch et al. 1991; Story et al. 1992), but even by using a membrane gas separator (Nakahara 1991) the determination was subject to interference by the $ArCl^+$ molecular ion. The removal of $40Ar35Cl^+$ interference was optimized (Sheppard et al. 1992; Vela et al. 1993) by coupling ion chromatography and ICP-MS detectors and lowered detection limits for As(III), As(V), DMA and MMA with the use of an He-Ar gas mixture as ionization source: they ranged between 0.032 and 0.080 ng for DMA and MMA respectively. More recently (Ding et al. 1995) a speciation of these compounds was obtained by micellar liquid chromatography coupled with ICP-MS detection. The method, based on micellar mobile phase (CTAB, propanol and borate buffer) and on a PRP-1 separation column coupled with the ICP-MS system, allowed linear dynamic ranges of three orders of magnitude and detection limits in the picograms range (90–300) and overcame the problems of chloride since it is not coeluted with any of the four arsenic species. Finally a detailed study must be mentioned on suitability of the ion-spray (IS) technique for arsenic speciation analysis in biological samples. A cation-exchange HPLC has been coupled with IS-MS-MS detection for analysis of oragnoarsenic species. Dual mode, elemental and molecular, analysis is presented using standard mixtures. Although detection limits are not as low as those obtained by HPLC-ICP-MS, the results indicate IS-MS-MS as a complementary technique to ICP-MS for speciation analysis (Corr and Larsen 1996).

Inorganic selenium species were determined by ion-pairing reversed-phase (silica C_{18}) microscale-liquid chromatography (eluent: methanol-water-TBA, flow rate 50 μl min^{-1}) and a direct injection nebulizer (DIN) coupled with ICP-MS (detection limits of 10–20 ng ml^{-1}) (Houck et al. 1991). A similar ion-pair chromatographic separation with an ICP-MS detector using an ultrasonic nebulizer gave detection limits between 0.17 and 0.76 ng ml^{-1} for TMSe, Se(IV) and Se(VI) (Yang and Jiang 1995). Se(IV) and Se(VI) species were also separated (Shum and Houck 1993) on an anion exchange microcolumn (eluent flow rate 100 μl min^{-1}) with a DIN-ICP-MS detection system. Isotope ratio measurements on chromatographically separated species of Se gave detection limits of 7–8 ng ml^{-1} for both of the species. A two-step eluent switching procedure (25 mM K_2SO_4 eluent switching to 200 mM after 200 s at flow rate 2.0 ml min^{-1}), with an anion exchange column, enabled the separation of inorganic selenium species in aqueous samples (Pitts et al. 1995). Detection of analytes, after online microwave reduction and hydride generation, was performed with an atomic fluorescence detector. The method provided detection limits of 0.2 and 0.3 ng ml^{-1} within 1.5 and 2.0% RSD for selenite and selenate respectively. Speciation of eight selenium compounds has also been obtained with a strong cation-exchange column by interfacing the chromatographic system with an ICP-MS by high pressure hydraulic nebulizer (Goessler et al. 1997).

Mercury species, methyl-, ethyl- and inorganic mercury, are both neutral and ionic, and their ion chromatographic separation as cysteine complexes was developed (Sarzanini et al. 1994). Eluent composition (acetic acid, sodium perchlorate and cysteine) was optimized with respect to the separation procedure and to the reductive reaction ($NaBH_4$) which permits the detection of mercury with cold vapour AAS. Online preconcentration procedures were also investigated using both C_{18} and ion exchange microcolumns. The detection limits, for 100 ml samples, were 2, 10 and 4 ng for Hg, CH_3Hg and C_2H_5Hg respectively. Enrichment and separation of methyl-, ethyl-, methoxyethyl-, ethoxyethyl-, phenyl- and inorganic Hg complexes were also performed with pyrrolidine dithiocarbamate (PDC). An RP C18 column was used coupled with an acetonitrile-water buffered eluent. Analytes were determined by ultraviolet, postcolumn oxidation, cold vapour atomic absorption spectrometry (UV-PCO-CVAAS) (Falter and Schöler 1994,1995). The preconcentration (300 ml samples) of mercury-chelates on a microcolumn (Hypersil-ODS RP C18) gave detection limits of 0.5 ng l^{-1}. ICP-MS detection through ultrasonic nebulization was enabled for methyl-, ethyl- and inorganic mercury by their separation on a C18 reversed phase column with a methanol-acetonitrile-2-mercaptoethanol eluent containing ammonium acetate (Huang and Jiang 1993). Detection limit values (0.4–0.8 ppb) 10 times lower than those obtained with LC-ICP-MS with a conventional nebulizer, and comparable to those for LC-ICP-MS with cold vapour generation, were obtained.

The determination of chromium speciation by ion-pair chromatography has been optimized (Posta et al. 1993). The separation of Cr(VI)-Cr(III) was performed on an RP C18 column by using a TBA-acetate, ammonium acetate, phosphoric acid and methanol based eluent. This eluent composition also enhanced the sensitivity of detection obtained by coupling HPLC to a flame AAS by a high pressure capillary with hydraulic high pressure nebulization. Detection limits of 0.02 and 0.03 mg l^{-1} for Cr(VI) and Cr(III) respectively were lowered to 0.5 μg l^{-1} for Cr(VI) after a preconcentration step. An anion chromatographic separation (polymer-based anion exchanger, eluent:

EDTA-oxalic acid) has been optimized by working at a 40 °C column temperature (Inoue et al. 1995). Detection was obtained by direct introduction of an eluate into an ICP-MS system. This method enabled 80–88 ng l^{-1} detection limits for Cr(III) and Cr(VI) respectively within a linear range from 0.5 to 5 000 μg Cr l^{-1} and simultaneous determination of chromium species and Mn, Fe, Ni, Cu, Mg and Ca in water samples. Total chromium, Cr(III) and Cr(VI) speciation was achieved (Powell et al. 1995) by coupling HPL anion-chromatography (eluent nitric acid) with direct injection nebulization and ICP-MS. The detection limits obtained were 30 and 60 ng l^{-1} for Cr(III) and Cr(VI) respectively without particular chromatographic approach. Mixed-mode columns were used (Byrdy et al. 1995), namely IonPac AS7, for Cr(III) and Cr(VI) separation with $(NH_4)_2SO_4$ eluent (pH 9.2). Cr(III) species was stabilized with EDTA before sample analysis and detection was performed both by ICP-AES and ICP-MS equipped with a high-performance interface and a concentric nebulizer. Relative detection limits were 0.40 ng l^{-1} for Cr(III) and 1.0 μg l^{-1} for Cr(VI) within a 4% RSD and a linear dinamic range from 3 to 600 μg l^{-1} and from 5 to 1 000 μg l^{-1} for Cr(III) and Cr(VI) respectively in aqueous media.

At the end of these examples it seems a matter of philosophy to define LC a separation technique coupled with a specific detector or a tool for on-line sample pretreatments (Buchberger and Haddad 1997).

References

Abollino O, Mentasti E, Porta V, Sarzanini C (1990) Immobilized 8-oxine units on different solid sorbents for the uptake of metal traces. Anal Chem 62:21–26

Abollino O, Sarzanini C, Mentasti E, Liberatori A (1993) Trace metal pre-concentration with sulfonated azo-dyes and ICP AES determination. Spectrochim Acta 49A:1411–1421

Alvarez-Zepeda A, Barman BN, Martire DE (1992) Thermodynamic study of the marked differences between acetonitrile - water and methanol - water mobile phase systems in reversed-phase liquid chromatography. Anal Chem 64:1978–1984

Branch S, Corns WT, Ebdon L, Hill S, O'Neil P (1991) Determination of arsenic by hydride-generation inductively-coupled plasma mass spectrometry using a tubular membrane gas - liquid separator. J Anal At Spectrom 6:155–158

Braverman DS (1992) Determination of rare-earth elements [in fly ash] by liquid-chromatographic separation using inductively-coupled plasma mass spectrometric detection. J Anal At Spectrom 7:43–46

Buchberger W, Haddad PR (1997) Advances in detection techniques for ion chromatography. J Chromatogr A 789:67–83

Byrdy FA, Olson LK, Vela NP, Caruso JA (1995) Chromium speciation by anion-exchange high-performance liquid chromatography with both inductively-coupled plasma atomic-emission spectroscopic and inductively-coupled plasma mass spectrometric detection. J Chromatogr 712:311–320

Caprioli R, Torcini S (1993) Determination of copper, nickel, zinc, cobalt and manganese in sea-water by chelation ion chromatography. J Chromatogr 640:365–369

Cardellicchio N, Cavalli S, Riviello JM (1993) Determination of cadmium and lead at μg l^{-1} levels in aqueous matrices by chelation ion chromatography. J Chromatogr 640:207–216

Cardellicchio N, Ragone P, Cavalli S, Riviello J (1997) Use of ion chromatography for the determination of transition metals in the control of sewage-treatment-plant and related waters. J Chromatogr A 770:185–193

Cassidy RM, Sun L (1993) Optimization of the anion-exchange separation of metal-oxalate complexes. J Chromatogr 654:105–111

Challenger OJ, Hill SJ, Jones P, Barnett NW (1992) Application of chelating exchange ion chromatography to the determination of trace metals in high ionic strength media. Anal Proc 29:91–93

Challenger OJ, Hill SJ, Jones P (1993) Separation and determination of trace metals in concentrated salt solutions using chelation ion chromatography. J Chromatogr 639:197–205

Comber S (1993) Determination of dissolved copper, nickel and cadmium in natural waters by high-performance liquid chromatography. Analyst 118:505–509

Corr JJ, Larsen EH (1996) Arsenic speciation by liquid chromatography coupled with ionspray tandem mass spectrometry. J Anal At Spectrom 11:1215–1224
Dasgupta PK (1992) Ion chromatography. The state of the art. Anal Chem 64:775A–783A
Dilli S, Haddad PR, Htoon AK (1990) Further studies of diethyldithiocarbamate complexes by high-performance liquid chromatography. J Chromatogr 500:313–328
Ding H, Wang J, Dorsey JG, Caruso JA (1995) Arsenic speciation by micellar liquid chromatography with inductively-coupled plasma mass-spectrometric detection. J Chromatogr 694:425–431
Dionex (1990) Determination of transition metals in complex matrices by chelation ion chromatography. Technical Note 25:1–17
Falter R, Schöler HF (1994) Interfacing high-performance liquid chromatography and cold-vapour atomic-absorption spectrometry with online UV irradiation for the determination of organic mercury compounds. J Chromatogr 645:253–256
Falter R, Schöler HF (1995) Determination of mercury species in natural waters at picogram level with online RP C18 preconcentration and HPLC-UV- PCO-CVAAS. Fresenius' Z Anal Chem 353:34–38
Farinas JC, Cabrera HP, Larrea MT (1995) Improvement in the ion-exchange-chromatographic separation of rare-earth elements in geological materials for their determination by inductively-coupled plasma atomic-emission spectrometry. J Anal At Spectrom 10:511–516
Gjerde DT, Wiederin DR, Smith FG, Mattson BM (1993) Metal speciation by means of micro-bore columns with direct-injection nebulization by inductively-coupled plasma atomic-emission spectroscopy. J Chromatogr 640:73–78
Goessler W, Kuehnelt D, Schlagenhaufen C, Kalcher K, Abegaz M, Irgolic KJ (1997) Retention behaviour of inorganic and organic selenium compounds on a silica-based strong-cation-exchange column with an inductively-coupled plasma mass spectrometer as selenium-specific detector. J Chromatogr A 789:233–245
Hill SJ, Bloxham MJ, Worsfold PJ (1993) Chromatography coupled with inductively-coupled plasma atomic-emission spectrometry and inductively-coupled plasma mass spectrometry. Review. J Anal At Spectrom 8:499–515
Houck RS, Shum SCK, Wiederin DR (1991) Frontiers in elemental analysis by mass spectrometry. Anal Chim Acta 250:61–70
Huang CW, Jiang SJ (1993) Speciation of mercury by reversed-phase liquid chromatography with inductively-coupled plasma mass-spectrometric detection. J Anal At Spectrom 8:681–686
Inoue Y, Sakai T, Kumagai HK (1995) Simultaneous determination of chromium(III) and chromium(VI) by ion chromatography with inductively-coupled plasma mass spectrometry. J Chromatogr 706:127–136
Janos P, Broul M (1992) Ion-exchange separation of metal cations on a dodecylsulfate-coated C18 column in the presence of complexing agents. Fresenius' Z Anal Chem 344:545–548
Jen JF, Chen CS (1992) Determination of metal ions as EDTA complexes by reversed-phase ion-pair liquid chromatography. Anal Chim Acta 270:55–61
Jones P, Nesterenko PN (1997) High-performance chelation ion chromatography. A new dimension in the separation and determination of trace metals. J Chromatogr A 789:413–435
Jones P, Challenger OJ, Hill SJ, Barnett NW (1992) Advances in chelating exchange ion chromatography for the determination of trace metals using dye-coated columns. Analyst 117:1447–1450
Jones P, Foulkes M, Paull B (1994) Determination of barium and strontium in calcium-containing matrices using high-performance chelation ion chromatography. J Chromatogr 673:173–179
Kobayashi S, Wakui Y, Kanesato M, Matsunaga H, Suzuki TM (1992) Chromatographic separation and inductively-coupled plasma atomic-emission spectrometric determination of the rare-earth metals contained in terbium. Anal Chim Acta 262:161–166
Kumar M (1994) Recent trends in chromatographic procedures for separation and determination of rare-earth elements. Analyst 119:2013–2024
LeGras CAA (1993) Simultaneous determination of anions and divalent cations using ion chromatography with ethylenediaminetetra-acetic acid as eluent. Analyst 118:1035–1041
López-Gonzálvez M, Gómez MM, Palacios MA, Cámara C (1996) Urine clean up method for determination of six arsenic species by LC-AAS involving microwave assisted oxidation and hydride generation. Chromatographia 43:507–512
Marina ML, Andrés P, Diéz-Masa JC (1993) Separation and quantitation of some metal ions by RP [reversed-phase]-HPLC using EDTA as complexing agent in mobile phase. Chromatographia 35:621–626
Miura J (1990) Determination of trace amounts of vanadium in natural waters and coal fly ash with 2-(8-quinolylazo)-5-(dimethylamino)phenol by reversed-phase liquid chromatography – spectrophotometry. Anal Chem 62:1424–1428
Nair LM, Saari-Nordhaus R, Anderson JM Jr (1994) Ion-chromatographic separation of transition metals on a polybutadiene maleic acid-coated stationary phase. J Chromatogr 671:43–49

Nakahara T (1991) Hydride generation techniques and their applications in inductively-coupled plasma atomic-emission spectrometry. Spectrochim Acta Rev 14:95–109
Ohashi H, Uehara N, Shijo Y (1991) Simultaneous determination of molybdenum, vanadium, gallium, copper, iron and indium as quinolin-8-olate complexes by high-performance liquid chromatography. J Chromatogr 539:225–231
Ohtsuka C, Matsuzawa K, Wada H, Nakagawa G (1991) Chromatographic behaviour of 2-(2-pyridylazo)-5-[N-(sulphoalkyl)amino]phenol chelates on hydrophobic stationary phases. Anal Chim Acta 252:181–186
Ohtsuka C, Matsuzawa K, Wada H, Nakagawa G (1992) Characterization of 2-(2-pyridylazo)-5-[N-(sulphoalkyl)amino]phenol derivatives as the pre-column chelating agent for trace metal determination by ion-pair reversed-phase liquid chromatography with spectrophotometric detection. Anal Chim Acta 256:91–96
Ohtsuka C, Matsuzawa K, Wada H, Nakagawa G (1994) Retention behaviour of metal chelates in ion-pair reversed-phase liquid chromatography as a function of mobile phase composition with methanol/water and acetonitrile / water mobile phases. Anal Chim Acta 294:69–74
Paull B, Jones P (1996) A comparative study of the metal selective properties of chelating dye impregnated resins for the ion-chromatographic separation of trace metals. Chromatographia 42:528–538
Paull B, Foulkes M, Jones P (1994a) Determination of alkaline earth metals in offshore oil-well brines using high-performance chelation ion chromatography. Anal Proc 31:209–211
Paull B, Foulkes M, Jones P (1994b) High-performance chelation ion-chromatographic determination of trace metals in coastal sea water using dye-impregnated resins. Analyst 119:937–941
Paull B, Fagan PA, Haddad PR (1996) Determination of calcium and magnesium in sea water using a dynamically coated porous graphitic carbon column with a selective metallochromic ligand as a component of the mobile phase. Anal Commun 33:193–196
Paull B, Nesterenko P, Nurdin M, Haddad PR (1998) Separation of metal ions using a polymeric reversed-phase column and a methylthymol blue containing mobile phase. Anal Commun 35:17–20
Pitts L, Fisher A, Worsfold P, Hill SJ (1995) Selenium speciation using high-performance liquid chromatography-hydride generation atomic fluorescence with online microwave reduction. J Anal At Spectrom 10:519–520
Porta V, Sarzanini C, Abollino O, Mentasti E, Carlini E (1992) Pre-concentration and inductively-coupled plasma atomic-emission-spectrometric determination of metal ions with online chelating ion exchange. J Anal At Spectrom 7:19–22
Posta J, Berndt H, Luo SK, Schaldach G (1993) High-performance flow flame atomic-absorption spectrometry for automated online separation and determination of chromium(III)/chromium(VI) and pre-concentration of chromium(VI). Anal Chem 65:2590–2595
Powell MJ, Boomer DW, Wiederin DR (1995) Determination of chromium species in environmental samples using high-pressure liquid chromatography direct injection nebulization and inductively-coupled plasma mass spectrometry. Anal Chem 67:2474–2478
Rauret G, Rubio R, Padró A (1991) Arsenic speciation using HPLC-hydride-generation ICP AES with gas-liquid separator. Fresenius' Z Anal Chem 340:157–160
Rey MA, Riviello JM, Pohl CA (1997) Column switching for difficult cation separations. J Chromatogr A 789:149–155
Riley JP, Taylor D (1968) Chelating resins for the concentration of trace elements from sea-water and their analytical use in conjunction with atomic absorption spectrophotometry. Anal Chim Acta 40:479–485
Robards K, Starr P, Patsalides E (1991) Metal determination and metal speciation by liquid chromatography. Analyst 116:1247–1273
Röllin S, Kopatjtic Z, Wernli B, Magyar B (1996) Determination of lanthanides and actinides in uranium materials by high-performance liquid chromatography with inductively-coupled plasma mass spectrometric detection. J Chromatogr 739:139–149
Rubio R, Padró A, Alberti J, Rauret G (1992) Speciation of organic and inorganic arsenic by HPLC – HG – ICP [hydride-generation inductively-coupled plasma]. Mikrochim Acta 109:39–45
Ryan E, Meaney M (1992) Determination of trace levels of copper(II), aluminium(III) and iron(III) by reversed-phase high-performance liquid chromatography using a novel online sample pre-concentration technique. Analyst 117:1435–1439
Sacchero G, Abollino O, Porta V, Sarzanini C, Mentasti E (1991) Reversed-phase ion-interaction chromatography of metal ions by EDTA pre-complexation. Chromatographia 31:539–543
Sarzanini C, Mentasti E (1997) Determination and speciation of metals by liquid chromatography. J Chromatogr A 789:301–321
Sarzanini C, Abollino O, Mentasti E (1992) Hydroxyazo-dyes in metal ions preconcentration by ion exchange. In: Slater MJ (ed) Ion Exchange Advances. Elsevier Applied Science, London, pp 279–286

Sarzanini C, Sacchero G, Aceto M, Abollino O, Mentasti E (1993a) Ion-interaction chromatographic studies on metal ions complexed with plasmocorinth B [C. I. Mordant Blue 13] dye. J Chromatogr 640:179–185
Sarzanini C, Sacchero G, Aceto M, Abollino O, Mentasti E (1993b) Ion-pair reversed-phase high-performance liquid chromatography for trace metal pre-concentration followed by ion-interaction chromatography. J Chromatogr 640:127–134
Sarzanini C, Sacchero G, Aceto M, Abollino O, Mentasti E (1994) Ion-chromatographic separation and online cold vapour atomic-absorption-spectrometric determination of methyl mercury, ethylmercury and inorganic mercury. Anal Chim Acta 284:661–667
Sheppard BS, Caruso JA, Heitkemper DT, Wolnik KA (1992) Arsenic speciation by ion chromatography with inductively-coupled plasma mass spectrometric detection. Analyst 117:971–975
Shum SCK, Houck RS (1993) Elemental speciation by anion-exchange and size-exclusion chromatography with detection by inductively-coupled plasma mass spectrometry with direct-injection nebulization. Anal Chem 65:2972–2976
Shum SCK, Pang HM, Houck RS (1992a) Speciation of mercury and lead compounds by microbore column liquid chromatography – inductively-coupled plasma mass spectrometry with direct injection nebulization. Anal Chem 64:2444–2450
Shum SCK, Neddersen R, Houck RS (1992b) Elemental speciation by liquid chromatography – inductively-coupled plasma mass spectrometry with direct-injection nebulization. Analyst 117:577–582
Sirén H (1991) Studies on metal complexes of nitroso-naphthol-sulphonates ion-paired with ammonium compounds, by UV/VIS-liquid chromatography with liquid-liquid extraction and on-line derivatization. Annales Academiae Scientiarum Fennicae, A, II. Chemia 233
Sirén H, Riekkola ML (1992) Determination of metal ions by online complexation and ion-pair chromatography. J Chromatogr 590:263–270
Siriraks A, Kingston HM, Riviello JM (1990) Chelation ion chromatography as a method for trace elemental analysis in complex environmental and biological samples. Anal Chem 62:1185–1193
Story WC, Caruso JA, Heitkemper DT, Perkins L (1992) Elimination of the chloride interference on the determination of arsenic using hydride-generation inductively-coupled plasma mass spectrometry. J Chromatogr Science 30:427–432
Sutton K, Sutton RMC, Caruso JA (1997) Inductively-coupled plasma mass-spectrometric detection for chromatography and capillary electrophoresis. J Chromatogr A 789:85–126
Timerbaev AR, Bonn GK (1993) Complexation ion chromatography-an overview of developments and trends in trace metal analysis. Chromatogr A 640:195–206
Tomlinson MJ, Wang J, Caruso JA (1994) Speciation of toxicologically important transition metals using ion chromatography with inductively-coupled plasma mass-spectrometric detection. J Anal At Spectrom 9:957–964
Uehara N, Katamine A, Shijo Y (1994) High-performance liquid-chromatographic determination of cobalt(II) as the 2-(5-bromo-2-pyridylazo)-5-diethylaminophenol chelate after pre-concentration with a cation-exchange resin. Analyst 119:1333–1335
van Elteren JT, Slejkovec Z (1997) Ion-exchange separation of eight arsenic compounds by high-performance liquid chromatography – UV decomposition – hydride generation – atomic fluorescence spectrometry and stability tests for food treatment procedures. J Chromatogr A 789:339–348
Vela NP, Olson LK, Caruso JA (1993) Elemental speciation with plasma mass spectrometry. Anal Chem 65:585A–597A
Voloschik IN, Litvina ML, Rudenko BA (1994a) Separation of transition and heavy metals on an amidoxime complexing sorbent. J Chromatogr 671:51–54
Voloschik IN, Litvina ML, Rudenko BA (1994b) Application of multi-dimensional liquid chromatography to the separation of some transition and heavy metals. J Chromatogr A 671:205–209
Yan B, Worsfold PJ (1990) Determination of cobalt(II), copper(II) and iron(II) by ion chromatography with chemiluminescence detection. Anal Chim Acta 236: 287–292
Yan D, Schwedt G (1990) Simultaneous ion chromatography of inorganic anions together with some organic anions and alkaline-earth-metal cations using chelating agents as eluents. J Chromatogr 516:383–393
Yang KL, Jiang SJ (1995) Determination of selenium compounds in urine samples by liquid chromatography – inductively-coupled plasma mass spectrometry with an ultrasonic nebulizer. Anal Chim Acta 307:109–115
Yi Z, Zhuang G, Brown PR, Duce RA (1992) High-performance liquid chromatographic method for the determination of ultratrace amounts of iron(II) in aerosols, rainwater, and sea-water. Anal Chem 64:2826–2830
Zappoli S, Morselli L, Osti F (1996) Application of ion-interaction chromatography to the determination of metal ions in natural water samples. J Chromatogr 721:269–277
Zhao Y, Fu C (1990) Pre-column chelation liquid-chromatographic separation and determination of trace vanadium, copper, cobalt, palladuim, iron and nickel. Anal Chim Acta 230:23–28

Chapter 19

PIXE Analysis for Trace Elements in Marine Environments

R. Cecchi · G. Ghermandi

19.1 Introduction

During the past decades considerable attention has been paid to the geochemical cycles of major and minor elements (and substances), and consequently to cycling models among different physically well-defined parts of the earth (or "boxes"), whose number depends on the previous knowledge of the way in which elements of interest are distributed about the earth's surface. For example, a global cycling model subdivides the earth in four boxes, namely land, atmosphere, ocean and sediments. From one box to another a substance is transferred by a transport path, that is, a directional property of the system depending on the physicochemical characteristics of the substance itself; for the ocean the transport paths are with the atmosphere (gases, dust and aerosol), from the land (rivers) and to the bottom sediments (sedimentation).

From this general point of view, to describe the behaviour of an element we need a set of differential equations (the mass balance for each box) whose number depends on the different phases (in the ocean the aqueous phase, the mineral suspension, the plankton), on the speciation and on the processes of removal or production in the system. One must also take into account the various transport agents controlling inputs and outputs of the element.

A very important point in such a cycling model is the knowledge of the above mentioned input and output terms, involving experimental evaluation of the fluxes. For diffusive processes it is important to measure concentration gradients; for advective ones, one must be able to determine the mean concentration of the substance under investigation. Consequently, in order to test a cycling model for the ocean in its various features, the crucial problem is the very high number of measurements involved. A fast analytical technique, with multi-elemental capability, covering a large concentration range, requiring generally little or no sample preparation and able to measure concentrations in different phases such as aerosols, waters, suspended inorganic particles and sediments, is obviously very useful.

PIXE (Proton Induced X-ray Emission) is an analytical method based upon X-ray spectrometry. Its intrinsic detection limits are not very much below 1 ppm (in weight) in respect to a given bombarded specimen. It offers its maximum sensitivity in the two atomic number (Z) regions ($20 < Z < 35$ and $75 < Z < 85$). Measurement errors are in the order of 10%, depending mainly on the target preparation procedure and on the slight variability of the proton flux. The PIXE technique allows fast (a few minutes), non destructive, highly sensitive simultaneous determination of a wide group of elements ($12 \leq Z \leq 85$), without high variations of sensitivity among different elements. The total element concentration is measured by PIXE, not distinguishing among different oxidation states. Adequate sample preparation procedures may be required in environmental PIXE applications, since solid samples have to be exposed to the proton beam.

The environmental studies also take a great advantage from the development of another variation of PIXE, called micro-PIXE, that in principle is analogous to the electron microprobe or Scanning Electron Microscope-Energy Dispersive X-ray Analyser (SEM-EDAX), where a proton accelerator replaces the electron gun. The proton beam can be scanned across the surface of the specimen, giving an elemental mapping of the tested area. The micro-PIXE detection limits are better by a factor of about 100 than the electron microprobe ones, and result of the order of 10 ppm of the exposed target: the small size of the irradiated area (a few of square micrometers) allows detection of absolute quantities of 10^{-15}–10^{-16} g.

The widest used application of PIXE analysis in the marine ecosystem consists of the determination of trace elements of interest in the atmospheric aerosol. In fact the membranes used in air samplers may be very suitable for immediate PIXE analysis, without further handling. In addition, micro-PIXE may be a very powerful tool in the investigation of major and trace elements in aerosol particles: the micro-PIXE maps of a portion of an aerosol deposit on a thin backing allows one to determine the element associations and their relative abundances, providing an immediate characterization of the studied aerosol.

Marine aerosol is strongly involved in an exchange of substances between sea and atmosphere, and further transfer to coastal areas. It is also well known that under favourable conditions, the finest fraction of aerosols may be transported far from the source, to remote areas: long-range transport of pollutants may follow.

Water sample analyses are more problematic. The seawater matrix may determine problems in the treatement of samples that contain several elements of interest at very low concentrations. Techniques of preconcentration are therefore necessary. In addition, solid targets have to be prepared for PIXE analysis. The preconcentration procedure has to be chosen, taking account of the main features of the samples: in sea water the concentration of the so-called major lighter elements are generally higher than the heavier ones; therefore it is necessary to employ selective preconcentration.

Bottom sediment samples may be analysed as such, or reduced to slices (if possible), powders or solutions. Destructive and non-destructive sample preparation procedures may be applied for bulk analysis by PIXE.

Other procedures are especially suited to extract the non reticular fraction from sediment samples. Even if satisfactorily applied, PIXE can not completely explain its capabilities in this field, given the high matrix effects. Pore waters extracted from bottom sediment cores may be instead usefully analysed by PIXE.

As it will be described in what follows, when coupled with adequate sample preparation methodologies, PIXE technique becomes a very powerful tool suitable for a wide investigation of a marine ecosystem.

19.2 The PIXE Technique and the Experimental Set-up

PIXE (Johansson and Campbell 1988; Johansson et al. 1995) is an analytical technique that uses the spectrometry of the characteristic X-rays emitted by target atoms ionized by accelerated particles, and detected by a suitable device (i.e. a Si (Li) X photon detector). The filling of produced vacancies induces the emission of photons, whose energies identify each particular atom (mainly K and L X-ray, M lines in a few cases, of different intensities given their relative emission rates). K and L or L and M lines from the same element may be simultaneously present in a PIXE spectrum.

The number of emitted X-ray photons for a given atom transition depends on its production cross section for protons at the energy of the used beam, on the relative intensity of the transition, and on the proton flux.

The PIXE measurement of a sample (target) is a spectrum (Fig. 19.1) that includes different contributions: the characteristic X-ray peaks, the continuum bremsstrahlung of beam particles (BP) and of secondary electrons (SEB) of the specimen atoms into the target matrix, and also the gamma rays from nuclear reactions (GB) induced by the beam.

The BP is responsible for the high energy background in a PIXE spectrum, and decreases with rising beam particle energy.

The SEB is the principal contribution to the low energy continuum X-ray background in PIXE. It increases with rising particle energy. The SEB emission has an unisotropic distribution in angle. The angle of 135° is generally preferred for X-ray detector placement in order to reduce SEB signal.

The interaction by Compton scattering of gamma radiations from nuclear reactions in the Si(Li) detector produces a flat background in the 1–30 keV region of the spectrum, worsening the detection limits of heavier elements. GB increases with rising particle energy.

The use of a 1–2 MeV beam of protons limits the GB while at the same time gives acceptable X-ray production cross section values. Other ways to reduce background

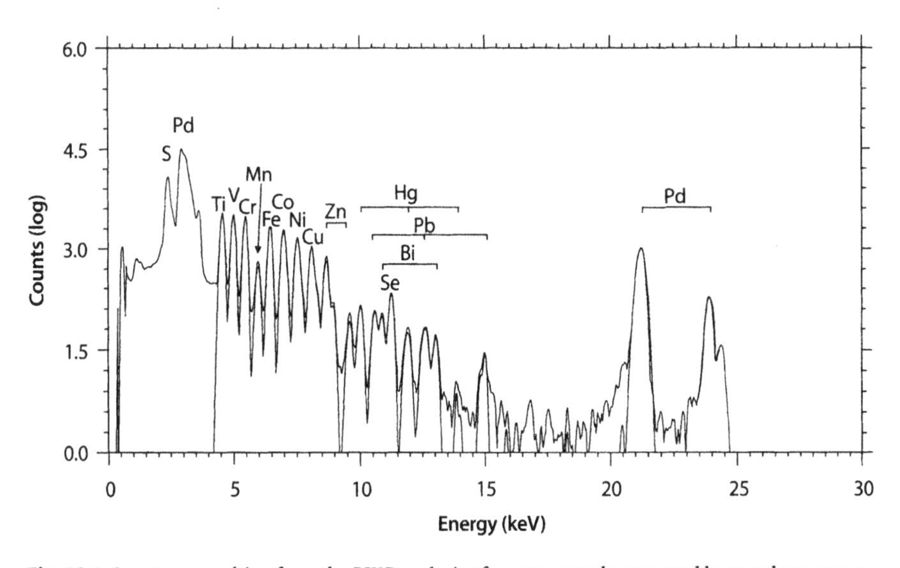

Fig. 19.1. Spectrum resulting from the PIXE analysis of a water sample, prepared by metal preconcentration as carbamate with Pd (as described in text). Some peaks from linear least-square fit (Cecchi et al. 1990a) are superimposed to the original measured spectrum before any data handling

effects are to use low Z and thin targets (of the order of 1 mg cm^{-2}, in which the energy loss of proton beam and the X-ray attenuation are negligible) which are suitable for limiting SEB and BP emissions (Ghermandi et al.1996).

The most widely used accelerators for PIXE experiments are single-ended Van de Graaff machines. The PIXE set-up used for environmental samples at the National Institute of Nuclear Physics (INFN) Laboratory of Legnaro (Padova, Italy) is here described (Aprilesi et al. 1984). The incident particles are protons accelerated to 1.8 MeV. The beam coming from the Van de Graaff is stabilized passing through a slit system into a deflector magnet, then driven in the line under high vacuum (10^{-6} Torr) by electrostatic equipments and lenses, up to the irradiation chamber. To obtain a homogeneous circular beam spot at the target, the beam is diffused through a Ni foil, 450 μm cm^{-2}, then carbon collimators pick off its central part (approximately 1.5 cm diameter). The Si(Li) detector (resolution 150 eV at 5.9 keV, solid angle between the target and the sensitive detector area 0.0198 sr) is placed at 135° from the beam and in front of the target. These measurement conditions limit the X-ray attenuation in the target. A Faraday cup is placed beyond the sample, at the end of the line, in order to measure the charge collected on the target. The resulting PIXE spectrum is collected in a multi channel analyser and then examined by suitable computer codes, in order to obtain the element concentration in the exposed target.

Mylar absorbers of various thickness or a funny filter (absorber with a central hole) may be placed in front of the detector. The mylar absorbers reduce the intensity of the emissions of exceptionally abundant light elements, as it may occur in sea and freshwater samples. Funny filters are preferable for aerosol specimens, because they balance the intensity between the low and high energy regions of the spectrum.

Pore water and mainly sediment samples, in which some medium-heavy elements are in general particularly abundant, may require the use of absorbers that selectively attenuate the most intense lines.

19.3 Sensitivity and Detection Limits of the Technique

PIXE intrinsic sensitivity is about 1 ppm of the exposed sample that, in a suitably thin target (for example 100 μg cm^{-2} of carbon), corresponds to very small amount (picograms) of an element. The sensitivity of the PIXE measure for a given atomic number depends on detector parameters, target composition, ion beam and measurement conditions (Cahill 1975). All these factors may be optimized to improve the technique's sensitivity. In a given set up and unvariable measurement conditions, the target features (thickness, characteristic ionization cross section of the elements both of the sample and of its backing) determine the behaviour of the yield versus atomic number. The highest sensitivity is obtained for $20 < Z < 40$ and $Z > 75$ and with low proton energies (1–3 MeV). These highest yield regions correspond to the detection limit minima. The sensitivity in a PIXE spectrum is determined by the signal/noise ratio of the detected emission. The detection limit is generally estimated as the number of counts in a peak N (>10) that result in significant respect to the noise (background counts N_b in the same spectral region):

$$N = 3\sqrt{N_b} \quad (19.1)$$

that corresponds to a confidence level of 99.86% for normally distributed measurements.

Representative curves of detection limits (Folkmann 1976) are reported in Fig. 19.2. It should be stressed that the detection limits have to be quoted with the used instrumental parameters and measurement conditions and with the exposed target features.

19.4 The PIXE Targets from Environmental Samples

It must be emphasized that in trace and ultratrace element analysis, sampling procedures and laboratory facilities and treatements must be adequate to eliminate any possible contamination. The target preparation should be performed in a "clean room" (wearing suitable suits and Low Density Polyethylene (LDPE) gloves), using high purity water (18 $M\Omega$ cm resistivity), high purity reagents as well as high purity salts and standard solutions. The sample containers, vessels and lab tools used should be made of conventional LDPE and Teflon (FEP) and cleaned by immersion in a succession of heated acid baths of increasing purity for several weeks (Boutron 1990). In addition, blank samples must be prepared in order to quantify any possible contamination during both the "in field" operation and the target preparation (Ghermandi et al. 1996).

PIXE has been shown to be a very suitable tool for the direct analysis of bulk (liquid matter or solid particles) aerosol specimens collected on filters, given both the PIXE sensitivity, but also the simplicity of preparing targets. The membranes used in air samplers (Nuclepore polycarbonate, cellulose and teflon filters) are generally suitable for immediate PIXE analysis. It follows that specially built samplers have been formed to produce aerosol deposits of suitable size for following PIXE analysis. In fact, if the

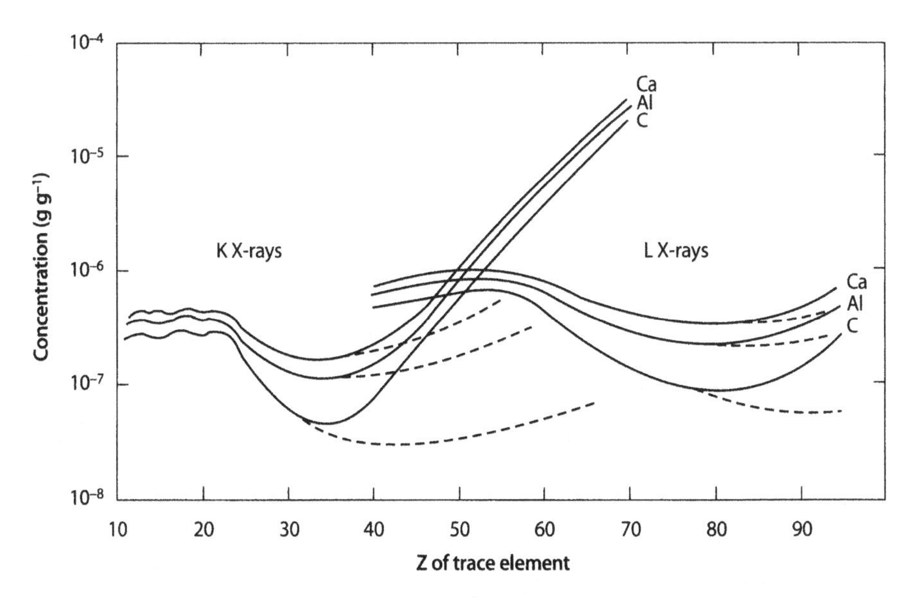

Fig. 19.2. Limits of detection evaluated for 1 mg cm^{-2} targets of carbon, aluminium and calcium, with 100 μC of 3 MeV protons (from Folkmann 1976, with permission of Plenum Press)

proton beam is larger than the deposit, it allows the integration of the whole specimen surface: the diameter of the proton beam impinging on the target is generally lower than 15 mm, while the commercially available air samplers collect a typically 47 mm diameter aerosol deposit on the backing.

An air sampler device may separate the aerosol in two or more size classes (Johansson et al. 1995). The dichotomous aerosol samplers ("Gent" stacked filter unit, IMPROVE dichotomous) separate by filtration coarse mode (2.0 μm < particle diameter < 10 μm) and fine mode fraction (<2 μm).

The cascade impactors provide the collection of aerosol particles by impaction on different stages. Some of these devices are especially suited for PIXE: the model I-1 PIXE International Cascade Impactor (PCI), with seven impaction stages and a backup filter stage, and the Small Deposit area Impactor (SDI). The SDI (Maenhaut et al. 1996) is a 12-stage, multinozzle device, with a deposit area with diameter less than 8 mm. The air flow rate is 11 l min^{-1}. The experimental aerodynamic cut diameters for the 12-stages SDI range from 8.5 to 0.045 μm, with 7 values under 1 μm.

Other impactors frequently coupled with PIXE are the rotating version of the Micro Orifice Uniform Deposit Impactor (MOUDI) and the DRUM sampler, that collect the aerosol particles on each stage along a rotating ring. The resulting specimen is a strip that, to optimize the PIXE analysis, should be mounted on a movable frame that adequately steps, on command, for proton irradiation of each increment.

PCI and SDI employ a 1.5 mm KIMFOL polycarbonate film as collection substrate, coated with vaseline (higher diameter stages) or paraffin (smaller diameter stages), or apiezon L in order to reduce particle bounce-off effects. The targets from all the SDI impaction stages allow the detection of 22 elements (Maenhaut et al. 1996), from Na to Pb (experimental conditions: 2.4 MeV proton beam, 150 nA beam current, 100 μC collected charge, funny filter absorber).

The quantitative PIXE analysis of aerosol samples are generally performed by comparison with external standard specimens (very thin, certified layers of elements evaporated onto suitable membrane). The INFN Laboratory at Legnaro, Padova (Italy) performs standard target by vaporization of elements onto polycarbonate (PC) backings.

The PIXE determination of elements suspended and/or dissolved in fresh and sea water involves greater difficulties in preparing targets, given the very low concentrations of several heavier elements of interest in presence of major lighter elements, and the need of solid targets for PIXE analysis. Different problems arise with ground and pore water. Suitable selective techniques of preconcentration are therefore necessary, also reducing the matrix effects.

Among the numerous treatments that use chelants to extract trace metals, a methodology has been optimized which performs the extraction of elements as carbamates, coprecipitated with a collector element on thin membrane filters, also involving minimal sample handling. The carbamate chelant complexes the trace elements in their more stable state, as they generally are in natural waters, and in their reduced form as ions, as they are carried by sample acidification (at the collection or during the pretreatment). The sample acidification also breaks chemical bonds of the majority of organic compounds. Obviously this procedure is not recommended for aqueous liquid waste samples, where other complexing agents like ethylenediaminetetraacetic acid (EDTA) may be discharged in significant concentrations, resulting in possible reduction of carbamate extraction efficiency.

The technique (Luke 1986) is based on the high stability of the carbamate of several elements and on the low solubility in water that is determined by pH. Keeping the pH values in a suitable range, a large number of carbamate elements precipitate and can be collected by filtration on a thin backing using a coprecipitant agent (that acts also as internal standard).

At pH 9, with sodium-diethyldithiocarbamate (Na DDTC) as chelant (Aprilesi et al. 1984) the largest number of elements, 41, precipitate (as carbamates mainly, as carbamates and hydroxides or as hydroxides alone). This coprecipitation effect represents an advantage when diluted solutions (under ppb (μg kg^{-1}) level) are treated, as is the case with open sea water, where the carbamate hydroxide precipitate may also act as a "scavenger" for constituents in solution. However it becomes a problem with fresh water or ground water, because, even if Na DDTC does not form chelates with alkaline earth metals, at pH 9 some major cations, like Ca, precipitate following the equilibra of their other salts. This phenomenon strongly affects the features of the resulting target.

A modified target preparation procedure was performed that takes place at pH 4 with the synergistic effect of two chelating agents, Na DDTC and ammonium-pirrolydine-dithiocarbamate (APDC), which also do not chelate the alkaline earths (Cecchi et al. 1987).

Palladium is used as the coprecipitant agent and internal standard, given the high stability of its carbamate complexes and for its rarity in the tested specimens. In the case of spectral interferences, Cu is used as the internal standard.

Precipitation is carried out as follows: 100 μg of Pd (from Pd solution) is added to the sample. An ammonia solution is added to keep the pH to 4. The carbamate elements are precipitated by adding 1 ml of a freshly prepared Na DDTC APDC (1% each one) solution which has been previously purified by shaking with an equal volume of freon (1,1,2-trichlorine-2,2,1-trifluorine ethane). The sample is allowed to stand for a certain time depending on the treated volume (from 30 minutes to 12 hours) and is then filtered through a Nuclepore PC membrane (0.4 μm pore size, 10 μm thickness, 1 mg cm^{-2} surface density and 10^8 pore cm^{-2}, filtration area of 0.79 cm^2). The possible non-homogeneity of the precipitate deposit can be integrated by performing PIXE measurements with a uniformly diffused beam larger than the filtration area.

This procedure has been widely tested and calibrated with mono and multi-element solutions (fresh and artificial sea water) and with reference materials for Ti, V, Cr, Mn, Fe, Co, Ni, Cu, Zn, Se, Hg, Pb, Bi, Mo, Y, Tl, Sb, Ag, Cd in the ppt-ppm range, with preconcentration of variable sample volumes, from 500 to 15 ml. The recovery efficiency (Cecchi et al. 1990b) remains generally higher than 80% for Fe, Co, Ni, Cu, Zn, Se, Hg, Pb, Ag, Cd, higher than 50% for Ti, V, Cr, Mo, Sb, and lower than 25% for Y and Tl.

The detection limits (for 15 μC of 1.8 MeV protons in an averaged blank spectrum for the calibrated K and L X-rays) are reported in Table 19.1. The values in the blank, with respect to the detection limits for the Nuclepore backing alone, are affected by the slightly increased target thickness and by the behaviour of each element as carbamate.

Satisfactory reproductiveness of the results have been verified, with measurement errors generally lower than 10%. The PIXE measurement error for these samples mainly depends on the uncertainty of the chemical treatement and on the variance of a suitable number of blank samples measured under the same experimental conditions; in minor part it depends on the uncertainty of the experimental setup condition.

Table 19.1. Detection limits for the calibrated K and L X-rays in the blank spectrum (averaged from several measured blank samples, prepared by element precipitation as carbamate on Nuclepore PC filters) for 15 μC of protons at 1.8 MeV

Element (detected by K line)	Detection limit ($ng\ cm^{-2}$)	Element (detected by L line)	Detection limit ($ng\ cm^{-2}$)
Ti	3.0	Sb	80.0
V	2.5	Hg	8.5
Cr	3.5	Tl	115.0
Mn	6.5	Pb	9.0
Fe	4.0	Bi	10.0
Co	3.0		
Ni	1.5		
Cu	2.0		
Zn	4.0		
Se	4.5		
Y	50.0		
Mo	40.0		
Ag	50.0		
Cd	50.0		

Bulk analysis of sediment and soil solid samples may be performed directly by PIXE or on samples previously reduced in slices, powders or solutions for further target preparations (Valkovic 1983).

Some procedures are especially suited to the extraction of the non-lattice-held fraction from sediment samples (Bernardi et al. 1988). The resulting liquid specimens are preconcentrated as Na DDTC and APDC carbamate on Nuclepore filters, using the procedure previously described for waters. Even if satisfactorily applied (the results are generally comparable with those obtained by Atomic Absorption Spectroscopy), PIXE can not completely explain its capabilities in these analyses. The matrix effects worsen the spectrum on each side of the most intense characteristic emissions, resulting in larger inaccuracies in the measurements.

19.5 The Proton Microbeam and its Applications

The best microbeam equipment (Johansson and Campbell 1988; Johansson et al. 1995) allows a spatial resolution of about 1 μm. This facility is provided defining the beam coming from the accelerator by a collimator (diameter range 10–100 μm) placed at the crossover point where the beam diameter has a minimum, then focusing the divergent beam. In the irradiation chamber a specimen may be scanned by the microbeam. A microscope allows the optical view of the sample. The X-ray detector and the electronic devices are the ones used for diffused beam PIXE set up.

The microbeam is widely used for mapping elemental concentrations, by moving the specimen (or the beam) in small steps, with positional reproducibility of the order of 1 μm, in two orthogonal directions or in linear scanning. A PIXE spectrum is recorded at each position.

Given its spatial resolution, the microbeam is not comparable with the electron microprobe resolution in surface analyses, while it is preferable for target thicker than some tens of micrometers, because the proton microbeam resolution is mantained (1–2 μm) for tens of micrometers of penetration depth.

The micro-PIXE sensitivity is of the order of 10^{-5} g g^{-1} of the exposed target. Given the small size of the exposed area (few micrometers diameter), the detection of absolute quantities of 10^{-16} g are allowed. The energy of the proton beam (some MeV) permits the detection of heavy elements by means of K lines. This sensitivity represents an advantage in trace element analysis with respect to the electron microprobe capabilities (10^{-3} g g^{-1}).

The application of micro-PIXE in the environmental field may be addressed to the investigation of major and trace elements in aerosol particles (such as marine aerosols), providing maps of a portion of an aerosol deposit on a thin backing. This technique is obviously not adequate for the examination of water samples, while it finds wide application in geology, for the immediate irradiation of slices of rocks and minerals.

19.6 PIXE Application in the Study of Pollutant Enrichment in Marine Aerosols

Long-range pollutant transport is of global importance in environmental research. Marine aerosol is a matrix strongly involved in exchange of substances between sea and atmosphere and in their further transfer to coastal areas, where, for example, vegetation damage may occur. It is also well known that, under favourable meteorological conditions, the finest fraction of aerosols may be transported far from the source to remote areas.

In the study of marine aerosol the role of many natural and man-made surfactants that spontaneously segregate at the sea surface-air interface has to be estimated. The marine aerosol is mainly generated by breaking wave events (Resch 1986). Air bubbles trapped by breaking waves, during the wave rolling motion, are particularly enriched in soluble surfactants. Marine spray is produced by the breaking of these bubbles, when they rise again at the sea surface. The bubble collapse involves films and jet formation (Blanchard 1975), particularly enriched of soluble surfactant matter. The jet microdrops are injected in the air at high levels (6–15 cm), with high probability of long life in the atmosphere. Recent studies (Cini et al. 1994) proposed a model of the enrichment partition of the surfactants in the jet drops, and the finest drops were shown to be the more enriched. The surfactant ability to interact with inorganic ions induces pollutant adsorption and their further long-range transport by means of jet drops.

The PIXE analysis of aerosol samples may be a very useful tool to better understand these processes, given the PIXE capability to detect the pollutant elements scavenged by the surfactants enriched in the aerosol jet drops.

An experiment is in running, in order to validate the model of the enrichment partition of the surfactants in the jet drops. The first step, performed up to now, consisted of analysis of samples obtained from an apparatus for artificial production of aerosol

(Cini et al. 1994) from Tyrrhenian Sea water. The approach involves the use of a multistage impactor, with size separation of the aerosol drops at the collection, and a measurement campaign performed in coastal areas near Livorno. In each experiment of artificial aerosol production, both the sea water before aerosolization and the produced aerosol and the sea water depleted after aerosolization were sampled. The samples were filtered through 0.45 µm pore size Nuclepore membranes. These filters are analysed by PIXE (using the described set-up of INFN Lab. of Legnaro), without any further treatement, and the element concentrations in the suspended particles have been measured. The soluted elements have been determined in targets obtained from liquid sample preconcentration, both by metal precipitation as carbamates, and also by a procedure of not boiling evaporation (Ghermandi et al.1996). The preliminary results are satisfactory: Ti, V, Cr, Mn, Fe, Co, Ni, Zn, Hg, Pb have been detected both in sea water and in aerosol samples, with enrichment factors of 10^3 in the aerosols. The elements detected are mainly associated to the suspended particles. It clearly demonstrates the efficiency of the aerosolization process in the concentration of pollutants, scavenged by the surfactant organic compounds in the subsurface marine layer.

19.7 The Study of the Venice Lagoon Ecosystem by Means of the PIXE Technique

The Venice Lagoon acts as a "reservoir" for municipal, industrial and agricultural wastes; the quality of its water has progressively deteriorated, with resultant damage to the ecosystem (decline in fish population, increase of eutrophication). The complicated water dynamics into the lagoon, whose effects are pronounced in the marginal areas of the northern basin, involve very long residence times of the pollutants.

For what concerns the inorganic pollutants, a constant monitoring and a study of their distribution processes within the Venice Lagoon has been performed by the authors for several years by means of the PIXE technique. About one thousand environmental specimens (water, sediment, pore water) were processed for this study, with the experimental setup performed at the INFN Laboratory of Legnaro (Padova, Italy).

19.7.1 The Fate of Pollutants Discharged into the Lagoon

The first objective was to describe and estimate the yearly discharge of fresh water and pollutants into the Lagoon, from the drainage basin (Bernardi et al.1983; Bernardi et al.1985).

Among the twentyfour identified streams from which water flows out from the drainage basin into the Lagoon, the eight most important natural and agricultural inputs were selected. Field measurement of hydrological and physicochemical parameters (in relation with tidal forcing) and water sample collection have been performed for some years at each of the eight channel outlets.

The investigation was limited to a group of more representative pollutant metals (Fe, Ni, Cu, Hg, Pb), and to some nutrients (ammonia, nitrite, nitrate and total orthophosphate). The heavy metal concentrations were measured by PIXE (Cecchi et al. 1983) in not prefiltered water specimens (soluted and suspended matter); the obtained total yearly inputs in the Lagoon are of 414.1 ton of Fe, 4.05 ton of Ni, 5.5 ton of Cu and 4.6 ton of Pb (Bernardi et al.1983; Bernardi et al.1985).

The evaluation of the mass budget of pollutant substances in the Lagoon requires also the estimation of the pollutant transfer from the Lagoon to the Adriatic Sea. It is really difficult, given the much higher water volume moved during each tidal cycle through the Lagoon inlets with respect to the net fresh-water flux from the drainage basin towards the sea. In addition, the pollutant distribution in the system is greatly affected by the complex processes of interaction among the biogeochemical phases in the Lagoon (surface water, bottom sediment, pore water and also living organisms).

Focusing attention on the distribution processes of heavy metals, a little shallow water area (Cona Marsh) has been identified in the bordering part of the northern Lagoon. This marsh constitutes the estuary of one of the eight fresh-water sources included in the previous study (Dese River). The investigations have been carryied on for some years in this test area, with systematic monitoring of surface water, bottom sediment, pore water and inorganic pollutant evaluation by means of PIXE.

19.7.2
Trace Element Distribution in Surface Bottom Sediment

The trace element distribution in surface bottom sediment (5 cm depth) of the test area was investigated because significant exchange processes take place at the sediment-water interface. In addition a part of the pollutant load may be remobilized from bottom sediments (given external factors that induce turbulence) reaching the water column again. The map of trace element distribution in surface bottom sediment of the Cona Marsh showed considerably lower concentrations near and in front of the mouth of a channel discharging into the marsh, with respect to greater distances or laterally away from the mouth. It is due to the more intense fresh stream from the channel. At both the sides of the mouth, on the contrary, insufficient currents determine the trace metal accumulation. The fine-grained fractions (that are the main carriers of metals among sediment grain sizes) are transferred far from the mouth (Bernardi et al. 1988).

The calculation of concentration gradients of elements among the upper sediment layer, pore water and bottom water allows for the evaluation of the fluxes of substances among these geochemical phases.

19.7.3
The Vertical Profile of Trace Elements in Pore Water

The trace element distribution in pore water, extracted by centrifugation (Iotti 1998) from sediment samples collected in Cona Marsh, has been investigated with the support of the PIXE technique. Pore water sample preparation has been performed in an inert N_2 atmosphere. Both the spatial variability of element concentrations and also their behaviour with depth have been tested in samples collected in sites that represent different conditions in the dynamics of the marsh waters. Several investigations were performed. At first pore water was extracted from two main depths alone: the surface (0–5 cm depth) and subsurface (5–10 cm) layer. Two groups of metals having a different spatial trend are distinguishable at the two sampling depths. Fe and Mn, and generally Cr, are enriched in the surface layer with respect to Cu, Ni and Zn (Zago et al. 1994).

A more detailed study (Cecchi et al. 1996) examined sediment cores (from the sediment-water interface to −34 cm depth) collected in the same sites previously sampled. These cores were sliced in specimens 3 cm thick. The pore water was extracted by centrifugation and analysed for trace metal, Eh, pH, H_2S and Carbon determination from six slices of each core (corresponding to depths of: 0/−3 cm, −3/−6 cm, −6/−9 cm, −13/−16 cm, −19/−22 cm, −31/−34 cm). Fe, Cr, Cu, V, Mn and Ni concentration shows a decrease in the values with the increasing depth, while H_2S, dissolved Carbon and the pH values have an opposite behavior. In addition, the vertical profiles from sites with comparable hydrodynamical characteristics exhibit common features, while other sites (close to the inner part of the marsh) affected by bioturbation and compaction phenomena provide less significant information.

A further investigation (Iotti 1998) has been preformed recently with core collection in three sites of the Lagoon with different environmental characters. The first site is in a small channel ("rio") in the center of the city of Venice. The second site is close to the Giudecca Island, in an area with relevant algae growth. The third one is located at the Malamocco port entrance. The sediment cores extracted from these sites had an average length of about 35 cm. They were sliced in order to obtain a continuous series of specimens 1 cm thick, that all were centrifugated for further PIXE analysis of pore water. AVS, H_2S, FeS_2 and SO_4^{2-} where also determined. The resulting vertical profiles of Fe, AVS, H_2S and SO_4^{2-} and the FeS_2 measurements in the same sediment samples show significantly the transition of iron (abundant in the surface sediment layer) to the sulfide phase. At the same time the sulfate was substituted by sulfide in the solid phase. Given that the sulfide is released more slowly than the metals, the metal concentration peaks generally occurred before (at lower depth) with respect to the peaks of H_2S, released by the solid phase. These resulting behaviours of metals, SO_4^{2-}, and H_2S (whose concentration in solution is very low) are in agreement with other studies about pore water (Carignan and Nriagu 1985).

19.7.4 Trace Elements in Surface Water and the Effect of Fresh and Salt Water Mixing

The surface water quality was analysed from 1984 both in the Cona Marsh and in the Dese River that discharge in the marsh. The seasonal behavior of the pollutant concentration, obtained from total water samples (i.e. soluted plus suspended matter), shows a maximum in June, probably depending on agricultural wastes.

Given that the particle settling is certainly the fundamental process in accumulation phenomena, we have investigated the metal distribution in suspended particulate, in order to evaluate the element flux from the water column toward the sediment. The first analyses were performed in 1988 in the test area. The trace metal concentration in the samples passed through 3 µm and 0.4 µm pore size filters were comparable, and equal to a low fraction of the total metal content. This indicates that the particles larger than 3 µm are mainly responsible for metal transfer in this ecosystem.

Field measurements of chemicophysical parameters, hydrodynamics and sample collection along the whole water column were performed in 1989 in two sites (one in a river and the other at the river mouth in the studied marsh), and completely covered a tide half-cycle. Filtration was performed in the field using 8 µm and 0.4 µm pore size filters. The metal distribution in the separated fractions of the suspended matter, obtained for the two sites, shows that in the marsh the metals are apparently bound to relatively coarse particles, since about one half of the total metal content was stopped by the 8 µm pore size filter, while in the river the metals are mainly carried by grains smaller than 8 µm. It is probable that part of the finer river load forms greater aggregates discharging in the marsh, where the deposition is favoured. In fact the grain-size distribution of the sediment in the northern marsh was centered around diameters a little smaller than 10 µm (Bernardi et al. 1988). Nevertheless, resuspension phenomena induced by water dynamics (Ghermandi et al. 1991) at the river mouth in the marsh are indicated by the comparison between the element concentration in the water column and the bottom current velocity along the axis of the mouth.

The processes occurring during salt- and fresh-water mix are already under investigation: these processes concern at first the soluted fraction in surface water. Two station sites along the Dese River (one site upstream characterized by fresh-water flow at the surface, and one downstream more subjected to the salt wedge) where chosen, and a third in a channel of the marsh, where a slack of hydrodynamics occurs. Physicochemical parameters measurement and water sample collection (with filtration operation in field (Zonta et al.1994)) were performed at the three sites during a tide half-cycle. Water dynamics were continously monitored.

The high concentration of suspended matter strongly affects and masks the behavior of the dissolved fraction (even if filtered samples were analysed). In spite of the difficulty to schematize the relationship between soluted trace metal distribution and physicochemical variables, the results confirm the significant correlation of salinity and pH with the trend of the ions (Ghermandi et al. 1993). The behaviour of both these variables strongly depends on the fresh and salt water mixing affecting the concentration of metals in solution. In addition each ion probably follows its particular pattern. On the other hand, given that the salinity gradient may represent a chemical boundary for dissolved metal circulation, the model of salinity vertical profile along the river and its branches towards the marsh is now under study.

This study demonstrates that the fate of the trace element load discharged in the marginal areas of Venice Lagoon is strongly affected by the complex water circulation and physicochemical processes in the water/sediment column. The main land-lagoon interface zones, as the tested marsh, operate as a filter for the metal load discharged from the drainage basin, and here transformation and accumulation of pollutants occur: the study of the quality of the water/sediment system in this areas is fundamental to the understanding of the future of the Venice Lagoon ecosystem.

References

Aprilesi G, Cecchi R, Ghermandi G, Magnoni G, Santangelo R (1984) Calibration and errors in the detection of heavy metals in fresh and sea waters by PIXE in the ppb-ppm range. Nucl Instr Meth 231 (B3) 1–3:172–176

Bernardi S, Costa F, Vazzoler S, Cecchi R, Ghermandi G, Magnoni G (1983) Transfer of fresh water and pollutants into the Lagoon of Venice. In: Proc. VI C.I.E.S.M. Workshop on Pollution of the Mediterranean, Cannes 1982, pp 73–78

Bernardi S, Costa F, Vazzoler S, Vincenzi Z, Cecchi R, Ghermandi G (1985) Lagoon of Venice: fresh water and pollutant transfer. In: Proc. VII C.I.E.S.M. Workshop on Marine Pollution in the Mediterranean Sea, Lucerna 1984, pp 91–97

Bernardi S, Costa F, Vazzoler S, Zonta R, Cecchi R, Ghermandi G (1988) Preliminary investigation on the distribution of heavy metals in surface sediments of the Cona tidal marsh (Venice Lagoon). Nuovo Cimento 11C 5–6:667–678
Blanchard DC (1975) Bubble scavenging and the water to air transfer of organic material in the sea. Adv Chem Ser 145:360–387
Boutron CF (1990) A clean laboratory for ultralow concentration heavy metal analysis. Fresenius' J Anal Chem 337:482–491
Cahill TA (1975) Ion-excited X-ray analysis of environmental samples. In: Ziegler JF (ed) New uses of ion accelerators. Plenum Press, New York, pp 1–71
Carignan R, Nriagu JO (1985) Trace metal deposition and mobility in the sediments of two lakes near Sudbury, Ontario. Geochim Cosmochim Acta 49:1753–1764
Cecchi R, Costa F, Galli G, Ghermandi G, Magnoni G (1983) Transfer of heavy elements into the sea: Detection by proton induced X-ray emission (PIXE). In: Proc. VI C.I.E.S.M. Workshop on Pollution of the Mediterranean, Cannes 1982, pp 349–353
Cecchi R, Ghermandi G, Calvelli G (1987) Ultratrace PIXE analysis in water with low selective metal preconcentration at various pH. Nucl Instr Meth B 22:460–464
Cecchi R, Ferrari MG, Ghermandi G, Rossi M (1990a) Spectre: A program for ultratrace PIXE analysis of thin targets. Nucl Instr Meth B 49:146–151
Cecchi R, Ghermandi G, Zonta R (1990b) PIXE analysis to study metal diffusion and sedimentation phenomena in Venice Lagoon. Nucl. Instr. Meth. B 49:283–287
Cecchi R, Ghermandi G, Zonta R (1996) Geochemical processes in Venice Lagoon by PIXE technique: an overview. Nucl Instr Meth B 109/110:407–414
Cini R, Desideri P, Lepri L (1994) Transport of organic compounds across the air/sea interface of artificial and natural marine aerosols. Anal Chim Acta 291:329–340
Folkmann F (1976) In: Meyer, Linker, Kappeler (eds) Ion beam surface layer analysis. Plenum Press, New York, pp 695–718
Ghermandi G, Cecchi R, Costa F, Zonta R (1991) Trace metal distribution in aquatic systems as studied by PIXE analysis of water and sediment. Nucl Instr Meth B 56–57:677–682
Ghermandi G, Campolieti D, Cecchi R, Costa F, Zaggia L, Zonta R (1993) Trace metal behaviour during salt and fresh water mixing in the Venice Lagoon. Nucl Instr Meth B 75:330–333
Ghermandi G, Cecchi R, Laj P (1996) Procedures of target preparation to improve PIXE efficiency in environmental research. Nucl Instr Meth B 109/110:63–70
Iotti F (1998) Caratterizzazione delle acque interstiziali estratte da sedimenti di fondale della Laguna di Venezia. Thesis, Modena University (Italy)
Johansson SAE, Campbell JL (1988) P.I.X.E: A novel technique for elemental analysis. John Wiley & Sons, Chichester
Johansson SAE, Campbell JL, Malmqvist KL (1995) Particle-induced x-ray emission (P.I.X.E). John Wiley & Sons, New York
Luke CL (1986) Determination of trace elements in inorganic and organic materials by x-ray fluorescence spectroscopy. Anal Chim Acta 41:237–250
Maenhaut W, Hillamo R, Makela T, Jaffrezo J-L, Bergin MH, Davidson CI (1996) A new cascade impactor for aerosol sampling with subsequent PIXE analysis. Nucl Instr Meth B 109/110:482–487
Resch F (1986) Oceanic air bybbles as generators of marine aerosol. In: Monham G, MacNiocaill G (eds) Oceanic whitecaps and their role in air-sea exchange processes. Reidel, Dordrecht, Boston, pp 101–112
Valkovic V (1983) Sample preparation techniques in trace elements analysis by X-ray emission spectroscopy. IAEA-TECDOC-300, Vienna
Zago C, Costa F, Zaggia L, Zonta R, Ghermandi G, Cecchi R (1994) Heavy metals distribution in pore water of the Cona marsh: A preliminary investigation on surface sediment. Nuovo Cimento 17C 4:503–510
Zonta R, Cecchi R, Costa F, Simionato F, Ghermandi G (1994) A filtration system for the size separation of fresh-water samples. Sc Total Environ 143:163–172

Chapter 20

Potentiometry for the Study of Acid-Base Properties of Sediments

V. Zelano · M. Gulmini

20.1 Introduction

The term sediments generally refers to both deposited solids and suspended ones: the latter are separated from dissolved components by filtration through a 0.45 μm membrane (Förstner 1989).

A sediment can be considered as a heterogeneous mixture of dissimilar particles, which are themselves a complex assemblage of different inorganic and organic components. Contaminants can be transferred from the water column onto the sediment through chemical and biological processes; sediment particles can then undergo a series of transportation, settling, resuspension and deposition, each accompanied by several chemical reactions which can result in a release of the associated pollutants. The sediment can therefore act both as a sink and as a non-point source for potentially toxic compounds or ions.

As a consequence, the need to understand the interaction that can take place between contaminants and sediments in natural water systems was brought to the attention of environmental chemists as a tool to determine pollutants' availability and to study the possibility of their release due to physicochemical modifications of the sediment sphere.

The general removal of a solute from the solution phase is called sorption. For heavy metals, the incorporation into the sediment is not generally due to precipitation, since their concentration is very low, and their solubility product is not reached. Therefore, their sorption onto the sediment is due to chemical and physical processes, which can be summarised as follows (Tessier and Campbell 1988):

- Adsorption and ionic exchange of ions bonded to particle surface, particularly onto clays, iron and manganese oxides and organic particles. This process is promoted by an increase of the pH value and a decrease of the ionic strength.
- Incorporation of cations into the crystal reticulate of secondary minerals such as carbonates and sulfides. This process is due to sulfate biological reduction and natural carbonation; it is promoted by an increase of pH and by favourable environmental conditions for biological activity.
- Coprecipitation with iron and manganese hydrous oxides. A decrease in the acidity and the oxidating conditions increase this possibility because the most insoluble oxides are the cations with an increased state of oxidation [Fe(III) and Mn(IV)].
- Incorporation in the crystal of the primary minerals, especially silicates created in the diagenesis processes.
- Complexation with inorganic and organic particles of biological or anthropical origin.

Hereafter we will deal with the last point, i.e. the formation of coordination bonds at the solid-water interface, with particular attention to the interaction between ions and protogenic groups on the sediment particles surface. The potentiometric approach for the characterisation of acid-base and ligand properties of sediment particles will be discussed.

20.2 Sediment: What is it?

Deposited sediments originate both from the settling of suspended particles and from the chemical and biological processes that take place on the sediment surface. Although sediment particles can be fairly different from one another, one can picture each particle as "onion-like": a matrix vehicle, generally siliceous, is covered with an array of compounds, such as carbonates, amorphous aluminosilicates, organic matter and iron and manganese oxides (Martin et al. 1987).

To simplify the discussion, it can be of great help to subdivide sediment components into inorganic and organic ones.

The main inorganic components, minerals or amorphous, are silicates (clay and non-clay), carbonates, iron and manganese oxides, phosphates and sulfides. Silicates derive mostly from fragmented rocks that are transported into the basin by wind and rain. Other inorganic components, namely carbonates, iron and manganese oxides and phosphates are formed by chemical (precipitation) and biological processes. Sulfides form on the bottom, anoxic layer of the sediment.

The organic component of the sediment derives from living processes of benthic organisms, mostly from accumulation and complex transformation of natural organic matter from the overlaying water (*NOM*).

NOM is a very complex mixture of compounds; because of the diversity of the natural processes of synthesis and degradation, the number of constituents of *NOM* is potentially infinite, and most of them have not been identified yet. Although there is little hope of completely separating and identifying each component of the *NOM*, a great amount of characterisation work has been completed. This work was reviewed in an excellent book by Buffle (Buffle 1988).

After sedimentation, the *NOM* from the water column can undergo significant modifications, since the chemical conditions in the sediment are very different from those immediately above the solid. The biochemical conditions existing at the sediment-water interface can be extremely variable, therefore it is very difficult to establish general rules concerning the composition of sediment *NOM*. In general, proteins, polysaccharides and amino-sugars are present, with a great variability in functional groups and molecular weight distribution.

To investigate the nature and role of the organic components of the sediment, they must be brought into solution by extraction from the solid, inorganic phase. The extract, that generally represents only a portion of the total organic content of the sediment, may be partitioned into fulvic and humic fractions, which are operationally defined on the basis of their different solubility in acids and bases. Taking into account the complexity of the initial mixture, humic and fulvic fractions are themselves heterogeneous and their composition may depend on the extraction condition used. Despite its complexity, a probable picture of the organic component of the sediment results in a three-dimensional, flexible structure of aliphatic, peptide and polysaccharide chains, containing a wide variety of oxygen, sulfur and nitrogen functional groups. These large, polyfunctional compounds are then adsorbed onto inorganic particles such as clays, oxides or carbonates.

In fact, a clear distinction between inorganic and organic particle components does not really exist, because of the close interactions that develop among the various sediment phases: for instance, a bit of silica covered with a thin film of humic material is considered essentially inorganic, but its properties will be those of the organic cover.

20.3 The Solid-Water Interface

The sediment particles are therefore covered with active sites that may adsorb protons, hydroxide ions, metal ions and ligands of the water column.

Until recently the adsorption phenomena were described using adsorption isotherms (Langmuir, Freundlich and other more complex ones), which give a useful indication of the capacity of retention of the surfaces, but the ideas upon which they were based did not permit the extension of their application to experimental situations which differed from the original ones (Davis and Kent 1990). For example, the Langmuir approach for adsorption of solute *C* onto surface *S*, develops as follows (Detenbeck and Brezonic 1991): taking into consideration the generic adsorption reaction

$$\mathrm{S} + \mathrm{C} \rightleftharpoons \mathrm{SC}$$

(S is a superficial adsorbent site) the mass law is expressed as

$$\frac{[\mathrm{SC}]}{[\mathrm{S}][\mathrm{C}]} = K_{\mathrm{ads}} = \exp\left(-\frac{\Delta G^0_{\mathrm{ads}}}{RT}\right)$$

and the molar balance for the superficial sites is

$$[\mathrm{S}]_{\mathrm{Tot}} \rightleftharpoons [\mathrm{S}] + [\mathrm{SC}]$$

By combining the two equations, we obtain

$$[\mathrm{SC}] = [\mathrm{S}]_{\mathrm{Tot}} \frac{K_{\mathrm{ads}}[\mathrm{C}]}{1 + K_{\mathrm{ads}}[\mathrm{C}]}$$

By defining $\Gamma = [\mathrm{SC}]$ / solid mass and $\Gamma_{\max} = [\mathrm{S}]_{\mathrm{Tot}}$ / solid mass and by substituting it in the preceding equation, the expression of the Langmuir isotherm is obtained:

$$\Gamma = \Gamma_{\max} \frac{K_{\mathrm{ads}}[\mathrm{C}]}{1 + K_{\mathrm{ads}}[\mathrm{C}]}$$

In recent years, the knowledge of the composition of the surface of sediment particles suggest treating their coordinating groups, both inorganic and organic, like their corresponding counterparts in solution, with a so-called surface complexation model (SCM) (Stumm et al. 1976; Davis et al. 1978; Schindler 1981; Davis and Kent 1990; Stumm 1992).

According to this model, reaction between solutes and functional groups on the solid surface are coordination reactions, which undergo the same concepts and mathematics of complexation reactions among solutes. All the surface ligand sites of the same type are considered identical, and the surface- and solution-phase interactions can be described by means of a set of linked equations, one for each reaction that takes place at the solid-solution interface. (Lekie 1988; Morel and Hering 1993; Stumm 1998)

Examples of surface reactions on oxides are:

- Acid base equilibria

$$\text{S-OH} \rightleftharpoons \text{S-O}^- + \text{H}^+ \qquad K = \frac{[\text{S-O}^-][\text{H}^+]}{[\text{S-OH}]}$$

$$\text{S-OH}_2^+ \rightleftharpoons \text{S-OH} + \text{H}^+ \qquad K = \frac{[\text{S-OH}][\text{H}^+]}{[\text{S-OH}_2^+]}$$

- Metal binding (the -OH-group may coordinate the metal ions from the solution)

$$\text{S-OH} + \text{Cu}^{2+} \rightleftharpoons \text{S-OCu}^+ + \text{H}^+ \qquad K = \frac{[\text{S-OCu}^+][\text{H}^+]}{[\text{S-OH}][\text{Cu}^{2+}]}$$

$$2\ \text{S-OH} + \text{Cu}^{2+} \rightleftharpoons (\text{S-O})_2\text{Cu} + 2\ \text{H}^+ \qquad K = \frac{[(\text{S-O})_2\text{Cu}][\text{H}^+]^2}{[\text{S-OH}]^2[\text{Cu}^{2+}]}$$

$$\text{S-OH} + \text{Cu}^{2+} + \text{H}_2\text{O} \rightleftharpoons \text{S-OCuOH} + 2\ \text{H}^+ \qquad K = \frac{[\text{S-OCuOH}][\text{H}^+]^2}{[\text{S-OH}][\text{Cu}^{2+}]}$$

Similarly, the metal ion of the oxide acts like a Lewis acid, and as such may substitute the hydroxyl group with other basic groups:

- Ligand exchange

$$\text{S-OH} + \text{L}^- \rightleftharpoons \text{S-L} + \text{OH}^- \qquad K = \frac{[\text{S-L}][\text{OH}^-]}{[\text{S-OH}][\text{L}^-]}$$

Many other surface species may be formed this way, including ternary and mixed complexes:

- Ternary surface complex formation

$$\text{S-OH} + \text{L}^- + \text{Cu}^{2+} \rightleftharpoons \text{S-L-Cu}^{2+} + \text{OH}^- \qquad K = \frac{[\text{S-L-Cu}^{2+}][\text{OH}^-]}{[\text{S-OH}][\text{L}^-][\text{Cu}^{2+}]}$$

Although the surfaces of aquatic particles contain functional groups whose acid-base or metal binding properties are similar to those of their counterparts in dissolved ligands, their behaviour toward coordinative properties can be different and difficult to understand due to the geometric restrictions imposed by the solid nature of surface. We can picture the functional groups on the sediment particle as if they were part of a flat, impenetrable surface, that can show an electrical charge due to chemical reactions or crystalline defects or adsorption of ionic surfactants. The electrical status of the surface may influence the reactions that can take place on the surface and, as a consequence, the related equilibrium constants. As such, the previously indicated constants should take into consideration a further term to take into account this electrostatic effect. Such a corrective term is expressed by considering the surface charge as a double electrical layer: the first stratum is modelled as a fixed charge on the particle surface, while the second one is diffused in the solution upon the surface (Gouy-Chapman Model) (Leckie 1988; Lumsdon and Evans 1995).

In agreement with the Gouy-Chapman theory, after having measured the surface charge density σ (C m^{-2}), we may calculate the surface potential Ψ (volt) using the equation:

$$\sigma = \sqrt{8RT\varepsilon\varepsilon_0 c10^3}\,\sinh\left(\frac{Z\Psi F}{2RT}\right)$$

where R = molar gas constant (8.314 J mol^{-1} K^{-1}); T = absolute temperature (K); ε = water dielectric constant (78.5 at 25 °C); ε_0 = vacuum permittivity (8.854×10^{-12} C J^{-1} m^{-1}); c = electrolyte molar concentration (M); Z = ion charge.

In order to consider the electric component, adsorption free energy ΔG^{o}_{ads} can be subdivided into an intrinsic free energy (ΔG^{o}_{int}) and a coulometric term (ΔG^{o}_{coul})

$$\Delta G^{0}_{ads} = \Delta G^{0}_{int} + \Delta G^{0}_{coul}$$

The electrostatic component is expressed as $\Delta G^{o}_{coul} = \Delta ZF\Psi_S$, where ΔZ is the variation of the charge of the surface species and Ψ_s is the surface electric potential (volt). Adsorption free energy is defined as

$$\Delta G^{0}_{ads} = -RT \ln K^{app}$$

where K^{app} is the apparent surface acidity constant.

By combining the previous equations, we obtain:

$$-RT \ln K^{app} = -RT \ln K^{int} + \Delta ZF\Psi_s$$

Substituting the expression for K^{app}, we obtain:

$$K_1^{int} = \frac{[\text{S-OH}][\text{H}^+]}{[\text{S-OH}_2^+]} \exp\left(-\frac{\Psi_s F}{RT}\right)$$

$$K_2^{int} = \frac{[\text{S-O}^-][\text{H}^+]}{[\text{S-OH}]} \exp\left(\frac{\Psi_s F}{RT}\right)$$

The exponential term may be considered as an activity coefficient (Dzombak and Morel 1990) which takes into account the effects of the surface charge on the apparent acidity constants.

Because of the very complex composition of sediments, the particle surfaces may have a wide variety of sorption characteristics; however, most of the particles are covered with a limited number of compounds, namely amorphous iron hydroxides and *NOM* (Lion et al. 1982; Luoma and Davis 1983; Tipping et al. 1990; De Wit et al. 1993a; De Wit et al. 1993b), which largely determine the complexing capacity of the solid, even if their relative abundance in the total sediment composition is rather low.

This prompted the study of sorption properties of well-characterized solid phases, in order to model the interaction between the sediment and the dissolved compounds as a sum of discrete contributes of the solid components. Particular attention was given to Fe and Mn oxides (Stumm et al. 1980; Benjamin and Leckie 1981; Farley et al. 1985) and organic matter (Ephraim and Marinsky 1986; De Wit et al. 1990; Puls et al. 1991; Masini 1993; Lead et al. 1994). Studies on oxides were generally performed with synthetic material and the properties of organic matter were determined after their extraction from natural sediment and purification.

Such an approach however, does not consider that the properties of a single component may be influenced by the presence of other components (Honeyman and Santschi 1982) and the extraction of a solid component from its natural matrix, i.e. extraction of organic matter (Templeton and Chasteen 1980), may lead to its modification, hence the model may not be representative of the real sediment behaviour.

A better approach could be the direct study of natural sediment ligand properties with as little manipulation as possible. For this reason, the interaction between ions and protogenic ligand groups on the sediment surface can be studied by alkalimetrically titrating suspensions containing the water-sediment system, carried out with potentiometric measurements. This technique still has a wide application in the study of the acid-base and ligand properties of a vast range of soluble compounds and it also appears promising for the study of heterogeneous systems (Gulmini et al. 1996; Gulmini et al. 1997). Although the method is only qualitatively correct at a molecular level, i.e. it is not sensitive to the detailed structure of the interfacial region and it does not distinguish between structurally different groups of the same type, it is possible to correctly describe proton-sediment interaction by curve-fitting, and, in some instances, it can be demonstrated that the sorption of metal ions is mainly due to interaction with the protogenic sites of the sediment surface.

20.4 Titration Procedure

Since the titration is performed on a heterogeneous system, we must take into consideration some procedural aspects in order to correctly carry out the titration.

As previously mentioned, the carbonates are a relevant component of the sediments, and their presence interferes with the titration. It is for this reason that it is preferable to eliminate them beforehand with a pH 4 acid treatment.

A second relevant aspect is the homogeneity of the sediment, upon which depends the possibility of performing reproducible titration on different subsequent aliquots. It is for this reason that it is better to use sediments with controlled granulometry, for instance, selecting the granulometric fraction <63 μm. This fraction is suggested for environmental analyses in the literature (Förstner 1989) because it has a similar composition as that of the suspended particulate, which is the primary cause of the transport of the pollutants within the water systems. It also presents a better surface/weight rate in comparison with the unsieved sediment and remains more easily dispersed in the solution.

As an index of the homogeneity of the sample, we may use the relative standard deviation percentual values (RSD%) calculated from the mean values of the parameters used in sediment characterization (metals, elemental analyses). Using as a reference the RSD% value obtained by the analysis of certified sediments, we may consider that the homogeneity of the sample is satisfactory when the RDS% value calculated from 5 replications is less than 2%.

To perform titrations, the pretreated sediment is placed in suspension in a solution containing HCl 8×10^{-3} M with a controlled ionic strength ($0.01 \le I \le 1$ M KCl). The solid is maintained in suspension by a magnetic stirrer and the temperature is kept constant at 25.0 ±0.1 °C by water circulation in the outer chamber of the titration cell. For each titration, different solid solution ratios, namely from 0.25–2% w/v were considered. The suspensions are then titrated from pH 2–10 with standard carbonate-free KOH. The same HCl solution is used to calibrate the electrodes (a glass electrode and a silver/silver chloride reference half-cell) in hydrogen ion concentration units. The experimental data are collected using an automatic titrator which includes an electrometer and a computer-controlled motor-driven piston burette. The titration program allows the evaluation of equilibrium potential values and determines the amount of titrant based on the actual buffering properties of the titrated solution, so that there is a difference in pH values of 0.05–0.08 between two successive readings. The EMF is considered to be stable when the variation is less than 0.05 mV min^{-1}.

Normally, for pH < 7, a constant signal is reached within 3 to 4 min with good reproducibility; in the alkaline region instead, a longer time is needed, and it is dependent upon both the pH as well as the quantity of the sediment: obviously, the most critical pH region is the unbuffered one between 7 and 10, and more time is needed for higher quantities of suspended sediment. For example, for an equal quantity of sediment (0.25%), a few minutes are needed when the pH is more than 11 in order to reach the equilibrium, while more than 80 min are needed in the titration zone with pH less than 10. This trend is shown in Fig. 20.1, where ΔE is the difference between the EMF measured at time t and that measured at time t_0, immediately after the addition of titrant.

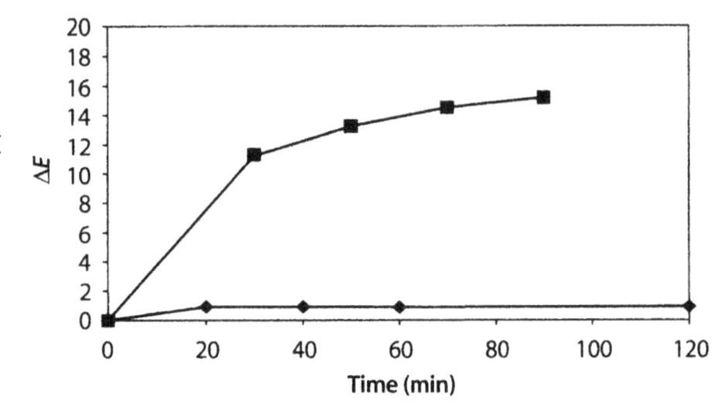

Fig. 20.1. Effect of pH on the equilibration time for 0.25% of suspended sediment. *Squares* refer to an initial pH value of 9.7, *rhombs* refer to an initial pH value of 11.1

The effect of the quantity of the sediment is seen in Fig. 20.2.

By correctly choosing the titration conditions, it is possible to reach the equilibrium for all the titration points and to obtain reproducible curves (Fig. 20.3).

By superimposing the titration graph with that of the reference solution (calibration), a first indication regarding sediment interaction with the acidic solution is obtained (Fig. 20.4). The difference between the curves depends upon the specific sediment properties, and it is proportional to the suspended quantity.

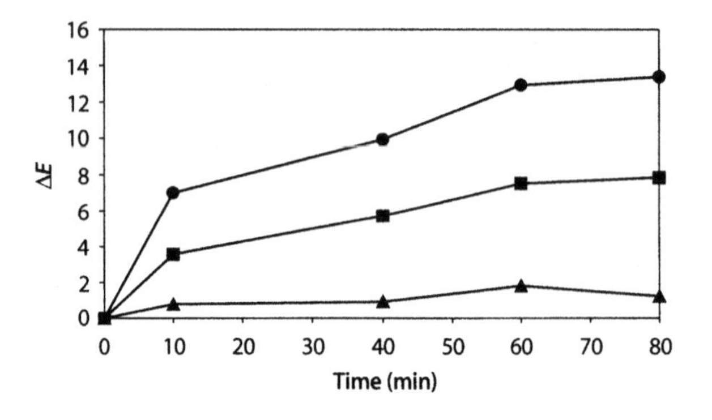

Fig. 20.2. Effect of suspended sediment quantity on the equilibration time for the initial pH value of 10.5. *Circles* refers to 1.5% of suspended sediment, *squares* refer to 1% and *triangles* refer to 0.25%

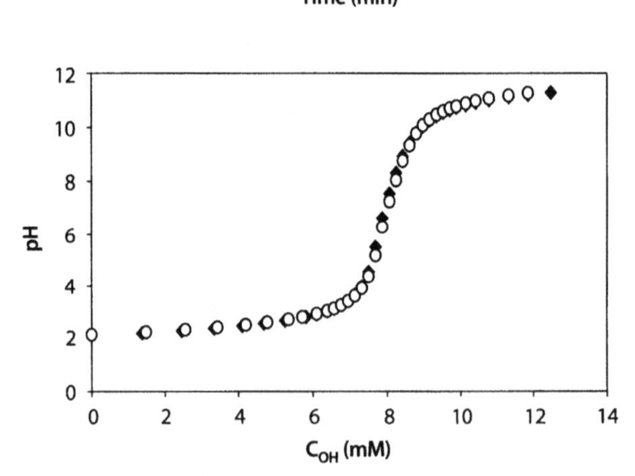

Fig. 20.3. Replicates for the titration of 1.25% of suspended sediment

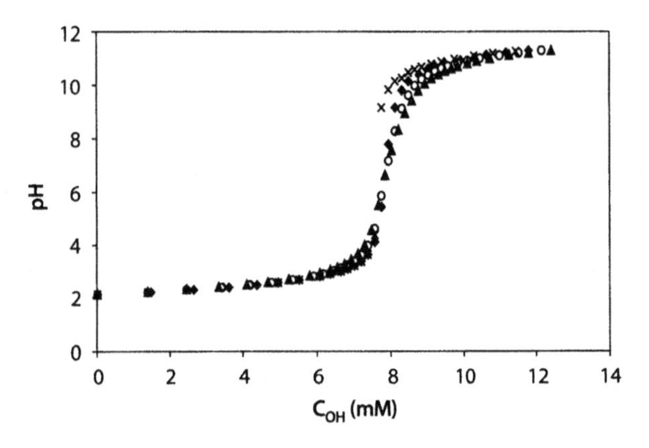

Fig. 20.4. Comparison among titration curves for increased quantities of suspended sediment. *Crosses*: 0%-reference; *rhombs*: 0.5%; *circles*: 1%; *triangles*: 1.25% of suspended sediment

Of the samples studied thus far, regardless of their origins, it was found that when less than 0.25% of suspended sediment was used, the titration curve does not differ from the reference curve, and therefore, this quantity is considered to be the minimum amount required for the study. On the other hand, for concentrations over 2%, the solid tends to deposit and does not remain homogeneously suspended in the solution.

The same titration procedure may be used to study the sediment interaction with metal ions. In this case, the initial solution also contains, in addition to strong acid and ionic medium, a known concentration of the ion under study.

20.5 Elaboration of the Potentiometric Data

The concentration of titratable sites present is unknown, but may be calculated from the data of the alkalimetric titrations on the solution-sediment system. This system is, in fact, a mixture of a strong free acid (HCl) and of a weak acid (sediment). The titration data may be elaborated to put into evidence the two inflection points, one of the strong acid (using the Gran plot method), and the other of total acid (derivative of the titration curve). The difference between the two values corresponds to the concentration of the acidic sites on the sediment.

To select the models which best explain the trend of the experimental data for the protonation and the complexation between the sediment and the copper(II) ion, the statistic software BSTAC (De Stefano et al. 1993) may be used. The software employs an iterative and convergent numeric method which is based upon the linear combination of the mass balance equations. Several chemical models, i.e. stoichiometry and stability constants of species that are possibly involved in the equilibrium, are loaded into the software to perform calculations. On the basis of the given models, the software indicates the most probable species and calculates their formation constants by minimising the sum of squared errors between calculated and experimental EMF values. Since there is a lack of information regarding titratable group concentration, this parameter is refined together with the protonation constants, and the calculated values are in good agreement (within 5%) with those calculated by the Gran plot and derivative curve.

The ionic strength and the ionic medium employed may influence, even significantly, the interaction between the ions in solution and the surface active centres, due to the nature of the sediment surface charges. In the elaboration of potentiometric data, we assume that the dependence on ionic strength of the formation constants for the interaction between hydrogen or copper(II) ions and the surface-active centres can be modelled by adopting the following equation:

$$\log\beta(I) = \log\beta(I') - z^*\left(\frac{\sqrt{I}}{2+3\sqrt{I}} - \frac{\sqrt{I'}}{2+3\sqrt{I'}}\right) + C(I - I') + D(I^{1.5} - I'^{1.5})$$

- $C = 0.10p^* + 0.23z^*$
- $D = -0.10z^*$
- $p^* = \Sigma p_{\text{reagents}} - \Sigma p_{\text{products}}$
- $z^* = \Sigma z^2_{\text{reagents}} - \Sigma z^2_{\text{products}}$

that is usually employed for homogeneous solutions (Daniele et al. 1985; Daniele et al. 1991); p and z are the stoichiometric coefficients and charges, respectively, $\beta(I')$ the formation constant at the selected reference ionic strength. In this case, the ligand corresponds to the donor groups present in each surface-active centre, and eventual reciprocal influence among the charges of donor group is neglected.

Ionic strength seems to play a marginal role in the reactions between sediment and solutions, and when some effects are seen, they are of the same size as those in homogeneous solutions. Figure 20.5 is such an example: titration curves of the same amount of sediment (0.75%) titrated at different ionic strengths (0.01 and 1 M) are shown.

This behaviour differs from that of inorganic oxide, whose titration curves are very different for different ionic strengths (Buffle 1988).

These results have prompted an elaboration of the data assuming that each donor group acts as a free ligand in solution and its charge is not significantly affected by charges present in the neighbouring active centres. The reliability of the foregoing assumption, even if a remarkable simplification in the interpretative model is introduced, is confirmed by the computer analysis of the data, since a good fit is obtained by simultaneously computing all the potentiometric readings, at 0.01, 0.1 and 1.0 M ionic strength. Of course, in the application of the chemical model for the dependence of formation constants on ionic strength, it also takes the ion-pair formation between either K^+ and negatively charged form(s) or chloride and positively charged form(s) of active centres into consideration. For instance, if the chemical model involves two types of surface-active centres, L_AH and $L_BH_2^+$, the formation of following ion-pair are considered:

$$K^+ + L_A^- \rightleftharpoons KL_A$$

$$K^+ + L_B^- \rightleftharpoons KL_B$$

$$L_BH_2^+ + Cl^- \rightleftharpoons L_BH_2Cl$$

First, the experimental data are elaborated in order to define the protonation constants and to evaluate the concentration of titratable sites present. After having defined the interaction with the proton, the data recorded on the sediment-proton-copper(II) system are elaborated to determine the complexes formation constants.

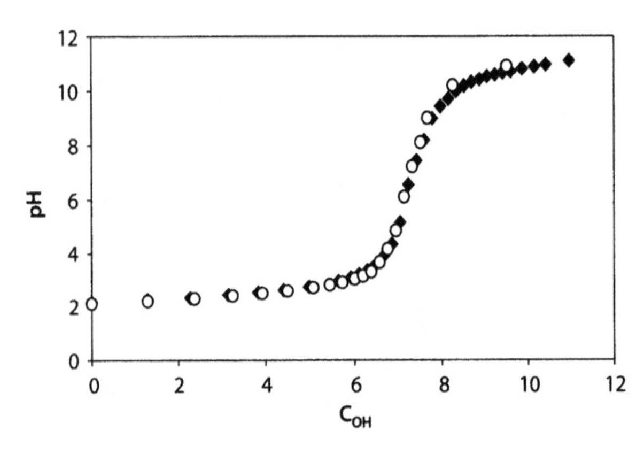

Fig. 20.5. Comparison between titration performed for 0.75% of suspended sediment at ionic strength of 0.01 M (*circles*) and 1 M (*rhombs*)

20.6 Results of the Potentiometric Approach

With the previously mentioned procedure some river and lake sediments were studied. The protonation constants and the copper(II) complexation constants were calculated, and the concentration of the superficial titratable sites were also estimated. The selected models for the sediments examined, indicated by a code, are reported in Table 20.1.

For instance, the computation for the sediment EN shows that all the potentiometric data are well interpreted by assuming the presence of two sediment components: one (L_A) contains the groups titratable in the acid region, and the other (L_B) the groups titratable in the alkaline one, which means that the acid-base properties of the sediment can be modelled as a mixture of a monoprotic and a diprotic acid. The concentration of the ligand sites L_A and L_B are expressed as moles of sites for gram of dry sediment.

Table 20.1. Formation constants for proton/sediment and for proton/copper(II)/sediment systems and estimate of the ligand sites (L) concentration on the sediment particle surface

Sediment	Equilibrium	Log *K*	Concentration (mol g^{-1})
MA	$H^+ + L_A^- \rightleftharpoons HL_A$	3.85 ±0.09	$L_A = 9.9 \times 10^{-5}$
	$H^+ + L_B^- \rightleftharpoons HL_B$	9.81 ±0.09	$L_B = 6.4 \times 10^{-5}$
	$H^+ + HL_B \rightleftharpoons H_2L_B^+$	7.16 ±0.04	
EN	$H^+ + L_A^- \rightleftharpoons HL_A$	5.05 ±0.06	$L_A = 2.63 \times 10^{-4}$
	$H^+ + L_B^- \rightleftharpoons HL_B$	10.17 ±0.07	$L_B = 5.05 \times 10^{-4}$
	$H^+ + HL_B \rightleftharpoons H_2L_B^+$	8.23 ±0.04	
	$L_B^- + Cu^{2+} \rightleftharpoons L_BCu^+$	9.20 ±0.08	
VG	$H^+ + HL_B \rightleftharpoons H_2L_B^+$	4.05 ±0.03	$L_A = 3.73 \times 10^{-5}$
	$H^+ + L_B^- \rightleftharpoons HL_B$	7.89 ±0.14	$L_B = 4.38 \times 10^{-5}$
	$H^+ + L_C^- \rightleftharpoons HL_C$	9.87 ±0.07	$L_C = 5.54 \times 10^{-5}$
	$L_AH + L_BH + Cu^{2+} \rightleftharpoons L_AL_BCu + 2H^+$	13.35 ±0.27	
SM	$H^+ + L_A^- \rightleftharpoons HL_A$	4.38 ±0.07	$L_A = 4.59 \times 10^{-5}$
	$H^+ + L_B^- \rightleftharpoons HL_B$	10.00 ±0.13	$L_B = 2.00 \times 10^{-4}$
	$HL_A + Cu^{2+} \rightleftharpoons L_ACu^+ + H^+$	2.78 ±0.08	
	$HL_B + Cu^{2+} \rightleftharpoons L_BCu^+ + H^+$	7.33 ±0.05	
	$HL_A + HL_B + Cu^{2+} \rightleftharpoons L_AL_BCu + 2H^+$	10.53 ±0.10	
TE	$H^+ + L_A^- \rightleftharpoons HL_A$	3.66 ±0.05	$L_A = 4.21 \times 10^{-5}$
	$H^+ + L_B^{2-} \rightleftharpoons HL_B^-$	10.89 ±0.18	$L_B = 1.29 \times 10^{-4}$
	$H^+ + HL_B^- \rightleftharpoons H_2L_B$	8.82 ±0.27	
	$L_AH + L_BH_2 + Cu^{2+} \rightleftharpoons CuL_AL_B^- + 3H^+$	14.45 ±0.07	

The validity of these models and of their related constants is supported by the good level of agreement between experimental and calculated potentiometric data. As an example, Table 20.2 reports the differences between calculated (E_{calc}) and experimental (E_{exp}) EMF values for some titrations of proton/sediment systems previously considered.

For proton/copper(II)/sediment systems, by using the concentration of ligand sites, the protonation and the complexation constants that were calculated by refining pH-metric data, it is possible to calculate the concentration of the copper(II) in solution for each pH value. These calculated values can be compared with those experimentally determined by FA/AAS for each titration point and the comparison may be considered a general validation for the potentiometric method used.

Should the concentration value determined by AAS of copper(II) in solution be inferior to that calculated from the potentiometric models, this would be an indication of the presence of non-protogenic sites which interact with the copper(II) ions. On the other hand, significant discrepancies in the opposite sense would indicate that the models deduced by the potentiometric method do not describe with reasonable accuracy the real interaction between the sediment and the investigated ions.

Table 20.2. Differences between calculated and experimental EMF values for some titrations taken as an example from those of MA and VG sediments

% w/v (Sediment)	pH_{mes}	$E_{calc}-E_{exp}$ (mV)	% w/v (Sediment)	pH_{mes}	$E_{calc}-E_{exp}$ (mV)
0.25	2.33	−1.22	1.5	2.36	1.76
(MA)	2.55	−0.56	(MA)	2.64	0.54
	2.79	−0.08		2.95	0.20
	3.02	0.24		3.29	−0.32
	3.31	0.24		3.70	−0.56
	3.89	0.05		5.51	0.44
	7.97	0.00		6.94	1.63
	9.45	1.01		8.52	−1.76
	10.06	0.42		9.70	0.07
	10.46	0.39		10.32	−0.38
1	2.82	−0.05	1.25	2.83	0.21
(VG)	2.91	0.02	(VG)	2.93	0.29
	3.03	0.18		3.04	0.23
	3.12	0.16		3.14	0.26
	3.25	0.19		3.27	0.18
	3.41	0.07		3.43	0.05
	3.63	−0.04		3.66	−0.03
	3.98	−0.07		3.98	−0.10
	4.59	−0.29		4.48	−0.82
	7.09	0.31		7.50	−0.41
	8.02	−0.44		8.13	−1.79
	8.87	−0.17		8.83	−1.26
	9.52	0.86		9.42	0.22
	9.89	0.37		9.78	0.2
	10.15	0.15		10.04	−0.26
	10.49	−0.39		10.41	−0.02
	10.61	−0.29		10.54	0.10
	10.70	−0.08		10.64	0.49
	10.82	−0.01		10.73	0.93

In reality, by comparing the potentiometric data with the spectrometric ones, a good agreement is obtained for all the sediment considered till now. For instance, Fig. 20.6 and Fig. 20.7 report the concentration of copper(II) not bonded to the sediment determined by FA/AAS and that is calculated from the potentiometric data for EN and VG sediments respectively.

The above expounded procedure, that considers the sediments as a mixture of simple ligands in solution, is able to correctly explain the trend of the potentiometric experimental data recorded on different solid solution ratios, both for proton/sediment and for proton/copper(II)/sediment systems. As a consequence, the calculated formation constants have a good level of accuracy, with low values of the related standard deviations. Moreover, the good agreement between copper(II) concentration in solution calculated with the models selected with the potentiometric technique and that are determined by AAS indicates both that the potentiometric procedure is successful in characterizing correctly and satisfactorily the titratable sites of the sediment particle surface, and that these sites are the main agents for the interaction between sediment and copper(II) in the pH range below metal precipitation.

An overview of the protonation constants reported in literature for humic and fulvic material of various origins (Buffle 1988; Masini 1993) and for Fe and Al hydrous oxides (Stumm et al. 1980) demonstrates that the values calculated for the considered

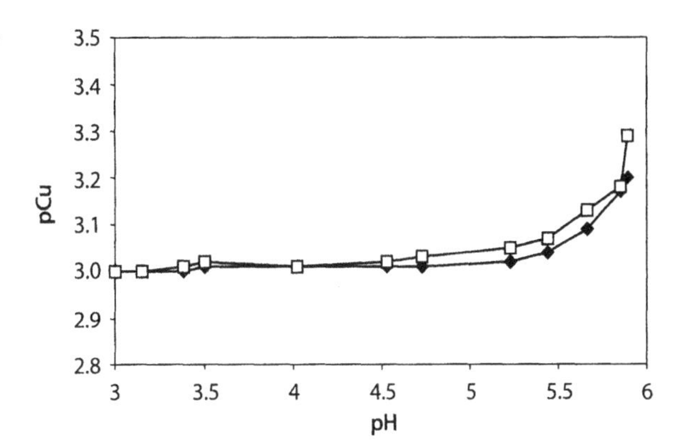

Fig. 20.6. Concentration of the non-bonded copper(II) for the EN sediment (0.5% of suspended sediment). *Squares*: values determined by AAS; *rhombs*: values calculated with the models selected by potentiometric titrations

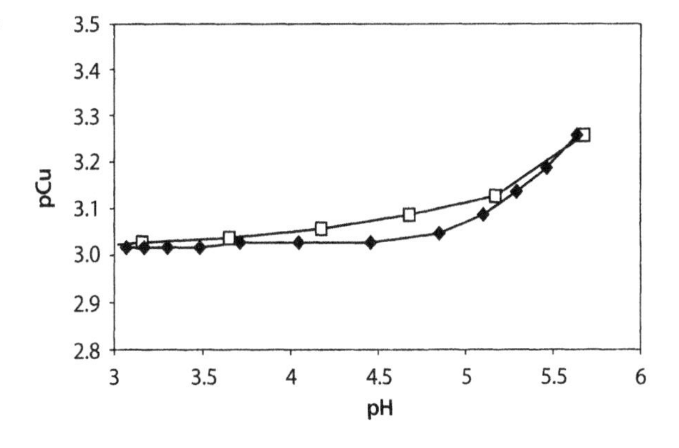

Fig. 20.7. Concentration of the non-bonded copper(II) for the VG sediment (0.5% of suspended sediment). *Squares*: values determined by AAS; *rhombs*: values calculated with the models selected by potentiometric titrations

sediments are in the range indicated for the assignment of the constants to natural organic and inorganic titratable groups.

References

Benjamin MM, Leckie JO (1981) Multiple site adsorption of Cd, Cu, Zn and Pb on amorphous iron oxyhydroxide. J Coll Inter Sci 79:209–221
Buffle J (1988) Complexation reactions in aquatic system. An analytical approach. Ellis Hornwood, Chichester
Daniele PG, De Robertis A, De Stefano C, Sammartano S, Rigano C (1985) On the possibility of determining the thermodynamic parameters for the formation of weak complexes using a simple model for the dependence on ionic strength of activity coefficients: Na^+, K^+ and Ca^{2+} complexes of low molecular weight ligands in aqueous solution. J Chem Soc Dalton Trans 2353–2361
Daniele PG, De Robertis A, De Stefano C, Sammartano S (1991) Ionic strength dependence of formation constants XIII. A critical examination of preceding results. In: Alegret S, Arias JJ, Barceló D, Casal J, Rauret G (eds) Miscel·lània Enric Casassas. Universitat Autònoma de Barcelona, Bellaterra, pp 121–126
Davis JA, Kent DB (1990) Surface complexation modelling in aqueous geochemistry. In: Hochella MF, White AF (eds) Mineral-water interface geochemistry. Mineralogical Society of America, Washington DC (Reviews in mineralogy, vol XXIII, pp 177–248)
Davis JA, James RO, Leckie JO (1978) Surface ionization and complexation at the oxide-water interface. I. Computation of electrical double layer properties in simple electrolytes. J Colloid Interface Sci 63:480–499
De Stefano C, Mineo P, Rigano C, Sammartano S (1993) Ionic strength dependence of formation constants. XVII. The calculation of equilibrium concentrations and formation constants. Ann Chim (Rome) 83:243–277
De Wit JCM, Van Riemsdijk WH, Nederlof MM, Kinniburg DG, Koopal LK (1990) Analysis of ion binding on humic substances and the determination of instrict affinity distributions. Anal Chim Acta 232:189–207
De Wit JCM, Van Riemsdijk WH, Koopal LK (1993a) Proton binding to humic substances. 1. Electrostatic effects. Environ Sci Technol 27:2005–2014
De Wit JCM, Van Riemsdijk WH, Koopal LK (1993b) Proton binding to humic substances. 2. Chemical heterogeneity and adsorption models. Environ Sci Technol 27:2015–2022
Detenbeck NE, Brezonik PL (1991) Phosphorus sorption by sediments from a softwater seepage lake. 1. An evaluation of kinetic and equilibrium models. Environ Sci Technol 25:395–402
Dzombak DA, Morel FMM (1990) Surface complexation modelling: Hydrous ferric oxide. John Wiley & Sons, New York
Ephraim J, Marinsky JA (1986) A unified physicochemical description of the protonation and metal ion complexation equilibria of natural organic acids (humic and fulvic acids). 3. Influence of polyelectrolyte properties and functional heterogeneity on the copper ion binding equilibria in an Armadale Horizons Bh fulvic acid sample. Environ Sci Technol 20:367–376
Farley KJ, Dzomback DA, Morel FMM (1985) A surface precipitation model for the sorption of cations on metal oxides.J Coll Inter Sci 106:226–242
Förstner U (1989) Contaminated sediments, lecture notes. In: Bhattacharij S, Friedman GM, Neugebauer HJ, Seilacher A (eds) Earth science, vol XXI. Springer, Berlin, Heidelberg, New York
Gulmini M, Zelano V, Daniele PG, Prenesti E, Ostacoli G (1996) Acid-base properties of a river sediment: Applicability of potentiometric titrations. Anal Chim Acta 329:33–39
Gulmini M, Zelano V, Daniele PG, Ostacoli G (1997) Acid-base and Copper(II) sorption properties of a natural lake sediment: Potentiometric and atomic absorption spectrometric characterisation. Anal Chim Acta 358:195–204
Honeyman BD, Santschi PH (1982) Metals in aquatic systems. Environ Sci Technol 16:862–871
Lead JR, Hamilton-Taylor J, Hesketh N, Jones MN, Wilkinson AE, Tipping E (1994) A comparative study of proton and alkaline earth metal binding by humic substances. Anal Chim Acta 294:319–327
Leckie JO (1988) Coordination chemistry at the solid-solution interface. In: Kramer JR, Allen HE (eds) Metal speciation. Theory, analysis and application. Lewis, Chelsea Michigan, USA, pp 41–69
Lion WL, Altman RS, Leckie JO (1982) Trace-metal adsorption characteristics of estuarine particulate matter: Evaluation of contribution of Fe/Mn oxide and organic surface coatings. Environ Sci Technol 16:660–666
Lumsdon DG, Evans LJ (1995) Predicting chemical speciation and computer simulation. In: Ure AM, Davidson CM (eds) Chemical speciation in the environment. Blackie A & P, London

Luoma SN, Davis JA (1983) Requirements for modeling trace metal partitioning in oxidized estuarine sediments. Mar Chem 12:159–181
Martin JM, Nierel P, Thomas AJ (1987) Sequential extraction techniques: Promises and problems. Mar Chem 22:313–341
Masini JC (1993) Evaluation of neglecting electrostatic interactions on the determination and characterization of the ionizable sites in humic substances. Anal Chim Acta 283:803–810
Morel FMM, Hering JG (1993) Principles and applications of aquatic chemistry. John Wiley and Sons, New York, pp 319–420
Puls RW, Powell RM, Clark D, Eldred CJ (1991) Effects of pH, solid/solution ratio, ionic strength and organic acids on Pb and Cd sorption on kaolinite. Wat Air Soil Poll 57–58:423–430
Schindler P (1981) Surface complexes at oxide water interfaces. In: Anderson MA, Rubin AJ (eds) Adsorption of inorganics at solid-liquid interfaces. Ann Arbor Science Publishers, Ann Arbor, pp 1–49
Stumm W (1992) Chemistry of the solid-water interface. Processes at the mineral-water and particle-water interface in natural ystems. John Wiley and Sons, New York
Stumm W (1998) The solid-water interface in natural systems. In: Macalady DL (ed) Perspectives in environmental chemistry. Oxford University Press, Oxford, pp 3–24
Stumm W, Hohl H, Dalang F (1976) Interaction of metal ions with hydrous oxide surfaces. Croat Chem Acta 48:491–499
Stumm W, Kummert R, Sigg L (1980) A ligand exchange model for the adsorption of inorganic and organic ligands at hydrous oxide interfaces. Croat Chem Acta 53:291–312
Templeton GD III, Chasteen ND (1980) Evaluation of extraction schemes for organic matter in anoxic estuarine sediments. Mar Chem 1:31–46
Tessier A, Campbell PGC (1988) Partitioning of trace metals in sediments. In: Kramer JR, Allen HE (eds) Metal speciation. Theory, analysis and application. Lewis, Chelsea Michigan, USA, pp 183–184
Tipping E, Reddy MM, Hurley MA (1990) Modelling electrostatic heterogeneity effects on proton dissociation from humic substances. Environ Sci Technol 24:1700–1705

Chapter 21

Chemometrics for Sampling and Analysis: Theory and Environmental Applications

M. Forina · S. Lanteri · R. Todeschini

21.1 Introduction

Chemometrics is the chemical discipline that uses mathematics, statistics, artificial intelligence and formal logic

- to design or select optimal experiments,
- to extract maximum relevant information from chemical measured or computed data, and
- to obtain knowledge about chemical complex systems, processes and products.

Chemometrics is characterized by:

- the data base,
- the tools,
- the models and
- the objective.

Data for chemometrics are complex, multivariate, in form of a matrix of many rows (samples, molecules), described by many columns (measured physical or chemical quantities, computed quantities) and frequently tridimensional (e.g. sampling sites, time, measured quantities). The below reported examples will generally use only two variables for representation purposes. Hundreds or thousands of variables often describe real objects: e.g. a spectrophotometer can give the absorbance at about one thousand wavelengths. GRID, a technique used to describe molecules by interaction potentials, can produce 10 000 or more descriptors for each studied molecule.

The variability of data between objects depends on many known or unknown factors and on measurement error (afterward simply "noise"). The information content of the data is closely related to the factor-dependent variability (afterward simply "variability"). When the noise is higher than the variability, data cannot contain information. The situation is that of "quality" – all objects are equal within the noise:

- Variability = Information
- Non-variability = Quality

The tools of chemometrics (Massart et al.1998; Meloun et al. 1992; Frank and Todeschini 1994) range from those of the base theory of probability (with particular attention to the special case of experimental data, characterized by the discontinuity due to the readability, very important with determinations close to the limit of quantization) to wavelet transforms, artificial neural networks, genetic algorithms, Kalman filtering, trilinear decomposition, etc. However, some tools have fundamental importance, and they can be introduced and used without a deep knowledge of the underlying mathematics. Chemists must understand the performances of these ones. The tools are equal to instruments with a lot of knobs: their settings must be selected with reference to the chemical problem and to the characteristics of chemical data.

The mathematical models obtained by the tools are soft models, without (or with low) theoretical validity. Sometimes they can help in theoretical development, but generally their function is to solve practical problems.

The final goal of chemometrics is the prediction of quantities or of qualities. A mathematical model for the prediction of a chemical quantity is a regression model; in the case of a quality it is a class-modelling model. All models must be carefully tested for their predictive ability, so that the result of the prediction can be given with a measure of its uncertainty. The techniques of predictive validation are a very important characteristic of chemometrics.

21.2 The Fundamental Tools of Chemometrics

21.2.1 Data Pretreatments (The Importance of Data Knowledge and of Problem)

An object is described by several variables, which are the result of the practical work of the chemist. These variables can be homogeneous (e.g. concentrations in the same unit), homogeneous and ordered (e.g. the absorbances at regularly spaced wavelengths) or non-homogeneous (e.g. concentration, time, temperature, absorbance). Moreover, each variable is characterized by its noise.

Data analysis is based on distances. For example, in the simple expression of the normal distribution of probability density:

$$f(x) = \frac{1}{\sqrt{2\pi}\sigma} e^{-\frac{(x-\mu)^2}{2\sigma^2}}$$

where the operator of the exponential is the square of a distance $(x - \mu)$ divided by the standard deviation σ, i.e. a distance standardised by means of a measure of the dispersion.

A fundamental pretreatment for non-homogeneous data is column *autoscaling*, i.e.

$$x_{\text{autoscaled}} = \frac{x_{\text{original}} - m_x}{s_x}$$

where m_x is the sample mean of a variable (column of the data matrix) and s_x is its sample standard deviation, hopefully due to variability, with a negligible contribution of standard deviation of noise, s_N. Autoscaling transforms original data so that all variables have the same mean (0) and the same standard deviation (1). So, all autoscaled variables have the same importance in regard to information, because they have the same dispersion. Then, the autoscaled variables can be divided by the standard deviation of the relative noise, s_N / s_x (weighting). So, the importance of a variable becomes proportional to the signal/noise ratio. Simple autoscaling must not be used with homogeneous data with almost constant noise. For example, consider absorbances at two wavelengths, with the same noise: the first has high absorption, the second has no absorption. Simple autoscaling gives the same importance to informative and the noisy absorbance. Weighting restores the original data. In this case an often-used transform is the simple subtraction of the column mean (*centring*). Therefore, it is evident that only the *knowledge of data* can suggest which pretreatment we have to use. A very simple mathematical operation as autoscaling can have pernicious effects when improperly used.

Frequently, in the case of homogeneous data, row profiles (percentages) are computed:

$$x_{\text{profile}} = \frac{x_{\text{original}}}{\sum\limits_{\text{Variables}} x_{\text{original}}}$$

This other simple mathematical operation cancels a part of the original information, so proportional objects become equal. Only the *knowledge of the problem* allows the decision about the adequacy of this pretreatment, since the knowledge of the problem is fundamental to select the suitable treatment in the case of homogeneous ordered variables, as those of spectra, chromatography, hyphenated techniques. Smoothing, first and second derivative, Fourier transforms, wavelet transforms, base line subtraction and row autoscaling (also known as Standard Normal Variate) are examples of important pretreatment tools.

21.2.2 Similarity and Clustering

Also similarity is a problem-dependent concept. Usually, two samples are considered similar when the difference in composition is small if compared to the variability. Fig. 21.1 shows nine samples described by two variables.

A current definition of the similarity between two objects i and j is:

$$s_{ij} = 1 - \frac{d_{ij}}{d_{\max}}$$

where the distance between the two objects (dependent on the pretreatment, obviously) is divided by the maximum inter-object distance of the data set, so that the farthest objects have similarity 0.

However, there is a second concept of similarity, based on the membership to a structure. Fig. 21.2 shows that object *2* is more similar to object *1* than to object *3*, because both *1* and *2* belong to the same structure.

As for objects, the similarity between variables can be defined. A usual definition is that similarity is measured by the $(1 + r) / 2$, where r is the linear correlation coefficient. So, the similarity is zero when the two variables are opposite, linked by a linear relationship with negative slope. A second definition is that the similarity is measured

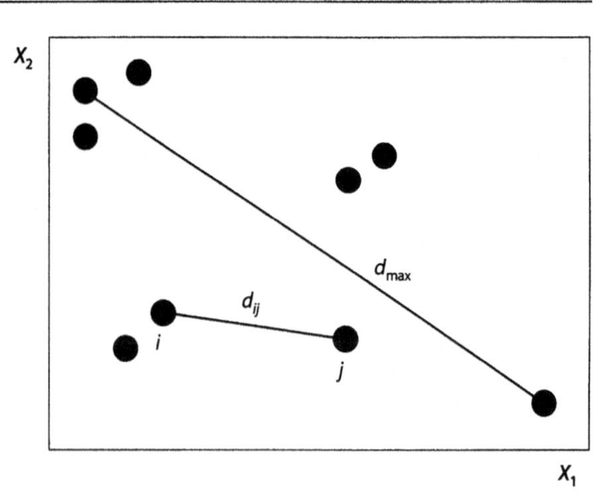

Fig. 21.1. Distances for similarities in clustering techniques

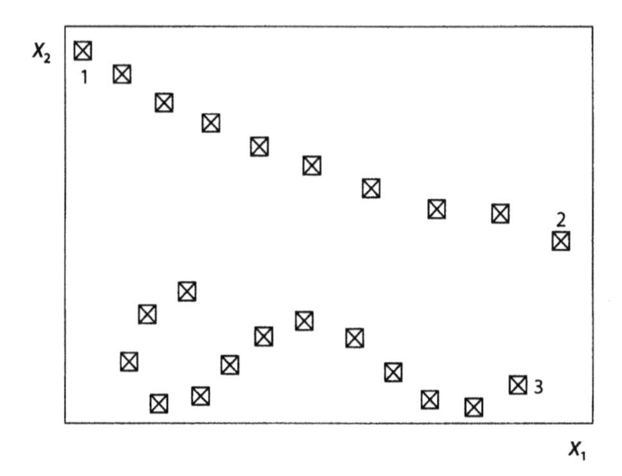

Fig. 21.2. Structures in the space of information

by the absolute correlation coefficient. Therefore, the two above opposite variables are considered with similarity 1. Also in this case the choice is problem-dependent.

Clustering techniques work on the similarities of objects or variables, to detect clusters of similar objects or of similar variables. Some clustering techniques build clusters progressively, starting from separated objects (or variables), joining the two most similar objects in a first cluster, which replaces the two objects and represents a single object for the next agglomeration. The typical output of these techniques is a dendrogram of similarities. An example of dendrogram will be discussed in the applications.

21.2.3 Principal Components

Principal Component Analysis (PCA) is surely the most important tool of chemometrics.

Remember that each object is described by many (V) variables, and these variables are more or less correlated. For example, the absorbances at two close wavelengths are highly correlated. Because of this correlation, in the *V*-dimensional information space the objects are not uniformly distributed, but they are grouped into a restricted area of the space; they form a structure.

Let us consider a very simple example, where the objects are solutions of the same compounds at different concentrations, described by the absorbance at two wavelengths:

$$a_{i1} = k_1 \, c_i + e_{i1}$$

$$a_{i2} = k_2 \, c_i + e_{i2}$$

The two absorbances are proportional to the concentration (with a noise component for each object i for the two absorbances). The space of information is two dimensional, but the two absorbances are closely correlated, because

$$a_{i2} = \frac{k_2}{k_1} c_i + e_i$$

so that (Fig. 21.3) the objects are on a straight line. The useful information has an unidimensional structure; the distance of an object from the structure is a consequence of the noise, i.e. the space outside the structure is that of useless information.

There are a lot of possible structures (linear, non-linear, clusters), but in real problems the V-dimensional space of the information can be always divided into an inner subspace of structured, useful information, and in an outer space of noise. Very frequently the dimension of the inner space is small, also when the dimension of the original information is very large.

PCA aims to detect the structure and to evaluate the number of dimensions of the inner space.

Principal components are the eigenvectors of centred data, often of autoscaled data. In a bidimensional example (Fig. 21.4) the first eigenvector E1 is the direction of maximum variance, with origin in the centroid. Its loadings (the direction cosines of the two axes) can be obtained as in the usual univariate regression, as the direction for

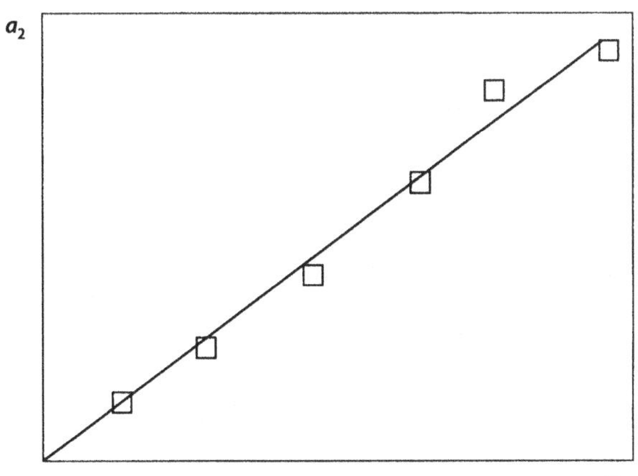

Fig. 21.3. Linear structure caused by correlated variables

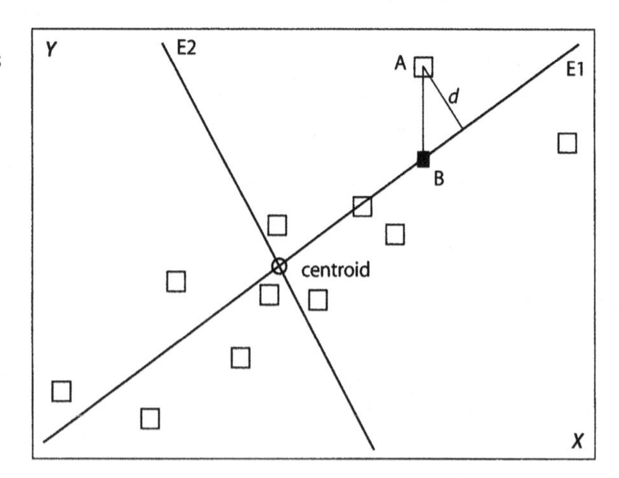

Fig. 21.4. Eigenvectors, E1 and E2. *d*: residual orthogonal to E1; *B*: position of object *A* predicted by the first Eigenvector in the case of unknown ordinate

which the sum of the squares of the residuals is minimum. The difference is that these residuals *d* are orthogonal, not vertical as in the usual regression. For this reason, the direction E2, orthogonal to E1, is the direction of minimum variance.

Because the residuals are distances it two dimensions, the pretreatment of data is very important in PCA. E1 is midway between the usual regression line of *Y* vs. *X* and that of *X* vs. *Y* with autoscaled variables.

Generally, PCA obtains a new set of variables from a set of *V* variables, by orthogonal rotation around the centroid. These new variables, the eigenvectors of centred data, are sorted from that with the maximum variance (the eigenvalue is the PC variance) to that with the minimum variance, according their information content. A very important feature of the principal components is that they are uncorrelated variables, so that each PC shows information not duplicated in other PCs. For this reason the first two or three PCs are used to visualise the maximum amount of information on a bi- or tridimensional plot. The space of the first PCs can be considered as the optimum window opened in the space of the total information to see the maximum amount of information. There are many mathematical tools to compute PCs, their *loadings*, the cosines of the angles of each PC with the original variables, and the *scores*, i.e. the coordinates of the objects in the space of PCs. An excellent technique is NIPALS algorithm, it can compute PCs in the case of missing data too. There are many techniques to detect the number of significant PCs, i.e. the dimensionality of the inner space. The most important one seems the technique of double-cross validation (DCV), based on predictive ability. According to a cancellation design, some data are deleted from the data set, and the first PC is computed by NIPALS algorithm. Suppose that the ordinate of point A in Fig. 21.4 was cancelled; object *A* is predicted in position *B* by the first component with an error on the cancelled ordinate. Significant components have predictive ability, in the sense that they improve the prediction of deleted data, referring to the prediction obtained by the centroid (no components) or by one component less for the next PCs.

Figure 21.5a shows a three dimensional space of information. The inner space is bidimensional, a plane. The direction orthogonal to this plane, the third component, has the minimum sum of the squares of the orthogonal residuals. The first two eigen-

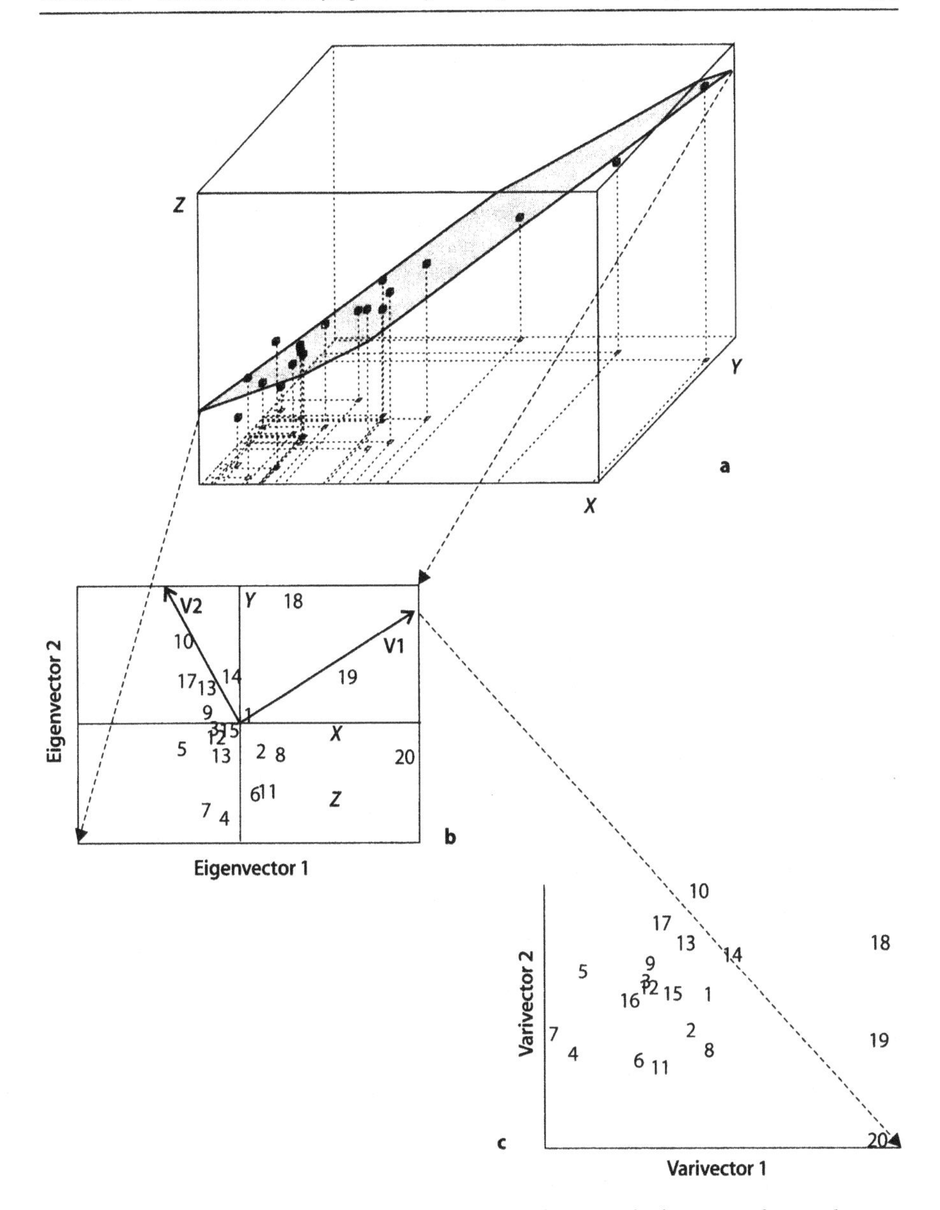

Fig. 21.5. a The space of original information; **b** The space of structured information, of PCs; **c** The space of rotated useful information

vectors define this plane, and they can be used to visualise the inner space, the relevant information (Fig. 21.5b).

Note that the three original variables are projected on the same plot according their loadings. Variable *X* contributes to component E1 (large positive loading of E1 on *X*), whereas its contribution to component E2 is negligible. On the contrary, *Y* contributes with large positive loading to component E2. *Z* contributes with positive loading to component E1 and with negative loading to component E2. The location of object *20* can be explained with very large values of X and Z. The location of object *5* is due to small values of all the original variables.

Moreover, three objects, namely *18*, *19*, *20*, are clearly separated from the cluster of the other objects. A new direction, *V1*, explains the difference between the three objects and the others. This direction, and the second direction *V2*, is obtained by further rotation with a technique of the family of varimax rotations, and for this reason they are called varivectors. As PCs, varivectors have their loadings of the original variables, so that the three objects are characterized by very large values of *X*, and slightly large values of *Y* and *Z*. Fig. 21.5c shows the projection of the objects on the two varivectors.

Factor Analysis (FA) works by means of further rotations (orthogonal or not) in the inner space, with the aim of helping in the interpretation and in the further use of the structured information. Varimax rotations are only one example of rotation techniques of FA.

21.2.4
Class-Modelling Techniques

This is a family of techniques used in classification problems (e.g. in quality control the classes are acceptable/rejected; in toxicology they are often toxic/non-toxic).

Class-modelling techniques compute a mathematical model of the studied class. In Fig. 21.6 the mathematical model of a class is a point, the centroid of the class. The mathematical model of the other class is a suitable range of its first principal component. An object falling exactly on the mathematical model can be considered an ideal object. The effect of some factors produces a more or less large distance from the mathematical model for the other objects. A statistical procedure computes the maximum permitted distance, the boundary of the class space. This maximum distance corresponds to a critical value of the used statistics. In the case of the centroid model (UNEQ method) the distance is a Mahalanobis distance, which takes into account the different variances of the variables and their correlation, and χ^2 statistics is used to compute the critical value. In the case of PC model (SIMCA method) the class boundary is obtained by the variance in the outer space, and by F statistics.

The class model accepts an object which falls within the class space, with a significance level decreasing with the distance from the class centroid. The class model does not accept an object which falls outside the class space. It can be considered as a non-typical object, or as an object very different from the other objects of the class according to its significance level.

In Fig. 21.6, object *A* fits both class models. This double possibility depends on the not perfectly separated two-class space. Object *A* can be considered more similar to class modelled by SIMCA. At first, it was considered an object of the class modelled by UNEQ, but the assignment can be modified, and the models computed again (refinement of class model).

21.2.5
Regression Techniques, Responses and Predictors

Regression techniques are used to model the relationship between one (or more) response variable and some predictor variables.

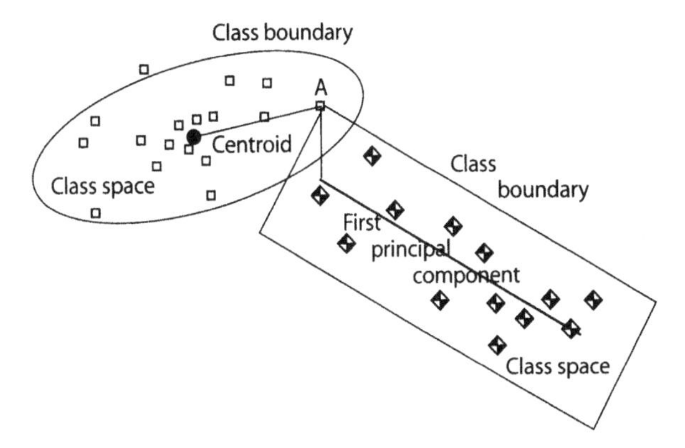

Fig. 21.6. Class models

A response is a quantity whose direct determination is impossible (e.g. a chemical quantity), time-consuming or expensive (e.g. octane number).

The fundamental type of regression is univariate regression, very familiar to chemists. A chemical quantity Y is indirectly evaluated, by means of the measure of a physical quantity X (generally X is used to denote the chemical quantity, and Y to denote the physical quantity, but the multivariate situation typical of chemometrics suggests the above notation). Some standards (Y known) are used; X is measured with these standards and the model is obtained by:

$$y = \frac{x - a}{b}$$

where a and b are the intercept and the slope of the regression line, the parameters of the model. This procedure is known as *calibration.*

In multivariate regression there are many predictors. The predictors can be:

- *Controllable quantitative continuous factors* (e.g. temperature, pressure, time, concentration of reactants; in the last case Y is the yield of a chemical reaction). These factors can be settled independently, so that they are independent variables. A wrong procedure can transform them into correlated variables (e.g. when the experimenter modifies two concentrations so that they are always equal).
- *Experimental quantities* (e.g. the absorbances in a NIR spectrum, as used in multivariate calibration). These predictors are naturally correlated, and the experimenter cannot settle their value.
- *Controllable qualitative factors* (e.g. the nature of a catalyst, of a solvent, of a substituent). There is the possibility to associate a conventional level to such factors (e.g. 0 for water, 1 for ethanol), but only for two level factors.

Carlsson (1992) introduced for these cases the idea of *principal properties.* More than one hundred solvents were described by some physical characteristics, from boiling point to dipolar moment. On the basis of the loadings of the descriptors, the first two PCs of autoscaled data were interpreted respectively as polarity and polarizability, the two principal properties (PP) of the solvents (Fig. 21.7).

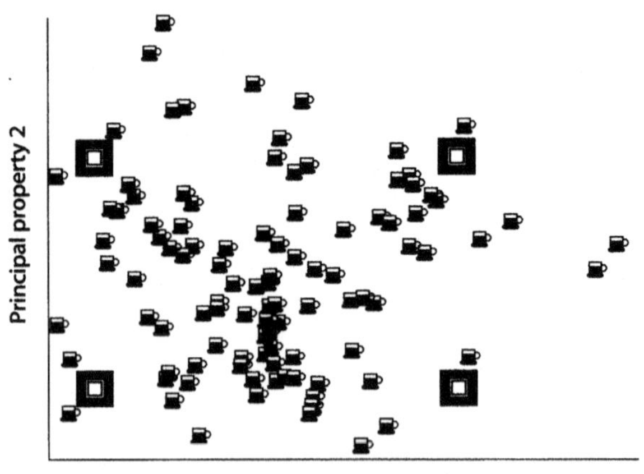

Fig. 21.7. Principal properties of solvents (adapted from Carlsson 1992)

To perform the experiments, the choice of four solvents, with characteristics close to those of the squares shown in the Fig. 21.7, allows exploring the main properties of the possible solvents with a reduced experimental cost. The four solvents will be described by the value of their PPs, which are independent variables.

A similar strategy can be used in the case of experimental data, too. Multivariate calibration requires standards, as univariate calibration. In many cases the standards are samples analysed for the response with a reference, a slow and expensive technique. To select the samples for calibration among many possible samples, the spectra of the candidates are recorded. The choice is the made on the first two or three PPs of the spectra. However, the used predictors are the absorbances, correlated variables, because the experience demonstrated that 2–3 PPs are not enough to retain all the useful information.

Sampling is one of the critical points of the analytical procedure. The number of sampling sites, the time frequency of the sampling and the related chemical analysis fix the cost of the collected information. Principal properties of sampling sites, determined by their position, wind, distance from possible pollution sources, can help very much to reduce the number of samples without loss of information.

21.2.6 Regression Techniques, Biased and Unbiased Techniques

The controllable factors are generally independent and their values can be selected so that they are also uncorrelated or with a small correlation. Mixtures are a typical exception: the fraction of each component is correlated to the other fractions (the sum is one). Mathematical transforms can be used in this case to obtain $N-1$ independent variables from N fractions.

The controllable factors are the domain of experimental design: a number of controllable factors, a reduced number of experiments, enough to compute the parameters of the regression model and to check the validity of the model. The regression technique used is ordinary least squares regression (OLS), the multivariate analogue of univariate least squares regression.

The main problems here are:

- the choice of the model,
- the design (number of experiments, values of the factors in each experiment) necessary to obtain the required information with a sufficient quality,
- the refinement of the regression model, and
- the validation of the model.

The strategy generally used in the chemical laboratory for univariate calibration is compared with that of experimental design in Fig. 21.8. In the first strategy, a number of equally spaced standards are used to cover the experimental domain of the response. Experimental design uses only two standards, at the boundary of the experimental domain, and a third standard at the centre of the domain to evaluate the quality of the model. Under the hypothesis of a linear model, two standards are sufficient to compute the parameters of the model. The quality of the model is measured by the residual of the third standard: generally, the standard deviation of the error is approximately known, so that it is possible to decide the adequacy of the model. Really, in the case of the first strategy, the underlying objective is complex: to check the linearity, to evaluate the parameters, to test the homoscedasticity of the error, and frequently the number of standards is not enough to answer to all these questions.

OLS has severe drawbacks in the case of experimental predictors, or generally when the number of predictors is large and the correlation is large too. First, when the number of predictors is higher than the number of objects or when the correlation is too large, the algorithm cannot be used (infinite solutions). Large correlations cause large uncertainty of the model coefficients (impossibility of evaluation of the relative importance of the predictors), and overfitting (use of the noise to minimise the sum of squares) and consequent poor prediction. A series of "biased" regression techniques have been developed. PCR (Principal Component Regression) uses the scores on PCs as predictors. The optimum number of PCs is obtained (Fig. 21.9) by predictive optimization: it defines the optimal complexity of the model. The first PCs generally carry good information, so that the predictive ability of the model increases. The last PCs with small eigenvalues are generally associated with the noise, so that their addition to the predictors produces a worsening of performances.

PLS (Partial Least Squares regression) is a very popular biased regression technique, with some advantages over PCR: one step instead of the two steps of PC computing and regression, the possibility to work with missing data too. PLS works with variables similar to PCs, called latent variables of PLS.

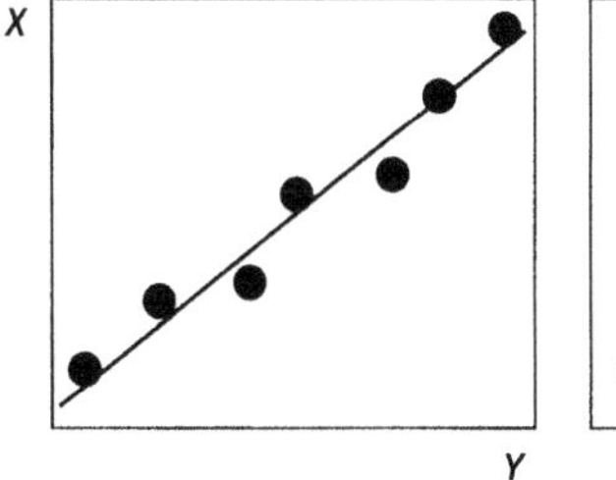

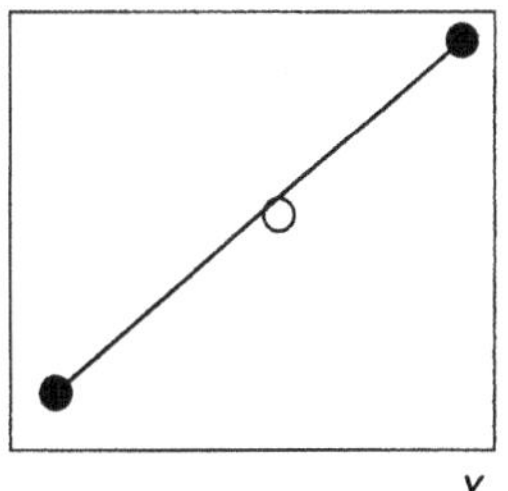

Fig. 21.8. Usual strategy for univariate calibration (*left*); strategy suggested by experimental design (*right*)

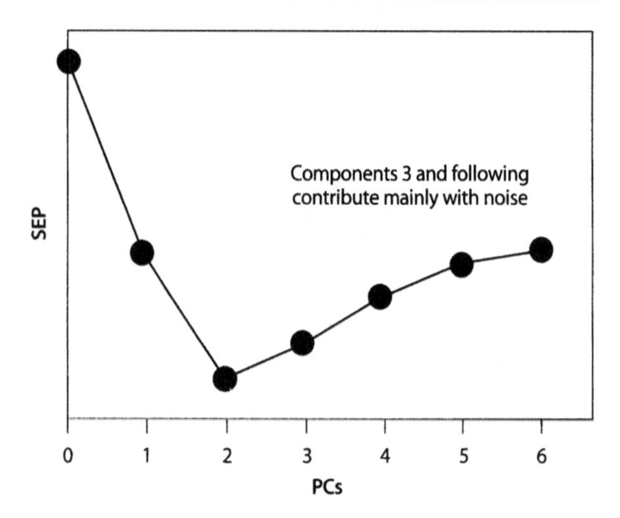

Fig. 21.9. Standard error of prediction (SEP) as a function of the complexity of the model, in the case of a biased regression technique as PCR

In spite of their relative insensitivity to noise, PCR and PLS produce bad results when few good predictors are joined with a lot of uninformative predictors. In these cases the model can be refined by elimination of noisy predictors. A number of techniques have been developed, generally based on the predictive optimization of the model.

The final regression model must be validated: its predictive ability is evaluated on samples never used in the development and in the refinement of the model (Lanteri 1992).

21.3 Applications

21.3.1 Example 1 – Rain Chemistry

All examples here used are presented in very rough form, with the aim of showing the use of the different techniques of chemometrics in the environmental studies. Details can be found in the original papers.

Data used in the first example (Knudson et al. 1977) refer to rain samples gathered during a single storm (24 hours) in north-west USA. Rain samples are considered an ideal means for sampling the chemical state of the atmosphere.

Figure 21.10 shows the contour maps of the volume of rain, the sampling sites and the positions of the two most important towns, Tacoma and Seattle. The direction refers to the predominant movement of the storm.

Figure 21.11 shows the contour maps for other two chemicals, copper and sulfate. The analysis of these maps can give useful information about the origin of pollution. However, a lot of work is necessary to compare all the maps. Moreover, details can be due to minor sources and to measurement errors.

Cluster analysis (Fig. 21.12) detects some "families" of chemical species, one with As, Cu, Sb and Cd, one with Na, Mg, Cl, Ca, K. H^+ and NO_3^- are very similar, too.

PCA analysis is more informative. Three components are significant. The space of the first two PCs (Fig. 21.13, about 60% explained variance) clearly shows that one object

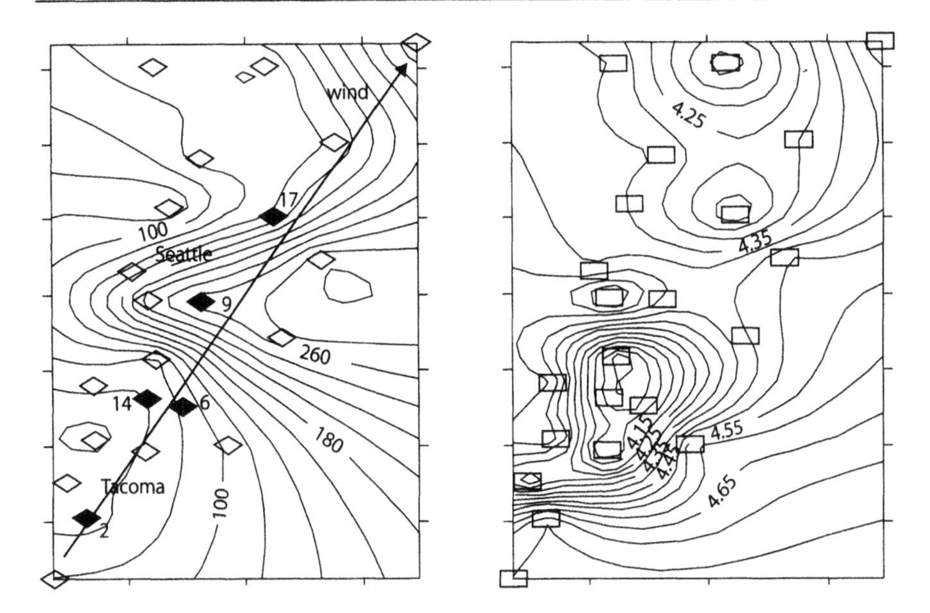

Fig. 21.10. Contour maps; *left*: volume of rain, sampling sites, direction of the storm; *right*: pH (adapted and recomputed from Knudson et al. 1977)

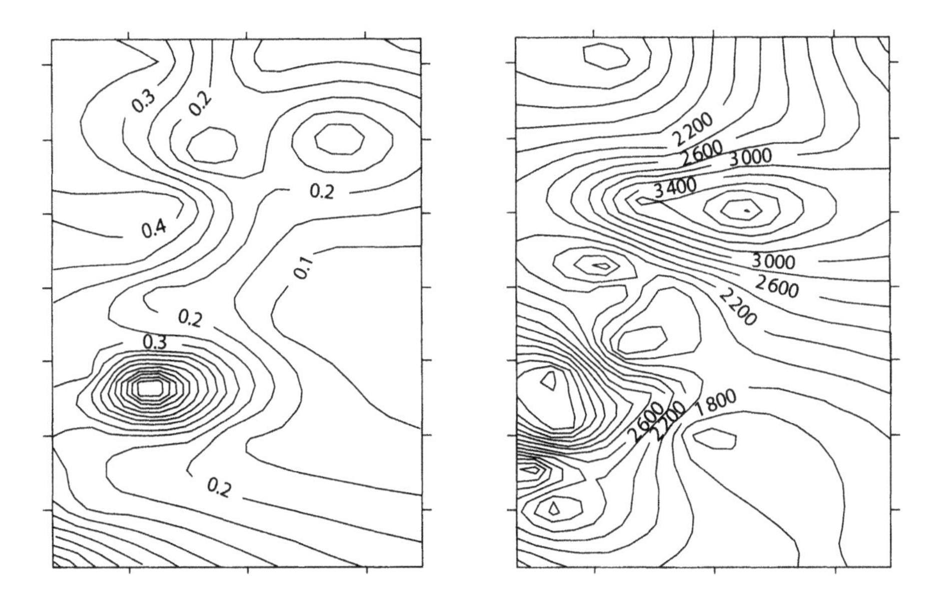

Fig. 21.11. Contour maps; *left*: Cu; *right*: sulfate(adapted and recomputed from Knudson et al. 1977)

(*Sample site 14*) is very different from others. The loadings of As, Cu, Sb, Cd and, less, H^+, are in the direction of object *14*, so that they indicate that high values of the concentration of these species correspond to object *14*. A second direction, almost orthogo-

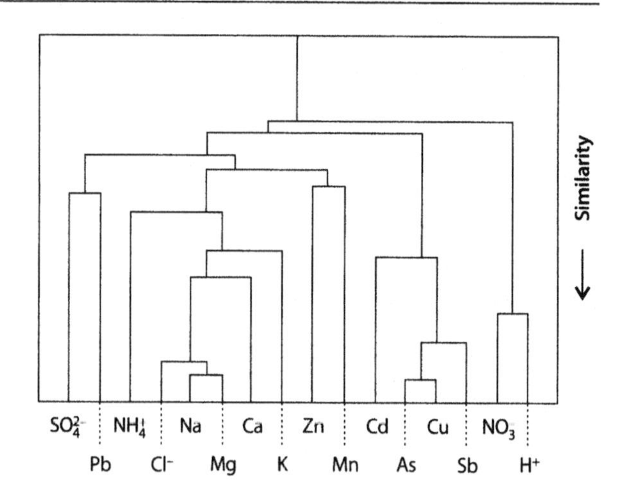

Fig. 21.12. Variable dendrogram

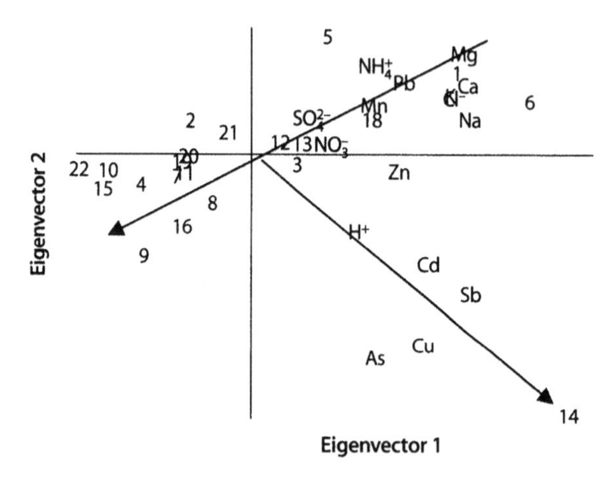

Fig. 21.13. Score and loading plot on the first two eigenvectors

nal to the first one, is characterized by high loadings of Na, Mg, Cl, Ca, K, the second group of variables detected by clustering. The dispersion on component 3 is almost fully due to H^+ and NO_3^-, corresponding to the three sampling sites with both high value of nitrate and of acidity (*Site 14* has a relatively high value of acidity but a low value of nitrate). Finally, components *4* and *5* are not so clear (the interpretation of non-significant components cannot give sure conclusions), except for the loading of Pb, connected with the two sampling sites in Seattle.

Varimax rotation is used to rotate the information in the space of the three PCs: the obtained varivectors represent fairly well the three clusters of variables detected by clustering analysis (Fig. 21.14 shows the loadings on the first two varivectors). The varivectors can be interpreted as the 3 relevant pollution sources (sea salt back-ground + urban sources, a copper smelter near Tacoma, urban sources), and the variscores can be used as the original variables in the contour plots, as those in Fig. 21.15.

Note that the farthest objects, namely *2* and *14*, can represent (Fig. 21.13) one direction; the second direction can be represented by objects *6* and *9*; finally objects *2* (again)

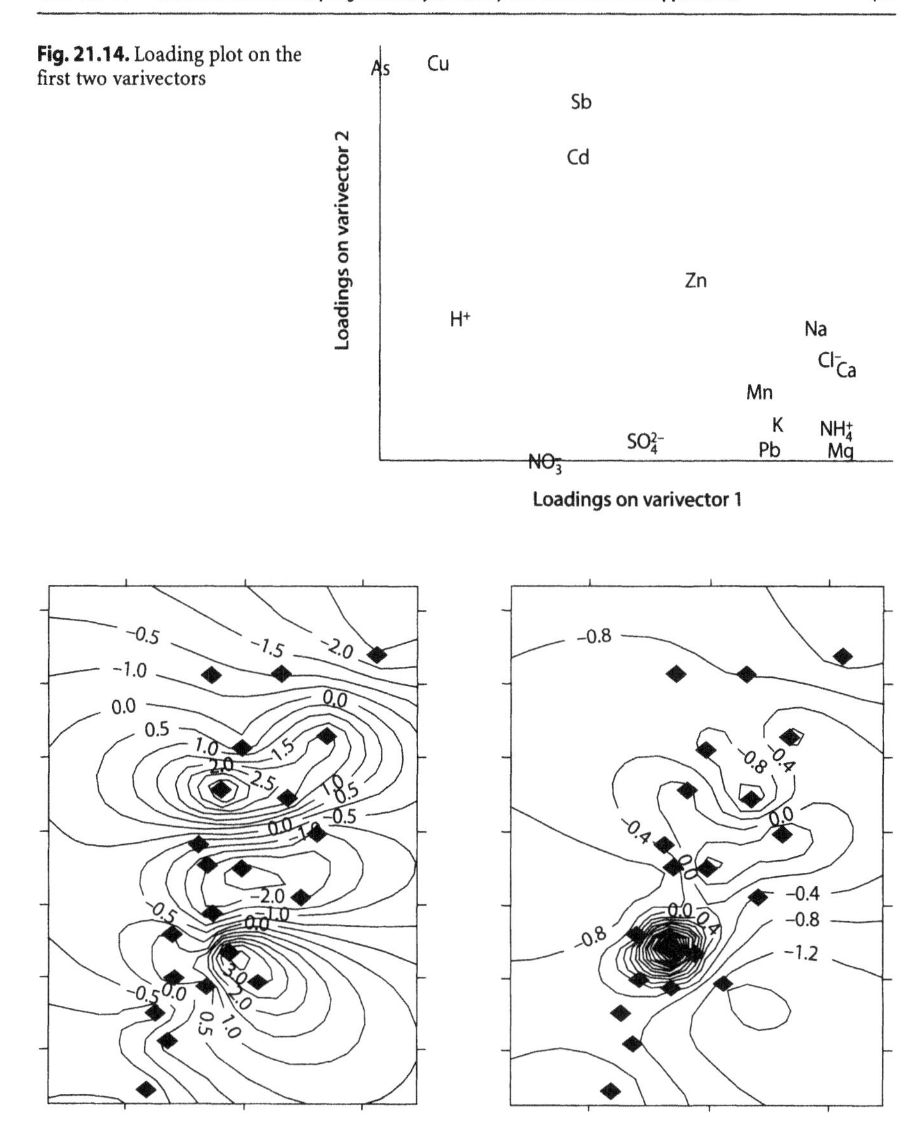

Fig. 21.14. Loading plot on the first two varivectors

Fig. 21.15. Contour maps; *left*: varivectors 1 and 2; *right*: varivectors 1 and 3

and *17* represent the third varivector. So the principles of design on principal properties suggest that 5 stations (instead of 22) can be used for an efficient monitoring of the three main pollution factors in the studied area. These stations are in the direction of the movement of the storm (Fig. 21.10). Stations *2* (south-west of Tacoma) and *6* (very close to the sea) can be considered as zero references for the three factors. Station *9*, in Seattle, represents the effect of urbanised and industrialised areas. Station *14* represents the effect of the Tacoma smelter. Finally, station *17*, downwind of Seattle, represents the effect of both industries and of areas of heavy suburban traffic towards Seattle with its high level of nitrate and lead.

21.3.2
Example 2 – Polycyclic Aromatic Hydrocarbons (PAHs)

The polluting effect of PAHs is closely related to their mutagenic and carcinogenic activity, so they can be used as indexes of pollution able to differentiate several environmental settings: urban, industrial and suburban.

Airborne particulate samples, used to assess the environmental contamination by PAHs, where collected (Lanteri et al. 1998) in thirteen sites of Genoa (the map of Genoa, with the approximate position of these sites is shown in Fig. 21.16) from December 1992 to December 1995. For each sample nine PAHs were measured: fluoranthene (FLT), pyrene (PY), benzo(a)anthracene (BaA), chrysene (CRY), benzo(b)fluoranthene + benzo(k)fluoranthene (BF), benzo(a)pyrene (BaP), indeno(1,2,3-cd)pyrene (IND), dibenzo(a,h)anthracene (DBA) and benzo(g,h,i)perylene (BghiP).

In Genoa a coal power plant and a large steel factory with coke ovens are present, they are important sources of combustion-derived PAHs, in addition to the pollution due to urban traffic, which is heavily present.

A class modelling analysis was performed. Samples were assigned to one of the two classes (urban or industrial pollution) on the basis of the position of the sampling site. Many samples of the "industrial" class were rather far from the model of the class, and close to the model of the class "urban." This fact was explained by the direction of wind. These samples were consequently assigned to the class "urban." The refined models were more compact of the original models, as shown in the plot on the two first princi-

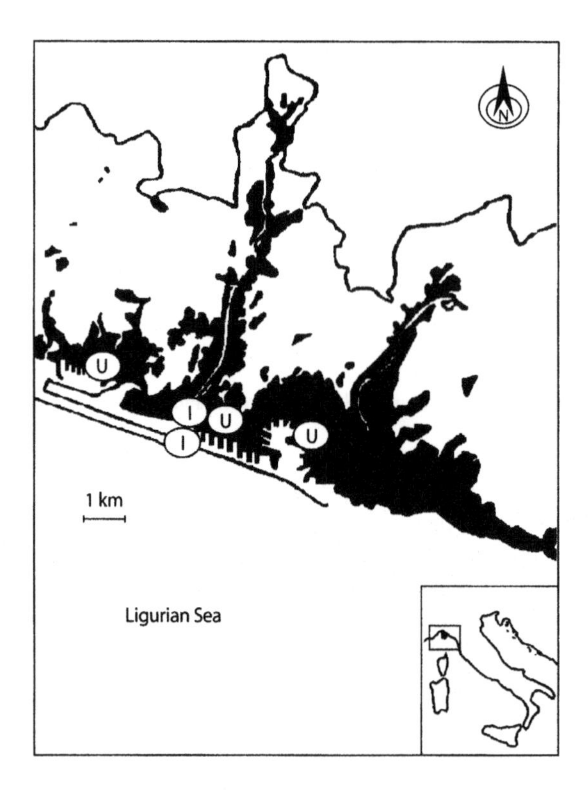

Fig. 21.16. Map of Genoa

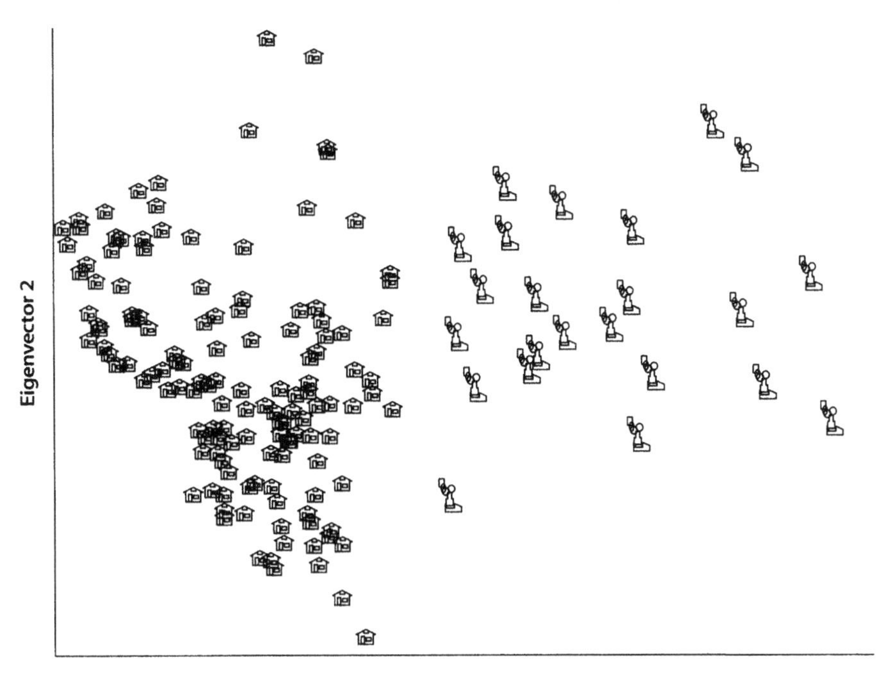

Fig. 21.17. Score plot on the first two eigenvectors

pal components (Fig. 21.17), where the first component is clearly the direction of the increasing pollution.

21.3.3 Example 3 – Toxicity

117 compounds were studied (Todeschini et al. 1997) for which Water Quality Objectives for the protection of aquatic life (WQO) have been produced (CSTE/EEC 1994). According with WQO, the compounds were grouped in five classes, by the use of a logarithmic scale of the toxicity, from WQO 10 $\mu g\ l^{-1}$ to 0.001 $\mu g\ l^{-1}$. The compounds were described by means of molecular descriptors (topological and structural descriptors). PCA of these descriptors explains about 60% of the total variance with two principal components. Dimensional descriptors mainly form the first component, while the second is related to shape and other structural features. The space of these two components can be divided in five sectors as Fig. 21.18 shows. Sector *I* includes chemicals belonging to the first class and some compounds of class 2 (relatively small compounds, as chlorinated aliphatics, benzene, some substituted benzenes). Sector *V* includes a single chemical of class 3, flucofenuron, with 6 F and 2 Cl, and the highest molecular weight in the data set. PCA can give a first qualitative evaluation of the relationship between toxicity and structure, to be possibly used with new chemicals of unknown toxicity.

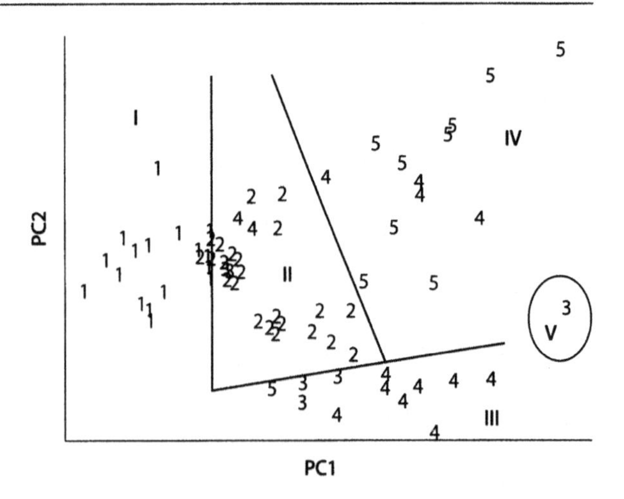

Fig. 21.18. Score plot on the first two eigenvectors

Regression analysis produces a refined model with only five descriptors and a good predictive ability, measured by a prediction error smaller than 0.5 logarithmic units over a range of about 4 orders of magnitude.

21.4 Conclusions

The tools of Chemometrics help to plan experiments, to improve the analytical procedures and to extract useful information from large databases. Simple tools, such as PCA, class-modelling techniques and biased regression techniques do not require a lot of mathematical knowledge, and for several complex problems can represent the only way to understand the studied system and to obtain models, perhaps with local validity, but high predictive ability.

References

Carlsson R (1992) Design and optimization in organic synthesis. Elsevier, Amsterdam

CSTE/EEC (1994) EEC Water Quality Objectives for chemical dangerous to aquatic environments. Rev Environ Contam Toxicol 137:1–91

Frank IE, Todeschini R (1994) The data analysis handbook. Elsevier, Amsterdam

Knudson EJ, Duewer DL, Christian GD (1977) Application of factor analysis to the study of rain chemistry in the Pudget Sound region. In: Kowalski BR (ed) Chemometrics: Theory and application. American Chemical Society, Washington D C (ACS Symposium Series 52, pp 80–116)

Lanteri S (1992) Full validation procedures for feature selection in classification and regression problems. Chemometrics and Int Lab Systems 15:159–169

Lanteri S, Pala M, Armanino C, Stella A(1998) Models of urban, industrial and suburban atmospheric pollution in Genoa. – IV Colloquium chimiometricum mediterraneum (Abstract book), Burgos 8–11 June 1998, p 104

Massart DL, Vandeginste BGM, Buydens LMC, De Jong S, Lewi PJ, Smeyers-Verbeke J (1998) Handbook of chemometrics and qualimetrics, Part A. Elsevier, Amsterdam

Meloun M, Militky J, Forina M (1992) Chemometrics for analytical chemistry, vol I: PC-aided statistical data analysis. Ellis Horwood, New York (Ellis Horwood Series in Analytical Chemistry)

Todeschini R, Vighi M, Finizio A, Gramatica P (1997) 3D-Modelling and Prediction by WHIM descriptors, Part 8: Toxicity and Physico-chemical properties of environmental priority chemicals by 2D-TI and 3D-WHIM descriptors. SAR and QSAR Environmental Research 7:173–193

Chapter 22

Chemometric Applications to Seawater Analysis

M.C. Gennaro · S. Angelino

22.1 Introduction

22.1.1 The Sea: A Very Complex Ecosystem

Seas and oceans have been defined as "*water bodies acting as natural reactors*" in which matter is constantly imported, exported and transformed. The main boundaries are the land, the atmosphere and the sediments; the inlet and the outlet pathways are rivers, channels and the air-water and water-sediment interfaces. Since a large number of physical, chemical, geochemical and biological processes are there taking place, the marine system is a very complex one. The chemical species undergo adsorption, desorption, solubilization, precipitation, oxido-reduction and settling processes which also govern speciation competitive equilibria; organisms give rise to mechanisms of production, uptake, excretion, decomposition and accumulation; sediments, particulate and colloids are involved in aggregation, fragmentation and desegregation phenomena, etc. Each of these processes can be individually described, but its role in the environment can be correctly evaluated only considering its connection with all the others simultaneously taking place.

The equilibria and natural cycles of biochemical transformation of the marine ecosystem are threatened by human impact and pollution, and oceans too often become the ultimate sink for many pollutants. Pollutants, when in the marine environment, can assume different forms and undergo different fates, due to the taking place of processes as adsorption onto suspended particles and sediments, scavenging by inorganic and organic particles, assumption by living organisms, mobilisation during regeneration and diagenesis, etc. Pollutant transport is strongly influenced by the behaviour of sediments and suspended matter, in turn driven by water dynamics and organic matter production and diagenesis; pollutants can be adsorbed and cumulate in sediments, be mixed by the physical action of waves and by the benthic organisms activity, be transported downward into the sediment or resuspended or solubilized in water. Suspended matter can be therefore considered as a sort of link among the different components of the system and it can be said that particles control a large part of the aquatic environmental chemistry.

The natural composition of water depends on many factors, such as time (variations within day, month, season, year), depth, tide and conditions of the main components of the ecosystem (water, sediments, organisms). This compositional variability must be distinguished from pollution induced variations (domestic and industrial sewage, leaching processes from agricultural activities, atmospheric precipitation, etc.).

In the study of eutrophycation processes the determination of nutrient-anions (the biogenic salts: NO_3^-, NO_2^-, phosphate, sulfate, silicate) is very important, since their amounts are related to the population of autotrophic organisms, which are mainly responsible for the production of organic matter from the inorganic. In particular NO_2^-, an intermediate stage in the nitrogen cycle, may occur in waters as a result of biological decompositions of proteinaceous material and is an index of organic pollution. In oxygen-poor conditions, autotrophic organisms are able to take oxygen from the reduction of NO_3^- to NO_2^-. Then, when nitrate is not available anymore, oxygen is taken from sulfate with production of sulfide, whose presence indicates therefore anoxic water conditions. It follows that nitrate and sulfide can not be simultaneously present in lagoon water, unless interchanges between different waters have taken place (Manahan 1994; Grasshoff et al. 1983; Riley 1989; Riley and Chester 1971; Carpenter and Capone 1984). These considerations give an example of the complex and competitive chemical and biological equilibria naturally present in sea-water and an idea of the difficulty of preparing good laboratory models or reference solutions. On the other hand, analytical chemistry, devoted to the determination both of natural constituents and xenobiotic compounds, requires the definition of a "reference status" of the water to be used during the development of new analytical methods and for speciation studies (De Stefano et al. 1994; Demianov et al. 1995).

Considering all the interactions and their implications for ocean management is a challenge, because they operate on global scales and often require long-term experiments, often done in remote locations. But a global view is emerging, thanks to increasing satellite coverage, new techniques and measurement campaigns. The large scale interactions between ocean biology and chemistry have been the focus of a recent issue of Science (Uppenbrink et al. 1998). This issue illustrates how bioinorganic chemistry and use of trace elements by marine organisms can affect the cycling of these elements, how recent satellite data and high quality monitoring allow a global assessment of primary production and a better view of carbon and nutrient cycling, and how long term studies in the Pacific Ocean are giving insight into the effects of climate change on marine ecosystem and chemistry. The awareness and availability of high-quality global data have begun to reveal the extent to which humans have altered the biology and the chemistry of the ocean. This increased awareness and the recognition of many remaining uncertainties prompted the United Nations to proclaim 1998 as the Year of the Ocean. Strong investment and the finest international scientific collaboration are needed worldwide to support pending scientific research: the aim is improving the health and the productivity of coastal oceans, sustaining ocean ecosystems, predicting climate variations and modernising ocean observational capabilities.

22.1.2
Chemometrics: A Helpful Tool in the Investigation of the Marine Ecosystem

The techniques and the methods for accurate and sensitive analytical determinations of chemical species in complex matrices have been greatly improved in the last years, and the instrumentation available can lead to the production in relatively low times of enormously large data sets. What can be further improved is the development of strategies able to extract and interpretate all the information contained in these data. The use of multivariate analysis methods able to simultaneously treat a large number of variables and their possible interactions seems the most suitable for a better knowledge of the marine system.

Literature offers examples of chemometric treatments that make it possible to obtain information from chemical data on the whole behaviour of sea waters and related systems (sediments, particulate and atmosphere). In fact the chemometric approach allows researchers to make a screening of the most significant variables and to find out their latent correlation, thus leading in general to a better knowledge of the ecosystem.

The data evaluated from samples of sea water, aerosol, particulate, sediments and marine flora and fauna collected all around the world and concerning both natural and anthropogenic components present in the sea system have been treated by pattern recognition methods such as: principal components analysis (PCA), factor analysis (FA), polytopic vector analysis (PVA), hierarchical (HCA) and non hierarchical (NHCA) clustering methods, fuzzy clustering methods (FCA), soft independent modelling of class analogy (SIMCA) classification method and non parametric k^*-nearest neighbour (k^*-NN).

Beside these kinds of applications, chemometric treatments are largely used also during the development of analytical methods. In particular techniques of experimental design and partial least squares (PLS) have been employed for the optimization of new methods for the determination of pollutants in complex matrices, as seawaters and sediments. Examples concerning the analysis of pesticides, PAHs, PCBs and other chlorinated compounds are reported.

22.2 Chemometric Applications to Marine Samples

22.2.1 Classification Techniques

22.2.1.1 *Sea Water*

Pollution from polychlorinated biphenyls (PCBs) was studied in seawater and particulate samples collected in the San Francisco Estuary (Jarman et al. 1997), the largest one on the coast of California: its drainage basin area having an inflow of 600 $m^2\,sec^{-1}$, mainly from the Sacramento and San Joaquin River. In addition to fresh-water inflow, ocean tides and winds bring ocean waters into the bay through the Golden Gate. The northern bay is a partially mixed estuary and the southern bay is a tidal lagoon: this distinction results in dramatically different biogeochemical processes in the two sections of the bay. The San Francisco Estuary is also a highly impacted area and several areas of the bay are highly industrialised. The matrix of the experimental data is formed by 14 samples and 32 PCB congeners for each of the six cruises performed over the years 1993–1995. Polytopic vector analysis (PVA) was designed to determine three important parameters in the analysis of mixtures: *(i)* the number of sources (end-members or chemical fingerprints) contributing to the mixture, *(ii)* the chemical composition of each-end member and *(iii)* the relative contribution of each-end member in each sample (Ehrich and Full 1987). PVA method is relatively new in environmental and chemometric literature but extensively used in geological sciences. It is a self-training classification method, because no training data set of known or suspected sources is required to solve a mixing model. With respect to PCA, PVA offers the advantage to report the final results as percentages rather than as abstract scores or loadings: this makes the end-member chemistry easier to interpret in a scientific context and allows the evaluation of concentration gradients by end-member mapping. Using the data of the cruise of April 1995, five separate PCB congener fingerprints were identified and their presence was correlated to the different locations in the bay. As concerns the different cruises, both within and between them, there is a large range of values, not correlated with seasonality.

On 26–28 November 1986, through a fracture in the Claymore pipeline close to its junction with the main Piper/Flotta oil line (approximately 10 km north of the Claymore platform), 2000–3000 t of crude oil leaked into the North Sea (Grahl-Nielsen and Lygre 1990). In the same period a typical industrial spill also happened: an overflow of a slop tank ended up in Byfjorden, Bergen. A study was carried out in 1990 for the identification of oil related to the two spills. The identification of samples from a spill becomes increasingly difficult the longer the oil has been exposed to light, air, water and microorganisms. A method based on the separation of the aliphatic fraction of the oil followed by GC-MS of polycyclic alkanes of low volatility was used for the analysis. In addition to samples from these two spills, samples of other North Sea crudes were analysed by PCA. The pairwise distances between the samples in the multivariate space was calculated for the most doubtful cases by the k^*-nearest-neighbour (k^*-NN) method. The plot of PC1 vs. PC2 and the k^*-NN results show significant difference between the samples of different origin.

By PCA and FA treatments (Machado 1993), the dependence on pH and copper(II) concentration of fluorescence signals of fulvic acids present in coastal marine waters collected in the Atlantic Ocean at Mindelo, Vila do Condo (20 km to the north of Porto) was evidenced.

With particular attention to sites devoted to hatcheries of edible mussels, the degree of pollution in the Gulf of Trieste (Northern Adriatic Sea), in particular in areas receiving industrial, agricultural and urban wastes, was studied with the help of chemometric methods (Reisenhofer et al. 1996). Four trace metals (Zn, Cd, Pb, Cu) and 5 nutrients (phosphate, silicate, NH_3, NO_2^-, NO_3^-) were determined in the filtered fraction of coastal surface water. Sampling was effected with a monthly frequency in the year 1993 in 6 representative stations to obtain information about the effect of possible seasonal trends and of the influence of meteorological factors. The use of PC method allows us to identify the sources and the typology of pollution and to discriminate between the quality of the water sampled at the different sites. In this area, largely inhabited and with low agricultural activity, the first and the most important factor resulted related to NO_3^- and silicate concentrations. This fact was attributed to the leachate of farmlands, whereas urban wastes (NH_3 from sewage and phosphate from domestic detergents) were related to the third factor. The heavy metal content was related to the second factor. It is worthwhile to mention the result obtained for Zn, which is not related to the other metal ions and seems therefore to have a different origin. A similar study was performed in the Marano lagoon: three sites were chosen in correspondence to hatcheries of edible clams to evaluate the environmental quality of the waters. The distribution of heavy metals (Cu, Pb, Cd, Zn, Cr, Mn, Ni) in sediments, in sea water at different depths, particulate matter and soft tissues of mollusks was evaluated. The results of HCA analysis and LDA (linear discriminant analysis) indicate a similar composition for the particulate matter and the sediments, while the composition of heavy metal in soft tissues reflects that of seawater body (Adami et al. 1998).

The composition of surface water (within 1 m depth) can result different for sampling sites even very close. This can happen in particular for lagoon water, as observed for two sites in Canal Grande in Venice (Gennaro et al. 1993), due to the lower circulation of waters, to the coastal geographical morphology and to the structure of the hydraulic net of channels differently affecting the cumulation processes and the circulation of clean water from the open sea. Also the effect of tide, of winds and of different sources of anthropogenic pollution play a role. Lagoon waters generally present larger problems of pollution, with respect to the open sea, and their composition can give useful information (Gennaro et al.1994; Gennaro et al.1995). At this purpose and to study also the seasonal dependence, monthly samples were collected in 1993 in three different stations characterized by different level of pollution (Marengo et al.1995). 13 variables (mainly inorganic anions) were determined. Unfortunately in 1993 the weather behaved atypically as concerns both rains and temperature and the raw data do not allow any direct correlation with the season, resulting predominant the differences due to the geographical positions. Only a PCA treatment permitted to characterize the samples as a function of seasonality and allowed to select the significant discriminating factors. The trend which is a function of the season can be envisaged in the plot of PC2 vs. PC3 and the discriminating variables resulted to be water pH, temperature, dissolved oxygen and phosphate and nitrate content. As a matter of fact, these species are the more affected by phytoplankton activity, that in turn depends on seasonality.

22.2.1.2
Seawater Vertical Distribution

As already mentioned, seawater composition is also a function of water depth. Studies of the vertical distribution can increase the information concerning the geochemical circumrotation of the elements. A study was devoted to find correlation between composition and depth in the open-sea region in the south-western part of the Black Sea, at about 200 miles from the town of Varana (Simeonov et al. 1993). The chemical components and the physical parameters collected for waters at different depths were studied by PCA, FA and hierarchical cluster analysis. The study was performed for 9 sampling depths and 22 variables, including NH_3, P, Zn, Mn, Cu, Fe, O_2, pH, salinity, alkalinity, Eh (mV), temperature and total suspended matter. The total variance could be explained by 5 PCs, being the 78.2% explained by the first two PCs. The plot of the loadings (PC1) reported in Fig. 22.1 permits the identification of two groups of data: the first includes the data for the samples collected between 0 and 100 m and the second one collects the data for the waters sampled between 125 and 200 m. The depth of 100 m seems to correspond to a variation in the composition that can be explained by two major effects:

i. the action of the wind able to mix the water mass of Black Sea up to this depth, and
ii. the effects of vertical density gradients.

These effects in turn cause accumulation of suspended matter at depths of about 100 m due to the retaining of suspensions by the more dense water layers. Additional

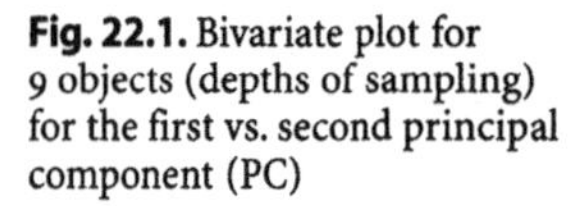

Fig. 22.1. Bivariate plot for 9 objects (depths of sampling) for the first vs. second principal component (PC)

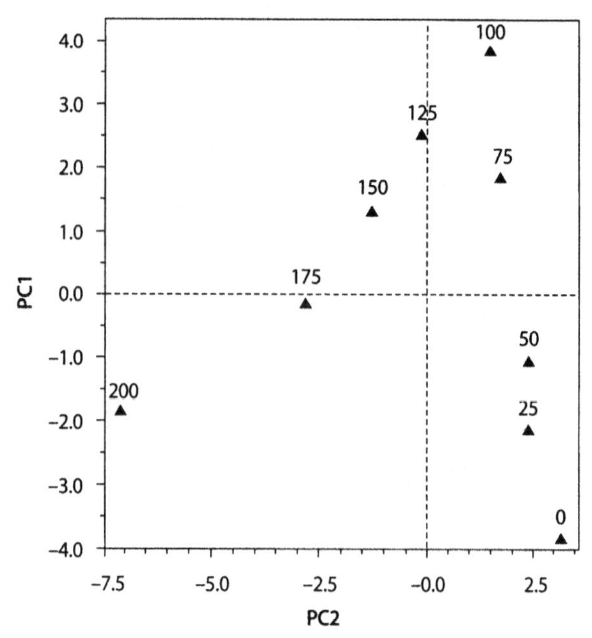

processes considered to explain the vertical distributions are adsorption, sedimentation, circulation of biogenic elements and bacterial activity. It was also observed that the vertical density gradient is closely connected with the effect of the hydrogen sulfide zone (below 100 m), characterized by additional redox influences, so that the change of the trend around 100 m (the aerobic zone) likely reflects the strong influence of the density gradient above and near the hydrogen sulfide zone. Both the plot of the loadings for the first two PCs and the dendrogram individuate four clusters: one includes the components whose values only smoothly decrease with increasing depth (pH, suspended matter, O_2, T), one includes the components with maximum concentration near to 100 m depth (Mn, P, Eh (mV)), the third contains the components whose concentration increases with depth (NH_3, Fe, Salinity, As) and the last one the components whose concentration increases in the H_2S zone (suspended As, Cu, Zn, alkalinity).

The vertical distribution was also studied in Sea of Japan, based on the determination of PCBs and nonylphenols (NoPHs) (Kannan et al.1998). The aim is not only to evaluate the extent of marine pollution but also to better understand the deep sea structure of that semi-enclosed "small ocean." The Sea of Japan (east sea) is a marginal sea in the northern Pacific Ocean: its average depth is 1700 m and the maximum is 4000 m. The water exchanges with the northern Pacific and with the Sea of Okhotsk is very limited because of sill depths of maximal 130 m. Using in-situ filtration-extraction techniques, two vertical profiles (deep water and shallow coastal water) and two-space-integrated surface profiles were taken. The data obtained by GC-MS were treated with PCA. A vertical structure in the water column was evidenced and compared with that of the open ocean. Figure 22.2 is a biplot of PC1 and PC2 of the correlation matrix of concentrations of PCBs in solution and suspension at the various depths of the Siribesi Trough. The similarity of composition results in neighbouring positions. Figure 22.2a

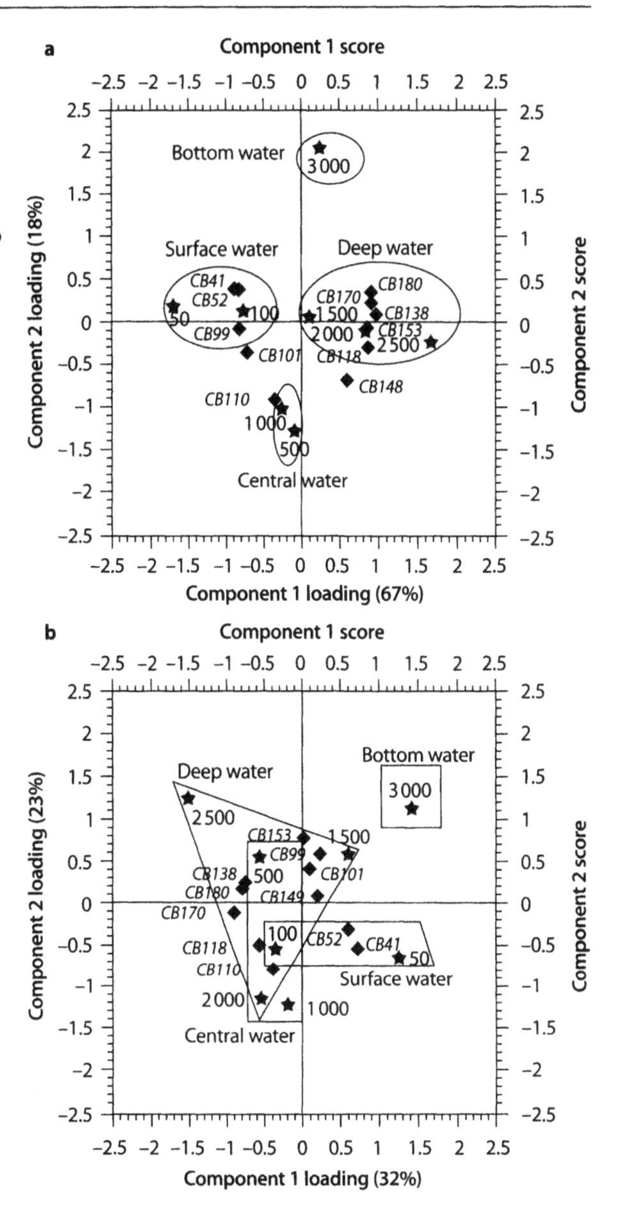

Fig. 22.2. Biplot of principal components 1 and 2 derived from the correlation matrix of ml % contribution of 11 dominant individual CBs in samples of the vertical profile in the Siribesi Trough **a** Solution; **b** Suspension. *Stars* represent scores, and the numbers refer to the depth of sampling in meters. The component loadings are represented by *rhombs*, along with corresponding CB numbers

shows that this applies to the sets of samples originating from: *(i)* 50 and 100 m, *(ii)* 500 and 1000 m, *(iii)* 1500, 2000 and 2500 m, while the composition at 3000 m sample can not be matched. PCA showed that a correlation exists between deep, bottom and surface waters of the nearby region. The water samples at different depths are characterized by different compositions of the CB mixture. In particular PC1, accounting for 67% of variability, gives the degree of chlorination. The loadings (the individual CBs) show that surface water is characterized by lower chlorinated CBs and that deep water is characterized by the higher chlorinated ones. As concerns the results obtained for suspended matter, a clear grouping can not be evidenced in the biplot of Fig. 22.2b. PC1 explains only 32% of the total variance, while PC2 contributes for an additive 22% and probably explains the composition of the biological material, the main carrier of the CBs. Figure 22.2a also confirms, as expected, that the anthropogenic contaminants in deeper waters have however their source in sea surface as confirmed by the generally higher concentration in intermediate-deeper waters than in the bottom ones. Since the composition of the CBs mixture in surface waters in this Japan region is quite variable, as shown by the plot of PC1 loading vs. PC2, correspondingly large differences were observed in the different water masses in the vertical profiles. Similar behaviour was observed for NoPHs too. The behaviour was explained by the taking place of mixing processes of different water bodies.

22.2.1.3
Marine Aerosol

In the North Sea, in the southern bight, aerosol samples were collected from an aircraft, in order to characterize aerosol particles for their content in metals (Cr, Zn, Pb, Ni, V, Sn, Cd, Cu) (Van Malderen et al. 1996). The input of heavy metals in the North Sea significantly increased due to the growing industrial activity and car traffic in European countries. Wet and dry deposition of pollutants from the atmosphere is a dominant transport system for a number of elements. To assess heavy metal concentrations and size distribution and to quantify the deposition fluxes of these elements, electron-probe X-ray microanalysis (EPXMA) was used, because it requires shorter sampling times (minutes rather than hours) with respect to the use of other analytical techniques. Thus it is possible to study short time variations in the atmospheric composition (for instance induced by changes of wind direction) and to overcome some problems of airborne sampling from aeroplanes. The aircraft flew tracks of about 110 km at six different heights, more or less evenly spaced between sea level and the inversion layer. The lowest track, 10–30 m above the sea level, was intended to assess particle resuspension by sea spray: airborne particulate matter was collected on 0.4 μm-filters. 96 samples were taken during 16 flights and 45 000 particles were analysed. More than 5 000 were found to contain significant concentrations of at least one of the following metals: Cr, Pb and Zn. Significant differences in abundance were evidenced in the aerosols, depending on the origin of the air masses: the highest values for Cr, Pb and Zn were always found under south-eastern wind directions. The chemometric treatment of the data set made use of a combination of hierarchical (dendrogram) HCA, non hierarchical (NHCA) and fuzzy clustering analysis (FCA). The dendrogram offers the advantage of an easy visualisation of the results but presents risks of misclassification, since it assumes a hierarchical structure (unlikely for atmospheric data sets). On the other hand, since NHCA methods would require a prior knowledge of the data structure, the results of HCA as the centroids for the NHCA were used. Both HCA and NHCA are hard clustering methods, according to which each sample can belong to only one cluster and can therefore lead to misclassifications when overlapping clusters are present. On the contrary, in FCA the possibility for each object to belong to each cluster is considered and its probability evaluated.

The results of the FCA individuated 5–7 major particle types. Those containing Cr, Pb or Zn can be, directly or indirectly, connected to emission processes of metallurgical industry, mainly present in the northern part of France, the German Ruhr area and industrial towns in the middle of the United Kingdom. A major part of Pb-rich particles was found to be associated with automotive exhaust. Less important sources of metal pollution were shown to be cement production and refuse incineration.

22.2.1.4
Water Microlayer

A study was performed in the North Sea (English Channel and Celtic Sea) to characterize individual particles and to find possible correlations between the sea surface microlayer and the bulk waters (Xhoffer et al. 1992). Sea surface microlayer (thickness between 0.1 and 3.0 μm) is characterized by chemical, physical and biological properties different from those of the water underneath. This layer is the main communication channel for transfer of matter between atmosphere and water. The transport between surface and sea body is due to processes of diffusion, convection, upwelling and rising of bubbles. The dispersion of the microlayer into the bulk waters occurs by sinking of particles and dissolution of water-soluble molecules, while bubble bursting and generation of aerosol by wind action are responsible for the transport into atmosphere. From the atmosphere matter is transferred to microlayer through wet or dry deposition processes. Many pollutants, such as petroleum hydrocarbons and chlorinated hydrocarbons, are concentrated in the microlayer together with proteinaceous material and organic acids that can provide complexing sites for many heavy metals thus often present in microlayer. Samples of microlayer and bulk water were collected in five positions in the North Sea: the data set contained more than 3 000 individual particles. The samples were analysed by electron-probe X-ray microanalysis (EPXMA) and laser microprobe mass analysis (LAMMA), techniques able to give indication if a specific element is uniformly distributed over all the particles. Eight different types of particles were individuated: aluminosilicate, Ca-rich aluminosilicates, Ca carbonates, silicates, organic particulate matter, Fe-rich particulate and Ti-rich particulate and a miscellaneous one. The results are compared with data from particles of atmosphere and river and related to the most probable sources: minerals, river waters, silicon-rich diatom, skeletal fragments of sponge, yeasts, bacteria, organic particulate (dead organism), skeleton of diatoms, coccolith, phytoplankton, etc. The concentration of dissolved silicate is lower at the surface and increases with depth, because of the decomposition of organisms. Biological activity in sea water generates an extremely high chemical mobility of silica, while calcite and aragonite are mainly present in oceans. Resuspension of a Ca-rich particulate fraction into marine atmosphere can occur during sea-spray processes and reactions with species S-rich (as SO_2, H_2SO_4 or dimethylsulfoxide) present in the atmosphere can take place with formation of $CaSO_4$. The variations of the particle types among the samples was studied by PCA. The first two PCs explain more than 94% of the total variance. The loadings (Fig. 22.3a) of the first PC explains the differences between aluminosilicates and organic and metal-rich particles, while PC2 shows high positive loading on both the Ca-rich and Ca-Si-rich particles types, that probably have the same source. The scores (Fig. 22.3b) and the dendrograms (Fig. 22.4) show good correlation between microlayer content and bulk seawater. For the five stations four clusters can be identified, composed as follows:

i. ML (microlayer) 3 and B (bulk water) 3
ii. ML4 + B4 + B5
iii. B1 + ML2 + B2
iv. ML1 + ML5

Except for the station 5, each microlayer and its corresponding bulk water sample are found in the same group, thus indicating a larger similarity between each microlayer sample and corresponding bulk sample than among the various bulk or microlayer samples.

Fatty acids have been analysed by GC and GC-MS in microlayer and subsurface water samples collected in the open western Mediterranean Sea (Sicre et al. 1988) in the area between Spain, France, Corsica and Sardinia. The concentration and composition of fatty acids present in various organisms have already been used as biological

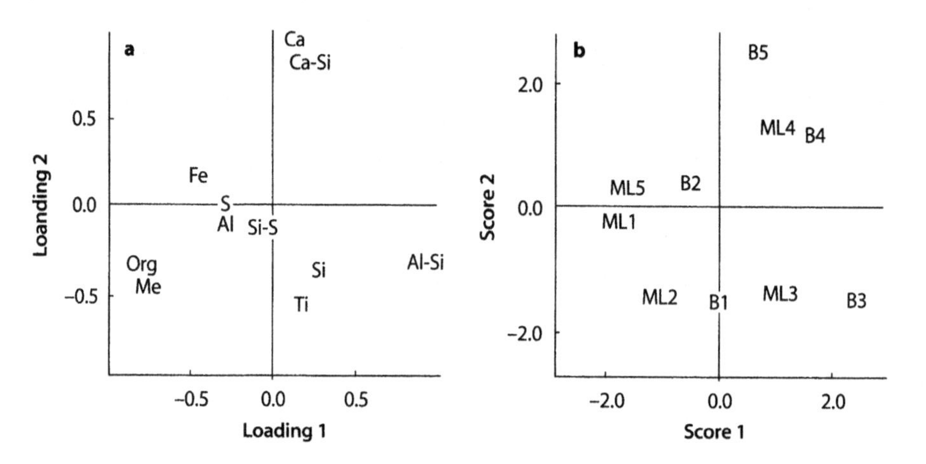

Fig. 22.3. a Loadings of the first two principal components obtained by PCA or EPXMA of North Sea microlayer and bulk water results; **b** Component scores of the first two principal components obtained from the same data set. Particle types: Org, organic; Me, metal-rich; Fe, Fe-rich; Ti, Ti-rich; S, S-rich; Al, Al-rich; Si, Si-rich; Ca-Si, Ca-rich aluminosilicate; Al-Si, aluminosilicate

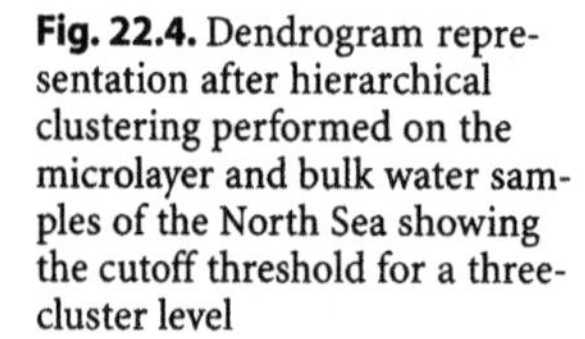
Fig. 22.4. Dendrogram representation after hierarchical clustering performed on the microlayer and bulk water samples of the North Sea showing the cutoff threshold for a three-cluster level

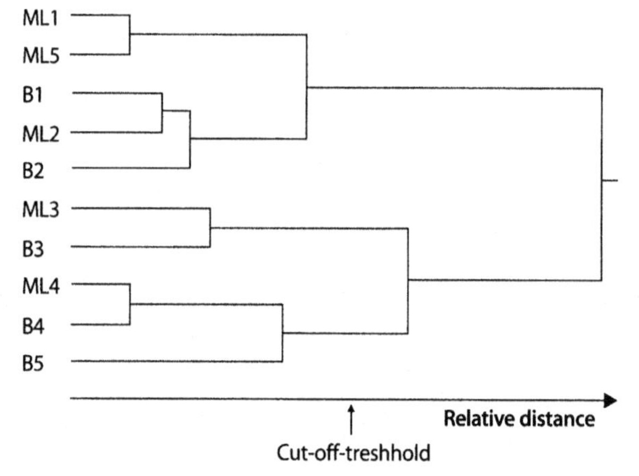

markers to individuate the sources and to assess the transformation processes of organic material in various environments. Subsurface (0.2 m depth) water samples were collected and several hundred data were treated by the multivariate method of correspondence factorial analysis (CFA) and non-parametric HCA. The chemometric treatments permitted to distinguish between different inputs in the dissolved and in particulate material and to classify the samples in two main groups: one contains planktonic markers and the second can further discriminate between a zooplanktonic and a microbiological pole.

PCA was employed to categorise sources of variability in rainwater data, and PLS (partial least-square) regression was used to compare the data for the same trace elements in crust, sea salt and smelter "signatures" (Vong 1993). 131 rainwater samples were analysed for eighteen trace elements, principally metals. PCA extracted three significant components: PC1 characterizes mainly the insoluble crustal elements, PC2 discriminates soluble sea salts and PC3 is mainly related to As, Sb, Cu and Se concentrations.

22.2.1.5
Particulate

In the northern Baltic Sea (Strandberg et al. 1998) the composition profile and the spatial differences of many pollutants that are globally distributed by transport through air and waters were studied, namely PCBs, hexachlorocyclohexanes (HCHs), hexachlorobenzene (HCBz), DDT and related metabolites, chlordane-related compounds (CHLs), dieldrin. The Baltic Sea contains high levels of halogenated organic compounds that, depending on their hydrophobicity and organic carbon-water partition coefficients, can be dissolved in the water phase or be associated with particles. Since atmospheric inputs often predominate in remote water areas, and since a limited biotransformation level takes places in particles, the composition of settling particulate matter can give indications of the situation of the air and of the contamination load of seawaters. The sampling of settling particulate matter was carried out in remote coastal and in offshore stations in the Bothnian Bay and in the Bothnian Sea, characterized by a large water exchange with open sea. Sediments were collected in coastal and in offshore stations. Surface bottom sediments were sampled in the coastal stations at Harufjarden (HF) and Simpnas (SN) in autumn and spring, to compare the contamination load at the coastal stations to that of surface bottom sediments. At one coastal station sediment samples (SED) were collected in autumn and in spring, in order to individuate seasonal variations. The levels of settling particulate matter (SPM) at the offshore stations were 10 times higher on the average than the coastal stations, likely because of an increased carbon content in the offshore settling particulate matter. The analysis was performed after cleanup procedures by GC/MS. The data set was then treated by PCA. Figure 22.5 reports a combined plot of scores (samples, in bold) and loadings (compound class, in italic) with the combined origins for the first two PCs. The plot permits one to study the distribution of the samples and to determine how the individual variables influence the samples. The profiles of the settling particulate matter from the four stations show great similarity, the major difference being the greater concentration of HCHs in the coastal stations (*SN* and *HF*), while PCBs are the most abundant pollutants at the offshore stations (*SR5* and *F9*). A more densely populated area might cause the elevated levels of DDTs in both stations. The plot also

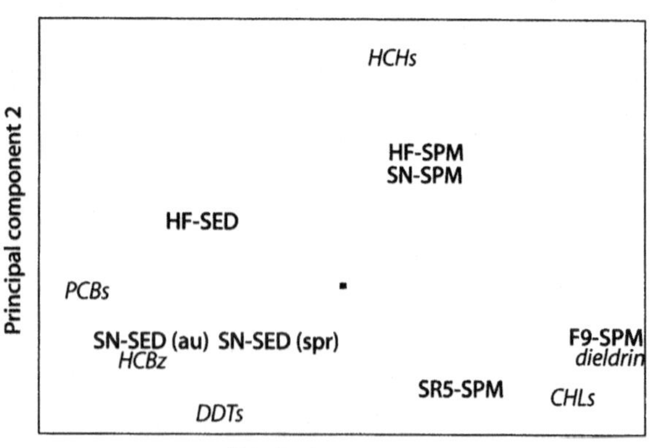

Fig. 22.5. Combined plot of PCA scores (describing samples; *bold*) and loadings (describing compounds; *italic*) for the first two principal components, explaining 24 and 22% of the variance, respectively

indicates that there are only minor seasonal differences, individuated by slightly higher DDT and CHL concentrations in the spring samples. The ratio (DDT metabolite/DDT) can be used to establish if the DDT emission occurred recently or in the past, since lower ratios indicate a recent use of technical DDT while a ratio larger than 3/1 is normally expected for aged mixtures in the environment. In the settling particulate the ratio matter shows unexpected values between 0.6 and 2.0 that seem to indicate a recent use, even though DDT has been banned in Scandinavia since the early 1970s. Since the atmosphere is an important pathway for global transport and deposition of organochlorine compounds, the result was attributed to the long-range atmospheric transport from distant developing countries where the pesticide is still in use.

22.2.1.6 *Sediments*

To collect information about the contamination of the Southern Baltic Sea, 20 samples of sediments and corresponding samples of surface water and of 5 m depth water were collected. Analyses were performed by HPLC for PAH and by GC for alkanes (Lamparczyk et al.1990). Multivariate analysis of PC was applied on the data matrices to find connection of the PC score values with the geography of the sampling area and collect suggestions on the usefulness of these pollutants to estimate marine contamination. As concerns sediments, the plot of PC1 vs. PC3 clearly distinguishes the samples from an industrial area concentrated in Gdansk and in the Pomeranian Bay from the samples collected in the open sea. Only one sample, collected in a site very close to the shore, behaves differently: but its similarity to the open sea is justified because in this Polish region industrial activity is very low. For surface water samples, the two first PC explain 76% of the total variability. Again, the samples from the industrial area are well separated from those from open sea, with the only exception of the Polish station. A similar behaviour is observed for the water sampled at 5 m depth. As for sediments, the effect of the relatively high PAH concentration is assigned to sedimentation of organic particles from the atmosphere. A study of vertical distribution, performed as a function of variations of salinity, showed that the vertical water exchange ratio is very high. It must be underlined that, even if the behaviour between surface and 5 m depth waters is similar, the PC score values are different, to indicate that, although an equilibrium ratio between PAHs in surface water and in 5 m depth water may exist, the PAH composition is also governed by specific processes occurring in each layer, as on the other hand evidenced by the vertical distribution. As concerns *n*-alkanes in surface water, PCA analysis shows no connection between the geography of sampling and the composition in *n*-alkanes in the samples. Also, for water from 5 m depth the distribution of the PC scores is not connected with the geography of sampling, but the PC score plots for surface water and water from 5 m depth differ significantly. Likely, the vast majority of *n*-alkanes due to their extremely low solubility in water are present in emulsion form or adsorbed on organic matter. For sediment samples, the main factor governing the composition and concentration of *n*-alkanes seems to be the depth from the bottom. In conclusion, in water and sediment samples no connection was found between variable patterns and the geography of sampling: these results suggest that *n*-alkanes cannot be used as a measure of long term anthropogenic contamination.

The concentration of 28 PAH was measured in superlayer and underlayer deep sea sediments collected at 15 sampling sites in the Gulf of Lion, in the Mediterranean Sea (Domine et al. 1994). For comparison, samples of sediments of the Rhone River were also collected. The Mediterranean Sea receives large amounts of toxic contaminants from inflowing rivers and the atmosphere as well as industrial and tourist activities along its coasts. In particular, PAHs are generated on land by the combustion of fossil fuels and wood. PAH-loaden particles can be transported by the wind to distant locations and are then removed from the atmosphere by rain and dry fall-out. Some particles are deposited on water surfaces and then incorporated in the sediments, others are redeposited on land and then delivered to natural waters by erosion and runoff, followed by the resuspension and transport of contaminated sediments. The results of GC and HPLC methods emphasise that contamination is relatively important in deep sea sediments, and it is mainly due to material coming from the Rhone River. The data were treated by means of the non linear mapping (NLM) method. The NLM method, particularly suitable for the analysis of large sets of data, is a non linear display method to represent the points of an *n*-dimensional space in a 2-D space, preserving as possible the interpoint distances. The NLM was performed both on the 15 sites with the PAH concentrations as variables and on the 28 compounds with the site concentrations as variables and the results interpreted by means of multiple graphical displays. It was possible to obtain a better knowledge of the PAH contamination of underwater canyons and deep sea fans in Gulf of Lion. It also allowed some correlation between the PAH structures and their origins and concentrations in the different samples. The use of NLM coupled with graphical tools has shown that it is possible to easily visualise the level of PAH contamination: lower concentrations are generally observed in the underlayer sediments, with the only exception of retene and perylene which only in fact have a biogenic origin.

22.2.1.7
Bioindicators

Information about marine ecosystem health can be obtained also from studies on the state of flora and fauna present. Thus species of algae, fish, turtles, etc. are used as bioindicators. For example, aquatic macrophytes and algae can be used as alternative indicators of water pollution, even if the increasing role of bioindication as an appropriate system for pollution control requires the identification of the aquatic plants being suitable as biological monitors. Their selective absorption of certain ions combined with their sedentary nature are good reasons for using hydrophytes as biological monitors: the ability of algae in coastal waters to cumulate metals is well known. A study was performed in Greece, in the Aegean Sea, in the neighbourhood of the Thessaloniki Bay and of the islands of Crete and Thira (Stratis et al. 1996). In 14 sites, 9 kinds of algae were sampled. Each sample was analysed for 5 metals (Cu, Zn, Cd, Pb, Mn). Both hierarchical and non-hierarchical approaches were used to detect similarities or dissimilarities in the bioindicating ability. The samples collected in biotopes characterized by the same kind of pollution can be found in one cluster, while in another bioindicators from two relatively unpolluted biotopes are present. This result suggests that most of the marine algae do not react specifically to a certain pollutant but to the total pollution caused by heavy metals. In this way algae are bioindicators of the total situation of the environment and not of the presence of a specific pollutant.

Also bivalves, such as oysters clams and blue mussels (*Mytilus edulis*), are able to accumulate pollutants in their tissue up to concentrations that far exceed the concentration in the ambient water. For this reason they can advantageously be used in investigations devoted to the distribution of pollutants in coastal areas, as in the study carried out on the coast of Norway (Kvalheim et al. 1983). Four categories of pollutants were considered: petroleum hydrocarbons, halogenated hydrocarbons, trace metals and radio nuclides. Polluted mussels were collected from a dock in the Bergen harbour and unpolluted blue mussel from a farm at Austevoll, at about 50 km (sea distance) south of Bergen in a relatively unfrequented fjord: in total 19 samples were collected. Geographical and seasonal variations in both mussels and environments make it difficult to use the levels of pollutants found in mussel tissues as a direct measure of the degree of environment pollution. The purpose is to establish whether the composition of the components naturally present in tissues is different for mussels collected in different places. To level out the short-time differences caused by the different situations (access to food, temperature, salinity, light), all the mussels were transferred to an aquarium with running clean seawater and maintained here for 16 weeks. Samples from muscle and gonad tissues were analysed by GC: in the chromatograms 60 peaks were selected for muscle and 56 from gonad tissue samples. The experimental data were treated by SIMCA, which clearly shows that the muscle samples of polluted and unpolluted mussels form well-described and nicely-separated models in the 60-dimensional space and shows clear differences between the mussels from the two sampling locations. The significant PCs for each class were evaluated and cross-validated. The final results show:

i. The samples from unpolluted mussels are closely grouped and well separated from the polluted mussel samples.
ii. The more polluted samples are more scattered than the unpolluted ones.
iii. Muscle tissues give a better distinction than gonad tissues.

It must be noticed that these differences are not of short-term character because, as mentioned, all the mussels sampled were kept in the same aquarium where they lived for 16 weeks before the analysis. Another consideration to be made is that the major difference between the two sampling locations is that one is polluted by urban fall and ship discharges and the other one is pristine. It is reasonable to believe that the difference in the sea conditions at the two locations is reflected by the difference in composition of the natural components in mussels. Further studies of similarities and differences among the components naturally present in samples from different unpolluted locations are desirable, to make the method suitable for monitoring coastal areas.

Mussels were also studied for heavy metal pollution samples collected in 12 stations in the eastern coast of the Bothnian Sea, adjacent to the town of Pori in Finland (Lindstrom et al. 1988). In this area the two main pollution sources are the local titanium oxide industry and the River Kokemaenjoki, responsible for both industrial and agricultural discharges. Aims of the study were: *(i)* to investigate how far metals reach from the Vuorikemia factory effluent, *(ii)* to distinguish the waters coming from the factory from those coming from the river and *(iii)* to study biological effects of pollution to mussels during a short (2–3 months) and a long (years) period. In both soft parts and shells, a selective fractionating procedure for the determination of Al, As, Cd, Co, Cr, Cu, Fe, Hg, Mn, Pb, Sb, Ti, V and Zn and the treatment of the chemical data with SIMCA method gave the best correlation to the pollution sources with data from the calcite shells. The soft tissue data of the caged mussels correlated well with the pollution load contributed by the factory. Ongoing pollution is best measured by soft tissue method while long term or past pollution is better characterized by data from the shells.

A major recent advance in environmental science has been the use of molecular markers for the identification of organic inputs in aquatic systems. Generally speaking, molecular markers are compounds whose structure can be related to specific sources; for example long-chain alkylbenzenes and trialkylamines can be related to contributions from anionic or cationic detergents, respectively. The usefulness of PCA and FA in environmental studies when using molecular markers for sample description was evaluated in a study involving the determination of aliphatic and chlorinated hydrocarbon, fatty acids, alcohols, chlorophylls and some detergents in water and in particulate (Grimalt and Olivè 1993). Seven sampling sites were selected in the delta of the Ebre River, in the Mediterranean Sea. Samples of surface sea water and of about 3 m depth were collected in Fangar and Alfacs Bays and in the Ebre River bed, as well as in the irrigation and in the drainage channels. The sampling was carried out every 2 months covering a period of 18 months. 40 samples were collected in total. The use of GC and GC-MS techniques for hydrocarbon, alcohol and fatty acids fractions allowed the identification and determination of 122 compounds and two standard industrial mixtures of PCBs, while by UV spectrophotometry chlorophyll pigments and phaeophorbide were determined. PCA analysis of the data shows that PC1 explains 34% of the variance and corresponds to a mixture of compounds originating from various sources. This indicates that there is a concurrence of biogenic and anthropogenic inputs, which is not surprising because the river is the major geographical source which determines the distribution in the delta. PC2 explains 14% of the variance and shows a rather complex distribution. As concerns the score plot, PC1 scores afford a good differentiation between the bay water and those of the river and channels, while PC2 scores separate the irrigation channel waters from those of the river and of the discharge channel. FA was performed in an iterative way, by replacing in each cycle the diagonal elements of the correlation matrix by the square communalities obtained from the factor loadings and continuing the process until the difference in the traces of two successive correlation matrices were smaller than a prefixed threshold. Nine factors were identified as the most reasonable estimate of the data. Most of the loadings are unspecific: so for instance chlorinated species exhibit a distribution behaviour that is roughly independent on their origin. In general, however, source inputs are better differentiated by FA than by PCA. However, since the data set used to describe the system has unsuitable dimension (102 variables for 40 samples), the data set was divided between biogenic and anthropogenic markers and each group was again evaluated with PCA and FA. PCA analysis did not lead to higher definitions of geochemical sources, while FA analysis for the biogenic subgroup evidenced four factors as the most likely to estimate the reduced data set: two main biogenic factors correspond to algae origins, the first related to diatom sources and the second to green algae and dinoflagellates, while two factors differentiate between river and irrigation channel samples. The characterisation of anthropogenic data are more difficult, also because the diversity of pollutant discharge sites in the Ebre Delta.

SIMCA, a principal component chemometric modelling program, was used to examine complex mixtures of PCB residues in fish and turtles (Stalling et al. 1987) obtained from Tinicum National Environmental Center in Philadelphia and to compare the results with Aroclor mixtures. The data set consists of three classes: Aroclor mixture (14 samples), turtles (7 samples) and fish (6 samples). The chemometric treatment of SIMCA gave three well differentiated classes, and PCA individuated 3 PCs able to explain more than 93% of the total variance. These results confirm that PCB residues in turtles and fish are markedly different from Aroclor standards or their mixtures, even though fish residue profiles were closer to Aroclor than the turtle residue.

More than 150 samples of particulate, sediments, isopods, zoo- and phytoplankton, mysidis, perch, fourhorned sculpin, herring, salmon, salmon trout and whitefish were collected in the autumn 1991 and spring 1992 in the Gulf of Bothnia, between the coasts of Sweden and Finland, to study the accumulation of PCBs and other organochlorine contaminants in benthic and pelagic marine food chains (Van Bavel et al. 1994). In the treatment of the data of PCBs for sediments and for the amphipod *Monoporeia affinis*, PCA was employed. The first two PCs describe 85% of the total variance: PC1 is able to describe for the same station a concentration gradient from spring (lower concentration) to autumn (higher concentration) and a gradually higher level from Hornslamnder station to the others. This difference was explained as an indication of different sources of pollution of one sampling site with respect to the other three: this site is located in a sub arctic region where run off water from the arctic region is collected. Large seasonal fluctuations in PCB levels are then generally observed from the autumn (high concentration) and spring (low) for both sediment and amphipod samples.

22.2.2 Optimization Methods

Chemometric methods can be applied to collect from the chemical data more information about the marine system, as in the examples reported above, as well as for the development and the optimization of adequate methods for sample pretreatment and analyte recovery. The large number of species present in sea water plays in fact a complex matrix effect, affecting the analysis of both natural and anthropogenic species. Many variables must be often simultaneously optimized in order to maximize the extraction yield or to minimise the matrix effect; methods of experimental design able to treat simultaneously the effect of the factors and of their interactions on the response are used.

In the optimization of supercritical fluid extraction (SFE) of methyl mercury in marine samples, the effect of 8 variables is considered: CO_2, flow-rate, density, extraction cell temperature, static extraction time, nozzle and trap temperature, amount of hydrochloric acid and contact time between acid and sample before the extraction (Cela-Torrijos et al. 1996). A 26-experiment folded Plackett-Burman factorial design ($2^8 \times 3 / 32 + 2$ central points) was used. The results suggested that the extraction cell temperature is the most significant variable. The method was validated with respect to three reference materials containing different methyl-mercury contents. Recoveries are comparable to the widely accepted extraction procedure, being always greater than 82% (for concentrations of the order of $\mu g\ g^{-1}$). Precision is slightly better.

Chemometrics was also employed to optimize procedures of auto-injection and of programs of column temperature in a GC method for the evaluation of neutral lipid carbon number profiles in marine samples (Yang et al.1996). Natural lipids including mono-, di-, and triglycerols, sterols, steryl esters, hydrocarbons, wax esters, free fatty acids and free aliphatic alcohols are important components of marine food webs. Triacylglycerols, for instance, constitute a major form of energy storage for marine plants and animals. Sterols regulate membrane functions and act as precursors for many metabolically active molecules. The composition of neutral lipid compounds in marine animals can easily change through diet: environmental factors and concentration ratios of neutral lipids subclasses are indices of their physiological conditions and thus of the health of seawaters. The method was applied to a series of marine samples including algae, bivalves, polychaetes, fish eggs and fish larvae to evidence neutral lipid carbon number distributions. A two-level multivariate analysis Latin square L8 was used to screen the number of the potentially important variables in the autosampler and in the injection temperature program. L8 (2^7) means that seven variables were scanned in eight runs each at two levels (0.3 and 0.7 of the full range respectively). In the second experiment the better point of each variable in the previous run served as the central value. New values of the two-level analyses were selected from the middle points between the better and worse points in the last run. Finally, the response-surface method was employed to locate the optimum.

A two-level fractional factorial experimental design involving six chromatographic factors was employed to optimize the ion-chromatographic determination of low concentrations of nitrate and phosphate in sea water. $2^{(6-2)} + 3$ (for the central points) experiments were performed (Dahllof et al. 1997). For the development of an SFE procedure for a series of PAH from spiked marine sediments, four variables (pressure, temperature, extraction fluid volume, methanol concentration) were optimized (Notar and Leskovsek 1997). The PAHs were subdivided in four groups, according to the number n of the aromatic rings in their structure ($n = 2–3, 4, 5, 6$). A five-level spherical factorial experimental design was employed: the lower levels of the parameters were set on the basis of the literature data and the upper according to the extraction instrument capability. The other three levels of the parameters were set on the basis of the spherical experimental design, when all the experiments chosen lie on the surface of a mathematic sphere (Fig. 22.6). As a result three five-level factorial spherical designs were obtained, each of them involving 15 experiments (45 in total). Through the use of hierarchical clustering analysis, a confirmation was searched as to whether the dividing of the PAH into 4 groups was appropriate: the results of the dendrogram show four major clusters representing the four groups.

A two-level orthogonal array design, OAD, was used to optimize sample preparation for the determination of 4-nonylphenol in sea water and sediments (Chee et al. 1996). Orthogonal array design presents both the advantages of Simplex and of factorial design design (Lan et al. 1994 a,b). In a two-level orthogonal array design with S factors, only $S + 1$ experiments are required as in the Simplex method, instead of 2^S. A two-level orthogonal array data set is a $(S + 1) \times S$ matrix, where S is the number of the columns which correspond to the factors and $S + 1$ the number of the rows that correspond to the experiments. OAD is used to assign factors to a series of experimental conditions; the significance of the different factor effects is evaluated by the analysis of the variance (ANOVA). The main effects of the factors and the two-variable interactions can be considered separately as different factors and estimated by OAD along with the corresponding linear graphs on triangular tables. OAD was also successfully used for the optimization of recovery in the solid-phase extraction (Wan et al. 1994a) and of liquid chromatographic analysis (Wan et al. 1994b) of pesticides in seawater.

For monitoring triphenyltin (tPhT) in sea water, a method (0.56 ng l^{-1} as detection limit) was developed by a partial least square (PLS) multivariate calibration approach, based on solid-phase extraction and fluorescence measurement, using flavonol as the fluorogenic reagent in micellar medium (Leal et al. 1997). The calibration model is built from seawater samples of known tPhT concentrations. The data set consisted of 150 fluorescence intensity data values for each of the nine excitation wavelengths (between 375 and 415 nm): the method was cross-validated by leave-one-out mode. The fluorescence given by sea water is due to naturally occurring compounds, such as chlorophyll and fulvic acid, as well as to the presence of anthropogenic compounds such

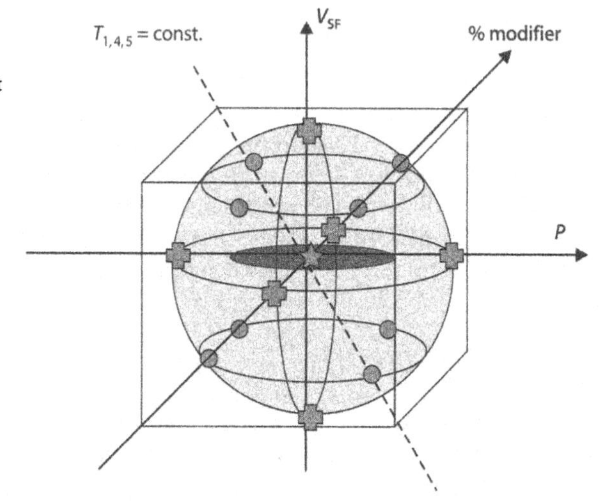

Fig. 22.6. Graphical presentation of five-level factorial spherical experimental design with 45 experiments labelled at different levels (*crosses*: first and fifth level, *circles*: second and fourth level, *star*: third level)

as PAHs. The PLS calibration method applied to fluorescence data overcomes the difficulties found in the determination, by univariate calibration of tPhT in samples containing high contents of organic compounds.

References

Adami G, Barbieri PL, Predonzani S, Reisenhofer E (1998) Heavy metals in surface sediments of Marano lagoon: A chemometric evaluation of their environmental relevance on clam hatcheries. Ann Chim (Rome) 88:709–720

Carpenter EJ, Capone DG (1984) Nitrogen in the marine environment. Verlag Chemie, Weinheim (Germany)

Cela-Torrijos R, Miguéns-Rodrìguez M, Carro-Dìaz AM, Lorenzo-Ferreira RA (1996) Optimization of supercritical fluid exctraction-gas chromatography of methyl mercury in marine samples. J Chromatogr A 750:191–199

Chee KK, Wong MK, Lee HK (1996) Optimization of sample preparation techniques for the determination of 4-nonylphenol in water and sediment. J Liq Chrom & Rel Technol 19:259–275

Dahllof I, Stevenson O, Torstensson C (1997) Optimising the determination of nitrate and phosphate in sea water with ion chromatography using experimental design. J Chromatogr A 771:163–168

Demianov P, De Stefano C, Gianguzza A, Sammartano S (1995) Equilibrium studies in natural waters: Speciation of phenolic compounds in synthetic seawater at different salinities. Environ Tox Chem 14:767–773

De Stefano C, Foti C, Sammartano S, Gianguzza A, Rigano C (1994) Equilibrium studies in natural fluids.Use of synthetic seawater and other media as background salts. Ann Chim (Rome) 84:159–175

Domine D, Devillers J, Garrigues P, Budzinski H, Chastrette M, Karcher W (1994) Chemometrical evaluation of the PAH contamination in the sediments of the Gulf of Lion (France). Sci Total Environ 115:9–24

Ehrich R, Full WE (1987) Use and abuse of statistical methods in earth sciences. In: Size W (ed), Oxford University Press, Oxford

Gennaro MC, Abrigo C, Saini G (1993) Simultaneous determination of nitrite, nitrate, iodide, chloride and sulfate in Venice lagoon water by ion-interaction reversed-phase HPLC and dual detection. Ann Chim (Rome) 83:125–146

Gennaro MC, Abrigo C, Giacosa D, Saini G, Avignone MT (1994) The role of tide on the composition of Venice lagoon., J Environ Sci Health 29:967–981

Gennaro MC, Abrigo C,Giacosa D, Saini G (1995) Determination of nitrite, nitrate, iodide, bromide, chloride, sulfate in Venice lagoon water by reversed-phase ion-interaction HPLC. The effect of the geographical position. J Environ Sci Health 30:675–687

Grahl-Nielsen O, Lygre T (1990) Identification of samples of oil related to two spills. Mar Pollut Bull 21:176–183

Grasshoff K, Ehrhardt M, Kremling K (1983) Method of seawater analysis. Verlag Chemie, Weinheim (Germany)

Grimalt JO, Olivé J (1993) Source input elucidation in aquatic systems by factor and principal component analysis of molecular marker data. Anal Chim Acta 278: 159–176

Jarman WM, Johnson GW, Bacon CE, Davis JA, Risebrought RW, Ramer R (1997) Levels and patterns of polychlorinated biphenyls in water collected from the San Francisco Bay and Estuary,1993–95. Fres J Anal Chem 359:254–260

Kannan N, Yamashita N, Petrick G, Duinker JC (1998) Polychlorinated biohenyls and nonylphenols in the sea of Japan. Environ Sci Technol 32:1747–1753

Kvalheim OM, Oygard K, Grahl-Nielsen O (1983) SIMCA multivariate data analysis of blue mussel components in environmental pollution studies. Anal Chim Acta 150:145–152

Lamparczyk H, Ochocka RJ, Grzybowski J, Halkiewicz J, Radecki A (1990) Classification of marine environment samples based on chromatographic analysis of hydrocarbons and principal component analysis. Oil & Chem Pollut 6:177–193

Lan WG, Wong MK, Chen N, Sin YM (1994a) Orthogonal array design as a chemometric method for the optimization of analytical procedures. Part 1. Two-level design and its application in microwave dissolution of biological samples. Analyst 119:1659–1667

Lan WG, Wong MK, Chen N, Sin YM (1994b) Orthogonal array design as a chemometric method for the optimization of analytical procedures. Part 2. Four-level design and its application in microwave dissolution of biological samples. Analyst 119:1669–1675

Leal C, Granados M, Beltran JL, Compano R, Prat MD (1997) Application of partial least squares multivariate calibration to triphenyltin determination in sea-water with excitation-emission matrix fluorescence. Analyst 122:1293–1298

Lindstrom R, Vuorinen A, Piepponen (1988) Monitoring environmental heavy metal pollution by selective fractionation method of mussel shells (*Mytilus edulis* L.). Heavy Met Hydrol Cycle 155–162
Machado AASC, Esteves da Silva JCG (1993) Factor analysis of molecular fluorescence data of marine and soil fulvic acids. Chemom Intell Lab Systems 19:155–167
Manahan S (1994) Environmental chemistry. Lewis Publishers, New York
Marengo E, Gennaro MC, Giacosa D, Abrigo C, Saini G, Avignone MT (1995) How chemometrics can helpfully assist in evaluating environmental data. Lagoon water. Anal Chim Acta 317:53–63
Notar M, Leskovsek H (1997) Optimisation of supercritical fluid extraction of polynuclear aromatic hydrocarbons from spiked soil and marine sediment standard reference material. Fres J Anal Chem 358:623–629
Reisenhofer E, Adami G, Favretto A (1996) Heavy metals and nutrients in coastal, surface seawaters (Gulf of Trieste, Northern Adriatic sea): An environmental study by factor analysis. Fres J Anal Chem 354:729–734
Riley JP (1989) Chemical oceanography, vol IX. Academic Press, London, New York
Riley JP, Chester R (1971) Introduction to marine chemistry. Academic Press, London, New York
Sicre MA, Paillasseur JL, Marty JC, Saliot A (1988) Characterization of seawaters samples using chemometric methods applied to biomarker fatty acids. Org Geochem 12:281–288
Strandberg B, van Bavel B, Bergqvist PA, Broman D, Ishaq R, Naf C, Pettersen H, Rappe C (1998) Occurrence, sedimentation and spatial variations of organochlorine contaminants in settling particulate matter and sediments in the Northern part of the Baltic Sea. Environ Sci Technol 32:1754–1759
Stratis JA, Simeonov V, Zachariadis G, Sawidis T, Mandjukov P, Tsakovski S (1996) Chemometrical approaches to evaluate analytical data from aquatic macrophytes and marine algae. Fres J Anal Chem 355:65–70
Simeonov V, Andreev G, Karadjov M (1993) Chemometric interpretation of analytical data for vertical distribution of some chemical components in sea water. Fres J Anal Chem 345:744–747
Stalling DL, Schwartz TR, Dunn III WJ, Wold S (1987) Classification of polychlorinated biphenyl residues: Isomers vs. homologue concentrations in modeling aroclors and polychlorinated biphenyl residues. Anal Chem 59:1853–1859
Uppenbrink J, Hanson B, Stone R, Malakoff D, Schmidt K, Irion R, Kaiser J, Costanza R, Andrade F, Antunes P, van den Belt M, Boersma D, Boesch DF, Catarino F, Hanna S, Limburg K, Low B, Molitor M, Pereira JG, Rayner S, Santos R, Wilson J, Young M, Falkowski PG, Barber RT, Smetacek V, Butler A, McGowan JA, Cayan DR, Dorman LM, Jickells TD, Delaney JR, Kelley DS, Lilley MD, Butterfield DA, Baross JA, Wilcock WSD, Embley RW, Summit M (1998) Special section chemistry and biology of the oceans. Science 281:189–230
Van Bavel B, Strandberg B, Lundgren K, Bergqvist PA, Zook D, Broman D, Naf C, Axelman J, Rolff C, Rappe C (1994) Levels of PCBs in the marine ecosystem of the Gulf of Bothnia: A chemometric comparison of different sampling sites. Organohalogen Compounds 20:547–552
Van Malderen H, Hoornaert S, Van Grieken R (1996) Identification of individual aerosol particles containing Cr, Pb, and Zn above the North Sea. Environ Sci Technol 30:489–498
Vong RJ (1993) Atmospheric chemometrics for identification of trace element sources in precipitation. Anal Chim Acta 277:389–404
Xhoffer C, Wouters L, Van Grieken R (1992) Characterization of individual particles in the North Sea surface microlayer and underlying seawater: Comparison with atmospheric particles. Environ Sci Technol 26:2151–2162
Yang Z, Parrish CC, Helleur RJ (1996) Automated gas chromatographic method for neutral lipid carbon number profiles in marine samples. J Chromatogr Sci 34:556–568
Wan HB, Lang WG, Wong MK, Mok CY, Poh YH (1994a) Orthogonal array designs for the optimization of solid-phase extraction. J Chromatogr A 677:255–263
Wan HB, Lang WG, Wong MK, Mok CY (1994b) Orthogonal array designs for the optimization of liquid chromatographic analysis of pesticides. Anal Chim Acta 289:371–380

Index

Symbols

A

B

D

E

I

J

K

L

M

N

O

P

T

X

Y

Z

Lightning Source UK Ltd.
Milton Keynes UK
UKOW06f1127101116
287325UK00005B/71/P